Nanomedicine and Cancer

Nanomedicine and Cancer

Editors

Rajaventhan Srirajaskanthan MBBS MD
Department of Gastroenterology
University Hospital Lewisham
London
UK

Victor R. Preedy PhD DSc
Professor of Nutritional Biochemistry
School of Medicine
King's College London
and
Professor of Clinical Biochemistry
King's College Hospital
UK

CRC Press
Taylor & Francis Group
an informa business
www.crcpress.com

6000 Broken Sound Parkway, NW
Suite 300, Boca Raton, FL 33487
270 Madison Avenue
New York, NY 10016
2 Park Square, Milton Park
Abingdon, Oxon OX14 4RN, UK

Science Publishers
Jersey, British Isles
Enfield, New Hampshire

Published by Science Publishers, an imprint of Edenbridge Ltd.

- St. Helier, Jersey, British Channel Islands
- P.O. Box 699, Enfield, NH 03748, USA

E-mail: info@scipub.net Website: www.scipub.net

Marketed and distributed by:

CRC Press
Taylor & Francis Group
an informa business
www.crcpress.com

6000 Broken Sound Parkway, NW
Suite 300, Boca Raton, FL 33487
270 Madison Avenue
New York, NY 10016
2 Park Square, Milton Park
Abingdon, Oxon OX14 4RN, UK

ISBN 978-1-57808-727-3

Cover Illustrations: Reproduced by kind courtesy of the undermentioned authors:

- Figure No. 4 from Chapter 2 by Conchita Tros de Ilarduya and Gloria Gonzalez-Aseguinolaza
- Figure Nos. 4 and 6 from Chapter 3 by Cheng-Lung Chen and Yang-Yuan Chen
- Figure Nos. 2 and 4 from Chapter 15 by Mansoor M. Amiji et al.

Library of Congress Cataloging-in-Publication Data

Nanomedicine and cancer / editors, Rajaventhan Srirajaskanthan, Victor R. Preedy.
p. ; cm.
Includes bibliographical references and index.
ISBN 978-1-57808-727-3 (hardcover)
I. Srirajaskanthan, Rajaventhan. II. Preedy, Victor R.
[DNLM: 1. Nanomedicine--methods. 2. Neoplasms. QZ 200]

616.99'401--dc23

2011043091

Printed in the United States of America

Preface

The nanosciences are a rapidly expanding field of research with a wide applicability to all areas of health. They encompass a variety of technologies ranging from particles to networks and nanostructures. For example, nanoparticles have been proposed to be suitable carriers of therapeutic agents whilst nanostructures provide suitable platforms for sub-micro bioengineering. However, understanding the importance of nanoscience and technology is somewhat problematical as a great deal of text can be rather technical in nature with little consideration to the novice. In this collection of books on ***Nanoscience Applied to Health and Medicine*** we aim to disseminate the information in a readable way by having unique sections for the novice and expert alike. This enables the reader to transfer their knowledge base from one discipline to another or from one academic level to another. Each chapter has an abstract, key facts, and a "mini-dictionary" of key terms and phrases within each chapter. Finally, each chapter has a series of summary points. In this book we focus on nanomedicine and naontechnology as applied to cancer. We cover for example, introductions to the field, general aspects, methods and techniques, nanocarriers for gene delivery, siRNA, nanostrategies, nanoprobes, magnetic nanoparticles, nanoimaging, nanoscaffolds, ethylcellulose nanoplatforms, nanorings, nanotubes, nanowires, nanobubbles multidrug resistance, tocotrienol, the microenvironment, and cancers of the lung, liver, head and neck and many more areas. Contributors to **Nanomedicine and Cancer** are all either international or national experts, leading authorities or are carrying out ground breaking and innovative work on their subject. The book is essential reading for oncologists research scientists, medical doctors, health care professionals, pathologists, biologists, biochemists, chemists and physicists, general practitioners as well as those interested in disease and nano sciences in general. **Nanomedicine and Cancer** is part of a collection of books on ***Nanoscience Applied to Health and Medicine***.

The Editors

Contents

Section 1: General Aspects, Methods and Techniques

1

Nanomedicine-based use of siRNA in Cancer

Guillermo N. Armaiz-Pena,[1,a] *Bulent Ozpolat,*[1] *Anil K. Sood*[2] and *Gabriel Lopez-Berestein*[3,*]

ABSTRACT

Small interfering RNA (siRNA) possess the ability to silence genes with a high degree of specificity. This ability has resulted in its development as a potential therapeutic tool for the treatment of cancer and other diseases. Although siRNA has the potential to be a powerful drug, its systemic delivery has remained a major obstacle. The main limiting factor preventing its use as a therapeutic agent is poor cellular uptake due to rapid siRNA degradation by serum nucleases and rapid clearance by the body. In order to overcome these factors, various approaches have been developed, although with different efficacy and safety profiles. Cationic liposomes have

[1]Department of Experimental Therapeutics, The University of Texas MD Anderson Cancer Center, 1515 Holcombe Blvd Unit 422, Houston, Texas 77030, USA.
[a]E-mail: gnarmaiz@mdanderson.org
[2]Departments of Gynecologic Oncology, Cancer Biology, Center for RNA Interference and Non-coding RNA, The University of Texas MD Anderson Cancer Center, 1515 Holcombe Blvd Unit 1362, Houston, Texas 77030, USA; E-mail: asood@mdanderson.org
[3]Departments of Experimental Therapeutics, Cancer Biology, Center for RNA Interference and Non-coding RNA, The University of Texas MD Anderson Cancer Center and the Nanomedicine and Biomedical Engineering Department, UT Health, Houston, Texas, 1515 Holcombe Blvd Unit 422, Houston, Texas 77030, USA; E-mail: glopez@mdanderson.org
*Corresponding author

List of abbreviations after the text.

been one of the most popular nanoparticles to deliver siRNA, but their effectiveness has been hampered by potential lung toxicity. In this review, we highlight the preclinical use of 1,2-dioleoyl-sn-glycero-3-phosphatidylcholine and chitosan nanoparticles loaded with siRNA. We have demonstrated that both approaches are more effective than other current technologies and naked siRNA for systemic delivery of siRNA into tumor tissues and its associated microenvironment. Furthermore, we present an overview of emerging technologies that are being developed to improve siRNA delivery into tumors.

INTRODUCTION

Discovery of siRNA as a Therapeutic Tool

The existence of RNA interference (RNAi) was first reported more than 20 years ago in *p. x hybrida* flowers (Napoli et al. 1990). It was shown that when the gene responsible for its pigmentation was overexpressed, the flower lost its endogenous color by suppressing both the transgene and the endogenous gene. Napoli and colleagues termed this process "cosuppression". More importantly, they observed that both genes were synthesized by the cell, but were silenced post transcriptionally. Years later, Fire and colleagues described the process of RNAi in animal cells when they discovered the ability of double stranded RNA to silence gene expression in the nematode worm C. elegans (Fire et al. 1998). Based on the substantial amount of data generated following this discovery, it is now believed that RNAi exists in all animals (Tokatlian and Segura 2010). The natural function of RNAi appears to be protection of the genome against invasion by mobile genetic material elements, such as transposons and viruses that produce aberrant RNA or dsRNA when they become active. Recently, the use of siRNA has rapidly become a powerful tool in silencing gene expression. Because of its high specificity and ability to silence genes that could not be targeted before, the process of RNA degradation has garnered a great deal of attention from the scientific community for its possible use as a therapeutic agent in various disease targets, including viral infections and cancer (Landen et al. 2005; Lu et al. 2010; Morrissey et al. 2005; Okumura et al. 2008).

A well-designed siRNA construct can be used as a therapeutic tool in order to utilize the cell innate RNAi machinery to silence any desired gene. This feature makes siRNA more attractive as a therapeutic tool than current drugs, such as small molecule inhibitors. SiRNA molecules possess the ability to knock down its target sequence in a very specific manner by direct homology dependent post-transcriptional gene silencing. In this process, RNAi is triggered by the presence of long pieces of double stranded RNA, which are cleaved by the endonuclease Dicer into 21–23 nucleotide long fragments called siRNA (Hannon and Rossi 2004). In practice, siRNA can be synthetically produced and then directly introduced into the cell, thus circumventing Dicer processing (Fig. 1). Argonaute 2, a multifunctional protein contained within the RNA-induced silencing

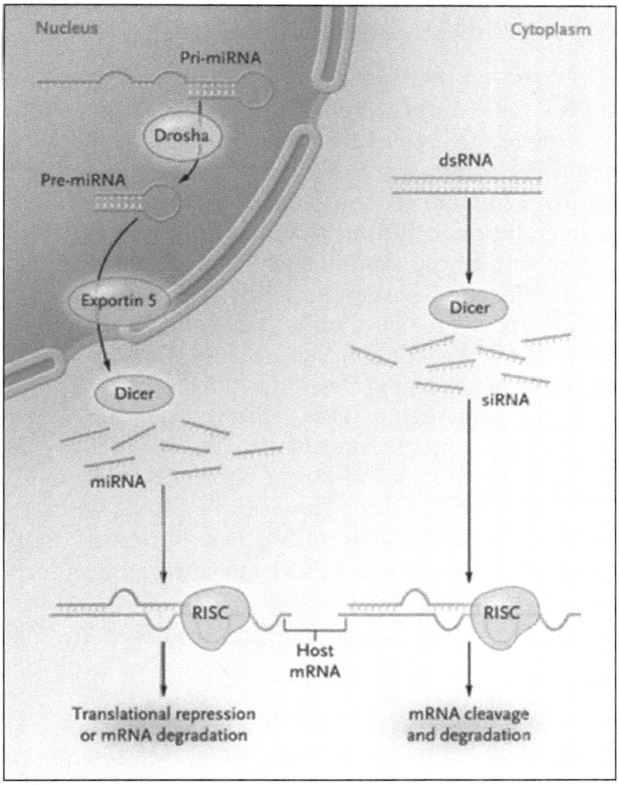

Fig. 1. Mechanism of RNA interference. Long precursor microRNA (miRNA) segments, called pri-miRNA, are first cleaved in the nucleus by Drosha, an RNase III endonuclease, into segments of approximately 70 nucleotides each (called pre-miRNA). Transportation into the cytoplasm by means of exportin 5 leads to cleavage by Dicer, another RNase III endonuclease, which produces mature miRNA segments. Host degradation of messenger RNA (mRNA) and translational repression occurs after miRNA binds to the RNA-induced silencing complex (RISC). Cytoplasmic long double-stranded RNA (dsRNA) is cleaved by Dicer into small interfering RNA (siRNA), which is incorporated into RISC, resulting in the cleavage and degradation of specific target mRNA. (Reprinted with permission from Merritt et al. Copyright 2008a Massachusetts Medical Society. All rights reserved).

Color image of this figure appears in the color plate section at the end of the book.

complex (RISC), unwinds the siRNA, after which the sense strand of the siRNA is cleaved. The RISC complex is activated and can selectively seek out and degrade mRNA that is complementary to the antisense strand. The cleavage of mRNA occurs at a position between nucleotides 10 and 11 on the complementary antisense strand, relative to the 5′-end (Hannon and Rossi 2004). The activated RISC complex can then move on to destroy additional mRNA targets, which further propagate gene silencing.

Challenges for Pre-clinical and Clinical use of siRNA

Since it has been shown that some tumors lack Dicer or express it at low levels, it is a preferable alternative to bypass Dicer processing (Merritt et al. 2008a). To achieve this, use of siRNA constructs that are no longer than 21 nucleotides is preferred. For any therapeutic application, siRNA use is recommended over longer constructs, since these can be potentially toxic, especially to liver function (Frantz 2006). Additionally, special attention is required when selecting potential siRNA sequences to avoid off-target and non-specific effects. These can be reduced or eliminated by avoiding certain immunogenic sequence motifs, and validating siRNA sequences using freely available search engines (for example see: http://www.dharmacon.com/DesignCenter/DesignCenterPage.aspx). For instance, by using the minimum effective dose, avoiding partially complementary sequences and immunogenic motifs can substantially diminish any possible toxicity and eliminate off-target effects. It is imperative to point out that one of the most valuable properties of siRNA when compared to antisense oligonucleotides, is that siRNA is significantly more potent at silencing a gene, positioning it as a more attractive option for therapeutic use (Ozpolat et al. 2010)

The first use of siRNA *in vivo* was accomplished by high pressure administration of naked siRNA in a large volume of physiological solution intravenously to mice (McCaffrey et al. 2002). Subsequently, siRNA has been shown to be effective in various physiological settings, such as the brain, eye and lung (de Fougerolles and Novobrantseva 2008; DiFiglia et al. 2007). Unfortunately, in order for siRNA to efficiently reach many tissues in the body, it will require an additional delivery system to facilitate delivery to cells. The major hurdle for the preclinical and clinical use of siRNA is poor cellular uptake, mostly due to degradation by nucleases or rapid renal clearance when administered systemically. Moreover, siRNA molecules are too big to cross the cell membrane and are negatively charged, leading to electrostatic repulsion from the membrane. Throughout the years, numerous siRNA transfection reagents have been developed for *in vitro* purposes, but these remain non-viable for *in vivo* use because of their innate toxicity and inefficient *in vivo* delivery of siRNA

(Bumcrot et al. 2006). Currently, there are multiple siRNA-based therapies that have entered into clinical trials (Tokatlian and Segura 2010). These therapies are designed to deliver siRNA locally, such as the intravitreal or intranasal routes and, as a result, there remains a void in the development of safe and effective nanoparticles that would make systemic delivery of siRNA a viable option. For the rest of this review, we will concentrate on the preclinical and clinical use of siRNA on the treatment of cancer. Other uses of siRNA, although important, are out of the scope of this review.

Approaches for Systemic siRNA Delivery

To develop a highly efficient vehicle for siRNA delivery into the tumor, certain conditions have to be met. First, these particles have to be relatively small (size no bigger than 1000 nm). Furthermore, these nanoparticles need to localize into the tumor vasculature and be able to extravasate into the tumor microenvironment to release the loaded siRNA into cells. Moreover, these nanocarriers need to be biocompatible, biodegradable, and lack immunostimulatory properties, and be able to bypass rapid hepatic or renal clearance. For example, the size, surface charge and hydrophobicity of the particle will determine its distribution pattern, safety, and uptake by cells and tissues. In addition, particle charge is important to its bioavailability, since negatively charged particles seem to be cleared faster than particles with a positive charge (Whitehead et al. 2009). Particles with diameter greater than 100 nm are quickly cleared by the reticuloendothelial system in the liver, spleen, lung and blood marrow, meanwhile particles smaller than 100 nm normally have an extended time in the circulation. Several techniques have been developed to extend the circulation time. For example, various studies have shown that PEG or non-anionic surfactants can substantially enhance the particle circulation time (Tokatlian and Segura 2010). Nanoparticles serving as delivery vehicles for siRNA present numerous advantages over naked siRNA delivery due to its ability to stabilize siRNA while delivering higher concentrations of siRNA directly into tumor sites. More importantly, some of these nanoparticles can be modified with high affinity ligands to specifically target siRNA directly into the tumor. Finally, these nanoparticles can serve to promote controlled release and when formulated correctly, they can provide a safe and reliable platform for siRNA delivery for treatment of cancer and other diseases.

Liposomes and Lipid-like Nanoparticles

This category of nanoparticles is comprised, among others, of neutral and cationic liposomes, micelles, emulsions and solid lipid nanoparticles. These nanoparticle systems have been shown to work more efficiently with siRNA compared to other types of particles, while the reason for this mechanistic difference is not clear (Tokatlian and Segura 2010). It is thought that weaker interactions between lipids and siRNA allows for faster decomplexation in the cytosol. This ability has made these particles the most widely used method for delivering therapeutic agents. Liposomes are composed of lipid bilayers that are formed upon the addition of water to dried phospholipid films. Therefore, they consist of a hydrophobic intra-membrane space and a hydrophilic interlayer space. They can be composed of naturally-derived phospholipids with mixed lipid chains or other surfactants. For example, one polar functional group can make up the outer part of the nanoparticle, meanwhile another polar functional group faces the interior hydrophilic core that contains the nucleic acid or desired drug to be carried. The addition of different particles, such as PEG, to its outer surface, can enhance its bioavailability leading to enhanced delivery into the tumor. Other strategies to enhance their safety profile and circulation half-life include the optimization of its lipid composition, siRNA to lipid ratio and reducing its size. Table 1 reviews several nanoparticles that have been utilized for siRNA delivery.

Charged liposomes

Since the discovery of the cationic lipid DOTMA (N-[1-(2,3-dioleyloxy) propyl]-N,N,N-trimethlylammonium chloride) and its ability to carry and deliver DNA and RNA into cells, liposomes have been used efficiently to deliver nucleic acids (Collins et al. 2004; Spagnou et al. 2004). Other cationic liposomes have been described since, being DOTAP (1,2-dioleoyl-3-trimethylammonium-propane) one of the highlights. The principle behind using cationic liposomes is that in order to achieve efficient delivery, the lipids form complexes with the negatively charged nucleic acid. There are various issues with this approach preventing its use with siRNA constructs, meanwhile their safety as a therapeutic agent remains under investigation. For example, positively charged liposomes have the ability to form strong interactions with negatively charged nucleic acids, effectively limiting its *in vivo* success because of low transfection rates due to its inability to release its payload.

Electrostatic interactions between siRNA and lipids may have profound effects on particle characteristics and behavior in biological fluids and tissues. Further complicating these interactions is that siRNA, unlike DNA, can not be condensed, causing particles to be too large and unstable, opening the door for potential enzymatic or physical degradation before being taken up by a cell. As a result of their positive charge they have a strong affinity towards serum proteins, lipoproteins and the extracellular matrix. These interactions can lead to accumulation at

Table 1. Selected nanoparticles that have been used for siRNA delivery.

Category of particle	Type, form and composition of carrier	Natural versus synthetic	siRNA/shRNA	Comments Lung toxicity
Lipid comlex	Cationic liposomes	Synthetic	siRNA	Biodegradable-nontoxic
	Neutral liposomes (DOPC)	Synthetic	siRNA	
	Lipoplexes	Synthetic	siRNA	
	Stable nucleic acid-lipid particles	Synthetic	siRNA	
Conjugated polymers	Polymer-functional peptides	Synthetic	siRNA	
	Polymer-lipophilic molecules (e.g., cholesterol)	Synthetic	siRNA	
	Polymer-PEG	Synthetic	siRNA	
Cationic Polymers	Chitosan	Natural	siRNA/shRNA	Biodegradable-nontoxic
	Atellocollagen	Natural	siRNA/shRNA	Biodegradable-nontoxic
	Pegylated	Synthetic	siRNA/shRNA	Cytotoxicity
	Cyclodextrin	Synthetic	siRNA/shRNA	
	Poly-L-lysine	Synthetic	siRNA/shRNA	

different tissues and cause the premature release of siRNA. These particles have the potential to activate the complementary system and they can also be rapidly cleared by macrophages of the reticuloendothelial system. In our hands, cationic liposomes, such as DOTAP, are mostly observed in or near the vasculature and accumulate at the liver and spleen, reducing its therapeutic value (Landen et al. 2005). Although liposomes have been among the most popular delivery particles for nucleic acid, still some concerns remain regarding their safety record when used as a therapeutic agent because there have been reports describing *in vivo* toxicity by cationic liposomes (Whitehead et al. 2009). For example, cationic liposomes caused dose-dependent toxicity and pulmonary inflammation, being multivalent cationic liposomes, such as Lipofectamine, more toxic than monovalent cationic lipids, such as DOTAP (Spagnou et al. 2004). These studies also revealed that the creation of reactive oxygen species was partially responsible for the observed toxicity. On the other hand, it has been shown that the administration of negatively charged liposomes did not result in any lung toxicity (Ozpolat et al. 2010). However, due to their negative charge it is expected that these particles will not be able to bind efficiently to siRNA and be quickly taken up by circulating macrophages, further reducing their effectiveness. Other questions still linger regarding the non-specific effects that a liposome/lipid-based system can induce. For example, they could affect the total cellular protein content potentially. Notwithstanding these issues, liposomes still hold promise for future clinical use, as evidenced by the FDA approval of PEGylated liposomes as a carrier of doxorubicin.

In addition to liposomal formulations, other types of lipid-based systems have been described for systemic siRNA delivery. For example, stable nucleic acid-lipid particles (SLN) and cationic solid lipid nanoparticles (SNALP) have been previously described (Kim et al. 2008a; Zimmermann et al. 2006). SNALPs made of cationic and fusogenic lipids have been shown to posses the ability to deliver siRNA and silence apolipoprotein B protein expression for up to 11 days (Zimmermann et al. 2006). It has been shown that SLN, when reconstituted from protein-free LDL was able to successfully deliver siRNA. These particles were prepared by utilizing a modified solvent-emulsification method composed of cholesteryl ester, triglyceride, cholesterol, dioleoyl phosphatidylethanolamine (DOPE), and 3-beta-[N(N′,N′-dimethylamino ethane)carbamoyl]-cholesterol (DC-cholesterol). Several studies have shown that these nanoparticles can effectively deliver siRNA that can efficiently silence its target gene, while being relatively stable in serum and causing minimal toxic effects (Kim et al. 2008a).

Neutral nanoliposomes

The delivery particle surface charge is one of the most important parameters that have to be taken into consideration, since it is going to determine how a particle interacts with its surroundings and other molecules. For example, a positive-charged nanoparticle, *in vitro*, could be very effective in delivering its load due to its interactions with the negative-charged cell membrane. On the other hand, in an *in vivo* setting a positive-charged nanoparticle can interact with negative-charged serum proteins in the blood losing its effectiveness. Additionally, siRNA nanocomplexes tend to aggregate when their net surface charge is positive. As a result, different strategies have been developed to prevent the aggregation of nanoparticles. Among these, the addition of PEG, sugar molecules (cyclodextrin), and hyaluronic acid (HA) to the nanoparticles have been the most studied (Tokatlian and Segura 2010). These additions resulted in the neutralization of the particle's charge, enhanced the particle circulation time and opened the door for more specific nanoparticle targeting by restricting non-specific interactions between the nanoparticle and the cell membrane.

To overcome the limitations posed by cationic liposomes, our group developed neutral nanoliposomes, such as 1,2-dioleoyl-sn-glycero-3-phosphatidylcholine (DOPC) in order to deliver siRNA more efficiently into tumor tissue (Fig. 2) (Landen et al. 2005). We have shown that DOPC can deliver siRNA into tumors more effectively than cationic liposomes, such as DOTAP, and naked siRNA. Moreover, we have shown the ability of DOPC-siRNA to silence several target genes important to cancer progression when dosed twice a week. More specifically, we have shown that DOPC/siRNA nanoparticles have the ability to silence their target for up to 4 days (e.g., EphA2, FAK, neuropilin-2, IL-8 or Bcl-2) (Gray et al. 2008; Halder et al. 2006; Landen et al. 2005; Merritt et al. 2008b). Furthermore, as a result of its robust target gene silencing, DOPC-siRNA particles have shown the ability to reduce tumor size in mice, growing orthotopic or subcutaneous models of human cancer. Throughout all the *in vivo* experiments using DOPC, we have not observed any detectable distress or toxicity and found the nanoparticles to be safe in mice and non-human primates (A.K. Sood and G. Lopez-Berestein, unpublished observations). Moreover, DOPC/siRNA nanoparticles are not toxic to any tissue tested, including fibroblasts, bone marrow and hematopoietic cells.

Other neutral particles have been developed for systemic and localized siRNA delivery. For example, polyelectrolyte complex micelles (PEC) with PEG or PEI have been utilized with modest success for siRNA delivery (Kim et al. 2008b). When administered intratumorally or intravenously,

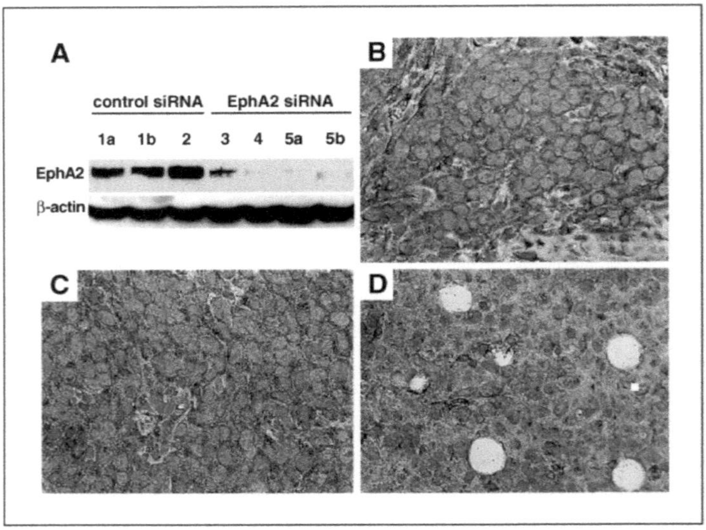

Fig. 2. *In vivo* down-regulation of EphA2 by siRNA/DOPC after systemic administration. (a) Western blot of lysate from orthotopic tumors collected 48 hours after a single administration of control siRNA (*lanes 1* and 2) or EphA2-targeting (*lanes 3–5*) siRNA, each complexed within DOPC. To control for sampling error, *lanes 1a* and *1b* are separate preparations from the same tumor treated with control siRNA. Similarly, *lanes 5a* and *5b* are separate preparations from the same tumor treated with EphA2-targeting siRNA. *Lanes 2* to *4*, additional tumor-bearing mice treated with control or EphA2-targeting siRNA-DOPC. Adjacent sections were stained by H&E to confirm the presence of tumor. (b) Immunohistochemical staining for EphA2 of tissue treated with control siRNA/DOPC. The typical cobblestone appearance of this overexpressed membrane-bound protein is noted. (c) Immunohistochemistry 48 hours after a single treatment of EphA2-targeting siRNA without a transfection agent ("naked") is shown and had no detectable affect on EphA2 expression. (d) Treatment of EphA2-targeting siRNA encapsulated within DOPC effectively down-regulated EphA2 expression 48 hours after a single dose. EphA2 expression is restored 1 week after a single treatment (not pictured). (b-d) Original magnification, ×400. (EphA2: Brown). (Reprinted with permission from the American Association for Cancer Research, Inc., from Landen et al. 2005; Fig. 4).

Color image of this figure appears in the color plate section at the end of the book.

they demonstrated the ability of silencing its target gene resulting in decreased tumor growth in mice. Another delivery vehicle that has been used is lipophilic siRNA conjugates (e.g., cholesterol conjugates) (Wolfrum et al. 2007). Cholesterol-siRNA and HDL complexes were shown to be more effective than cholesterol-siRNA alone at silencing its target gene *in vivo*. The main problem with these approaches was that relatively high doses were administered, therefore increasing the chances of toxicity and non-specific effects.

Polymeric Delivery Systems

Polymers with a linear or branched structure can be used as siRNA delivery particles due to their ability to interact and condense nucleic acids into stable nanoparticles. Polymers are known to stimulate nonspecific endocytosis and endosomal escapes. Polymeric particles that are used for siRNA delivery can be biodegradable and be attained from natural sources (e.g., chitosan, atelocollagen) or be of synthetic origin (e.g., PEI, PLL, cyclodextrin). As carriers, polymers can stabilize siRNA, preventing its degradation and allow for extended circulation times in the bloodstream. For example, PEI has been a popular nanocarrier for genetic material delivery, but recently it has been used to deliver siRNA (Ozpolat et al. 2010). Atelocollagen and cyclodextrin particles have been shown to have few side effects and the ability to carry siRNA to target sites (Whitehead et al. 2009). Furthermore, when used to deliver siRNA, their administration resulted in increased cellular uptake and prolonged release of genes and oligonucleotides.

Recently, we have also developed chitosan nanoparticles for systemic delivery of siRNA into both tumor cells and tumor-associated vasculature (Fig. 3) (Lu et al. 2010). Their small size, slight positive charge and high incorporation efficiency of siRNA made them a desirable carrier. Our results demonstrated that siRNA was efficiently delivered into the tumor, but interestingly enough we observed substantial siRNA delivery to tumor-associated endothelial cells, suggesting potential applications for targeting the tumor vasculature (Lu et al. 2010). Chitosan particles containing Hs Ezh2 and Ms Ezh2 were highly effective in silencing EZH2 expression and the combination of both particles was effective in reducing ovarian tumor growth (Lu et al. 2010). Even though, we noted siRNA uptake in various organs, very little siRNA was taken up by macrophages. Furthermore, we did not observe any toxicity of safety issues with chitosan-siRNA nanoparticles.

SiRNA Modifications

Throughout their existence, humans have developed various defense mechanisms to defend themselves against exogenous siRNA. In the hope of evading these mechanisms, the siRNA molecule can be modified. For example, numerous siRNA sequences can nonspecifically activate the immune system by inducing the Toll-like receptor 7 pathway (Whitehead et al. 2009). Studies have shown that the addition of 2′-O-methyl modifications into the sugar structure of various nucleotides within both the sense and antisense strands can diminish this effect (Chiu and Rana 2003; Harborth et

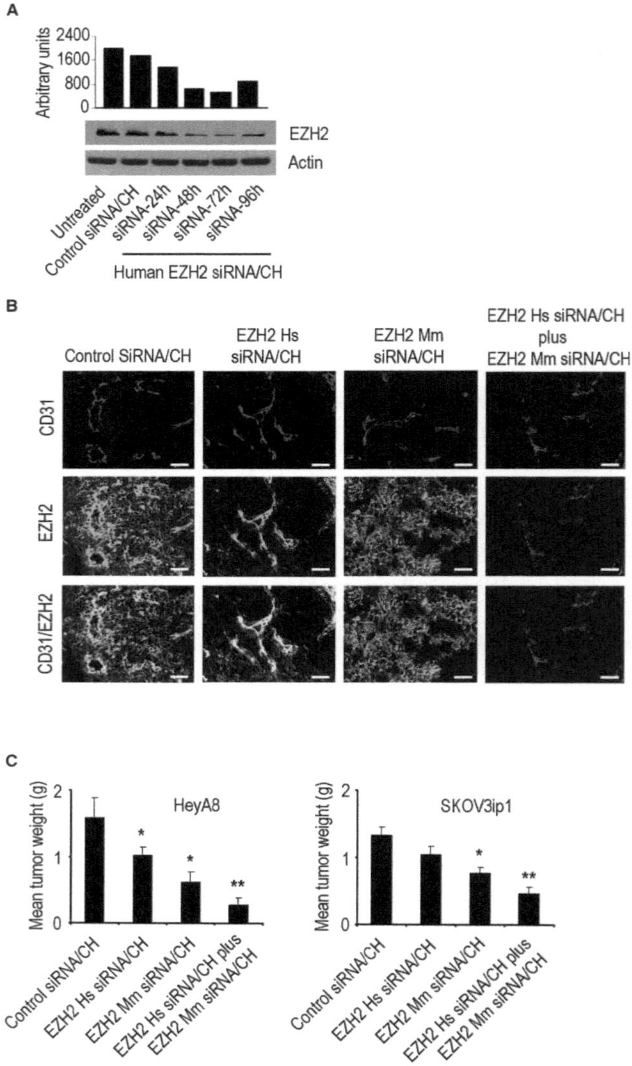

Fig. 3. Effects of *ezh2* gene silencing on *in vivo* ovarian cancer growth. (a) Western blot of lysates from orthotopic tumors collected 24, 48, 72, and 96 hr after a single injection of control siRNA/CH or human (*ezh2* Hs siRNA/CH). (b) *Ezh2* gene silencing in HeyA8 tumor as well as tumor endothelial cells. Tumors collected after 48 hr of single injection of control siRNA/CH, *ezh2* Hs siRNA/CH, or *ezh2* Mm siRNA/CH and stained for EZH2 (green) and CD31 (red). The scale bar represents 50 μm. (c) Effects of *ezh2* Hs siRNA/CH or *ezh2* Mm siRNA/CH on tumor weight in orthotopic mouse models of ovarian cancer. Error bars indicate SEM. $^{*}p < 0.05$; $^{**}p < 0.001$. (Reprinted with permission from Elsevier, Inc., from Lu et al. 2010. Fig. 5).

Color image of this figure appears in the color plate section at the end of the book.

al. 2003). Other modifications to the siRNA developed to help stabilize the molecule include the addition of phosphorothioate backbone linkages at the 3′-end of the RNA strands. This approach reduces the susceptibility of siRNA to exonucleases. Addition of alternative 2′ sugar modifications (e.g., a fluorine substitution) can lead to increased resistance to endonucleases (Chiu and Rana 2003; Harborth et al. 2003). It is important to point out that chemical changes or additions to the siRNA molecule may result in off-target effects, greatly diminished siRNA activity, increased production of toxic metabolites, and reduced therapeutic index.

Target-specific siRNA-nanoparticle

Another venue being investigated to enhance siRNA delivery into the tumor is the development of target-specific nanoparticles. By targeting the tumor cell and its microenvironment, we can increase the therapeutic window of any nanoparticle by increasing its bioavailability at target tissues and eliminating non-specific delivery that would result in reduced toxicity. To add the ability of active targeting of tumor cells to nanoparticles, a coupling ligand must be added to the exterior surface. For example, functional peptides, lipophilic molecules, PEG, and aptamers have been used for this purpose (Park et al. 1995; Stephenson et al. 2003). Our group has shown that nanoliposomes can be passively delivered to tumors due to the enhanced permeability and retention effect caused by the tortuous endothelium of the tumor microenvironment (Landen et al. 2005). Ligand-targeted liposomal nanoparticles and imaging agents have been shown to enhance binding while improving the drug therapeutic window.

Various approaches have been developed for tumor specific targeting. Among them, folate receptors, transferrin receptors, and alphaV/beta3 integrin have been used as targets, meanwhile high affinity ligands include folate, antibodies, antibodies parts (Fab) and RGD (Arg-Gly-Asp) peptides. Recently, our group developed an RGD peptide-labeled chitosan nanoparticle to directly target tumoral cells (Fig. 4) (Han et al. 2010). We showed that RGD-CH-NP loaded with siRNA substantially increased intratumoral siRNA delivery in an orthotopic ovarian cancer model, in addition to efficient silencing of several genes (POSTN, FAK and PLXDC1) (Han et al. 2010). Furthermore, we showed that this approach resulted in significant inhibition of tumor growth. As a whole, ligand conjugations will allow to target nanoparticles actively to tumor tissue leading to enhanced delivery and increased therapeutic index when delivering chemotherapy agents, peptides and siRNA.

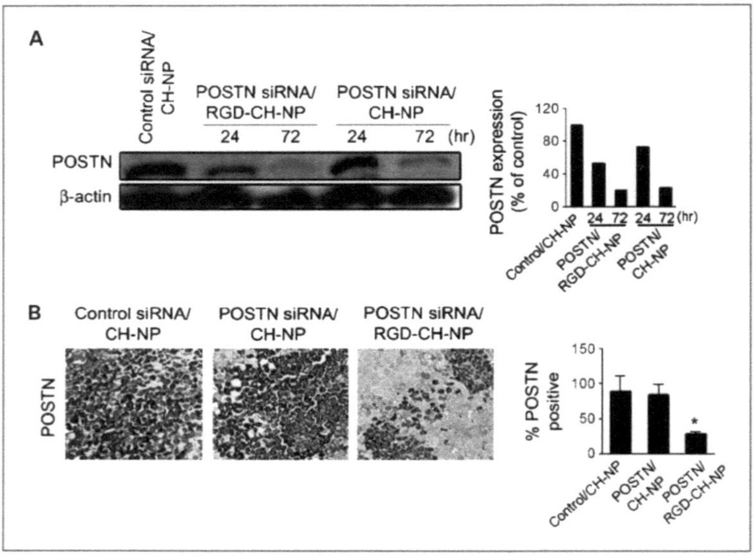

Fig. 4. Effect of *POSTN* downregulation following i.v. injection of POSTN siRNA/RGD-CH-NP into SKOV3ip1-bearing mice. (a) Western blot analysis was done for *POSTN* expression in tumor tissue (20 µg of protein used). Quantitative differences were determined by densitometry analysis. (b) *POSTN* expression in tumor tissues was assessed by immunohistochemistry at 24 hours. All of these analyses were recorded in five random fields for each slide and quantitative difference was determined by positive/negative expression of cells for staining (magnification, ×100). Error bars, SE; *, $P < 0.05$. (Reprinted with permission from the American Association for Cancer Research, Inc., from Han et al. 2010. Fig. 4).

Color image of this figure appears in the color plate section at the end of the book.

Preclinical Approaches

During the last five years our group has developed various nanoparticles designed to deliver siRNA systemically (Han et al.; Landen et al. 2005; Lu et al. 2010; Tanaka et al. 2010). We have reported the efficacy of DOPC nanoliposomes, chitosan nanoparticles and mesoporous silicon particles for delivery of siRNA in orthotopic models of ovarian cancer in nude mice. As a proof of principle study, we administered systemically DOPC/EphA2 nanoparticles and demonstrated that EphA2 gene and its protein could be completely silenced for up to four days (Landen et al. 2005). Based on these experiments, in all siRNA therapy studies we treat mice twice weekly. The therapeutic efficacy of DOPC/EphA2 siRNA nanoparticles was tested on two orthotopic ovarian cancer models. We showed that targeting EphA2 alone diminished tumor growth in both models compared to control groups (Landen et al. 2005). Moreover, when DOPC/EphA2 siRNA was combined with paclitaxel we observed a significant reduction in tumor growth compared to control siRNA and paclitaxel alone (Landen et al. 2005). Previously, we have used different gene targets in various preclinical cancer models in mice (Ozpolat et al. 2010). For example, we have shown that DOPC/siRNA could effectively downregulate its target and lead to decreased tumor growth in several tumor models, including pancreatic cancer, melanoma, liver, colorectal and breast cancer (Ozpolat et al. 2010). We have recently reported the development of a multistage vector composed of mesoporous silicon particles loaded with DOPC containing siRNA (Fig. 5) (Tanaka et al. 2010). As a proof of concept, this novel delivery method resulted in EphA2 gene silencing for at least 3 weeks in orthotopic ovarian cancer models following a single administration. A single dose of this delivery system had a therapeutic effect similar to DOPC/siRNA administered twice a week, proving to be a more efficient method to silence gene expression.

We developed and characterized chitosan nanoparticles for systemic delivery of siRNA into both tumor cells and tumor-associated vasculature (Lu et al. 2010). We selected chitosan nanoparticles due to their small size, slight positive charge, and high incorporation efficiency of siRNA. siRNA/chitosan was delivered into the tumor-associated endothelial cells, suggesting potential applications for targeting the tumor vasculature(Lu et al. 2010). To test its therapeutic efficacy we used Ezh2 siRNA directed to either the human or mouse sequence. First, after intravenous injection of either control siRNA/CH, Ezh2 Hs siRNA/CH, Ezh2 Ms siRNA/CH or the combination of Ezh2 targeted siRNAs into orthotopic ovarian cancer models we showed that Ezh2 siRNA/CH nanoparticles could silence ezh2 gene and protein expression for up to four days (Lu et al. 2010). Next, we assessed the effect of Ezh2 siRNA/CH nanoparticles on *in vivo* tumor growth. In all orthotopic ovarian cancer models tested treatment with Ezh2 Mm siRNA/CH resulted in a significant decrease in tumor burden, meanwhile Ezh2 Hs siRNA/CH as a single agent had modest effects on tumor growth. We observed the greatest reduction when we combined Ezh2 Hs siRNA/CH and Ezh2 Mm siRNA/CH nanoparticles (Lu et al. 2010). Additionally, we have developed an Arg-Gly-Asp (RGD) peptide-labeled chitosan nanoparticle (RGD-CH-NP) as a novel tumor targeted delivery system for siRNA (Han et al. 2010). We reported that RGD-CH-NP loaded with siRNA significantly increased selective intratumoral delivery of various cancer promoting genes in orthotopic ovarian cancer models of ovarian cancer leading to a significant inhibition of tumor growth (Han et al. 2010).

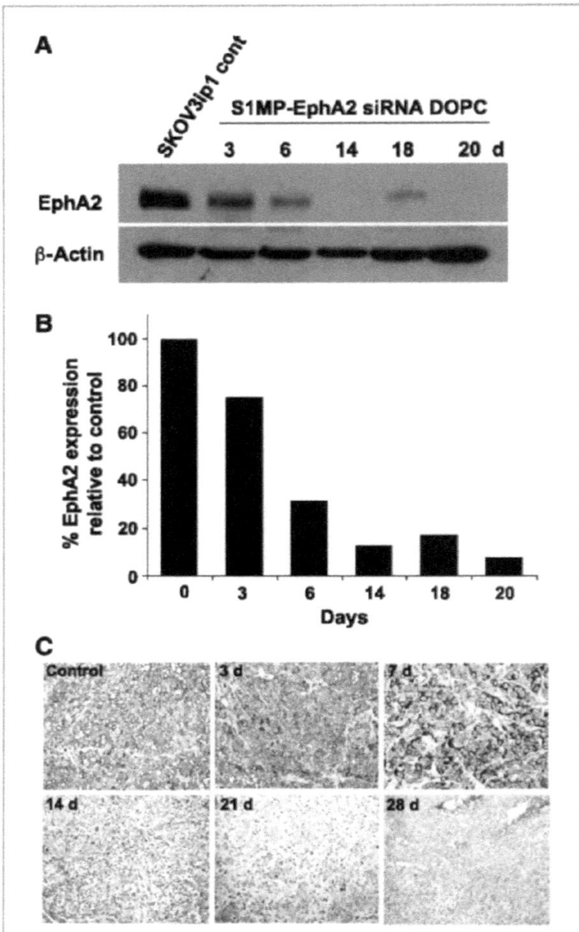

Fig. 5. Systemic delivery of siRNA-DOPC using S1MP results in long-lasting *in vivo* gene silencing. The mice (three mice per time point) bearing SKOV3ip1 orthotopic ovarian tumors were injected with S1MP-EphA2-siRNA-DOPC or left untreated. (a) The tumors were harvested at the indicated time points for Western blot to measure EphA2 expression levels. Thirty micrograms of tumor lysate were separated on a 10% SDS-PAGE and transferred on to a polyvinylidene difluoride membrane. The membrane was incubated with anti-EphA2 antibody overnight at 4°C. The membrane was tested for β-actin to confirm equal loading. (b) Densitometric analysis was performed to normalize EphA2 expression by β-actin. Data were expressed as % of normalized value to the untreated group. (c) Immunohistochemical analysis of EphA2 expression in the SKOV1ip3 tumor. Images were taken at original magnification of ×400. (Reprinted with permission from the American Association for Cancer Research, Inc., from Tanaka et al. 2010. Fig. 2).

Color image of this figure appears in the color plate section at the end of the book.

Clinical Approaches

Currently, there are several clinical trials using siRNA for therapeutic gene silencing. The majority of these trials rely on forms of localized siRNA delivery or chemically modified or naked siRNA. The most prominent clinical trials are focusing on age-related macular degeneration, which is a leading cause of blindness. However, currently there is, to the best of our knowledge, one anti-cancer siRNA therapy (Calando Pharmaceuticals' CALAAA-01) that has entered to phase 1 trial (Tokatlian and Segura 2010).

Final Remarks

As the field moves forward, we are certain that novel nanoparticles ranging from polymers, lipids, and conjugated molecules will be developed to improve and maximize the therapeutic use of siRNA. As we learn more about cancer and its microenvironment, we will be able to exploit that knowledge in order to treat this disease better. For example, novel carrier nanoparticles that specifically target tumor cells or its associated vasculature with minimal delivery to non cancerous tissue can be developed, effectively minimizing side effects and improving its therapeutic range. In brief, since systemic siRNA delivery to tumor remains the most significant barrier in the development of siRNA as a therapeutic modality, new technologies have to be developed. A variety of nanotechnology platforms, as discussed in this review, hold potential to achieve this goal successfully. These and other novel technologies, currently under development, are leading the way to better and safer nanoparticles to deliver siRNA and offer hope for better treatment options for cancer patients.

Applications to Areas of Health and Disease

In the last decade, siRNA has rapidly become a powerful tool for silencing gene expression due to its high degree of specificity leading scientists to target genes that could not be targeted before. Although the clinical application of naked siRNA appears to be limited, its use as a therapeutic agent is greatly expanded when siRNA is loaded to a nanoparticle and administered systemically. This approach has garnered substantial attention from the scientific community for possible use to treat several diseases, including viral infections and cancer. At this time, various RNAi-based drugs are being developed for the treatment of age-related macular degeneration and for respiratory syncytial virus. We have developed various siRNA/nanoparticles and tested their efficacy in various preclinical cancer models. We have shown that siRNA-induced silencing of genes important in cancer progression leads to significant reduction of tumor growth in mouse models of cancer.

Key Facts of siRNA (196 words)

- RNA inhibition of protein expression was discovered more than twenty years ago when Jorgensen and colleagues were trying to improve the color of petunia flowers by introducing pigment-producing genes into the plant. Instead of improving the flower's color, they found that the flowers had white patterns or were completely white. This process was known at that time as post-transcriptional gene silencing.
- In 1998, Fire and Mello observed in *c. elegans* that double-stranded RNA was the source of sequence-specific inhibition of protein expression, which they called "RNA interference".
- In 2001, Elbashir and colleagues demonstrated that 21-nucleotide siRNA duplexes specifically suppressed expression of endogenous genes in several mammalian cell lines.
- A year later, McCaffrey and colleagues showed that transgene expression could be suppressed in mice by synthetic siRNA and by shRNA transcribed *in vivo* from DNA templates.
- Four years later, Zimmermann and colleagues demonstrated that siRNA delivered systemically in a liposomal formulation can result in gene silencing in non-human primates, supporting the potential use of RNAi molecules for therapy.
- The 2006 Nobel Prize in medicine was awarded to Fire and Mello for their discovery of RNA interference gene silencing by double stranded RNA.

Definition of Key Terms (250 words)

RNA interference (RNAi): A biological pathway in eukaryotic cells by which a short RNA fragment is able to induce the elimination of mRNA that contains a complementary sequence.

Small interfering RNA (siRNA): RNA fragments that are approximately 21–23 nucleotides long and capable of inducing sequence-specific elimination of complementary mRNA.

Liposomes: A drug delivery vehicle composed of lipids. These nanoparticles are either unilamellar (one set of head-groups) or multilamellar (two or more sets of head-groups). Liposomes are used to deliver hydrophilic or hydrophobic drugs, depending on their structure and composition.

DOPC: A type of liposome. It has a neutral charge and has been proven to be effective in delivering its payload.

Chitosan: A linear polysaccharide normally found in the exoskeleton of crustaceans and the walls of fungi. Recently, it has been used in different areas of the biomedical field, such as a delivery nanoparticle.

RGD: Peptide composed of the three amino acids "Arginine-Glycine-Aspartic Acid". It serves as a recognition sequence for integrin binding to many extracellular matrix proteins allowing for better cancer cell targeting.

EphA2: Tyrosine kinase receptor in the ephrin family that plays a key role in neuronal development. It has been shown to be involved in the formation of tumoral vasculature and promotion of tumoral invasion.

Ezh2: A member of the polycomb-group proteins that has been implicated in the progression and metastasis of several cancers. Polycomb-group proteins are negative regulators of gene expression and are involved in the stable transmission of the repressive state of their target gene.

Summary Points

- SiRNA can silence gene expression with high specificity by inducing degradation of its complementary mRNA sequence.
- Naked siRNA has been used to silence topical or local gene expression, but its systemic delivery remains a challenge due to the lack of safe and effective nanoparticles.
- Cationic liposomes have been among the most popular delivery particles for nucleic acids, but major concerns remain regarding their safety for *in vivo* use.
- Neutral liposomes, such as DOPC, deliver siRNA efficiently and safely into the tumor tissue leading to decreased tumor growth in pre-clinical cancer models.
- Polymeric carriers, such as chitosan, increased tumor and tumor-associated endothelial cell siRNA uptake resulting in the reduction of tumor growth.
- Chemical modifications of the siRNA molecule can help stabilize the molecule, but can increase the risk for off-target effects, diminished activity and increased toxicity.
- The addition of ligands that target the tumor cell and its microenvironment to a nanoparticle can increase its therapeutic window by increasing its bioavailability at target tissues and eliminating non-specific delivery.
- At this time, there are several clinical trials using siRNA for therapeutic gene silencing, but just one developing it as an anti-cancer approach.

Acknowledgements

Portions of this work were supported by the NIH (P50 CA083639, P50 CA098258, CA128797, RC2GM092599, U54 CA151668), the Ovarian Cancer Research Fund, Inc. (Program Project Development Grant), the Zarrow Foundation, the Kim Medlin Fund, and the Laura and John Arnold Foundation.

Abbreviations

RNAi	:	RNA interference
siRNA	:	small interfering RNA
DOPC	:	1,2-dioleoyl-sn-glycero-3-phosphatidylcholine
dsRNA	:	double stranded RNA
RISC	:	RNA-induced silencing complex
PEG	:	polyethylene glycol
DOTMA	:	N-[1-(2,3-dioleyloxy)propyl]-N,N,N-trimethylammonium chloride
DOTAP	:	1,2-dioleoyl-3-trimethylammonium-propane
FDA	:	Food and Drug Administration
SLN	:	nucleic acid-lipid particles
SNALP	:	cationic solid lipid nanoparticles
LDL	:	low-density lipoprotein
DOPE	:	dioleoyl phosphatidylethanolamine
HA	:	hyaluronic acid
PEC	:	polyelectrolyte complex micelle
PEI	:	polyetherimide
HDL	:	high density lipoprotein
PLL	:	poly-L-lysine
Hs	:	human
Ms	:	mouse
RGD	:	Arg-Gly-Asp peptide

References

Bumcrot, D., M. Manoharan, V. Koteliansky and D.W. Sah. 2006. RNAi therapeutics: a potential new class of pharmaceutical drugs. Nat. Chem. Biol. 2: 711–719.

Chiu, Y.L., and T.M. Rana. 2003. siRNA function in RNAi: a chemical modification analysis. RNA 9: 1034–1048.

Collins, C.D., C.A. Pedersen, P.J. Schneider, A.S. Miller, S.J. Sierawski and R.K. Roux. 2004. Effect on amphotericin B lipid complex use of a clinical decision support system for computerized prescriber order entry. Am. J. Health Syst. Pharm. 61: 1395–1399.

de Fougerolles, A., and T. Novobrantseva. 2008. siRNA and the lung: research tool or therapeutic drug? Curr. Opin. Pharmacol. 8: 280–285.

DiFiglia, M., M. Sena-Esteves, K. Chase, E. Sapp, E. Pfister, M. Sass, J. Yoder, P. Reeves, R.K. Pandey, K.G. Rajeev, M. Manoharan, D.W. Sah, P.D. Zamore and N. Aronin. 2007. Therapeutic silencing of mutant huntingtin with siRNA attenuates striatal and cortical neuropathology and behavioral deficits. Proc. Natl. Acad. Sci. USA 104: 17204–17209.

Elbashir, S.M., J. Harborth, W. Lendeckel, A. Yalcin, K. Weber and T. Tuschl. 2001. Duplexes of 21-nucleotide RNAs mediate RNA interference in cultured mammalian cells. Nature 411: 494–498.

Fire, A., S. Xu, M.K. Montgomery, S.A. Kostas, S.E. Driver and C.C. Mello. 1998. Potent and specific genetic interference by double-stranded RNA in Caenorhabditis elegans. Nature 391: 806–811.

Frantz, S. 2006. Safety concerns raised over RNA interference. Nat. Rev. Drug. Discov. 5: 528–529.

Gray, M.J., G. Van Buren, N.A. Dallas, L. Xia, X. Wang, A.D. Yang, R.J. Somcio, Y.G. Lin, S. Lim, F. Fan, L.S. Mangala, T. Arumugam, C.D. Logsdon, G. Lopez-Berestein, A.K. Sood and L.M. Ellis. 2008. Therapeutic targeting of neuropilin-2 on colorectal carcinoma cells implanted in the murine liver. J. Natl. Cancer Inst. 100: 109–120.

Halder, J., A.A. Kamat, C.N. Landen Jr., L.Y. Han, S.K. Lutgendorf, Y.G. Lin, W.M. Merritt, N.B. Jennings, A. Chavez-Reyes, R.L. Coleman, D.M. Gershenson, R. Schmandt, S.W. Cole, G. Lopez-Berestein and A.K. Sood. 2006. Focal adhesion kinase targeting using *in vivo* short interfering RNA delivery in neutral liposomes for ovarian carcinoma therapy. Clin. Cancer Res. 12: 4916–4924.

Han, H.D., L.S. Mangala, J.W. Lee, M.M. Shahzad, H.S. Kim, D. Shen, E.J. Nam, E.M. Mora, R.L. Stone, C. Lu, S.J. Lee, J.W. Roh, A.M. Nick, G. Lopez-Berestein and A.K. Sood. 2010. Targeted gene silencing using RGD-labeled chitosan nanoparticles. Clin. Cancer Res. 16: 3910–3922.

Hannon, G.J., and J.J. Rossi. 2004. Unlocking the potential of the human genome with RNA interference. Nature 431: 371–378.

Harborth, J., S.M. Elbashir, K. Vandenburgh, H. Manninga, S.A. Scaringe, K. Weber and T. Tuschl. 2003. Sequence, chemical, and structural variation of small interfering RNAs and short hairpin RNAs and the effect on mammalian gene silencing. Antisense Nucleic. Acid. Drug. Dev. 13: 83–105.

Kim, H.R., I.K. Kim, K.H. Bae, S.H. Lee, Y. Lee and T.G. Park. 2008a. Cationic solid lipid nanoparticles reconstituted from low density lipoprotein components for delivery of siRNA. Mol. Pharm. 5: 622–631.

Kim, S.H., J.H. Jeong, S.H. Lee, S.W. Kim and T.G. Park. 2008b. Local and systemic delivery of VEGF siRNA using polyelectrolyte complex micelles for effective treatment of cancer. J. Control Release 129: 107–116.

Landen, C.N., A. Chavez-Reyes Jr., C. Bucana, R. Schmandt, M.T. Deavers, G. Lopez-Berestein and A.K. Sood. 2005. Therapeutic EphA2 gene targeting *in vivo* using neutral liposomal small interfering RNA delivery. Cancer Res. 65: 6910–6918.

Lu, C., H.D. Han, L.S. Mangala, R. Ali-Fehmi, C.S. Newton, L. Ozbun, G.N. Armaiz-Pena, W. Hu, R.L. Stone, A. Munkarah, M.K. Ravoori, M.M. Shahzad, J.W. Lee, E. Mora, R.R. Langley, A.R. Carroll, K. Matsuo, W.A. Spannuth, R. Schmandt, N.B. Jennings, B.W. Goodman, R.B. Jaffe,

A.M. Nick, H.S. Kim, E.O. Guven, Y.H. Chen, L.Y. Li, M.C. Hsu, R.L. Coleman, G.A. Calin, E.B. Denkbas, J.Y. Lim, J.S. Lee, V. Kundra, M.J. Birrer, M.C. Hung, G. Lopez-Berestein and A.K. Sood. 2010. Regulation of tumor angiogenesis by EZH2. Cancer Cell 18: 185–197.

McCaffrey, A.P., L. Meuse, T.T. Pham, D.S. Conklin, G.J. Hannon and M.A. Kay. 2002. RNA interference in adult mice. Nature 418: 38–39.

Merritt, W.M., Y.G. Lin, L.Y. Han, A.A. Kamat, W.A. Spannuth, R. Schmandt, D. Urbauer, L.A. Pennacchio, J.F. Cheng, A.M. Nick, M.T. Deavers, A. Mourad-Zeidan, H. Wang, P. Mueller, M.E. Lenburg, J.W. Gray, S. Mok, M.J. Birrer, G. Lopez-Berestein, R.L. Coleman, M. Bar-Eli and A.K. Sood. 2008a. Dicer, Drosha, and outcomes in patients with ovarian cancer. N. Engl. J. Med. 359: 2641–2650.

Merritt, W.M., Y.G. Lin, W.A. Spannuth, M.S. Fletcher, A.A. Kamat, L.Y. Han, C.N. Landen, N. Jennings, K. De Geest, R.R. Langley, G. Villares, A. Sanguino, S.K. Lutgendorf, G. Lopez-Berestein, M.M. Bar-Eli and A.K. Sood. 2008b. Effect of interleukin-8 gene silencing with liposome-encapsulated small interfering RNA on ovarian cancer cell growth. J. Natl. Cancer Inst. 100: 359–372.

Morrissey, D.V., J.A. Lockridge, L. Shaw, K. Blanchard, K. Jensen, W. Breen, K. Hartsough, L. Machemer, S. Radka, V. Jadhav, N. Vaish, S. Zinnen, C. Vargeese, K. Bowman, C.S. Shaffer, L.B. Jeffs, A. Judge, I. MacLachlan and B. Polisky. 2005. Potent and persistent *in vivo* anti-HBV activity of chemically modified siRNAs. Nat. Biotechnol. 23: 1002–1007.

Napoli, C., C. Lemieux and R. Jorgensen. 1990. Introduction of a Chimeric Chalcone Synthase Gene into Petunia Results in Reversible Co-Suppression of Homologous Genes in trans. Plant Cell 2: 279–289.

Okumura, A., P.M. Pitha and R.N. Harty. 2008. ISG15 inhibits Ebola VP40 VLP budding in an L-domain-dependent manner by blocking Nedd4 ligase activity. Proc. Natl. Acad. Sci. USA 105: 3974–3979.

Ozpolat, B., A.K. Sood and G. Lopez-Berestein. 2010. Nanomedicine based approaches for the delivery of siRNA in cancer. J. Intern. Med. 267: 44–53.

Park, J.W., K. Hong, P. Carter, H. Asgari, L.Y. Guo, G.A. Keller, C. Wirth, R. Shalaby, C. Kotts, W.I. Wood, et al. 1995. Development of anti-p185HER2 immunoliposomes for cancer therapy. Proc. Natl. Acad. Sci. USA 92: 1327–1331.

Spagnou, S., A.D. Miller and M. Keller. 2004. Lipidic carriers of siRNA: differences in the formulation, cellular uptake, and delivery with plasmid DNA. Biochemistry 43: 13348–13356.

Stephenson, S.M., W. Yang, P.J. Stevens, W. Tjarks, R.F. Barth and R.J. Lee. 2003. Folate receptor-targeted liposomes as possible delivery vehicles for boron neutron capture therapy. Anticancer Res. 23: 3341–3345.

Tanaka, T., L.S. Mangala, P.E. Vivas-Mejia, R. Nieves-Alicea, A.P. Mann, E. Mora, H.D. Han, M.M. Shahzad, X. Liu, R. Bhavane, J. Gu, J.R. Fakhoury, C. Chiappini, C. Lu, K. Matsuo, B. Godin, R.L. Stone, A.M. Nick, G. Lopez-Berestein, A.K. Sood and M. Ferrari. 2010. Sustained small interfering RNA delivery by mesoporous silicon particles. Cancer Res. 70: 3687–3696.

Tokatlian, T., and T. Segura. 2010. siRNA applications in nanomedicine. Wiley Interdiscip Rev. Nanomed. Nanobiotechnol. 2: 305–315.

Whitehead, K.A., R. Langer and D.G. Anderson. 2009. Knocking down barriers: advances in siRNA delivery. Nat. Rev. Drug Discov. 8: 129–138.

Wolfrum, C., S. Shi, K.N. Jayaprakash, M. Jayaraman, G. Wang, R.K. Pandey, K.G. Rajeev, T. Nakayama, K. Charrise, E.M. Ndungo, T. Zimmermann, V. Koteliansky, M. Manoharan and M. Stoffel. 2007. Mechanisms and optimization of *in vivo* delivery of lipophilic siRNAs. Nat. Biotechnol. 25: 1149–1157.

Zimmermann, T.S., A.C. Lee, A. Akinc, B. Bramlage, D. Bumcrot, M.N. Fedoruk, J. Harborth, J.A. Heyes, L.B. Jeffs, M. John, A.D. Judge, K. Lam, K. McClintock, L.V. Nechev, L.R. Palmer, T. Racie, I. Rohl, S. Seiffert, S. Shanmugam, V. Sood, J. Soutschek, I. Toudjarska, A.J. Wheat, E. Yaworski, W. Zedalis, V. Koteliansky, M. Manoharan, H.P. Vornlocher and I. MacLachlan. 2006. RNAi-mediated gene silencing in non-human primates. Nature 441: 111–114.

2

Nanocarriers for Gene Delivery

Conchita Tros de Ilarduya[1] and *Gloria Gonzalez-Aseguinolaza*[2,*]

ABSTRACT

Gene therapy (GT) using nanocarriers has the potential to have a revolutionary impact on cancer treatment. Non-viral GT has been largely ignored in the past due to inefficient gene delivery, toxicity and short-term gene expression. However, recent advances in nanoparticle technology and in the genetic material to be transferred are dramatically changing the popularity of this gene transfer methodology. New nanoparticles with better DNA compaction capacities, endosome escaping properties and nuclear import signals have recently been developed. Furthermore, the versatility of nanoparticles engenders optimism since may well yield active, highly-targeted delivery vectors suitable for *in vivo* applications in cancer GT. In this chapter, the authors review recent advances in gene delivery for cancer therapy using nanoparticles. In particular, the authors focus on the results obtained in clinical trials and experimental, clinically-relevant animal models.

[1]Department of Pharmacy and Pharmaceutical Technology, School of Pharmacy, University of Navarra, Scieces Building. C/Irunlarrea 1, 31008 Pamplona, Spain; E-mail: ctros@unav.es

[2]Division of Hepatology and Gene Therapy, Center for Investigation in Applied Medicine (CIMA), University of Navarra, Av Pio XII 55, 31008 Pamplona, Spain;
E-mail: ggasegui@unav.es

*Corresponding author

List of abbreviations after the text.

INTRODUCTION

Cancer is among the top three "killers" in modern society. The current treatment options for most solid tumours are still surgical resection combined with radiation or chemotherapy to reduce incidence of tumour recurrence and obtain optimum local disease control. This therapy has significant toxicities and side effects are common. Moreover, the efficacy of the treatment is not always guaranteed. As an alternative, GT is becoming a promising option in cancer treatment. GT is based on the introduction of genetic material into cells with the aim of correcting a genetic defect or producing a therapeutic protein. In the beginning, GT was focused on the treatment of monogenic hereditary diseases, but it was rapidly applied to acquired-diseases. In fact, cancer research is currently the area of study in which most GT clinical trials are being performed (Fig. 1) (Gillet, Macadangdang et al. 2009). Cancer GT has been intensively developed using viral and non-viral vectors. Although viruses are incredibly efficient gene delivery agents, clinical trials are often limited by several concerns, e.g., toxicity, immunogenicity, inflammatory properties, production and packing problems and high cost. Synthetic or non-viral vectors show a higher bio-safety profile, are more cost-effective and are easier to manufacture than viral vector systems. However, development of non-viral vectors for *in vivo* gene delivery, especially for clinical applications, has also suffered from problems including toxicity, low gene transfer efficiency, short-term gene expression and poor stability *in vivo*. Therefore, it is difficult to find a single method that meets all the conditions for ideal gene transfer and vector expression. Among non-viral gene delivery vehicles, one of the most promising is nanoparticles. In general, genetic nanoparticles are materials

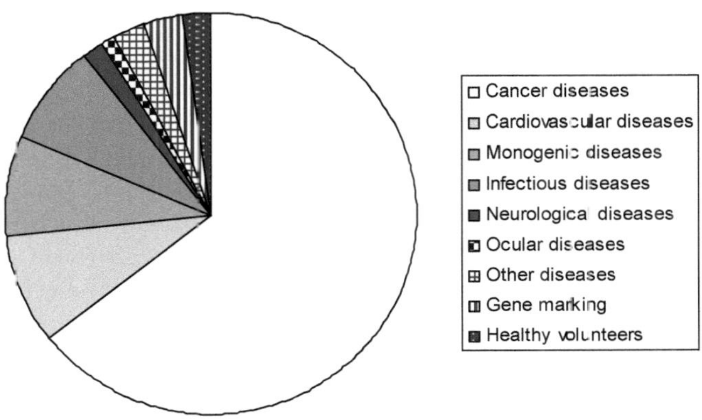

Fig. 1. Clinical targets for GT. Information obtained from The Journal of Gene Medicine Clinical Trial site. http://www.wiley.com/legacy/wileychi/genmed/clinical/

that bind electrostatically to DNA or RNA, condensing the genetic material into nanoscale dimension complexes between 1 and 100 nm in diameter. These particles protect the genetic material from degradation and allow the genetic material to enter cells. Because of the size and supramolecular structure of nanoparticles, technologies utilising nanoparticles have a great potential to improve cancer therapy. A primary attribute of nanoparticle delivery systems is their potential to specifically enhance the accumulation of the therapeutic agent in tumour cells over healthy tissues. Furthermore, in order to increase the transfection efficiency, non-viral delivery systems have been engineered to mimic viral delivery systems by incorporation of a condensing or encapsulation agent to protect the DNA from nucleases, targeting ligands to reduce nonspecific uptake, endosomal release agents to avoid lysosomal degradation and nuclear localisation signals to enhance nuclear uptake. There are a number of approaches that can be undertaken to increase the clinical efficacy and effectiveness of transgene delivery including improvement of the vector efficiency, enhancement of the vector's cell-targeting capabilities and development of transgenes with enhanced tumour-killing characteristics. In this chapter, we review some of the properties of nanoparticles as gene delivery vectors and the material most commonly used to produce genetic nanoparticles for cancer treatment, followed by a review of recent advances in the use of nanoparticles in gene delivery for cancer therapy.

PROPERTIES OF NANOPARTICLES AS GT VEHICLES FOR CANCER

The use of Nanoparticles to Circumvent Biological Barriers in Tumours

One of the most important obstacles to cancer GT is the presence of several barriers that impede direct and systemic access to the specific site of action. Effective cancer GT requires the successful transport of a transgene from the site of administration to the nuclei of the target cells. Biodistribution barriers for systemic gene delivery include interactions with blood components and nonspecific uptake by macrophages. Once in the blood stream, transgenes are exposed to serum inactivation and degradation by nucleases. Thus, effective vectors need to be capable of protecting DNA, rendering it inaccessible to these degradative enzymes; this protection can be achieved by encapsulation in nanoparticles. At the cell surface, the rate of entry into cells occurs relatively slowly and varies with cell type. After cellular uptake of the gene delivery systems by endocytosis, endosomal release is another critical barrier that affects the efficiency of gene transfer. Most DNA is retained in the endosomes and eventually degraded or inactivated by lysosomal enzymes. A number of strategies have been developed to enhance endosomal release. One uses fusogenic peptides or lipids to disrupt the endosomal membrane. Another strategy depends on using gene delivery systems with high buffering capacities, known as "proton sponges", which are presumably able to reduce the acidification of the endosome, resulting in swelling and membrane rupture (see below). Following release from the endosome, the transgene is transported to the nucleus through the cytoplasm, where it can be exposed to degradation by cytosolic nucleases. The incorporation of short peptide sequences known as "nuclear localisation signals" into delivery vectors has been shown to enhance DNA delivery to the nucleus via active transport along microtubules. Following nuclear uptake, the transgene is transcribed into messenger RNA (mRNA), which is exported to the cytoplasm and translated to the desired protein (Fig. 2).

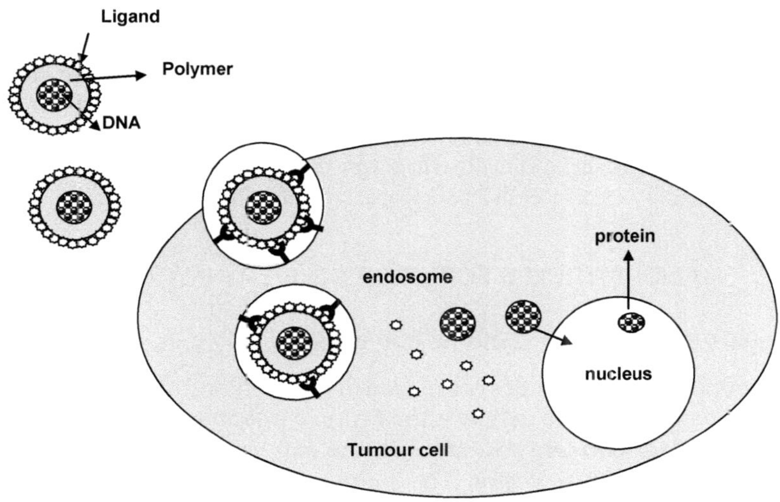

Fig. 2. Illustration of the uptake of nanoparticles, via endocytosis, into the tumour cells. Gene delivery is a complex process with many possible rate-limiting steps. After the rupture of the endosomal membrane, the nanoparticle cargo is delivered into the cytosol and enters the nucleus. After cellular uptake by endocytosis, the gene carrier must traverse the expansive and crowded cytoplasm to reach the nucleus.

In addition to those barriers, other biological barriers may challenge the efficiency of cancer GT. One of the characteristics of tumour tissue is the defective vascular architecture created caused by the rapid vascularisation necessary to serve fast-growing cancers. In contrast to normal tissue, tumours contain a high density of abnormal blood vessels that are dilated and poorly differentiated, with chaotic architecture and aberrant branching. Subsequently, various functions of the tumour vasculature were found to be impaired, which resulted in an increased accumulation of macromolecules. These findings support the use of nanoparticles in tumour therapy as carriers because the particles will passively accumulate in solid tumours after their intravenous administration.

Targeting Delivery

It is common knowledge that one major difficulty during cancer treatment is to destroy tumour cells without destroying normal tissues. Several approaches to improving the selective toxicity of anticancer therapeutics are presently being pursued. The most commonly used method is antibody- or ligand-mediated targeting of anticancer therapeutics. The basic principle that underlies ligand-targeted therapeutics is that the selective delivery of antineoplastic drugs to cancer cells or cancer-associated tissues, such as tumour vasculature, can be enhanced by associating the drugs with molecules that bind to antigens or receptors that are either uniquely expressed or overexpressed in target cells, when compared with normal tissues antigen. This technique allows the specific delivery of drugs to cancer cells. Nanoparticles are sufficiently large and flexible to contain multiple targeting ligands allowing for the specific engulfment of the nanoparticles by cancer cells (Wang and Thanou 2010).

NANOPARTICLES FOR GENE DELIVERY IN CANCER

Nanoparticles Based on Cationic Polymers

Polymers are large molecules composed of repeating subunits. For effective polymer-mediated gene delivery, the cationic polymer carrier has to fulfil a series of drug delivery functions in the extracellular and intracellular transport of the DNA vector. The polymer has to compact DNA into particles of virus-like dimensions, which can circulate though the blood to reach the target tissue. It also has to protect the DNA from degradation and against undesired interactions with the biological environment. Furthermore, the polymer should facilitate target cell binding and internalisation, ideally in a target cell-specific manner, as well as promote endosomal escape, trafficking through the cytoplasmic environment, localisation to the nucleus and vector unpacking. In addition, the polymer should be nontoxic, non-immunogenic, and biodegradable. In reality, no polymer is able to achieve all the desired extracellular and intracellular delivery functions; therefore, additional functional elements have to be included in the polyplex formulation. The majority of polymers used in gene delivery are synthesised by chemical methods or obtained from natural sources.

Polyethylenimine (PEI)

PEI is by far the most commonly used non-viral vector for gene and oligonucleotide delivery in animals (Demeneix and Behr 2005). PEI is a linear or branched polymer that has a high cationic charge, which promotes efficient DNA concentration and entry into cells. Furthermore, PEI nanoparticles have endosomolytic properties. Due to the presence of many protonatable amino nitrogen atoms, which make the polymer an effective proton sponge, PEI can cause the endosome to rupture through osmotic swelling, resulting in the release of PEI-DNA complexes into the cytoplasm. PEI can be covalently modified with different molecules, allowing the targeting of nanoparticles to specific cells or tissues. However, systemic delivery of PEI complexes has also been shown to result in passive tumour targeting (Kircheis, Sch et al. 1999).

Poly(lactic-co-glycolic acid) (PLGA)

PLGA is a widely-used polymer for preparing nanoparticles with a long-standing track record in biomedical applications. PLGA nanospheres have been suggested to be a good gene delivery carrier because of their safety, biocompatibility and well-documented ability to maintain sustained drug release. These DNA/RNA loaded nanoparticles extravasate through tumour vasculature, delivering their DNA cargo into the cells via an enhanced permeability and retention (EPR) effect, thereby increasing the therapeutic effect of the DNA loaded nanoparticles (Zou, Liu et al. 2009).

Dendrimers

Dendrimers are synthetic, branched macromolecules that form tree-like structures. Different types of dendrimers can be synthesised based on the core structure used during the polymerisation process. They are synthesised as well-defined spherical structures ranging from 1 to 10 nanometers in diameter and enter the cells by endocytosis, after which the DNA gets transported into the nucleus. The synthetic versatility of dendritic molecules has enabled the synthesis of a wide array of DNA binders and delivery vehicles, each with different advantages. The flexibility to modify and adapt dendrimers to meet the needs of therapeutic applications for pathological conditions, such as cancer or tumours, is one of the important characteristics of this technology. A major advantage of dendrimers for *in vivo* applications is their ability to protect DNA from the action of DNAase found in serum (Koppu, Oh et al. 2010).

The core structure most commonly found in the literature to form dedrimers for GT applications is Polyamidoamine (PAMAM). PAMAM dendrimers have shown promise for biomedical applications because they can be easily conjugated with targeting molecules, imaging agents, and drugs, have high solubility in water, have well-defined chemical structures, are biocompatible, and are rapidly cleared from the blood. More importantly, extensive work has been completed with PAMAM dendrimers, and *in vivo* studies have shown no evidence of toxicity when administered intravenously.

Chitosan

Chitosan is a mucopolysacharide closely related to cellulose. Chitosan is obtained by deacetylation of chitin, the major compound of exoskeletons in crustaceans. Because of its low production costs, biodegradability, biocompatibility, and recent FDA approval, the pharmaceutical and food applications of chitosan have increased remarkably over recent years. By coacervation between the positively charged amine groups on the chitosan and negatively charged phosphate groups on the DNA, chitosan-DNA nanoparticles have been reported to form easily (Bowman and Leong 2006).

Protein and peptide-based polymers

Peptide based nanoparticles consist of consecutive basic amino acids, like polylysine. Polylysine peptides can be modified by the inclusion of a sequence of histines (KHKHKHKHKK), which gives the polymer endosomolytic properties or a "proton-sponge" effect, as described for PEI-based nanoparticles. An Arg-Gly-Asp (RGD) peptide ligand and polylysine K(18) fusion peptide have been used to encapsulate plasmids. Protein based nanoparticles include, but are not limited to, chromatin components such as histones and protamine (Ezzat, El Andaloussi et al. 2010).

Protein/peptide-based polymers can be designed to incorporate a variety of functionalities, including response to microenvironmental stimuli, controlled biodegradation, and presentation of informational motifs for cellular and subcellular interactions. Biologically synthesised polymers do not contain toxic monomer residues or solvents and depending on their structure, can biodegrade to nontoxic amino acids that can be eliminated at a controlled rate from the body. However, limitations in the cloning and expression of cationic polymers in gene delivery must be overcome before these systems are considered successful systemic gene carriers.

Nanoparticles based on Liposomal Formulations

Liposomes consist of a lipid bilayer that encapsulates an aqueous phase containing the drugs. Positively charged lipids capture plasmids and deliver DNA into cells very efficiently. Generally, this is a simple procedure that requires mixing the cationic lipids with the DNA and adding them to the cells, resulting in the formation of aggregates composed of DNA and cationic lipids. The composition of liposomes may play an important role in their interactions with cells. The size of the liposomes and the type of cells are fundamentally important for efficient capture. Generally, liposomes are taken up by various endocytosic processes. However, rapid plasma clearance and high toxicity has limited the use of liposomes *in vivo*.

Nanoparticles based on Targeted Multifunctional Nanomedicine Platforms

Multifunctional nanoparticles are formed by the combination of several materials. Most of them contain materials that permit genetic material compaction, tumour-specific delivery and in many cases, the imaging of the delivery process (Fig. 3). These nanomaterials are characterised by increase serum stability and selective internalisation by cancer cells and are expected to be widely exploited as multifunctional delivery vehicles for cancer therapy. Some multifunctional nanoparticles have been already tested for GT transfer in tumour treatments and show very promising

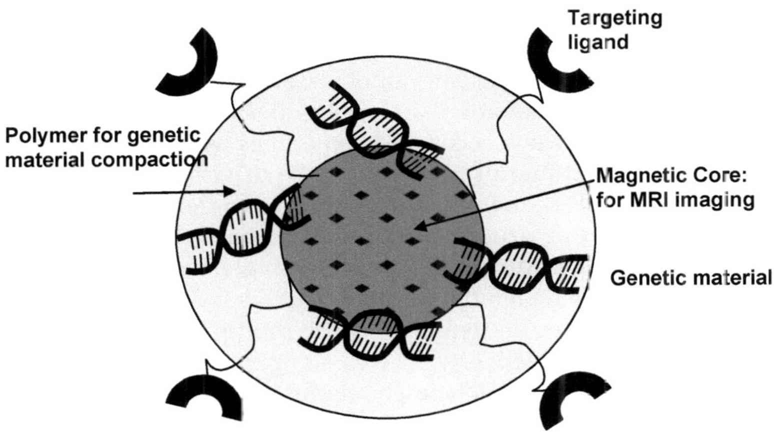

Fig. 3. Schematic representation of a targeted multifunctional nanomedicine platform.

results (Davis, Zuckerman et al. 2010). Some examples of this particular type of nanoparticles found in the literature are the following:

- Superparamagnetic Iron Oxide (SDION) nanoparticles, for magnetic resonance imaging (MRI) combined with Poly(Propyleneimine) generation 5 dendrimers (PPI G5), Poly(ethylene glycol) (PEG) coating and cancer specific targeting moiety (LHRH peptide) (Shen, Chen et al. 2010).
- Hollow manganese oxide nanoparticles (HMON) for MRI combined with surface functionalisation using 3,4-dihydroxy-l-phenylalanine (DOPA) and a therapeutic monoclonal antibody, herceptin, which selectively targets cancer cells (Bae KH et al. 2011).
- Iron oxide nanoparticle cores for MRI, coated with a copolymer of chitosan, polyethylene glycol (PEG), and polyethylenimine (PEI) and labelled with chlorotoxin (CTX) to target glioma cells (Veiseh, Kievit et al. 2009).
- Eight-armed polyethylene glycol (EAP) and low molecular weight (LMW) PEI (EAP-PEI) copolymer with a peptide capable of selectively binding fibroblast growth factor receptor (FGFR) on tumour cell membranes (Li D et al. 2010).
- Superparamagnetic iron oxide (SDION) nanoparticles for MRI, coated with the dye Cy 5.5 for near-Infra Red optical imaging and peptides that specifically target uMUC-1, to shuttle siRNA to human breast tumours (Kumar M et al. 2010).

THERAPEUTIC STRATEGIES FOR THE TREATMENT OF CANCER USING NUCLEIC ACID-LOADED NANOPARTICLES

Immunotherapeutic Approaches using Nanoparticles

Most of the existing cancer immunotherapeutic strategies are based on the systemic administration of a bolus dose of cytokines; for this reason, side effects always occur and sometimes are so intense that the patients must discontinue therapy before the drugs have had a change to eradicate the cancer. The use of nanocarries to deliver genes coding for immunotherapeutic drugs allows for an increase in the concentration of the drugs inside the tumour while decreasing systemic concentration and side effects, thus widening the therapeutic window. The IL-12 gene has been used for cancer treatment with multiple different types of nanoparticles. Aerosol delivery, as well as intravenous administration of PEI-IL12 complexes, has shown significant antitumor properties against lung cancer and lung metastases in a mouse model by activating cytotoxic T lymphocytes and NK cells and promoting a TH1 type of

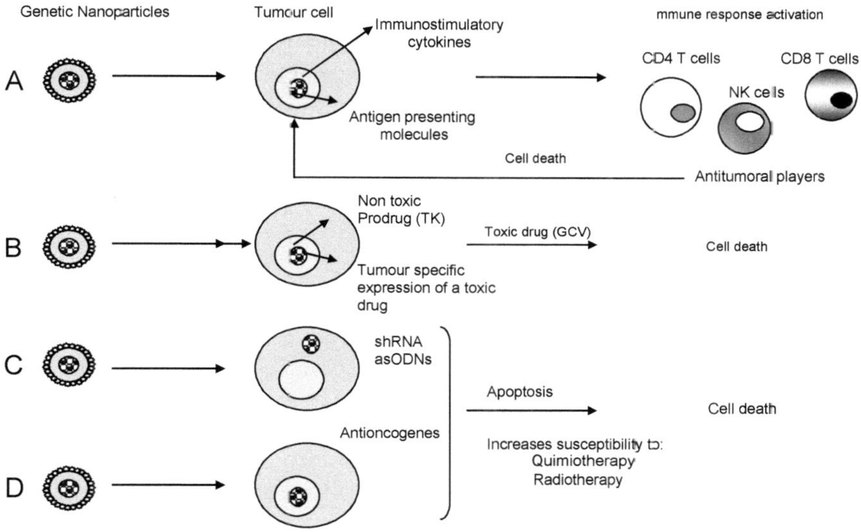

Fig. 4. Therapeutic strategies against cancer using nucleic acid-loaded nanoparticles. A. Immunotherapy. B. Suicide GT. C. Expression of genetic molecules that inhibit the expression of oncogenes. D. Expression of antioncogenes.

response (Hallaj-Nezhadi S et al. 2010). The antitumor efficacy of IL-12 coformulated with chitosan was assessed in mice bearing established colorectal (MC32a) and pancreatic (Panc02) tumours (Zaharoff, Hoffman et al. 2009). Weekly intratumoral (i.t.) injections caused complete tumour regression in 80% to 100% of the mice. Asialoglycoprotein receptor-targeted cationic nanoparticles composed of PLGA and DOTAP also showed a significant inhibition of tumour growth in a mouse model of hepatocellularcarcinoma.

In 2005, Schatzlein and his team used dendrimer nanoparticles to insert the gene carrying instructions for the production of a toxic protein called tumour necrosis factor α (TNFα) into cancer cells in mice. The results were impressive; in a mouse xenograft model of human epidermoid tumours (A431), the tumours shrank, and the mice survived significantly longer than those treated with just the gene or dendrimers alone (Koppu, Oh et al. 2010).

IFNα is a cytokine with strong antiproliferative and immunostimulatory activities. Recently, a copolymer consisting of eight-armed PEG (EAP) and PEI, coupled to a peptide that selectively bound fibroblast growth factor receptor (FGFR), MC11, was used to deliver a plasmid expressing IFNα into HepG2-bearing mice, resulting in a significant antitumoral effect.

A different strategy to increase the immunogenicity of a tumour cell is to increase the expression of antigen-presenting molecules, which are normally downregulated after cancer development. Liposomes expressing Human leukocyte antigen (HLA)-B7/beta-2 microglobulin (Allovectin-7) were used in the treatment of patients with melanoma (Table 1). Phase I and II trials have shown local and systemic response to these tumours, with an excellent safety profile. Currently, a Phase III trial has completed enrolment and is set to determine the safety and efficacy of treatment in comparison with standard chemotherapy (Chowdhery R 2011).

Suicide GT

Suicide GT consists of the expression of molecules that specifically induce the death of tumour cells, which can be achieved by the use of tumour specific promoters. In some cases, such genes encode enzymes that convert a prodrug into a toxic metabolite. Thymidine kinase (TK) from the herpes simplex virus is the best-characterised prodrug activation gene and converts ganciclovir (an antiviral drug) into a toxic phosphorylated compound that inhibits both nuclear and mitochondrial DNA synthesis. Several studies have demonstrated the efficacy of the TK expression system in the treatment of cancer in animal models. More importantly, liposomes carrying a plasmid expressing HSV-TK have been injected in patients with glioblastoma. After ganciclovir administration, two of eight patients exhibited a more than 50% reduction of tumour volume, and six of eight patients showed focal treatment effects (Voges, Reszka et al. 2003).

An antitumoral effect against different cancers was observed using other prodrug activating enzymes delivered via nanoparticles such as cytosine deaminase, which turns antifungal 5-fluocytosine (5-FC) into 5-fluorouracil, and *Escherichia coli* purine nucleoside phosphorylase, which converts fludarabine into a diffusible toxic compound (Li, Yu et al. 2005).

There is no doubt that two of most significant results in this field are the use of PEI-based gene delivery of diphteria toxin expressed under the control of H19 gene promoter for the treatment of human bladder carcinoma (Sidi, Ohana et al. 2008) and liposome-based gene delivery of a mutant of the proapoptotic gene BikDD (Table 1). The H19 gene is significantly expressed in 85% of bladder cancers. Following successful use of PEI-diphtheria toxin delivery for the treatment of this cancer in a rat model, two male human patients with recurrent bladder cancer were treated, both patients showed > 85% tumour remission (Sidi, Ohana et al. 2008).

BikDD is a mutant of the proapoptotic gene Bik designed to make cancer cells destroy themselves. Targeted expression of a therapeutic BikDD through intravenous delivery of liposomes carrying the BikDD gene effectively and preferentially killed cancer cells, but not normal cells, and has been shown to significantly repress growth of lung, liver, and pancreatic tumours, prolonging survival in multiple mouse models (Sher,

Table 1. Clinical trials against cancer using nanoparticles.

nanoparticle	gene	route of delivery	indication	clinical trial	ref
liposomes	HLA-B7/Beta 2-Microglobulin	intratumoural	Malignant melanoma	Phase I/II	Chowdhery, 2011
liposomes	Fus-1	intravenously	stage IV lung cancer	Phase I	Gopalan, 2004
liposomes	(HSV-TK)	intratumoural	Glioblastoma	Phase I/II	Voges, 2003
PEI	diphteria toxin	intravesical instillations	bladder carcinoma	Phase II	Sidi, 2008
liposomes	BikDD		Advanced Pancreatic Cancer	Approved-not yet active	Sher, 2009
liposomes	siRNA targetind bcr-abl	subcutanous/intravenous	Philadelphia CML	Phase I	Koldehoff, 2007
multifunctional nanoparticle	siRNA targeting RRM2	systemic	melanoma	Phase I	Davis, 2010

Tzeng et al. 2009). A phase I clinical trial using these nanoparticles for the treatment of advanced pancreatic cancer will be started soon (Table 1).

Nanocarriers for Delivering Antisense Oligonucleotides

The development of antisense Oligodeoxynucleotides (asODNs) that hybridise to and downregulate target mRNAs encoding the proteins that contribute to malignant transformation or that increase sensitivity to conventional treatment has a sound rationale, but has had an overall limited clinical success in cancer due to insufficient intracellular delivery. Different types of nanoparticles have been used to deliver asODNs. Transferrin conjugated lipopolyplex (TF-LPs) nanoparticles have been used to deliver an antisense asODN to the R2 subunit of ribonucleotide reductase (R2s-RR), which has been shown to contribute to chemoresistance in acute myeloid leukaemia (AML) (Jin, Liu et al. 2010). Chemoresistant AML cells treated with these TF-LPs-asODN-R2s-RR are sensitised to the chemotherapy agent cytarabine. PAMAM dedrimers have been used to deliver ODNs into breast cancer. During *in vivo* experiments in a human breast tumour xenograft mice model, PAMAM showed an accumulation of asODN that targetted VEGF and inhibited tumour vascularisation (Wang, Zhao et al. 2010). Furthermore, a PAMAM dendrimer was employed as a carrier to co-deliver antisense-miR-21 oligonucleotide (as-miR-21) and 5-fluorouracil (5-FU) to human glioblastoma cells. The co-delivery of as-miR-21 significantly improved the cytotoxicity of 5-FU and dramatically increased the apoptosis of U251 cells, while the migration ability of the tumour cells was decreased (Mei, Ren et al. 2010).

Table 2. Nanocarriers for gene delivery in cancer.

1. Genetic nanoparticles are materials that bind electrostatically to DNA or RNA, condensing the genetic material into nanoscale dimensions complexes between 1 and 100 nm in diameter.
2. Nanoparticles are based on cationic polymers (natural or artificial), liposomes, or a combination of different materials (multifunctional platforms).
3. One distinct characteristic of nanoparticles delivery systems is their potential to enhance the tumour accumulation of the therapeutic agent in tumour cells than in healthy tissues.
4. Nanoparticles can be easily modified, allowing the addition of ligands for cell specific gene delivery and targeting.
5. Today, the most promising nanoparticles for gene delivery are targeted multifunctional nanomedicine platforms.
6. Nanoparticles carrying different types of genetic material have shown significant antitumoural effects in a diverse variety of cancer animal models.
7. Genetic nanocarriers have reached the clinical oncology level with promising results.

This table lists the key facts of nanocarriers for gene delivery in cancer, including the particular properties of nanoparticles as gene delivery vectors.

Nanocarriers to Deliver small interfering RNA (siRNA)

Small interfering RNA (siRNA), short double-stranded RNA composed of 19–23 nucleotides, has received significant attention in the field of biomedical research and drug development. They have been shown to induce sequence-specific degradation of a complementary mRNA by a mechanism called RNA interference (RNAi), leading to knocking down of a target protein at a post-transcriptional level. New targets for RNAi-based cancer therapy are constantly emerging from the increasing knowledge of key molecular pathways important for carcinogenesis. Nevertheless, *in vivo* delivery of siRNA remains a crucial challenge for therapeutic success. In fact, the use of siRNA in clinical applications has been restricted by its poor intracellular uptake, failure to escape endosomes, and rapid enzymatic degradation in the bloodstream (Wang SL 2010). However, in order to overcome these obstacles, various cationic polymers, lipids, and peptides have been utilised to produce compact polyelectrolyte complexes with siRNA. Polyethylenimine (PEI) alone or complexed with polyethylene glycol- or chitosan has been one of the most popular cationic polymers used in non-viral GT due to its high transfection efficiency and superior ability to escape from endosomes. Many experimental data have shown the significant antitumor effect of siRNA-loaded nanoparticles.

More importantly, siRNA-loaded nanoparticles have spread into the arena of clinical oncology with promising results. A patient with recurrent Philadelphia chromosome-positive myeloid leukaemia (CML) resistant to conventional treatments was treated using liposomes loaded with siRNA, which targeted the bcr-abl oncogene, together with imatinib (Table 1). Liposomes loaded with siRNA were administered intravenously and subcutaneously three times. Furthermore, at the time of the first administration, one extramedullar CML node was also treated. The treatment was well tolerated with no significant side effects, and although a significant reduction on bcr-abl expression was only detected after the first administration, a reduction in size of treated and non-treated CML nodes was observed, (Koldehoff, Steckel et al. 2007) indicating a therapeutic effect. However, the exact mechanism for this therapeutic effect is not clear.

A most important achievement in this field has been reported recently: a phase I clinical trial involving the systemic administration of siRNA-loaded nanoparticles to patients with melanoma. The siRNA was targeted to the M2 subunit of ribonucleotide reductase (RRM2) using a multifunctional nanoparticle consisting of a cyclodextrin-based polymer, a human transferrin protein as the targeting ligand, and PEG (Table 1). The authors showed a reduction of the RRM2 protein and mRNA, and a dose dependent accumulation of the nanoparticles in the tumour (Davis,

Zuckerman et al. 2010). However, the results of this study (i.e., safety and efficacy) will not be available until later in the year. Furthermore, phase II and phase III clinical trials are necessary to prove that this method is efficacious in "fighting cancer".

Nanoparticle to Deliver Antioncogenes

Many types of cancers evolve as a result of loss of or mutation in the tumour suppressor genes that act as negative regulatory genes; these genes can overcome the uncontrolled cell proliferation driven by activated oncogenes or other stimulatory factors. Transduction of these tumour suppressor genes may result in dormancy of the tumour cells or cell death, also known as "apoptosis". An example of such a gene is p53, which has been studied extensively in cancer GT. Several nanocarriers have been used to restore p53 expression in tumours. PEI-mediated p53 gene transfer in mice bearing head and neck squamous cell carcinoma (HNSCC) xenografts has been shown to increase Bax and P21 proapoptotic proteins, to repress Bcl2, activate caspase-3, and cleave PARP, resulting in significant antitumoral effects (Ndoye, Dolivet et al. 2006). Furthermore, it has been shown that the restoration of p53 expression using PEI restored chemosensitivity and increased apoptosis, resulting in a 70% reduction in tumour size in an orthotopic bladder cancer model (Sweeney, Karashima et al. 2003).

An elegant strategy to restore p53 expression was developed by Shen et al. in which they used superparamagnetic nanoparticles (SDION) carrying the adenoviral gene E1A. Intratumoral delivery of SDION-EIA increased p53 expression and inhibited HER-2/neu expression, enhancing the radiosensitivity of human cervical cancer in xenograft mice (Shen, Chen et al. 2010).

Another example of the use of tumour suppressor gene expression to fight cancer is Bax, a pro-apoptotic molecule that functions as a tumour suppressor; Bax GT has been examined for the treatment of various cancers. The antitumour effect of Bax mRNA transfer by cationic liposomes was tested in human malignant melanoma cells in nude mice; treatment with Bax-loaded liposomes induced a significant reduction of tumour volume and was more efficient than the administration of naked DNA (Okumura, Nakase et al. 2008).

The tumour suppressor gene FUS1 is frequently inactivated early in the development of lung cancer; FUS1 mediates apoptosis in cancer cells, but not normal cells, through its interaction with Apaf1. Liposomes loaded with a FUS1-expressing plasmid were injected intravenously in stage IV progressive lung cancer patients after cisplatin combination chemotherapy, and there was no significant drug related toxicity (Gopalan, Ito et al. 2004). The median survival time for all patients was 14.6 months, which compares favourably to the seven-month median survival time for patients receiving second line therapy. However, since the last results published in 2004, no more information about this trial has been reported.

PUMA is induced by the p53 tumour suppressor and other apoptotic stimuli; it is a principal mediator of cell death in response to diverse apoptotic signals. Gene delivery of PEI/PUMA in oral cancer xenografts induced apoptosis and resulted in significant reductions (approximately 60%) of tumour growth *in vivo*. Furthermore, PEI-mediated PUMA GT prolonged the survival of animals with orthotopic oral cancers.

Conclusion

Because cancer is ultimately a genetic disease, it has long been the hope of researchers to use GT to attack tumours where they might be most susceptible. However, the lack of safe and efficient systemic gene delivery vectors has largely limited the potential of GT in the clinical setting. The use of nanotechnology has many advantages in cancer GT. With their small size, nanotechnology platforms can enter tumour vasculature, avoiding the existing physical barriers. In addition, functionalisation with hydrophilic polymers/oligomers can offer a long circulation half-life and prolong the exposure time of the tumour tissue to anticancer agents. Furthermore, inclusion of tissue recognition residues such as antibodies, lectins and ligands specific for cancer cells can help nanotechnology platforms to achieve tumour cell-targeting. Because most nanotechnology platforms are synthetic, there is less chance that they will provoke a harmful immune response. However, challenges still remain in the development and application of nanotechnology platforms in cancer therapy. Nanoparticles that target cells with a higher gene transfer efficiency need to be developed. Furthermore, more tests are needed to confirm the safety of use of the particles in humans. However, new information about the molecular mechanisms involved in cancer development and the discovery of new forms of therapeutic genetic material is increasing the number of genetic therapeutic options. In particular, since the discovery 12 years ago that double-stranded RNA can silence genes in a targeted manner, researchers have hailed RNAi as a powerful approach in creating new and potent medicines. In brief, nanotechnology platforms hold the promise to change the practice of oncology fundamentally, allowing easy and effective targeted therapies.

Applications to Areas of Health and Disease

Nanomedicine, as a newly emerging scientific field, has become one of the most studied subjects in the world. Biomedical application of nanotechnology is a rapidly developing area that promises new prospects in the improvement of diagnosing and treating human diseases. The ability to incorporate genes into a functionalised nanoparticle demonstrates a new era in pharmacotherapy in which genes can be selectively delivered to tissues or cells. As shown in this chapter, nanoparticles loaded with different genetic materials show promising results in the treatment of very different types of cancer in animal models. But more importantly, clinical trials are being conducted to demonstrate the safety and potential efficacy of these GT vectors in humans.

Summary Points

- Nanocarries are relatively easy to manipulate and allow the introduction of ligands that target the particle to specific cells and selectively attack the cancer cells, saving the normal tissue.
- Nanoparticles are, in theory; safer than recombinant viral vectors for gene delivery, but the transduction efficiency is significantly lower; however, the development of multifunctional gene nanocarriers that increase the transfer of DNA to the cell nucleus presents a very promising strategy.
- Gene delivery using nanoparticles is a promising strategy for the treatment of tumours when using siRNA as therapeutic agent, in particular.
- More tests are needed to confirm the safety of use of these particles in humans, but because they are synthetic, there is less chance that they will provoke a harmful immune response.

Key Terms

- **Gene Therapy** is a therapeutic strategy based on the introduction of genetic material into cells with the aim of correcting a genetic defect or producing a therapeutic protein.
- **Genetic Nanoparticles** are materials that bind electrostatically to DNA or RNA, condensing the genetic material into nanoscale dimension complexes between 1 and 100 nm in diameter.
- **Targeting**: Delivery of genetic material to specific targets (cells or tissues).
- **Genetic Immunotherapy**: Treatment of a disease by inducing, enhancing, or suppressing an immune response using gene transfer strategies.
- **Suicide Gene therapy**: consists of the expression of genes that induce the death of cells directly or alter the administration of a drug.
- **Gene Silencing**: Downregulation of target mRNAs that contribute to cancer development using genetic molecules like antisense Oligodeoxynucleotides or interference RNAs.

- **Small interfering RNA (siRNA):** Short double-stranded RNA composed of 19–23 nucleotides that induce sequence-specific degradation of a complementary mRNA.

Abbreviations

asODNs	:	Antisense Oligodeoxynucleotides
CTX	:	chlorotoxin
DOPA	:	3,4-dihydroxy-l-phenylalanine
EAP	:	Eight-armed polyethylene glycol
EPR	:	enhanced permeability and retention
FGFR	:	fibroblast growth factor receptor
GCV	:	Gancyclovir
GT	:	Gene Therapy
HMON	:	Hollow manganese oxide nanoparticles
LMW	:	low molecular weight
MRI	:	magnetic resonance imaging
PAMAM	:	Polyamidoamine
PEG	:	Poly(ethylene glycol)
PEI	:	Polyethylenimine
PLGA	:	Poly(lactic-co-glycolic acid
PPI G5	:	Poly(Propyleneimine) generation 5 dendrimers
RNAi	:	RNA interference
SDION	:	Superparamagnetic Iron Oxide
siRNA	:	Small interfering RNA
TK	:	Thymidine kinase from herpes simplex virus

References

Bae, K.H., C. Kim and T.G. Park. 2011. Surface functionalized hollow manganese oxide nanoparticles for cancer targeted siRNA delivery and magnetic resonance imaging. Biomaterials 32: 176–184.

Bowman, K., and K. W. Leong. 2006. Chitosan nanoparticles for oral drug and gene delivery. Int. J. Nanomed. 1: 117–128.

Chowdhery, G.R. 2011. Immunologic therapy targeting metastatic melanoma: Allovectin-7(®). Immunotherapy. 3: 17–21.

Davis, M.E., J.E. Zuckerman, C.H. Choi, D. Seligson, A.C. Tolcher, A. Alabi, Y. Yen, J.D. Heidel and A. Ribas. 2010. Evidence of RNAi in humans from systemically administered siRNA via targeted nanoparticles. Nature 464: 1067–70.

Demeneix, B., and J.P. Behr. 2005. Polyethylenimine (PEI). Adv. Gen. 53: 217–30.

Ezzat, K., S. El Andaloussi, R. Abdo and U. Langel. 2010. Peptide-based matrices as drug delivery vehicles. Cur. Pharm. Design 16: 1167–78.

Gillet, J.P., B. Macadangdang, R.L. Fathke, M.M. Gottesman and C. Kimchi-Sarfaty. 2009. The development of gene therapy: from monogenic recessive disorders to complex diseases such as cancer. Meth. Mol. Bio. 542: 5–54.

Gopalan, B., I. Ito, C.D. Branch, C. Stephens, J.A. Roth and R. Ramesh. 2004. Nanoparticle based systemic gene therapy for lung cancer: molecular mechanisms and strategies to suppress nanoparticle-mediated inflammatory response. Tech. Can. Res. Treat. 3: 647–57.

Hallaj-Nezhadi, S.L.F., and C. Dass. 2010. Nanoparticle-mediated interleukin-12 cancer gene therapy. J. Pharm. Pharm. Sci. 13: 472–485.

Jin, Y., S. Liu, B. Yu, S. Golan, C.G. Koh, J. Yang, L. Huynh, X. Yang, J. Pang, N. Muthusamy, K.K. Chan, J.C. Byrd, Y. Talmon, L.J. Lee, R.J. Lee and G. Marcucci. 2010. Targeted delivery of antisense oligodeoxynucleotide by transferrin conjugated pH-sensitive lipopolyplex nanoparticles: a novel oligonucleotide-based therapeutic strategy in acute myeloid leukemia. Mol. Pharm. 7: 196–206.

Kircheis, R., S. Schuller, S. Brunner, M. Ogris, K.H. Heider, W. Zauner and E. Wagner. 1999. Polycation-based DNA complexes for tumor-targeted gene delivery *in vivo*. J. Gen. Med. 1: 111–120.

Koldehoff, M., N.K. Steckel, D.W. Beelen and A.H. Elmaagacli. 2007. Therapeutic application of small interfering RNA directed against bcr-abl transcripts to a patient with imatinib-resistant chronic myeloid leukaemia. Clin. Exp. Med. 7: 47–55.

Koppu, S., Y.J. Oh, R. Edrada-Ebel, D.R. Blatchford, L. Tetley, R.J. Tate, and Duf. 2010. Tumor regression after systemic administration of a novel tumor-targeted gene delivery system carrying a therapeutic plasmid DNA. J. Control. Rel. 143: 215–21.

Kumar, M., Y.M. Dai, G.A. Moore and Z. Medarova. 2010. Image-guided breast tumor therapy using a small interfering RNA nanodrug. Can. Res. 70: 7553–7561.

Li, D., Y. Ping, F. Xu, H. Yu, H. Pan, H. Huang, Q. Wang, G. Tang and J. Li. 2010. Construction of a star-shaped copolymer as a vector for FGF receptor-mediated gene delivery *in vitro* and *in vivo*. Biomacromol. 11: 2221–2229.

Li, S., B. Yu, P. An, G. Chen, W. Lu, H. Cai, W. Guo and F. Zuo. 2005. Combined liposome-mediated cytosine deaminase gene therapy with radiation in killing rectal cancer cells and xenografts in athymic mice. Clin. Can. Res. 11: 3574–3578.

Mei, M., Y. Ren, X. Zhou, X.B. Yuan, L. Han, G.X. Wang, Z. Jia, P.Y. Pu, C.S. Kang and Z. Yao. 2010. Downregulation of miR-21 enhances chemotherapeutic effect of taxol in breast carcinoma cells. Tech. Can. Res. Treat. 9: 77–86.

Ndoye, A., G. Dolivet, A. Høgset, A. Leroux, A. Fifre, P. Erbacher, K. Berg, J.P. Behr, F. Guillemin and J.L. Merlin. 2006. Eradication of p53-mutated head and neck squamous cell carcinoma xenografts using nonviral p53 gene therapy and photochemical internalization. Mol. Ther. 13: 1156–1162.

Okumura, K., M. Nakase, M. Inui, S. Nakamura, Y. Watanabe and T. Tagawa. 2008. Bax mRNA therapy using cationic liposomes for human malignant melanoma. J. Gen. Med. 10: 910–917.

Shen, L.F., J. Chen, S. Zeng, R.R. Zhou, H. Zhu, M.Z. Zhong, R.J. Yao and H. Shen. 2010. The superparamagnetic nanoparticles carrying the E1A gene enhance the radiosensitivity of human cervical carcinoma in nude mice. Mol. Can. Ther. 9: 2123–2130.

Sher, Y.P., T.F. Tzeng, S.F. Kan, J. Hsu, X. Xie, Z. Han, W.C. Lin, L.Y. Li and M.C. Hung. 2009. Cancer targeted gene therapy of BikDD inhibits orthotopic lung cancer growth and improves long-term survival. Oncogene 28: 3286–3295.

Sidi, A.A., P. Ohana, S. Benjamin, M. Shalev, J.H. Ransom, D. Lamm, A. Hochberg and I. Leibovitch. 2008. Phase I/II marker lesion study of intravesical BC-819 DNA plasmid in H19 over expressing superficial bladder cancer refractory to bacillus Calmette-Guerin. J. Urol. 180: 2379–2383.

Sweeney, P., T. Karashima, H. Ishikura, S. Wiehle, M. Yamashita, W.F. Benedict, R.J. Cristiano and C.P. Dinney. 2003. Efficient therapeutic gene delivery after systemic administration of a novel polyethylenimine/DNA vector in an orthotopic bladder cancer model. Can. Res. 63: 4017–4020.

Veiseh, O., F.M. Kievit, J.W. Gunn, B.D. Ratner and M. Zhang. 2009. A ligand-mediated nanovector for targeted gene delivery and transfection in cancer cells. Biomaterials 30: 649–657.

Voges, J., R. Reszka, A. Gossmann, C. Dittmar, R. Richter, G. Garlip, L. Kracht, H.H. Coenen, V. Sturm, K. Wienhard, W.D. Heiss and A.H. Jacobs. 2003. Imaging-guided convection-enhanced delivery and gene therapy of glioblastoma. Ann. Neur. 54: 479–487.

Wang, M., and M. Thanou. 2010. Targeting nanoparticles to cancer. Pharm. Res. 62: 90–99.

Wang, P., X.H. Zhao, Z.Y. Wang, M. Meng, X. Li and Q. Ning. 2010. Generation 4 polyamidoamine dendrimers is a novel candidate of nano-carrier for gene delivery agents in breast cancer treatment. Can. Let. 298: 34–49.

Wang, S.L., H. Yao and Z.H. Qin. 2010. Strategies for short hairpin RNA delivery in cancer gene therapy. Exp. Opin. Biol. Ther. 9: 1357–1368.

Zaharoff, D.A., B.S. Hoffman, H.B. Hooper, C.J. Benjamin Jr., K.K. Khurana, K.W. Hance, C.J. Rogers, P.A. Pinto, J. Schlom and J.W. Greiner. 2009. Intravesical immunotherapy of superficial bladder cancer with chitosan/interleukin-12. Can. Res. 69: 6192–6129.

Zou, W., C. Liu, Z. Chen, N. Zhang. 2009. Studies on bioadhesive PLGA nanoparticles: A promising gene delivery system for efficient gene therapy to lung cancer. Int. J. Pharm. 370: 187–195.

3

Cancer Cell Photothermolysis and Gold Nanorod Surface Plasmons

Cheng-Lung Chen[1,a] and *Yang-Yuan Chen*[1,b]

ABSTRACT

Gold nanorods (GNRs) exhibit excellent two-photon photoluminescence (TPPL) and photothermal effect due to their characteristic surface plasmon resonance (SPR), making them attractive for potential applications in cancer diagnosis and therapy. Because of their anisotropic shape, the optical absorption efficiency of longitudinal absorption band in GNRs is strongly dependent on the polarization direction of the laser beam. Regarding this, radially polarized light is anticipated to excite randomly oriented GNRs with higher efficiencies, leading to more pronounced photothermal effect and photoluminescence. The therapeutical dynamics engendered by this photothermal effect is rather beyond simple hyperthermia: the laser-induced local superheating of GNRs induces observable internal explosions within cells, resulting in significant cavitations in the vicinity of GNR clusters, and thus, compromises cell membrane integrity. In this scenario, the concept of photothermolysis is more appropriate in describing such selective heating of, and confined

[1]Institute of Physics, Academia Sinica, Taipei, Taiwan, Republic of China.

[a]E-mail: aabbss@phys.sinica.edu.tw

[b]E-mail: cheny2@phys.sinica.edu.tw

List of abbreviations after the text.

thermal damages to, targeted absorber hosts. Some research results further demonstrated that the induction of apoptosis due to GNRs mediated photothermolysis is possible, and the required laser power density/energy fluence is far below that of necrosis. To explain the challenge of future clinical applications, two remarkable examples of *in vivo* cancer cell photothermolysis are invoked; particularly, the important relation between GNR uptake amounts and GNR-based photothermolysis. Indeed, in the context of photothermal therapies, to achieve excellent selective tumor tissue destruction and higher therapeutic efficacy within the safety criterion, GNRs with active targeting abilities are definitely a prerequisite.

INTRODUCTION

Cancers are still among the leading causes of death in the world. Despite the significant efforts invested in various therapeutic strategies such as radiation therapy, surgery, chemotherapy, targeted therapy, and combined treatments, limited efficacy of present curative methodologies remains a hurdle to overcome. For example, inadequate drug dosages reaching the tumor and the growing phenomena of drug resistance (Wang et al. 2009). Recently, the application of heat to eliminate or restrain specific cancer cells is proposed as an encouraging approach in optimizing cancer therapy. This non-invasive technique is generally referred to as hyperthermia or thermotherapy (Wust et al. 2002), in which biological tissues are exposed to elevated temperatures to promote protein denaturation together with membrane disruption, leading to cell death. The key facts of thermotherapy are listed in Table 1. Although this type of treatment has been under investigation for a period, the main challenge of this therapy is to achieve selective application of the required energy to destroy tumors thermally.

To overcome the foregoing difficulties, several groups have proposed, and investigated, the feasibility of therapies that incorporates deep penetrating near infrared (NIR) lasers and plasmon-resonant nanoparticles (O'Neal et al. 2004). The advantage of these nanoparticles is their remarkable capacity to absorb light in near-infrared region. Moreover, these optical properties can be finely tuned by adjusting the nanoparticles size, shape, and dielectric medium, bringing about a diversity of applications in biological imaging and photothermal therapy. Among the variety of prospective photothermally active nanomaterials, in particular GNRs hold an interest because of their unique optical properties and high efficiency in converting photon energy to heat. Indeed, the use of GNRs has demonstrated considerable boost in cancer cell ablation. However,

Table 1. Key Facts of thermotherapy.

1. In the70s, pre-clinical investigations revealed that once the temperature of a defined region is above 43 degree Celsius, tumor cells are easily destroyed, while the injury to normal cells is minimal.
2. Depending on the extent of cancer metastasis, the treated area can be local, regional and whole-body heating.
3. The method to increase temperature can be achieved by a variety of applicators, such as lasers, alternating magnetic fields, radio frequency waves.
4. The resulting temperature distribution is greatly influenced by the power density of these applicators, energy absorption and thermal conductance of the tissues.
5. The therapeutic efficacy is dependent upon the temperature achieved during the treatment, and also incubation period and cell types.
6. If the hyperthermia process is not well managed, possible complications or side effects, for example blisters, burns, discomfort or nausea will occur.

This table lists the key facts of thermotherapy including the basic principle, methods and safety. (Adapted from Wust P. et al., Lancet Oncol. 2002; 3: 487–497, Copyright © 2002 with permission from Elsevier).

the approach is by no means perfect, and improvements are continuously sought to optimize the laser dosimetry and targeted delivery. Furthermore, due to its strong two-photon photoluminescence, GNRs are attractive candidates as contrast agent for imaging of cancer cells. Their surfaces can be easily modified with various targeting ligands, tailoring biological functions for site-specific therapy of tumors and diseased tissues; a scheme realized through the ligand's multivalent interactions with tissue-specific receptors. These extraordinary features are opening up diverse opportunities in biological imaging, detections and photothermal therapy (Huang et al. 2009).

In this chapter, we will first summarize the distinctive optical and thermal properties that are highly associated with surface plasmon resonance of GNRs. We then introduce the basic principle of cell photothermolysis and discuss the photothermal approach, presenting also, the underlying relationship between cell death and laser energy in GNR-mediated photothermolysis. Lastly, we describe several significant examples in animal studies involving the utilization of GNRs as photothermal agents.

GENERAL FEATURES OF PLASMONIC GOLD NANORODS

Surface Plasmon Resonance Absorption

When a metal nanoparticle is irradiated with electromagnetic radiation, if the frequency matches the material characteristic resonant frequency, all the 'free' electrons within the conduction band of the particle will undergo an in-phase oscillation with the frequency of the radiation. The oscillating electromagnetic field induces a polarization of the free conduction electrons with respect to the nanoparticle's ionic metal core; the effect, in turn, establishes a restoring Coulomb force. In this manner, a dipolar oscillation of the electrons is created: a phenomenon generally known as surface plasmon resonance (Bohren and Huffman 1983), as shown in Fig. 1. Meanwhile, this SPR sets off, through dissipation, a strong absorption of radiation; the effect results in a broad absorption band centered in a certain optical absorption spectrum. These pronounced absorptions are characteristics of metallic nanoparticles. Having said that, we will see that, long GNRs, being strong absorbers of near infrared radiation, are the most prospective candidates as medical photothermal convertors.

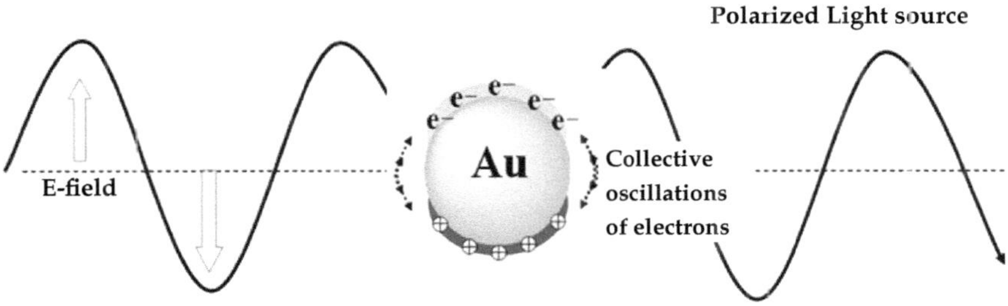

Fig. 1. The schematic illustration of plasmon oscillation in a gold nanoparticle under light irradiation. The resonance of electrons across the particle can be induced by the electromagnetic field of the light. (Unpublished material of Cheng-Lung Chen).

SPR is sensitively dependent on the geometry, size, structure, metal dielectric property, and the surrounding medium (Huang et al. 2009). Indeed, the geometric influence is very well exemplified by GNRs: observed shifts in GNRs' surface plasmon absorption spectrum are generally associated with changes in the nanorod aspect ratio *R* (length/width). A typical absorption spectrum and transmission electron micrographs of GNRs with different aspect ratios are shown in Figs. 2 and 3. GNRs have two surface plasmon absorption bands in the spectrum: a weak absorption band in the visible region around 520 nm due to the transverse electronic oscillation; and a much stronger absorption band in the near infrared region due to the longitudinal electronic oscillation (Link et al. 1999). While the transverse band absorption is rather indifferent to geometric variations, the longitudinal absorption band is very sensitive to aspect ratio. With the increase in aspect ratio, the absorption maximum of the GNRs is red-shift from the mid-visible wavelength, 600 nm ($R = 1.9$), to the near infrared region 900 nm ($R = 5$). This aspect ratio dependence of SPR absorption is very well studied, and is in accord with Gans theory on the scattering of radiation by small particles (Link et al. 1999). Owing

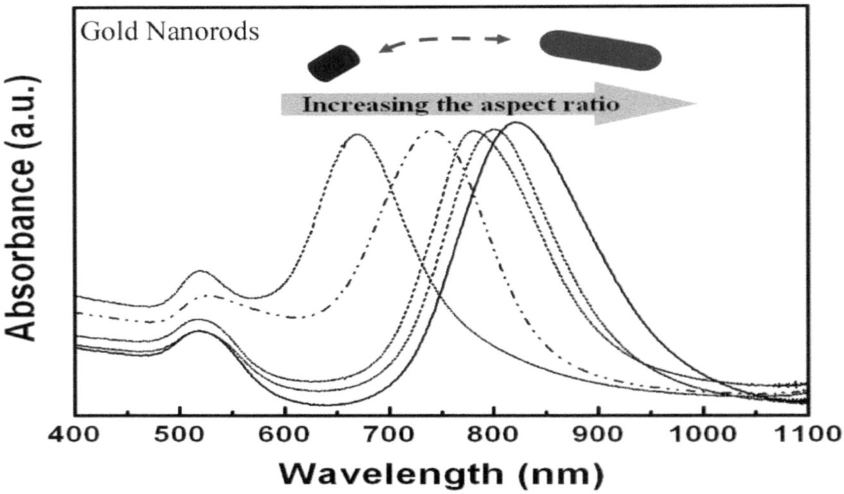

Fig. 2. UV–visible absorption spectra of gold nanorods with different aspect ratios. There is clearly a red shift in the longitudinal peak position as the aspect ratio of the nanorods increases. From left to right, the aspect ratios of curve are from 2.6 to 4.3. (Unpublished material of Cheng-Lung Chen).

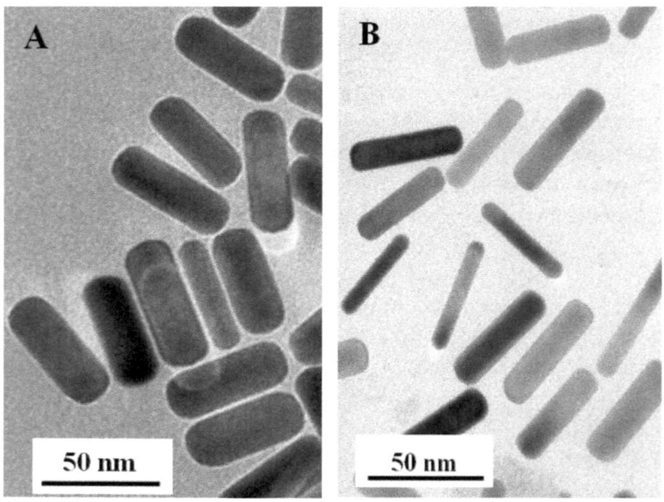

Fig. 3. Transmission electron micrographs of gold nanorods with different aspect ratios. (A) 3.1 ± 0.3; (B) 4.3 ± 0.5. (Unpublished material of Cheng-Lung Chen).

to the tunability of its optical properties, through varying the aspect ratio, GNRs outcompete spherical nanoparticles as prospective candidates for biomedical applications.

Two-photon Photoluminescence (TPPL)

The mechanism of two-photon photoluminescence in GNRs is generally described in a three-step process: (i) electrons are first excited by two-photon absorption from occupied *d* band to the unoccupied states of *sp* band; thus generating electron-hole pairs, (ii) the electrons lose partial energy to the phonon lattice via electron-phonon scattering processes, (iii) the electron-hole recombination, resulting in photoemission (Wang et al. 2005; Bouhelier et al. 2005). The TPPL intensities of GNRs are found several-folds higher than that of dye molecules; this enhancement is attributed to a resonant coupling with localized surface plasmons. Although gold nanoparticles also possess such kind of phenomena due to their well-defined SPR, the TPPL of GNRs is more attractive due to their longitudinal plasmon modes at near-infrared region.

In applications, the highly efficient TPPL of GNRs can serve as bright contrast agents for biological imaging. Such application has several advantages (Durr et al. 2007; Weissleder 2001): (1) TPPL can be excited at NIR range (750~900 nm), the window of greatest transmittivity through biological tissue; (2) the two-photon excitation can extend the maximum depth of imaging up to 75 μm in a tissue phantom; (3) the TPPL can be spectrally distinguished from tissue autofluorescence; (4) the excited power density for TPPL imaging is much below the damage threshold of biological tissues.

Polarization-dependent Optical Properties

Due to the anisotropic shape, the optical absorption and even photoluminescence of GNRs are strongly dependent on the relative orientation between longitudinal axis of GNRs and the polarization direction of laser beam. This intriguing optical property of GNRs was confirmed experimentally by Perez-Juste et al. (2005) in composite thin films of aligned GNRs. When polarized light is parallel to the long axis of aligned GNRs, the longitudinal plasmon mode is effectively activated. Whereas, only the transverse mode is excited by laser light with polarization which is perpendicular to the long axis. The progressive change of absorption in both plasmon modes was observed with varying intermediate polarization angles from ninety to zero degrees (Perez-Juste et al. 2005). No such kind of polarization dependence of absorption in ensemble GNRs with random orientation due to the spatial average on all directional configurations (Muskens et al. 2008).

Photothermal Effect

The appeal of GNRs is the capacity to convert the absorbed light into heat via non-radiative processes. El-Sayed's group and other workers have used ultrafast dynamic techniques and computational model to study this interesting heating process in nanoscale region (Link and El-Sayed 2003; Ekici et al. 2008). It can be understood that as the surface plasmons of GNRs are excited to gain kinetic energy upon irradiation with ultrafast laser pulses matching to the collective excitation wavelength of surface electrons, these hot electrons are relaxed through electron-electron scattering process during 10–50 fs. At this timescale, there is, practically, no energy exchange happening between phonons and electrons; thus, the electrons remain at elevated energy levels. The subsequent energy dissipation to the phonon bath—occurring around 1 ps later via electron-phonon scattering—results in homogeneous heating of GNRs. As the GNRs temperature increases, energy transfers from GNRs to their surrounding medium through phonon–phonon coupling. Thermal equilibrium is achieved within 100 ps to 1 ns, diminishing the sharp temperature discontinuity near the interface of the GNRs and the aqueous solution. Therefore, at low incident power density with low repetitive rate or photon intensity, heat dissipation is comparatively slow and observable in temperature change; the effect is similar to that of local hyperthermia. At high photon densities, however, the fast repetitive absorption of photons by GNRs can exceed the rate of heat diffusion; this superheating brings an extreme discontinuity in local temperature within a very short time, causing an extreme temperature rise. The process induces explosive evaporation of gold atoms of GNRs, it also enables the generation of shock waves, stress transients, and local high pressure (Letfullin et al. 2006).

CANCER CELL PHOTOTHERMOLYSIS

Basic Principle

Photothermolysis is a technique that depends on selective absorption of a short radiation pulse to produce heat at certain targeted tissue locations (Anderson and Parrish 1983). The main requirement is that the target areas should be bestowed with greater thermal energy than their surrounding tissues. This is achieved through implanting optical absorbers. During exposure to radiation, the optical energy absorbed by optical absorbers is converted into heat. After appropriate laser irradiation duration, the temperature in the target area will surpass the thermal denaturation threshold while the farther surrounding tissue remains well below this temperature. The schematic representation of selective photothermolysis is illustrated in Fig. 4. By controllable target selectivity, the confinement of damage can be achieved with microbeam techniques without necessary precise aiming.

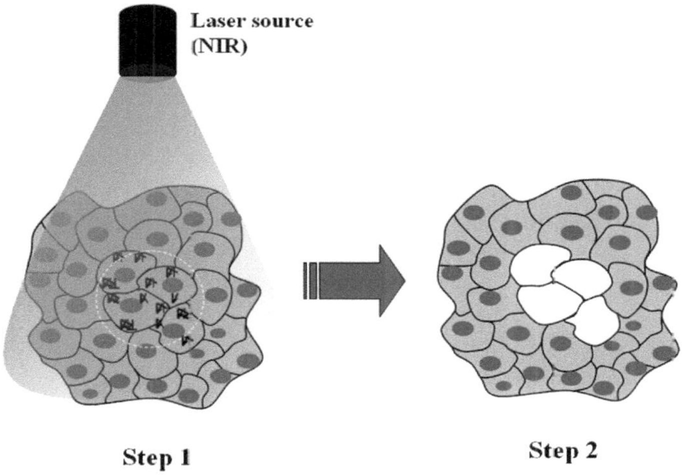

Fig. 4. The schematic representation of selective cancer cell photothermolysis. Step 1: Optical absorbers are specifically targeted to cancer cells (circle area), and exposed to the light. Step 2: Cancer cells (white area) are selectively destroyed without injuring surrounding normal cells. (Unpublished material of Cheng-Lung Chen).

Photothermolysis of Cancer Cells with Gold Nanorods

GNRs, with their remarkable opto-thermal conversion, are recently considered as promising photothermal target mentioned above, and have thenceforth been widely incorporated in cancer cell photothermolysis. The first example of using GNRs in eliminating cancer cells by NIR irradiation was reported by Huang et al. (2006). They showed that as the cells incubated with GNRs were exposed to laser irradiation, different laser power/energies were observed to induce photothermal destruction among malignant and nonmalignant cells (Huang et al. 2006). The energy needed to cause destruction of these cells was also dependent on the uptake quantity of GNRs in cells. Tong et al. (2007) attributed such kind of cell death as a result of plasma membrane disruption (Tong et al. 2007). In our previous *in situ* real-time investigations (Chen et al. 2010), polystyrenesulfonate polymer modified GNRs were ingested through endocytosis, which is a fast and general process for delivering clusters of GNRs from the cell membrane into the cytoplasm. The ingested GNRs are initially trapped in endosomes and eventually, after several hours of incubation, reside in lysosomes (final degrading organelles). It is worth noting that this normal biological process will congregate GNRs into larger clusters which have been demonstrated to increase the efficiency of the laser-induced photothermolysis (Zharov et al. 2005).

Due to the excellent photoluminescence of GNRs, the GNRs' location within the cell can be observed with precision, and hence correctly targeted with laser irradiation. For the deposited energy fluence of femtosecond-pulsed laser as high as 75 mJ/cm^2, cells with GNRs incur serious damages with observable internal explosion phenomenon occurring upon excitation (Fig. 5). In such a circumstance, the quantity of GNRs within a cell is affected insignificantly. The formation of characteristic cavities could reach as large as 6 μm, and cavities were especially pronounced at GNR cluster locations (cluster size: 2~3 μm). Theoretical modeling indicates that the presence of GNR clusters may provide synergistic enhancement of bubble formation and photothermolysis (Zharov et al. 2005). Indeed, such laser-induced bubble formations do play an important role in the perforation, or sudden rupture, of plasma membrane. The process of GNRs-mediated photothermolysis is depicted in Fig. 6. In order to gain deeper insights into the whole cell destruction process, the progression to death under the exposure of different energy fluences can be simultaneously monitored using fluorescent dyes—typically YOPRO-1 (green coloring) and propidium iodide (red coloring)—to qualitatively indicate the cell membrane integrity: variations in the morphology and the coloring of the nucleus provide vital information (Chen et al. 2010).

With a continuous decrease in energy fluences, the extent of internal explosions becomes more moderate as oppose to those caused by higher energy levels. However, the symptoms of oncosis were still observed, and

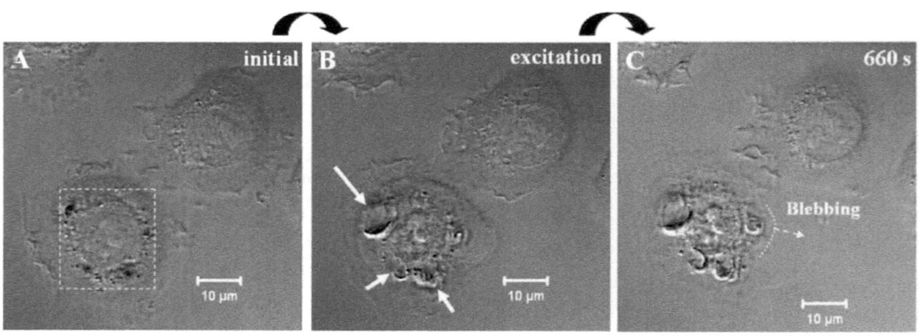

Fig. 5. Optical images of photothermolysis of the EMT-6 breast tumor cell triggered by gold nanorods under the energy fluence of 75 mJ/cm^2. (A) The laser activation area is marked with white dash line, and the black dots within this scope is the nanorod clusters. (B) Cells with nanorods exposed to femtosecond-pulsed laser irradiation produced several cavities (indicated by arrows). (C) Membrane exhibited perforation and blebbing. The blebbing trace is shown by dash line. (Unpublished material of Cheng-Lung Chen).

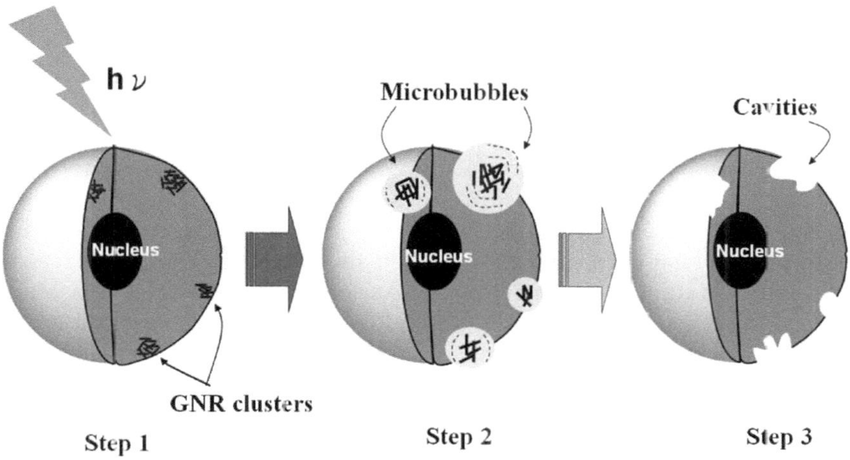

Fig. 6. A schematic diagram explaining steps of gold nanorod-mediated cancer cell photothermolysis. Step 1: Clusterization of nanorods and irradiation with a laser pulse. Step 2: Laser-induced thermal explosion due to superheating of nanorods and hence microbubble formation. Step 3: Cellular membrane of the cell is severely damaged and finally results in necrosis. (Unpublished material of Cheng-Lung Chen).

the total transiting time to cell necrosis was relatively longer. In fact, the lesions of plasma membrane resulting from different energy fluences are different and led to dissimilar destruction of cells. All of these processes can be confirmed by scanning electron microscopy and fluorescent dyes dynamics. When the laser energy fluence was reduced to 18 mJ/cm^2, the uptake quantities of GNRs by cells were quite crucial to the cell damage. The detailed mechanism will be discussed in the next section.

On the other hand, most of the GNRs are randomly oriented in cells. A more energy-efficient illumination method was proposed by Li et al. They demonstrated that radially polarized laser beam can effectively damage cancer cells with a much lower energy fluence than that required for a linearly polarized beam, which is far below the medical safety level and renders the photothermal therapy medically safer (Li et al. 2008).

Cell death Associated with Laser Dosimetry

Maximizing the therapeutic efficacy of photothermolysis with optimum laser dosimetry is a crucial issue for clinical applications. It is necessary to fully understand the mechanism underlying the cell death process during the photothermal treatments. In general, cell death is mainly discussed dichotomously as either apoptosis or necrosis (Majno and Joris 1995). Recent studies on GNR-based photothermolysis reveal when cancer cells labeled with GNRs were subjected to different severity of damage, it would lead to diverse injuries and hence cell death. In our previous work, when the deposited energy fluence is larger than required for elevating temperature above the melting threshold, thermal explosion is realized if heat is generated too fast to diffuse away (Chen et al. 2010; Ekici et al. 2008; Letfullin et al. 2006). Thus, in effect, GNRs can function as "nano-bombs" to damage intracellular organs or metabolism processes through laser-induced micro-explosions within the cells. The vital cellular components and even the outer membrane may be fatally cracked by this sudden local heating or the accompanied shock waves. That is the reason why the cells died completely in the crumbled form within a very short time. When exposing the cells to much lower energy fluence, 18 mJ/cm^2, the injuries become highly correlated with the uptake quantity of GNRs within the cell. It was observed that intracellular damages from small-scale explosions within the cells with a few GNRs ($N\sim$ 10–30 clusters) were not severe enough, likely to give rise to unrepairable wounds—the cell seemed pretty much healthy and continued to survive. With regard to cells with relatively higher population of GNR clusters ($N\sim$ 60–100), even the intracellular explosions did not immediately wound the plasma membrane. Nevertheless, the later intracellular complex responses, for example, the level of lysosomal rupture and hence an increase in the total volume of acidic compartments within the cell might be the factors that eventually engenders leaking of plasma membrane (Ono et al. 2003). The detailed mechanism of this process is still under investigation. This study primarily explores the relationship between the laser dosimetry and the progression to cell destruction. The operating energy fluence is still too high, and is unfavorable for apoptosis induction.

Studies with regard to controlled cell death in photothermolysis have recently been reported by two groups (Li and Gu 2010; Tong and Cheng 2009). For instance, Li et al. prepared functionalized GNRs to efficiently target HeLa cells and manipulated the appropriate energy fluence of a femtosecond laser to induce apoptosis of cells. They found that a low laser power of 1 mW (27.8 W/cm^2) is sufficient to induce the apoptosis of the cells after 10 raster scans (~6.3 mJ/cm^2). This energy fluence is indeed much lower than our work (18 mJ/cm^2). Cell necrosis was evidently observed, if the cells were scanned 30 times under this power density. Similarly, increasing the laser power up to 2 mW (55.6 W/cm^2) and above, only one raster scan is required to induce perforation of cell membrane. It is clear that the required power density/energy fluence for apoptosis induction is much less than that for necrosis. Meanwhile, the other study conducted by Tong et al., also showed the same behavior in the GNRs-mediated photothermolysis of macrophages. Obviously, cell death in GNRs-mediated photothermolysis is controllable through appropriately tuning the laser energy power level.

In vivo Tumor Targeting and Selective Photothermal Therapy

Since 2006, the applications of plasmonic GNRs in photothermal therapies of cancer cells have been progressively reported. The capability of exploiting GNRs for *in vivo* photothermal therapy has recently been reported by El-Sayed's research group (Dickerson et al. 2008). In their studies, the PEGylated GNRs were directly injected into mice. After GNRs accumulated in tumors, either through intravenous or subcutaneous injections, the tumors are easily recognized from NIR laser transmission imaging. As these tumors were exposed to a continuous-wave laser at 808 nm with a power density of 1W/cm^2 for 10 min, a dramatic size decrease in tumors was observed for both injection treatments. In contrast, the untreated tumors kept growing at a rapid rate. Although GNRs exhibit promising inhibition of tumor growth, the amount of GNRs targeting to the tumors is another major concern in enhancing the efficacy of photothermal therapy; the development of safe and highly targeting GNRs is essential in future clinical applications. In this respect, the polyamidoamine dendrimer-modified GNR (dGNR) conjugated with arginine-glycine-aspartic acid (RGD) peptides is one successful example, presenting highly efficient tracking and destructive effects on the melanoma tumor cells and vascular cells inside tumor tissues (Li et al. 2009). In their experiments, the RGD functionalized GNRs (RGD-dGNRs) that possessed active targeting abilities, and pure dGNRs that served as passive targeting groups were injected into mice with tumors via tail vein, respectively. The examination showed that RGD-dGNRs could rapidly distribute into the whole body of mice loaded with tumors through the circulatory system; it then gradually accumulated in the tumor tissues as time went by. On the other hand, for passive targeting group, although the dGNRs could accumulate in the tumors due to enhanced permeability and retention, the amount of GNRs in tumor location was much less than that of active targeting. After laser irradiation treatments, the average tumor size of mice with RGD-dGNRs (active targeting) became smaller and gradually disappeared in a span of several weeks. As opposed to this, the photothermal therapy of passive targeting groups only slightly inhibited tumor growth and showed lower therapeutic efficiency. This encouraging study literally confirms the significance of functionalized GNRs in the applications of tumor targeting and selective photothermal therapy.

Applications to Areas of Health and Diseases

Traditional cancer therapies lack invariably in specificity. Both cancerous and healthy cells in the body are usually affected by conventional anticancer drugs, and uncomfortable symptoms are unavoidable; for example, liver, kidney and cardiac damage, hair loss and nausea (Wang et al. 2009). Although these methods do help patients extend their life to some extent, the subsequent side effects also gradually threaten their health. Therefore, much effort is continually devoted in developing a more efficient and tolerable therapy. Recent advances in nanotechnology opens up opportunities in overcoming the above-mentioned predicaments. In this regard, plasmonic GNRs are prospected to provide revolutionary improvement on the diagnosis, as well as precise therapy of cancers. The ability to rapidly and sensitively detect malignant cells, pathogenic bacteria, DNA and antigens is extremely imperative in preventing the proliferation of cancer or diseases (Huang et al. 2009; Goodrich et al. 2010). Furthermore, GNRs can be easily modified with a variety of functional molecules to actively target to cancerous areas thus enhancing the efficacy of photothermolysis. Most importantly, the efficacy depends significantly on laser excitation power and optical absorbers rather than the characteristic properties of specific cells. Thus, in the foreseeable future, GNRs-based photothermolysis will surely become an indispensable tool in disease prevention and therapy.

Key Terms

- **Gold nanorod**: a rodlike gold nanostructure, which has many attractive applications in disease diagnosis and therapy due to its unique optical and thermal properties.
- **Hyperthermia:** it is, in the context of cancer therapy, a methodology in which tumor cells are eliminated through the application of heat.
- **Apoptosis:** a process in which cells play an active role in their own death (suicide) and characterized by cytoplasmic shrinkage and karyorhexis. The importance of this process in a living body is to govern the proliferation of cells and avoid cancerous mutations.
- **Necrosis:** a type of uncontrolled cell death resulting from acute cellular injury; it is characterized by cellular/organelle swelling, or blebbing, due to the perforation of the cell membrane. Plasma membrane disruption is usually the key characteristic that differentiates necrosis from apoptosis in cell death.
- **Active targeting:** efficient tumor cells targeting can be achieved by antibody- or ligand-conjugated nanocarriers, which will allow functionalized nanocarriers to "recognize" cells having specific receptors.

Summary

- GNRs exhibit two characteristic surface plasmon resonance absorption bands due to its anisotropic shapes. The longitudinal absorption band is highly associated with the aspect ratio of nanorods, and can be finely tuned to NIR region where the optical transmission through tissues is at its maximum.
- The excellent two-photon photoluminescence and photothermal effect of GNRs have rendered them as promising theranostic agents for biological imaging and photothermal therapy.
- The optical absorption efficiency of the longitudinal band in GNRs is strongly dependent on the polarization direction of the laser beam. Radially polarized light is thus suggested to excite randomly oriented GNRs for achieving higher efficiency in photothermal effect and photoluminescence.
- The notion of cancer cell photothermolysis of GNRs is rather beyond conventional hyperthermia. The laser-induced superheating of GNRs can induce observable internal explosion phenomena, lead to significant cavities near the GNR cluster positions, and thus compromise cell membrane integrity.
- The induction of necrosis or apoptosis in GNRs-mediated photothermolysis can be finely controlled by manipulating the laser dosimetry.
- Two examples of *in vivo* cancer cell photothermal therapy have indicated that the uptake amounts of GNRs in cells play an important role in GNRs-based photothermolysis. GNRs with active targeting abilities can exhibit excellent selective destructive effects on tumor tissues and offer greater therapeutic efficacy within the safety criterion.

Acknowledgements and Permissions

Financial supported by the National Science Council of the Republic of China (Grant No. NSC 99-2120-M-001-001) is greatly acknowledged. We deeply appreciate to all of our colleagues and collaborators, their names appear in the cited literatures. Particularly, we would like to thank Ms. Ling-Ru Guo, Ms. Candy I Chen Tsai and Mr. Cheng-Kuang Huang for their important contributions to the research effort. We also like to thank the following publisher and journal for permission to reuse previously published material:

1. *Biomaterials*, 31, 2010, C.L. Chen, L.R. Kuo, C.L. Chang, Y.K. Hwu, C.K. Huang, S.Y. Lee, K. Chen, S.J. Lin, J.D. Huang, Y.Y. Chen, "*In situ* real-time investigation of cancer cell photothermolysis mediated by excited gold nanorod surface plasmons," pp. 4104–4112. Copyright © 2010, with permission from Elsevier.
2. *Lancet Oncol*. 3, 2002, P. Wust, B. Hildebrandt, G. Sreenivasa, B. Rau, J. Gellermann, H. Riess, R. Felix, P.M. Schlag, "Hyperthermia in combined treatment of cancer," pp. 487–497. Copyright © 2002, with permission from Elsevier

Abbreviations

GNR	:	Gold nanorod
SPR	:	Surface plasmon resonance
TPPL	:	Two-photon photoluminescence
NIR	:	Near infrared
PEG	:	Polyethylene glycol
RGD	:	Arginine-glycine-aspartic acid
dGNR	:	Dendrimer-modified GNR

References

Anderson, R.R., and J.A. Parrish. 1983. Selective photothermolysis: precise microsurgery by selective absorption of pulsed radiation. Science 220: 524–527.

Bohren, C.F., and D.R. Huffman. 1983. Absorption and Scattering of Light by Small Particles. Wiley, New York.

Bouhelier, A., R. Bachelot, G. Lerondel, S. Kostcheev, P. Royer and G.P. Wiederrecht. 2005. Surface plasmon characteristics of tunable photoluminescence in single gold nanorods. Phys. Rev. Lett. 95: 267405–267408.

Chen, C.L., L.R. Kuo, C.L. Chang, Y.K. Hwu, C.K. Huang, S.Y. Lee, K. Chen, S.J. Lin, J.D. Huang and Y.Y. Chen. 2010. *In situ* real-time investigation of cancer cell photothermolysis mediated by excited gold nanorod surface plasmons. Biomaterials 31: 4104–4112.

Dickerson, E.B., E.C. Dreaden, X. Huang, I.H. El-Sayed, H. Chu, S. Pushpanketh, J.F. McDonald and M.A. El-Sayed. 2008. Gold nanorod assisted near-infrared plasmonic photothermal therapy (PPTT) of squamous cell carcinoma in mice. Cancer Lett. 269: 57–66.

Durr, N.J., T. Larson, D.K. Smith, B.A. Korgel, K. Sokolov and A. Ben-Yakar. 2007. Two-photon luminescence imaging of cancer cells using molecularly targeted gold nanorods. Nano Lett. 7: 941–945.

Ekici, O., R.K. Harrison, N.J. Durr, D.S. Eversole, M. Lee and A. Ben-Yakar. 2008. Thermal analysis of gold nanorods heated with femtosecond laser pulses. J. Phys. D: Appl. Phys. 41: 185501–185511.

Goodrich, G.P., L. Bao, K. Gill-Sharp, K.L. Sang, J. Wang and J.D. Payne. 2010. Photothermal therapy in a murine colon cancer model using near-infrared absorbing gold nanorods. J. Biomed. Opt. 15: 018001–018008.

Huang, X., I.H. El-Sayed, W. Qian and M.A. El-Sayed. 2006. Cancer cell imaging and photothermal therapy in the near-infrared region by using gold nanorods. J. Am. Chem. Soc. 128: 2115–2120.

Huang, X., S. Neretina and M.A. El-Sayed. 2009. Gold nanorods: from synthesis and properties to biological and biomedical applications. Adv. Mater. 21: 1–31.

Letfullin, R.R., C. Joenathan, T.F. George and V.P. Zharov. 2006. Laser-induced explosion of gold nanoparticles: potential role for nanophotothermolysis of cancer. Nanomedicine 1: 473–480.

Link, S., M.B. Mohamed and M.A. El-Sayed. 1999. Simulation of the optical absorption spectra of gold nanorods as a function of their aspect ratio and the effect of the medium dielectric constant. J. Phys. Chem. B 103: 3073–3077.

Link, S., and M.A. El-Sayed. 2003. Optical properties and ultrafast dynamics of metallic nanocrystals. Annu. Rev. Phys. Chem. 54: 331–366.

Li, Z., P. Huang, X. Zhang, J. Lin, S. Yang, B. Liu, F. Gao, P. Xi, Q. Ren and D. Cui. 2009. RGD-conjugated dendrimer-modified gold nanorods for *in vivo* tumor targeting and photothermal therapy. Mol. Pharm. 7: 94–104.

Li, J.L., and M. Gu. 2010. Surface plasmonic gold nanorods for enhanced two-photon microscopic imaging and apoptosis induction of cancer cells. Biomaterials 31: 9492–9498.

Li, J.L., D. Day and M. Gu. 2008. Ultra-low energy threshold for cancer photothermal therapy using transferrin-conjugated gold nanorods. Adv. Mater. 20: 1–6.

Majno, G., and I. Joris. 1995. Apoptosis, oncosis, and necrosis. An overview of cell death. Am. J. Pathol. 146: 3–15.

Muskens, O.L., G. Bachelier, N.D. Fatti, F. Vallee, A. Brioude, X. Jiang and M.P. Pileni. 2008. Quantitative absorption spectroscopy of a single gold nanorod. J. Phys. Chem. C 112: 8917–8921.

O'Neal, D.P., L.R. Hirsch, N.J. Halas, J.D. Payne and J.L. West. 2004. Photo-thermal tumor ablation in mice using near infrared-absorbing nanoparticles. Cancer Lett. 209: 171–176.

Ono, K., S.O. Kim and J. Han. 2003. Susceptibility of lysosomes to rupture is a determinant for plasma membrane disruption in tumor necrosis factor alpha-induced cell death. Mol. Cell Biol. 23: 665–676.

Perez-Juste, J., B. Rodriguez-Gonzalez, P. Mulvaney and L.M. Liz-Marzan. 2005. Optical control and patterning of gold-nanorod-poly(vinyl alcohol) nanocomposite films. Adv. Funct. Mater. 15: 1065–1071.

Tong, L., Y. Zhao, T.B. Huff, M.N. Hansen, A. Wei and J.X. Cheng. 2007. Gold nanorods mediate tumor cell death by compromising membrane integrity. Adv. Mater. 19: 3136–3141.

Tong, L., and J.X. Cheng. 2009. Gold nanorod-mediated photothermolysis induces apoptosis of macrophages via damage of mitochondria. Nanomedicine 4: 265–276.

Wang, X., Y. Wang, Z. Chen and D.M. Shin. 2009. Advances of cancer therapy by nanotechnology. Cancer Res. Treat. 41: 1–11.

Wang, H.F., T.B. Huff, D.A. Zweifel, W. He, P.S. Low, A. Wei and J.X. Cheng. 2005. *In vitro* and *in vivo* two-photon luminescence imaging of single gold nanorods. Proc. Natl. Acad. Sci. USA 102: 15752–15756.

Weissleder, R. 2001. A clearer vision for *in vivo* imaging. Nat. Biotechnol. 19: 316–317.

Wust, P., B. Hildebrandt, G. Sreenivasa, B. Rau, J. Gellermann, H. Riess, R. Felix and P.M. Schlag. 2002. Hyperthermia in combined treatment of cancer. Lancet Oncol. 3:487–497.

Zharov, V.P., R.R. Letfullin and E.N. Galitovskaya. 2005. Microbubbles-overlapping mode for laser killing of cancer cells with absorbing nanoparticle clusters. J. Phys. D: Appl. Phys. 38: 2571–2581.

4

Tocotrienol Loaded Lipid Nanoparticles in Cancer

Sami Nazzal[1,*] and *Paul W. Sylvester*[2]

ABSTRACT

Tocopherols and tocotrienols represent the two subgroups that constitute the vitamin E family of compounds. Although chemically very similar, tocotrienols were shown to possess significantly greater antiproliferative and apoptotic activity than tocopherols in cancer cells using treatment doses that have little or no effect on normal cell growth or viability. Furthermore, studies have demonstrated that combined treatment of tocotrienols with other traditional chemotherapeutic agents results in a synergistic anticancer response. However, the potential therapeutic benefits of these lipophilic molecules are diminished by their poor solubility in aqueous biological media, low bioavailability, and non-specific targeting when given orally or parenterally. To circumvent these limitations, advanced drug delivery technologies have

[1]Associate Professor of Pharmaceutics, Department of Basic Pharmaceutical Sciences, College of Pharmacy, University of Louisiana at Monroe, 1800 Bienville Drive, Monroe, LA 71201, USA; E-mail: nazzal@ulm.edu

[2]B.J. Robison/Pfizer Endowed Professor of Pharmacology, Department of Basic Pharmaceutical Sciences, College of Pharmacy, University of Louisiana at Monroe, 1800 Bienville Drive, Monroe, LA 71201, USA; E-mail: sylvester@ulm.edu

*Corresponding author

List of abbreviations after the text.

been developed and used to maximize drug bioavailability. The emergence of therapeutic and economic incentives for the development of novel technologies for the delivery of lipophilic drugs has generated considerable interest in lipid-based formulations due to their inherent biocompatibility, scalability, and low production cost. The contemporary view on the use of lipid-based formulations for the oral and parenteral delivery of tocopherols and tocotrienols is discussed in this chapter. In this context, emphasis has been placed on two lipid-based nanotechnologies; self-emulsifying drug delivery systems and solid lipid nanoparticles. These nanosystems have considerable therapeutic potential in the treatment or prevention of cancer due to their capacity to entrap high doses of tocopherols and tocotrienols.

INTRODUCTION

Vitamin E is a generic term that refers to family of eight natural compounds that are further divided into two subgroups called tocopherols and tocotrienols (McIntyre et al. 2000). Tocopherols are commonly found in high concentrations in a wide variety of foods, whereas tocotrienols are relatively rare and found in appreciable levels only in a few specific vegetable fats (Table 1), such as palm oil (Cottrell 1991). Tocopherols and tocotrienols have the same basic chemical structure characterized by a long phytyl chain attached at the 2-position of a chromanol ring structure (Fig. 1).

Table 1. Vitamin E Levels (mg/L) in Common Dietary Oils.

Dietary Oil	Total Tocopherol (mg/L)	Total Tocotrienol mg/L	Total Vitamin E (mg/L)
Palm	152	738	890
Coconut	11	25	36
Coco Butter	198	2	200
Corn	800	0	802
Cottonseed	776	0	776
Peanut	367	0	367
Olive	51	0	51
Safflower	801	0	807
Soybean	958	0	958
Sunflower	546	0	546
Lard	19	7	26

This table lists the total amount of tocopherols and tocotrienols present in common dietary oils. The selected information shown in this table was obtained from reference Cottrell (1991).

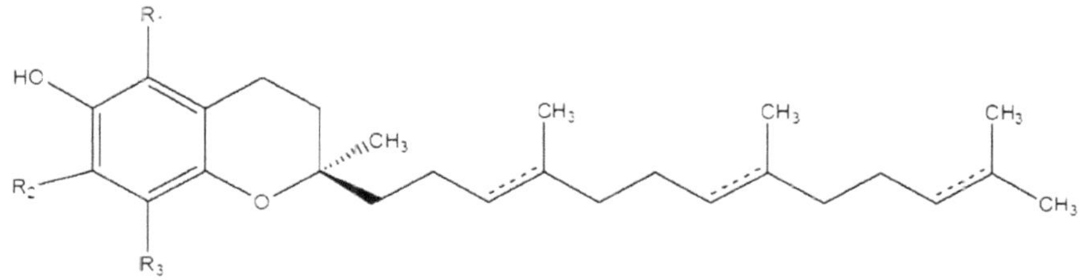

Compound	R_1	R_2	R_3	Phytyl chain
α-tocopherol	CH_3	CH_3	CH_3	Saturated
γ-tocopherol	H	CH_3	CH_3	Saturated
δ-tocopherol	H	H	CH_3	Saturated
α-tocotrienol	CH_3	CH_3	CH_3	Unsaturated
γ-tocotrienol	H	CH_3	CH_3	Unsaturated
δ-tocotrienol	H	H	CH_3	Unsaturated

Fig. 1. Generalized chemical structure of Vitamin E. Vitamin E refers to family of eight natural compounds that are divided into two subgroups called tocopherols and tocotrienols. Both tocopherols and tocotrienols have similar structure characterized by a phytyl side chain attached to a chromane ring. The difference between the individual isoforms lays in the degree of methylation of their chromane ring and the saturation of the phytyl chain (Reprinted from Colloids and Surfaces A: Physicochemical and Engineering Aspects, 353(1), H. Ali, A.B. Shirode, P.W. Sylvester and S. Nazzal. Preparation and *in vitro* antiproliferative effect of tocotrienol loaded lipid nanoparticles, 43–51, 2010a, with permission from Elsevier).

However, tocopherols have a saturated, while tocotrienols have an unsaturated phytyl chain. Individual tocopherols and tocotrienols include α-, β-, γ-, and δ-isoforms that differ from each other based on the number and position of methyl groups attached to their chromanol ring. Evidence suggests that phytyl chain saturation versus unsaturation and/or chromane ring methylation is critical in determining the differential biopotency demonstrated by individual vitamin E isoforms (Hayes et al. 1993). Another important chemical feature is the phenolic hydroxyl group located on the number 6 carbon of the chromane ring, which functions as the reactive site mediating antioxidant activity in both vitamin E subgroups. Direct comparisons between the two subgroups showed that tocotrienols were significantly more potent in suppressing growth and inducing cell death than tocopherols (McIntyre et al. 2000). Before the 1980s the majority of vitamin E research focused on the actions of α-tocopherol and very little was known about the therapeutic value of tocotrienols. However, since that time a great deal of information has accumulated regarding the health-related biological properties of tocotrienols. In addition, studies are rapidly identifying and characterizing the mechanism of action(s) mediating these effects of tocotrienols, and thereby providing the necessary information required for understanding the potential health benefits that might be obtained from tocotrienol dietary supplementation.

Applications to Area Of Health and Disease

Recently, a large body of research has accumulated demonstrating that tocotrienols display significantly more potent antioxidant, anticancer, anticholesterolemic, antiatherosclerotic, antihypertensive, anti-inflammatory, and neuroprotective activity than tocopherols (Hayes et al. 1993; Sylvester and Theriault 2003; Sylvester and Shah 2005). The exact reason why tocotrienols are more potent or possess unique biological actions as compared to tocopherols is presently unknown. However, studies have shown that tocotrienols can modulate intracellular signaling pathways that affect cellular proliferation, differentiation and survival. In cancer studies, the antiproliferative effects of tocotrienols appear to involve the suppression of multiple hormone- and growth factor-receptor mitogenic signaling pathways (Sylvester 2007), whereas tocotrienol-induced programmed cell death (apoptosis) is mediated by the activation of specific intracellular cysteine proteases (caspases) through various cell specific mechanisms (Sylvester 2007). Other studies have shown that the use of tocotrienols in combination therapy may not only enhance therapeutic responsiveness in a variety of pathologies, but also significantly reduce the adverse side effects associated with high dose monotherapy of many traditional medications (Wali and Sylvester 2007). These investigations have provided much information regarding the potential health benefits that might be gained from tocotrienol dietary intake or supplementation. It is now clearly evident that vitamin E is no longer synonymous with α-tocopherol and that other isoforms, particularly tocotrienols, play important and specific roles in the maintenance of good health and prevention of disease.

TOCOTRIENOLS AND CANCER

The antiproliferative effects of tocotrienols against various types of cancer cells are associated with its action to suppress mitogenic signaling initiated by hormones and growth factors receptor activation. Tocotrienols inhibit EGF receptor family member mitogenic signaling (Samant and Sylvester 2006). The EGF family is comprised of four related receptors defined as the EGF receptor (ErbB1/HER1), ErbB2/, ErbB3/HER3, and ErbB4/

HER4 (Normanno et al. 2006). Activation of the EGF receptor tyrosine kinase activity results in the recruitment and phosphorylation of many intracellular substrates (Normanno et al. 2006). A major downstream signaling pathway activated by EGF receptor is the Ras/Raf/MEK/ERK pathway, also known as the MAPK pathway (Matar et al. 2004). Activation of Ras initiates a multistep phosphorylation cascade that leads to the activation of ERK1, and ERK2 (Matar et al. 2004). Another important target in EGF receptor signaling is phosphatidylinositol 3-kinase and the downstream protein-serine/threonine kinase, Akt. Akt activation then triggers a cascade of responses from cell growth, and activation of important transcription factors, such as NFkB, which play an important role in tumor cell survival and metastasis (Matar et al. 2004).

ORAL DELIVERY OF TOCOTRIENOLS

Tocotrienols have poor aqueous solubility, which results in low therapeutic concentration in the plasma when given orally (Sylvester and Shah 2005). The absolute oral bioavailability of α-tocotrienols in rats was found to be approximately 28% whereas that of δ-tocotrienol and γ-tocotrienol was only 9% (Yap et al. 2003). To improve their bioavailability tocotrienols were formulated into SEDDS and SMEDDS. SEDDS and SMEDDS are isotropic mixtures of oil, surfactant, co-surfactant, co-solvent, and drug that form fine oil-in-water emulsion when introduced into an aqueous medium under gentle agitation (Nazzal et al. 2002). SEDDS and SMEDDS differ in the particle size and optical clarity of the resultant dispersion. SEDDS dispersions are typically opaque with particle sizes >100 nm whereas SMEDDS form transparent microemulsions with a droplet size < 100 nm (w/w). The best-known example of a successfully marketed SMEDDS formulation is the Neoral® cyclosporine formulation, which is composed of corn oil derived glycerides, polyoxyl 40 hydrogenated castor oil, propylene glycol, ethanol, and α-tocopherol as an antioxidant. Tocopherols are frequently used at low concentrations in lipid-based formulations as antioxidants or to aid in the solubilization of drugs.

Drugs formulated into SEDDS and SMEDDS usually show superior performance when taken orally. For example, when evaluated in healthy human volunteers, tocotrienol SEDDS formulations resulted in higher level of absorption and led to at least a two fold increase in the oral bioavailability of tocotrienols compared to a non self-emulsifying soya oil based formulation (Yap and Yuen 2004). Co-administration of tocotrienols with food was also found to increase their relative bioavailability in humans by more than two folds (Yap et al. 2001).

Although the increase in tocotrienols solubility when administered with food or formulated into SEDDS was found to significantly enhance their bioavailability, tocotrienol absorption was shown to be incomplete (Yap et al. 2001, 2003). Furthermore, *in vitro* analysis of tocopherol cellular uptake by Caco-2 cells revealed that the uptake of tocotrienols was concentration-dependent (Abuasal et al. 2010). These findings suggested the existence of a carrier-mediated process that is involved in the oral absorption of tocotrienols. Two carriers; NPC1L1 and R-B1 were thought to play a role in the transport of tocopherols (Abuasal et al. 2010). To further delineate the role of these carriers in the transport of tocotrienols, *in vitro* inhibition studies were carried out using ezetimib, an inhibitor of NPC1L1, and BLT1, a chemical inhibitor of SR-B1. Inhibition studies with BLT1 indicated that SR-B1 does not play a major role in the transport of γ-tocotrienol whereas inhibition studies with ezetimib showed that γ-tocotrienol is a substrate for NPC1L1, which is present on the enterocyte brush-border membrane to facilitate cholesterol absorption (Abuasal et al. 2010). These studies confirmed that the absorption of tocotrienols is a saturable carrier mediated process, which may explain the poor bioavailability of tocotrienols when taken orally.

PARENTERAL DELIVERY OF TOCOTRIENOLS

Parenteral administration of drugs is an alternate delivery method that could be employed to overcome the low oral bioavailability of tocotrienols. Parenterally administered cytotoxic agents however have posed significant challenges due to their poor specificity, high toxicity, and susceptibility to induce drug resistance. They often extensively and indiscriminately bind to body tissues and serum protein and only a small fraction of the drugs reach the tumor site, which may reduce their therapeutic efficacy and increase their systemic toxicity (Ratain and Mick 1996). While cytotoxic drugs should ideally only kill cancer cells, they are also toxic to non-cancerous cells, especially to the rapidly dividing cells, such as bone marrow cells and the cells of the gastrointestinal tract (Tipton 2003). These normal tissue toxicities, some of which are life-threatening, occur even when standard therapeutic doses of anticancer drugs are administered.

The systemic toxicity of cytotoxic drugs may be circumvented with the use of particulate drug carrier systems, such as nanoparticles, that could selectively deliver cytotoxic drugs to tumor cells without harming normal cells. Such selective extravasation and retention is possible due to the unique structural features of many solid tumors. It is well documented that tumors are often associated with a defective and leaky vascular architecture as a result of the poorly regulated nature of tumor angiogenesis. Actively growing tumors form new blood vessels in order to obtain nutrients and sustain their growth. These newly formed blood vessels are usually defective and have fenestrae ranging in size from 100 to 600 nm (Gullotti and Yeo 2009). Their discontinuous and leaky microvasculature allows the nanoparticles along with the entrapped drugs to readily penetrate into tumors. Furthermore, the interstitial fluid within a tumor is usually inadequately drained by a poorly formed lymphatic system. As a result, submicron-sized particles may preferentially extravasate and accumulate into the tumors. This is often referred to as the "enhanced permeability and retention" effect (Matsumura and Maeda 1986). Passive tumor targeting utilizing the EPR effect could therefore be achieved by properly designed nanoparticles such as solid lipid nanoparticles (SLN, also referred to as lipospheres or solid lipid nanospheres).

SLN are aqueous colloidal particles that exhibit size range between 50 and 1000 nm. The matrix of these particles is comprised of biodegradable and biocompatible solid lipids (Ali et al. 2010a, c). SLN were first introduced in the early 1990s. Since then SLN have attracted increasing interest as a carrier system for therapeutic applications and are considered emerging alternative carriers to colloidal systems for controlled and targeted drug delivery. Like other types of drug carrier used for cytotoxic drug delivery, such as polymeric micelles and liposomes, SLN have the advantages of physical stability, protection of labile drugs from degradation, and controlled release (Wissing et al. 2003). At the same time, SLN avoid some of the drawbacks associated with those drug delivery systems. SLN for example, have fewer storage and drug leakage problems compared to systems such as liposomes (Muller et al. 2000).

TOCOTRIENOL LIPID NANOPARTICLES

Due to their lipophilic cores, SLN were shown to be ideal carriers for tocopherols and tocotrienols (Ali et al. 2010a, c). The physicochemical properties and the antitumor activity of SLN loaded with tocotrienol rich fraction of palm oil (TRF) were previously reported in detail (Ali et al. 2010a, c). TRF is a mixture of α-tocotrienol, γ-tocotrienol, δ-tocotrienol, α-tocopherol, and other tocotrienol-related compounds. SLN loaded with TRF were prepared by the melt emulsification technique using poloxamer 188 as the surfactant and glycerol tristearate, glyceryl behenate, glycerol palmitostearate, or cetyl palmitate as the lipid cores (Ali et al. 2010a, c). TRF was incorporated into SLN by substituting 10% to 50% of the lipid core with TRF. To minimize surfactant's potential cytotoxicity, a surfactant to lipid ratio of 0.5:1 was used to prepare the nanoparticles. The entrapment of TRF within the SLN was confirmed by DSC and ^{1}H-NMR. Nonetheless, from the NMR signals it was also concluded that the mobility of the SLN increased with an increase in TRF concentration with a corresponding decrease in particle size, which suggested that a portion of the TRF, which exceeded the holding capacity of the solid lipids, was confined to the outer surface of the nanoparticles. Such arrangement is expected to have occurred during the preparation of the nanoparticles as evident by their long-term stability. No significant change in particle size ($p>0.05$) was observed after 6 months of storage when compared to the freshly prepared samples. After 6 months, the average size of the nanoparticles increased marginally from 148 to 163 nm, from 88 to 99 nm, from 71 to 83 nm, and from 121 to 135 nm for SLN made from CET, COMP, PREC, and DYN, respectively, in which 50% of the lipid core was substituted with TRF. Furthermore, high entrapment efficiency was observed for the SLN, which was verified by HPLC analysis (Ali and Nazzal 2009). Even after one month of storage at controlled room temperature, 90–110% of TRF was retained within the SLN, which also reflected good chemical stability of the nanoparticles. Long-term stability of the nanoparticles was further confirmed by particle size homogeneity and the absence of visible phase separation.

To investigate their potential use for cancer therapy, TRF-SLN were evaluated for their antiproliferative effects against the highly malignant +SA mammary epithelial cells (Ali et al. 2010a). In these studies, the antiproliferative effect of the TRF SLN was compared to unloaded (blank) SLN and TRF in solution bound to albumin (TRF/BSA, Fig. 2). Unloaded SLN did not show any significant effect on cell proliferation. The viability of the cells was above 95%, which suggested negligible effect on cell viability and the suitability of SLN as drug carriers. On the other hand, a significant decrease in cell viability was observed when the cells were treated with TRF/BSA and TRF SLN. Treatment with 1–7 µM of TRF as either TRF/BSA or TRF SLN significantly inhibited +SA cell growth in a dose-responsive manner (Fig. 2).

Similar results on the viability of neoplastic +SA mammary cells in the presence of TRF/BSA were reported by Wali and Sylvester (2007). Tocotrienols were shown to display potent antiproliferative and apoptotic activity against neoplastic mammary epithelial cells at treatment doses that have little or no effect on normal cell growth and function (Samant and Sylvester 2006). These effects appeared to be mediated by a reduction in the PI3K/PDK-1/Akt signaling, an important pathway associated with cell proliferation and survival in neoplastic mammary epithelial cells. While TRF has an inherent antiproliferative activity against neoplastic +SA mammary epithelial cells, entrapping TRF within SLN was found to potentiate this effect. IC_{50} values for the TRF SLN (average 2.0 µM) were lower than the IC_{50} value for the TRF/BSA solution (2.7 µM). These

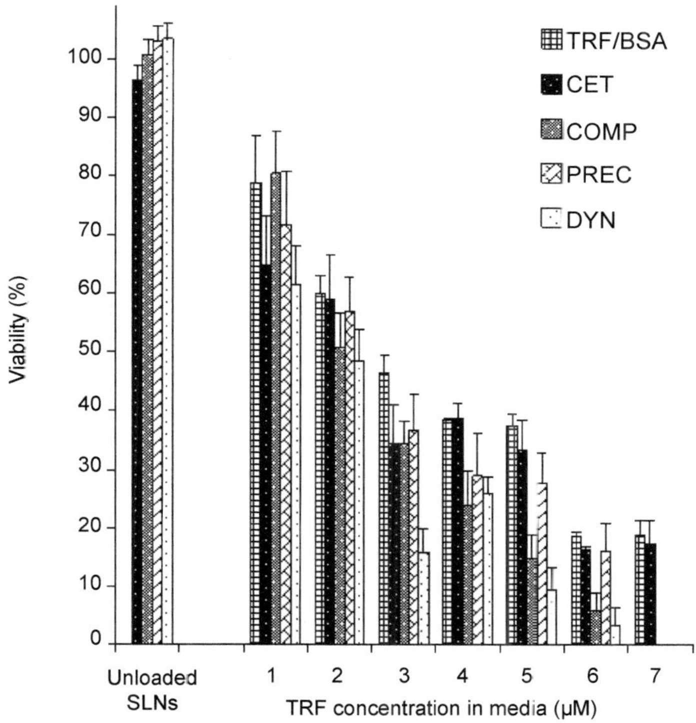

Fig. 2. Antiproliferative effects of unloaded SLN and TRF SLN on neoplastic +SA mammary epithelial cells. The figure shows the antiproliferative effects of unloaded (blank) SLN and SLN loaded with TRF on neoplastic mammary epithelial cells. Also shown is the antiproliferative effect of the TRF/BSA solution. Cells were initially plated at a density of $5x10^4$ cells/well (6 wells/group) in 24-well plates and exposed to formulation-supplemented media for a 4-day treatment period. Viable cell number was determined using the MTT colorimetric assay. Vertical bars indicate the mean cell count + SE (n=6). (Reprinted from Colloids and Surfaces A: Physicochemical and Engineering Aspects, 353(1), H. Ali, A.B. Shirode, P.W. Sylvester and S. Nazzal. Preparation and *in vitro* antiproliferative effect of tocotrienol loaded lipid nanoparticles, 43–51, 2010, with permission from Elsevier).

results suggested that entrapment of TRF within SLN increased their potency, which was attributed to enhanced nanoparticle uptake and their internalization by the cells, thereby potentiating their effect (Ali et al. 2010a). On the other hand, secondary or adjuvant lipids, which were used to construct the lipid core during the preparation of the nanoparticles, did not have a direct effect on cellular viability.

In another study (Fu et al. 2009), TRF was entrapped in transferrin-bearing vesicles made from a mixture of sorbitan monostearate (Span® 60), cholesterol, and a lanolin derivative (Solulan™ C-24). Transferrin was used as a targeting ligand due to the over-expression of transferrin receptor in a high percentage of human cancers (Fu et al. 2009). While transferrin receptors are also expressed in some rapidly dividing healthy tissues, selective targeting of the nanoparticles was achieved by a combination of active targeting, based on the use of ligands, such as transferring, and passive targeting, based on the accumulation of the nanoparticles in tumor tissue due to the enhanced permeability and retention (Matsumura and Maeda 1986). The mean diameter of these targeted tocotrienol lipid nanoparticles was 137 nm. When tested *in vitro*, transferring bearing vesicles led to at least 2-fold higher TRF uptake compared to control vesicles and more than 100-fold improved cytotoxicity in A431 (epidermoid carcinoma), T98G (glioblastoma) and A2780 (ovarian carcinoma) cell lines compared to TRF solution. TRF entrapped in transferring bearing nanoparticles was also tested *in vivo* using a murine xenograft model. When given by intravenous tail vein injection to immunodeficient BALB/c mice implanted with A431 cancer cells, TRF nanoparticles led to a marked tumor regression and improvement of animal survival. Mice treated with the nanoparticles had 19 days of extended survival when compared to untreated mice (Fu et al. 2009). These studies provided proof of concept that entrapment and delivery of TRF loaded lipid nanoparticles could be used as an alternate strategy for effective anti-cancer therapy.

TOCOTRIENOL-DRUG COMBINATIONS: SYNERGISTIC ANTICANCER EFFECTS

Tocotrienols were found to potentiate the anticancer effects of many drugs. For example, γ-tocotrienol treatment was recently found to potentiate the anticancer effects of the tyrosine kinase inhibitors erlotinib and gefitinib (Bachawal et al. 2010). Tyrosine kinase inhibitors, which target EGF receptors, have been developed for the treatment of cancer. However, the clinical usefulness of such agents as monotherapy has been found to be limited due to the ability of cancer cells to circumvent the actions of these drugs through the cooperation of different EGF receptor family members (Sylvester 2007). As a result, recent research has focused on the effect of combinational treatments directed against multiple EGF receptors. For example, combined treatment with subeffective doses of erlotinib (0.25 μM) or gefitinib (0.5 μM) with subeffective doses of γ-tocotrienol (0.5–3.0 μM) synergistically inhibited growth and induced apoptosis in mammary tumor cancer cells (Bachawal et al. 2010). Since the use of agents that target only one member of the EGF receptor family have shown limited success in the clinical treatment of cancer, these results strongly suggest that combined treatment of γ-tocotrienol with EGF receptor tyrosine inhibitors targets multiple EGF receptors and provides synergistic inhibition of mammary tumor cell growth and induces apoptosis (Bachawal et al. 2010). Taken together, combination therapy of tocotrienols with receptor tyrosine kinase inhibitors may greatly improve therapeutic responsiveness in breast cancer patients.

Tocotrienols were also found to potentiate the anticancer activity of statins (Wali and Sylvester 2007). Stains, such as simvastatin, constitute a class of drugs widely used for the treatment of hypercholesterolemia by suppressing the biosynthesis of cholesterol in the liver. Statins share structural resemblance with 3-hydroxy-3-methylglutaryl-coenzyme A and competitively inhibit HMGCoA reductase, the rate limiting enzyme in the mevalonate pathway for cholesterol synthesis (Fig. 3) (Wali and Sylvester 2007). In addition, the mevalonate pathway produces abundant non-sterol products that are required for isoprenylation of key signaling proteins in the Ras family (Ras, Rap, and Rab, etc.). Isoprenylation results in the anchoring of these proteins to the interior of the plasma membrane located in the proximity of membrane bound receptors, which when activated by their appropriate receptor ligand initiate Ras protein downstream mitogenic signaling pathways vital for cellular growth and survival. Interestingly, the mevalonate pathway is constitutively active in many types of cancer cells due to elevated and unregulated HMGCoA reductase activity (Mo and Elson

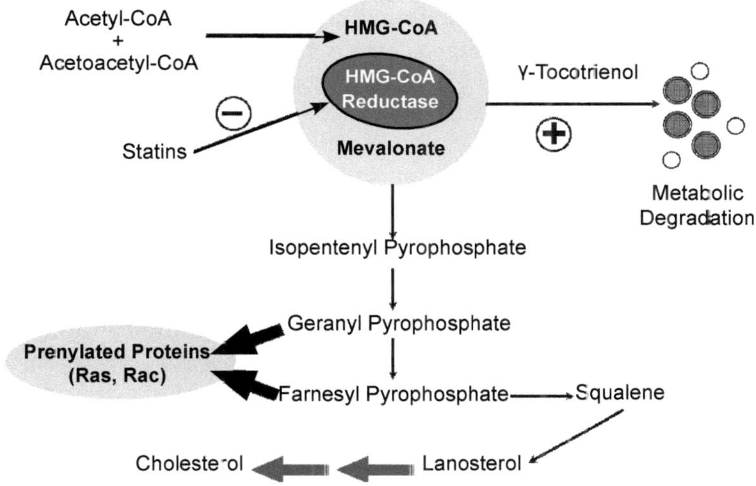

Fig. 3. The role of γ-tocotrienol and statins in attenuating 3-hydroxy-3methyl-glutaryl-coenzyme A (HMGCoA) reductase activity and downstream progression of the mevalonate synthesis pathway. A diagram showing the mevalonate synthesis pathway and the mechanism by which γ-tocotrienol and statins attenuate the activity of the enzyme 3-hydroxy-3methyl-glutaryl-coenzyme A (HMGCoA). (Unpublished material of the authors).

2004). Therefore, statins have also been shown to act as potent anticancer agents by blocking the mevalonate pathway and subsequent isoprenylation of Ras family proteins (Wali and Sylvester 2007). Unfortunately, clinical trials in prostate, breast, colorectal, ovarian, and lung cancer patients determined that high dose treatment of statins was associated with severe myopathy (Thibault at al. 1996). These findings indicate that statin monotherapy is too toxic to be used in the treatment of cancer.

Similarly, γ-tocotrienol treatment was shown to increases the intracellular metabolism and destruction of HMGCoA reductase and significantly decreases cholesterol synthesis (Fig. 3) (Parker et al. 1993). Since statins and γ-tocotrienol suppress HMGCoA reductase activity through independent mechanisms, studies were conducted to determine if combined treatment with these agents would produce additive or synergistic anticancer effects. Initial studies examined the growth inhibitory effects of low dose treatment with various statins alone and in combination with γ-tocotrienol against the highly malignant mouse mammary tumor cells (Wali and Sylvester 2007). Results from these studies showed that treatment with 2–8 μM simvastatin, lovastatin or mevastatin alone significantly inhibited EGF-dependent mammary tumor cell growth (Wali and Sylvester 2007). In addition, treatment with growth inhibitory doses (2–8 μM) of simvastatin, lovastatin, or mevastatin was not found to be cytotoxic to +SA mammary tumor cells. Subsequent studies showed that combined treatment with a subeffective dose (0.25 μM) of simvastatin, lovastatin or mevastatin with subeffective doses (0.25–2.0 μM) of γ-tocotrienol resulted in a synergistic inhibition of mammary tumor cell growth (Wali and Sylvester 2007). These findings may have very important clinical implications in that combined low dose statin and γ-tocotrienol treatment may not only provide synergistic anticancer activity in the treatment of breast cancer, but may also be devoid of the adverse effects and severe myotoxicity that is associated with high dose statin use. In addition, other experimental findings showed that treatment with subeffective doses of γ-tocotrienol or various statins alone had no effect on EGF-dependent mitogenic signaling in mammary tumor cells. However, when these subeffective doses of γ-tocotrienol and simvastatin, lovastatin or mevastatin were used in combination, a marked suppression in the relative intracellular levels of phosphorylated (activated) p44 MAPK, p54/46 JNK, p38 and Akt was observed (Wali and Sylvester 2007). The observed synergistic antiproliferative effects of combined low dose treatment of γ-tocotrienol and individual statins against mammary tumor cells strongly suggest that combined therapy with these agents may provide significant health benefits in the prevention and/or treatment of breast cancer in women, while at the same time avoiding myotoxicity that is associated with high dose statin monotherapy.

TOCOTRIENOL-SIMVASTATIN LIPID NANOPARTICLES

An obstacle to successful cancer therapy via a combined tocotrienol/simvastatin (SIM) treatment lies in the challenge of delivering the required therapeutic concentration of the drugs to the tumor cells while minimizing non-specific toxicity resulting from their systemic administration. This could be circumvented by passively targeting simvastatin and tocotrienols co-encapsulated in the form of lipid nanoparticles to tumor cells. Due to their lipophilic matrix, SLNs are ideal for the administration of lipophilic drugs, such as SIM and tocotrienols. The selective extravasation and retention of the nanoparticles in tumors is expected to potentiate the effect of tocotrienol while minimizing the side effects associated with the systemic administration of simvastatin.

In order to incorporate a drug into SLN it should however possess a sufficiently high solubility in the lipid core of the nanoparticles. While SIM could be dissolved in molten lipids such as COMP, SIM crystals were readily seen in the re-solidified COMP/SIM blends when inspected by polarized light microscopy (Ali et al. 2010b). Consequently, liquid oils such as tocopherols and tocotrienols were added to the solid lipid in order to enhance SIM payload in the nanoparticles and prevent drug precipitation. Tocotrienols and tocopherol introduce imperfections to the ordered crystalline structure of the solid lipid, thereby facilitating drug incorporation into the nanoparticles (Ali et al. 2010b). The addition of tocotrienols to SLN is advantageous because they (a) act as solubilizers, thereby enhancing the capacity of the nanoparticles to accommodate a larger quantity of a drug and (b) potentiate the therapeutic efficacy of the co-encapsulated drugs.

A combined TRF/SIM lipid nanoparticles were prepared by high-shear homogenization followed by ultrasonication technique during which 50% (by weight) of the lipid phase (i.e., COMP) was substituted with either TRF or α-tocopherol (Fig. 4). Complete solubilization of SIM in the lipid cores was confirmed by DSC and XRD studies, which indicated that SIM was molecularly dispersed in the lipid matrix. The average particle size of the nanoparticles was approximately 100 nm with SIM entrapment efficiency greater than 99%. No significant change in particle size and entrapment efficiency was observed after 6 months of storage at controlled room temperature. The release of SIM from the nanoparticles was characterized by an initial fast release within the first 10 hours followed by a slower and controlled release. The cumulative percentage of SIM released in the buffered dissolution media ranged from 25% to 51% over the 48 hr test period (Ali et al. 2010b). Similar release pattern was reported for other SLN formulations, whereby an initial fast release is followed by a slow and incomplete release of the drug (Wong et al. 2007). Nonetheless, it has

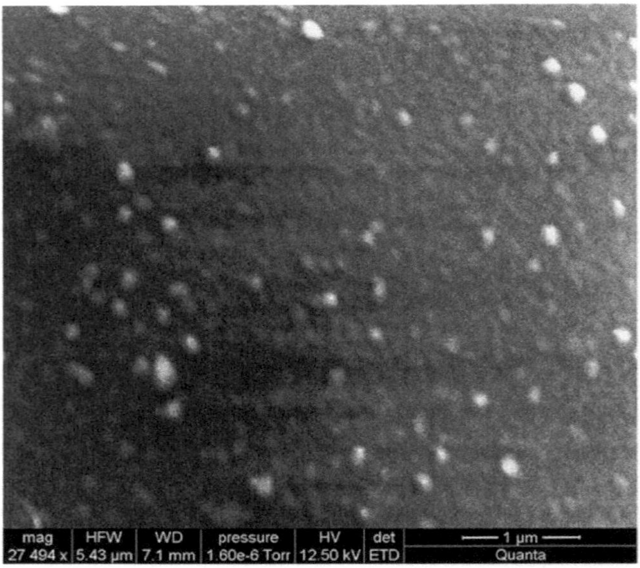

Fig. 4. Representative Cryo-SEM image of SIM/TRF SLN. Microscopic image of SLN composed of simvastatin (1 mM), TRF (5 mM), poloxamer 188 (0.125% w/v), and glyceryl behenate (0.25% w/v). (Reprinted from International Journal of Pharmaceutics, 389(1-2), H. Ali, A.B. Shirode, P.W. Sylvester and S. Nazzal. Preparation, characterization, and anticancer effects of simvastatin-tocotrienol lipid nanoparticles, 223–231, 2010, with permission from Elsevier).

also been shown that SLN formulations in which only a fraction of the drug is released are more cytotoxic than the free drug, thus indicating an increase in the cytotoxicity of drugs incorporated within the nanoparticles (Wong et al. 2007).

To confirm that the antiproliferative activity of TRF and SIM was retained when they were co-encapsulated within SLN, *in vitro* cell viability studies were carried out using neoplastic +SA mammary epithelial cells (Ali et al. 2010b). The effects of SLN at various doses of α-tocopherol and TRF with or without SIM on cell proliferation over a 4 day culture period are shown in Fig. 5.

α -tocopherol based SLN were used as negative control. As expected, α-tocopherol-SLN did not exhibit any significant effect on cellular viability within a concentration range from 0.25 to 8.0 µM, indicating absence of cytotoxic effect. Furthermore, α-tocopherol based SLN did not exhibit anticancer activity as evident from their high IC_{50} value (17.7 µM). While treatment with high doses of α-tocopherol may induce anticancer effect, the delivery of such doses is impractical. However, addition of SIM to the α-tocopherol based SLN, however, significantly inhibited +SA cell growth

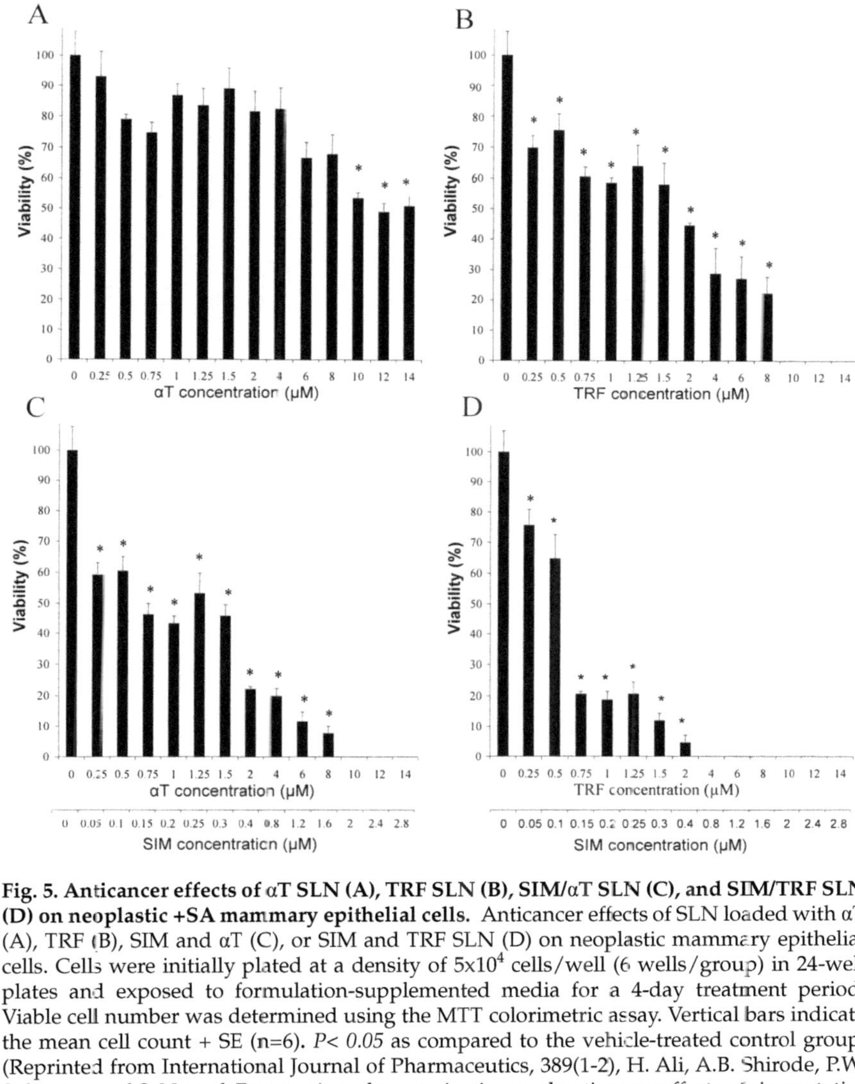

Fig. 5. Anticancer effects of αT SLN (A), TRF SLN (B), SIM/αT SLN (C), and SIM/TRF SLN (D) on neoplastic +SA mammary epithelial cells. Anticancer effects of SLN loaded with αT (A), TRF (B), SIM and αT (C), or SIM and TRF SLN (D) on neoplastic mammary epithelial cells. Cells were initially plated at a density of 5×10^4 cells/well (6 wells/group) in 24-well plates and exposed to formulation-supplemented media for a 4-day treatment period. Viable cell number was determined using the MTT colorimetric assay. Vertical bars indicate the mean cell count + SE (n=6). *P< 0.05* as compared to the vehicle-treated control group. (Reprinted from International Journal of Pharmaceutics, 389(1-2), H. Ali, A.B. Shirode, P.W. Sylvester, and S. Nazzal. Preparation, characterization, and anticancer effects of simvastatin-tocotrienol lipid nanoparticles, 223–231, 2010, with permission from Elsevier).

in a dose-responsive manner. Similarly, when α-tocopherol was substituted with TRF, the cellular viability decreased significantly with TRF treatment from 0.25 to 14 μM. The antiproliferative effect of the SLN was further potentiated when the cells were treated with co-encapsulated SIM/TRF nanoparticles. In comparison, when α-tocopherol was replaced with TRF,

IC_{50} decreased significantly to 1.5 μM. When SIM was incorporated into the TRF nanoparticles at a SIM to TRF ratio of 1:5, the IC_{50} further decreased to 0.52 μM. The observed increase in potency suggested that activities of both drugs were preserved when they were incorporated into SLN and substantiated the previous findings in which the synergistic effect of TRF and SIM was reported.

Key Facts

- Vitamin E is a generic term that refers to a family of eight natural compounds that are further divided into two subgroups called tocopherols and tocotrienols.
- Although chemically very similar, tocotrienols were shown to possess significantly greater anticancer activity than tocopherols using treatment doses that have little or no effect on normal cells.
- The anticancer effects of tocotrienols were first discovered in studies investigating the role of high dietary fat intake on cancer development in which high dietary intake of tocotrienol rich palm oil was found to suppress carcinogen-induced tumor growth in experimental animals.
- Tocotrienols were shown to have a synergistic anticancer activity when combined with many anticancer drugs suggesting that combined therapy with these drugs may provide significant health benefits in the prevention and/or treatment of cancer meanwhile minimizing the side effects associated with these drugs when given alone.
- Tocotrienols are poorly soluble in water resulting in low therapeutic concentration in the blood, a drawback that may be circumvented by encapsulating tocotrienols within lipid nanoparticles.

Summary Points

- Tocotrienols, in contrast to tocopherols, were shown to display potent antiproliferative and apoptotic activity both *in vitro* and *in vivo* against a wide variety of cancer cell types at treatment doses that have little or no effect on normal cell growth and function.
- Numerous studies have demonstrated that combined treatment of tocotrienol with other drugs often result in a synergistic inhibition in cancer cell growth and viability. These studies strongly suggest that tocotrienols may provide significant benefit in the prevention and treatment of cancer when used either alone as monotherapy or in combination with other anticancer agents.
- Tocotrienols are poorly absorbed when taken orally. Therefore, a simple oral consumption of tocotrienols is insufficient to maintain therapeutic levels in the blood. When formulated as SEDDS, the bioavailability

of tocotrienols was found to increase due to the formation of nanodispersion in the luminal contents of the gastrointestinal tracts and the subsequent increase in tocotrienol dissolution in the digestion media.

- While reformulating tocotrienols may improve their oral bioavailability, tocotrienols absorption is mediated by a saturable carrier that limits the extent of their absorption. Therefore, the delivery of tocotrienols with the aid of lipid nanoparticles through the parenteral route was considered as an alternate and efficient approach for the treatment of cancer.
- Preliminary studies with SLN and lipid vesicles showed promising results in cell culture and animal models. Lipid nanoparticles are safe and stable and have the capacity to accommodate a high dose of tocotrienols alone or in combination with other drugs, thereby paving the way for a potentially new era in the prevention and treatment of cancer.

Definitions

Antiproliferative: The ability of a drug to suppress cell growth.

Angiogenesis: The formation of new blood vessels.

Enhanced Permeability and Retention (EPR): A term used to describe the preferential accumulation of nanoparticles in tumors. This selective accumulation and retention of nanoparticles is due to the unique structural features of many solid tumors. Tumors are often associated with a defective and leaky vascular architecture as a result of the poorly regulated nature of tumor angiogenesis. Actively growing tumors form new blood vessels in order to obtain nutrients and sustain their growth. These newly formed blood vessels are usually defective and have fenestrae ranging in size from 100 to 600 nm, which allow the nanoparticles to readily penetrate and accumulate in tumors. Furthermore, the interstitial fluid within a tumor is usually inadequately drained by a poorly formed lymphatic system.

Extravasation: Leakage of drugs into the surrounding tissues.

Fenestræ: Small pores in the epithelial cells that allow for rapid exchange of molecules between blood vessels and surrounding tissue.

Parenteral: Taken into the body by means other than oral intake, usually by injection in a muscle or vein.

Self-Emulsified Drug Delivery Systems (SEDDS): Lipid formulations that are ideally composed of a blend of drug, oil, surfactant, co-surfactant, and co-solvent that spontaneously form fine oil-in-water emulsion when introduced into an aqueous medium under gentle agitation.

Solid Lipid Nanoparticles (SLN): Colloidal particles made from biodegradable and biocompatible solid lipids, that exhibit size range between 50 and 1000 nm.

Acknowledgments

This work was performed in the College of Pharmacy at the University of Louisiana at Monroe, Monroe, LA and supported in part by grants from First Tech International Ltd.

Abbreviations

+SA	:	Soft Agar positive
^{1}H-NMR	:	Proton Nuclear Magnetic Resonance
Akt	:	Protein kinase B
BSA	:	Bovine Serum Albumin
CET	:	Cetyl palmitate
COMP	:	Glyceryl behenate (Compritol® 888 ATO)
Cryo-SEM	:	Cryogenic Scanning Electron Microscopy
DSC	:	Differential Scanning Calorimetry
DYN	:	Glycerol tristearate (Dynasan® 118)
EGF	:	Epidermal Growth Factor
EPR	:	Enhanced vascular Permeability and Retention
HER	:	Human Epidermal Growth factor
HMG-CoA	:	3-hydroxyl-3-methyl glutaryl Coenzyme A
HPLC	:	High Performance Liquid Chromatography
IC_{50}	:	Dose resulting in 50% cell growth inhibition
JNK	:	c-Jun N-terminal kinase
MAPK	:	Mitogen-Activated Protein Kinase
MTT	:	3-(4, 5-dimethylthiazol-2yl)-2, 5-diphenyl tetrazolium bromide
NFkB	:	Nuclear Factor Kappa B
NPC1L1	:	Niemann-Pick C1-like 1 transporter
PI3K	:	Phosphatidylinositol-3-Kinase
PREC	:	Glycerol palmitostearate (Precirol® ATO 5)
SE	:	Standard Error
SEDDS	:	Self-Emulsifying Drug Delivery Systems
SIM	:	Simvastatin
SIM/αTSLN	:	Simvastatin and α tocopherol loaded solid lipid nanoparticles
SIM/TRF SLN	:	Simvastatin and tocotrienol rich fraction loaded solid lipid nanoparticles
SLN	:	Solid Lipid Nanoparticles

SMEDDS	:	Self-Microemulsifying Drug Delivery Systems
SR-B1	:	Scavenger Receptor Class B type 1
TRF	:	Tocotrienol-Rich-Fraction of palm oil
TRF SLN	:	Tocotrienol rich fraction loaded solid lipid nanoparticles
XRD	:	X-ray Diffraction
α	:	Alpha
αT SLN	:	α Tocopherol loaded solid lipid nanoparticles
β	:	Beta
γ	:	Gamma
δ	:	Delta

References

Abuasal, B., P.W. Sylvester and A. Kaddoumi. 2010. Intestinal Absorption of _-Tocotrienol Is Mediated by Niemann Pick C1-Like 1: *In Situ* Rat Intestinal Perfusion Studies. Drug Metab. Dispos. 38(6): 939–945.

Ali, H., A.B Shirode, P.W. Sylvester and S. Nazzal. 2010a. Preparation and antiproliferative effect of Tocotrienol loaded lipid nanoparticles. Colloids Surf. A 353(1): 43–51.

Ali, H., A.B. Shirode, P.W. Sylvester and S. Nazzal. 2010b. Preparation, characterization, and anticancer effects of simvastatin-tocotrienol lipid nanoparticles. Int. J. Pharm. 389(1-2): 223–231.

Ali, H., K. El-Sayed, P.W. Sylvester and S. Nazzal. 2010c. Molecular Interaction and Localization of Tocotrienol-Rich Fraction (TRF) Within the Matrices of Lipid Nanoparticles: Evidence Studies By Differential Scanning Calorimetry (DSC) and Proton Nuclear Magnetic Resonance Spectroscopy (^{1}H NMR). Colloids Surf. B. 77: 286–297.

Ali, H., and S. Nazzal. 2009. Development and validation of a reversed-phase HPLC method for the simultaneous analysis of simvastatin and tocotrienols in combined dosage forms. J. Pharm. Biomed. Anal. 49(4): 950–6.

Bachawal S.V., V.B. Wali and P.W. Sylvester. 2010. Enhanced antiproliferative and apoptotic response to combined treatment of gamma-tocotrienol with erlotinib or gefitinib in mammary tumor cells. BMC Cancer 10: 84.

Cottrell, R.C. Introduction: nutritional aspects of palm oil. 1991. Am. J. Clin. Nutr. 53(suppl): 989S–1009.

Fu, J.Y., D.R. Blatchford, T. Laurence and C. Dufès. 2009. Tumor regression after systemic administration of tocotrienol entrapped in tumor-targeted vesicles. J. Control Release 140(2): 95–9.

Gullotti, E., and Y. Yeo. 2009. Extracellularly Activated Nanocarriers: A New Paradigm of Tumor Targeted Drug Delivery. Molecular Pharmaceutics 6(4): 1041–1051.

Hayes, K.C., A. Pronczuk and J.S. Liang. 1993. Differences in the plasma transport and tissue concentrations of tocopherols and tocotrienols: observations in humans and hamsters. Proc. Soc. Exp. Biol. Med. 202: 353–359.

Matar, P., F. Rojo, R. Cassia, G. Moreno-Bueno, S. Di Cosimo, J. Tabernero, M. Guzman, S. Rodriguez, J. Arribas, J. Palacios and J. Baselga. 2004. Combined epidermal growth factor receptor targeting with the tyrosine kinase inhibitor gefitinib (ZD1839) and the monoclonal antibody cetuximab (IMC-C225): superiority over single-agent receptor targeting. Clin. Cancer Res. 10: 6487–6501.

Matsumura, Y., and H. Maeda. 1986. A new concept for macromolecular therapeutics in cancer chemotherapy: mechanism of tumoritropic accumulation of proteins and the antitumor agent SMANCS. Cancer Res. 6: 193–210.

McIntyre, B.S., K.P. Briski, A. Gapor and P.W. Sylvester. 2000. Antiproliferative and apoptotic effects of tocopherols and tocotrienols on preneoplastic and neoplastic mouse mammary epithelial cells. Proc. Soc. Exp. Biol. Med. 224: 292–301.

Mo, H., and C.E. Elson. 2004. Studies of the isoprenoid-mediated inhibition of mevalonate synthesis applied to cancer chemotherapy and chemoprevention. Exp. Biol. Med. (Maywood) 229: 567–585.

Müller, R.H., K. Mader and S. Gohla. 2000. Solid lipid nanoparticles (SLN) for controlled drug delivery—a review of the state of the art. Eur. J. Pharm. Biopharm. 50: 61–177.

Nazzal, S., I.I. Smalyukh, O.D. Lavrentovich and M.A. Khan. 2002. Preparation and *in vitro* characterization of a eutectic based semisolid self-nanoemulsified drug delivery system (SNEDDS) of Ubiquinone: Mechanism and progress of emulsion formation. Int. J. Pharm. 235: 247–265.

Normanno, N., A. De Luca, C. Bianco, L. Strizzi, M. Mancino, M.R. Maiello, A. Carotenuto, G. De Feo, F. Caponigro and D.S. Salomon. 2006. Epidermal growth factor receptor (EGFR) signaling in cancer. Gene. 366: 2–16.

Parker, R.A., B.C. Pearce, R.W. Clark, D.A. Gordon and J.J. Wright. 1993. Tocotrienols regulate cholesterol production in mammalian cells by post-transcriptional suppression of 3-hydroxy-3-methylglutaryl-coenzyme A reductase. J. Biol. Chem. 268: 11230–11238.

Ratain, M.J., and R. Mick. 1996. Principles of pharmacokinetics and pharmacodynamics. pp. 123–142., In: R.L. Schilsky, G.A. Milano, and M.J. Ratain [eds.] 1996. Principles of Antineoplastic Drug Development and Pharmacology. Marcel Dekker, New York, USA.

Samant, G.V., and P.W. Sylvester. 2006. γ-Tocotrienol inhibits ErbB3-dependent PI3K/Akt mitogenic signaling in neoplastic mammary epithelial cells. Cell. Prolif. 39: 563–574.

Sylvester, P.W. 2007. Vitamin E and apoptosis. Vitam. Horm. 76: 329–356.

Sylvester, P.W., and A. Theriault. 2003. Role of Tocotrienols in the Prevention of Cardiovascular Disease and Breast Cancer. Curr. Top. Nutraceutical Res. 1: 121–135.

Sylvester, P.W., and S.J. Shah. 2005. Mechanisms mediating the antiproliferative and apoptotic effects of vitamin E in mammary cancer cells. Front. Biosci. 10: 699–709.

Thibault, A., D. Samid, A.C. Tompkins, W.D. Figg, M.R. Cooper, R.J. Hohl, J. Trepel, B. Liang, N. Patronas, D.J. Venzon, E. Reed and C.E. Myers. 1996. Phase I study of lovastatin, an inhibitor of the mevalonate pathway, in patients with cancer. Clin. Cancer Res. 2: 483–491.

Tipton, J.M. 2003. Side effects of cancer chemotherapy. pp. 561–580. In: R.T. Skeel [ed.] 2003. Handbook of Cancer Chemotherapy. Lippincott Williams & Wilkins, Philadelphia, USA.

Wali, V.B., and P.W. Sylvester. 2007. Synergistic antiproliferative effects of gamma-tocotrienol and statin treatment on mammary tumor cells. Lipids 42: 1113–1123.

Wissing, S.A., O. Kayser and R.H. Muller. 2004. Solid lipid nanoparticles for parenteral drug delivery. Adv. Drug Deliv. Rev. 56: 1257–1272.

Wong, H.L., R. Bendayan, A.M. Rauth, Y. Li and X.Y. Wu. 2007. Chemotherapy with anticancer drugs encapsulated in solid lipid nanoparticles. Adv. Drug Deliv. Rev. 59(6): 491–504.

Yap, S.P., K.H. Yuen and A.B. Lim. 2003. Influence of route of administration on the absorption and disposition of alpha-, gamma- and delta-tocotrienols in rats. J. Pharm. Pharmacol. 55(1): 53–8.

Yap, S.P., K.H. Yuen and J.W. Wong. 2001. Pharmacokinetics and bioavailability of alpha-, gamma- and delta-tocotrienols under different food status. J. Pharm. Pharmacol. 53: 67–71.

Yap, S.P., and K.H. Yuen. 2004. Influence of lipolysis and droplet size on tocotrienol absorption from self-emulsifying formulations. Int. J. Pharm. 281(1-2): 67–78.

5

Utilization of Magnetic Nanoparticles for Cancer Therapy

Deryl L. Troyer[1,*] and *Stefan H. Bossmann*[2]

ABSTRACT

Magnetic nanoparticles are a powerful tool in cancer imaging, drug delivery, magnetic hyperthermia, immunotherapy, and detection of circulating tumor cells. Magnetic hyperthermia depends upon intratumoral injection of MNPs or various methods of targeting, including use of external magnetic fields, or the attachment of antibodies, peptides or other ligands that are tumor specific. Recently, stem cells have been used as a sort of 'Trojan horse' for targeted delivery of MNPs and magnetic hyperthermia in preclinical models. The working principle behind MHT is that of exposing MNP to an alternating magnetic field (A/C-MF or MF). The heat generated depends upon the frequency, amplitude and time of AMF exposure, as well as specific absorption rates (SAR) of the MNPs. SAR is very dependent on shape, size, dispersity and chemical type of the MNPs. There are many reports in the literature of successful magnetic hyperthermia in animal models

[1]Department of Anatomy and Physiology, 228 Coles Hall, Kansas State University, Manhattan, KS 66506, USA; E-mail: troyer@vet.ksu.edu

[2] Department of Chemistry, 213 CBC Building, Kansas State University, Manhattan, KS 66506, USA; E-mail: sbossman@ksu.edu

*Corresponding author

List of abbreviations after the text.

as well as human patients. Magnetic hyperthermia is capable of powerful immune enhancement in addition to its ability to cause cancer cell destruction due to elevated intratumoral temperatures. Magnetic hyperthermia shows synergy with other methods of cancer treatment (such as chemotherapy, radiation, immunotherapy, or gene therapy). It is rapidly developing into a powerful alternative to classic treatment approaches.

INTRODUCTION

Magnetic nanoparticles (MNPs) are the subject of intense research focusing on synthesis methods, characteristics, and potential for functionalization (Latorre and Rinaldi 2009). They have enormous potential for tumor therapy due to hyperthermia generated within the tumors when subjected to an external A/C-magnetic field, or as imaging agents via MRI. Magnetic hyperthermia (MHT) is attractive as a cancer therapy because it is associated with fewer off-target effects than many standard cancer treatments. MHT is most effective in synergy with chemo, radiation, immunotherapy, or gene therapy (Gazeau et al. 2008). In this review, we will focus primarily on MHT, with mention of imaging when used in conjunction with this modality. It is remarkable that the same nanoparticle could have such potential for both imaging and therapy. We will also mention possible targeting strategies for improved MHT.

Key Facts About the History of Hyperthermia

- Hyperthermic cancer therapy was first used in Egypt over 4000 years ago by applying hot oil to tumors.
- Since then, total body hyperthermia has been used with varying success, but has side effects.
- Tiny magnetic particles implanted in tumors will generate heat within the tumor when the area is put in an alternating magnetic field.

Key Facts Pertaining to Renewed Interest in Hyperthermia

- The potential use of hyperthermia to treat tumors has great appeal because it is a physical treatment and thus has fewer side effects than chemotherapy or radiotherapy.
- The hyperthermia can be repeated without development of drug resistance.
- Hyperthermia can be combined with other forms of therapy.
- Emerging targeting methods will allow better localization of hyperthermia to tumors.

Key Facts Pertaining to Circulating Tumor Cells (CTCs)

- Some cancer cells may be present in the blood of cancer patients that are extremely rare and thus difficult to identify.
- Detection of such cells may enable earlier intervention or more accurate assessment of treatment effectiveness.
- Magnetic nanoparticles that have proteins attached that recognize the circulating tumor cells can be used to separate them from other blood cells when a magnet is applied.

MAGNETIC HYPERTHERMIA

Mechanism

The working principle behind MHT is that of exposing MNP to AMF after they have somehow reached a tumor (Jordan et al. 1993; Jordan et al. 2001). Three major factors involved in MHT are the field generator, which generates the AMF, the induction coil, which defines the area of high field strength, and the MNPs, which generate heat when exposed to an AMF. The heat generation depends on the frequency and amplitude of the magnetic field, and the time exposed to AMF, as well as and on Curie temperature and Specific Absorption Rates (SAR), which is strongly dependent on the chemical type of MNP and its shape, size, dispersity (mono or poly) (Bossmann 2009).

MNPs create heat by hysteresis loss when they are exposed to AMF. In multi-domain particles the dominant heating mechanism is hysteresis loss due to the movement of domain walls. In single domain magnetic nanoparticles, Néel and/or Brownian relaxation are observed. Which of these two processes is dominant depends on MNP size and the viscosity of the surrounding medium (Kotitz et al. 1995; Pakhomov et al. 2005). In smaller magnetic nanoparticles, Néel relaxation (random flipping of the spin without rotation of the particle) is the main contribution to A/C-magnetic heating in tissue. Brownian relaxation (rotation of the entire nanoparticle) occurs in larger nanoparticles. It must be noted that Brownian relaxation is very dependent on the viscosity of the surrounding medium. It is greatly diminished if the nanoparticles are bound to biological structures or immobilized otherwise (Glöckl et al. 2006). Therefore SAR measurements performed in a solvent may be accurate in predicting the efficacy of MNPs in biological tissue. The transition between the two mechanisms occurs between 5–12 nm for various materials, but it also varies with frequency (Mornet et al. 2004).

Magnetic heating takes place by power absorption of MNPs in an AMF. The important factor for magnetic heating experiments is the ***specific absorption rate*** or SAR, which is determined by SAR = $C_c\ \Delta T\ \Delta t^{-1}$ where C_c is the specific heat capacity of the sample and T and t are the temperature and time, respectively. SAR (or SLP (specific loss power)) is very sensitive to material properties, medium and, AMF frequency(Fortin et al. 2007) (Fig. 1). The state of the art in hyperthermia is characterized by the use of very different AMF characteristics. Therefore, comparisons between the results obtained by the numerous research groups in the field are close to impossible. Thus, hyperthermia research would greatly profit from the use of standard AMF conditions for treating surface tumors and deep-seated tumors.

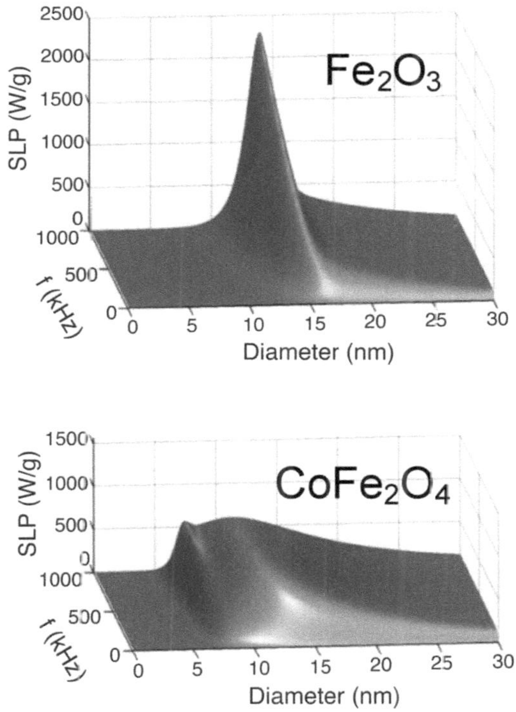

Fig. 1. Specific loss power (SLP) of monodisperse particles of maghemite (Fe_2O_3) and cobalt ferrite ($CoFe_2O_4$) as a function of particle diameter and magnetic field frequency. Particle were dispersed in water. The field amplitude was H = 24.8 kA/m. In water, the SLP is mostly due to Brownian relaxation at particle diameters above 9nm ($CoFe_2O_4$) and 14nm (Fe_2O_3). The Neel contribution is highly peaked as a function of particle size (at f = 1MHz, its maximum is found at 7nm ($CoFe_2O_4$) and 14nm (Fe_2O_3). (Reproduced with permission from (Fortin et al. 2007) Copyright: American Chemical Society 2007).

The mass magnetization of magnetic materials is an important factor to determine their performance in hyperthermia. Iron oxide MNPs are commonly used for MHT in animals because of their biodegradability and relatively low toxicity(Gordon et al. 1979). The mass magnetizations of the bulk materials are in the range of 80 (γ-Fe_2O_3, maghemite)(Zhao et al. 1997) to 92 Am^2kg^{-1} (Fe_3O_4, magnetite)(Wang et al. 2007). Other metal oxides show comparable bulk magnetizations as well, for instance $CoFe_3O_4$ (71 Am^2kg^{-1}). Metallic iron (218 Am^2kg^{-1}) and cobalt (161 Am^2kg^{-1}) feature higher mass magnetizations than metal oxides. Metallic nickel (57 Am^2kg^{-1}) is clearly inferior (Huber 2005). Alloys of magnetic elements (Fe/Co, Fe/Ni)(Huber 2005; Wilcoxon and Abrams 2006) or of iron with noble elements (e.g., Fe/Pt) (Sun et al. 1999) have been investigated.

The magnetic properties of the ferromagnetic materials arise from the alignment of unpaired electron spins with the external magnetic field. In bulk ferromagnetic materials, magnetic domains, are of nanometer scale. Below 10–20 nm, depending on the material, single domain nanocrystals are more stable than nanocrystals featuring more than one magnetic domain. The result is superparamagnetic materials, which are characterized by a dependence of the magnetization on temperature. The absence of a net magnetization in the absence of a field is due to the random orientation of the various particles' spins (Shapiro et al. 2006). Although superparamagnetic MNPs do not exhibit hysteresis effects at the slow timescales of typical magnetic experiments, they do show hysteresis in high-frequency A/C-fields because the nanoparticles are unable to reorient at the rate of oscillation of the field.

Their large magnetic susceptibilities make MNPs excellent heat generators in AMFs. For instance, Feridex® consists of a γ-Fe_2O_3-core of 4–5 nm in diameter and a dextran coating. The mass magnetization M_s of this material has been determined to be approximately 70 Am^2kg^{-1}, which is very close to the bulk magnetization. Examples for metallic MNPs with high mass magnetizations are polymer-stabilized Fe(0)-MNPs (172 Am^2kg^{-1}) and Fe/Pt-alloy MNPs (200 $Am^{-2}kg^{-1}$ (Sun et al. 1999). In general, when Fe(0)-nanoparticles are synthesized, they always require protection either by ligands (organic thiols, amines, dopamine-based ligands), polymers (polystyrene-sulfonate, polyacrylic acid, poly-N-isopropyl-acrylaminde etc.) (Huber 2005), and/or inorganic shells (e.g., Fe/Au (Ravel et al. 2002; Chen et al. 2003) or Fe/Fe_3O_4 (Zeng et al. 2007). We have found that iron core, iron oxide shell MNPs are efficient heat generators when small amounts are transplanted into mice and subjected to A/C magnetic field (Fig. 2).

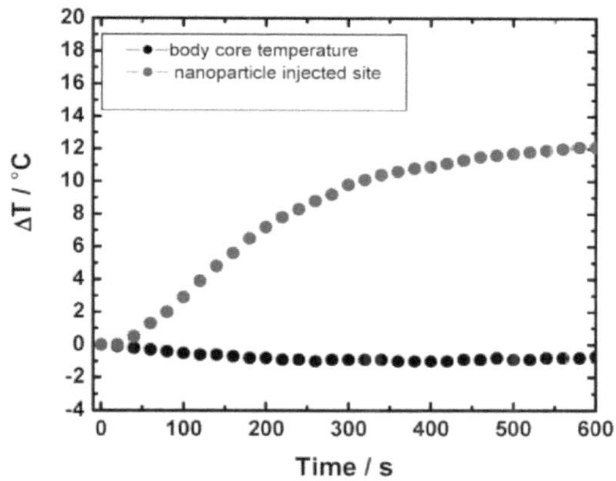

Fig. 2. Temperature change observed with magnetic hyperthermia. MNP were implanted into mouse subcutaneous tissue and exposed to AMF (red). The body core temperature (black) of the mouse is shown in comparison (Balivada et al., BMC Cancer, permission not required for reproduction).

Preclinical Cancer Therapy

Early work: The use of magnetic particles for generating heat within tumors began when Gilchrist and colleagues used micrometer sized particles to generate heat in canine lymph nodes (Gilchrist et al. 1957). This innovative idea was that lymphatic metastasis of colon cancer could be treated with heat by using microscopic ferromagnetic particles to embolize in lymph nodes draining primary tumors followed by the application of an external magnetic field to cause hysteric heating of the particles. Gordon et al. (1979) continued this innovation more than twenty years later when they used dextran magnetite NPs for hyperthermia after systemic application (Gordon et al. 1979). Jordan et al. in 1997 described direct injection of dextran-coated MNPs (Jordan et al. 1997). Since then, many examples of magnetic hyperthermia in animal models of human cancers have been published (Thiesen and Jordan 2008).

Examples of preclinical success: Most MHT preclinical studies have utilized the intratumoral injection of milligram amounts of iron followed by A/C MF exposure. Cationic liposomes loaded with magnetite were administered into subcutaneous gliomas in rats. After A/C MF exposure, 90% of the tumors had regressed almost completely (Shinkai et al. 1999). Magnetoliposomes coated with antibody were injected into subcutaneous gliomas in nude mice, which were then exposed to an A/C MF, with arrest of tumor growth (Le et al. 2001). Many other reports have detailed significant anti-tumor effects, in some cases very pronounced, with intratumoral injection of MNPs and A/C MF. These include animal models of melanoma (Ito et al. 2003), breast cancer (Jordan et al. 1997; Jordan et al. 1999; Hilger et al. 2001; Ito et al. 2003), primary prostate cancer (34–37) (Johannsen et al. 2005; Kawai et al. 2006), prostate tumors in bone (Kawai et al. 2008), brain cancer (Ohno et al. 2002; Jordan et al. 2006).

Immunmostimulation via MHT in preclinical models: There is now compelling evidence that MNPs with external alternating magnetic field exposure are potent inducers of 'heat immunotherapy'. A major reason for this is the upregulation of heat shock protein expression with eventual enhanced MHC class I-dependent tumor antigen presentation and antitumor cell mediated immunity (Krishnamachari et al. 2011). Hsp70, Hsp90 and glucose-regulated protein 96 (GRP-96) are all important proteins in the immune-enhancement in combination with MHT (Ito et al. 2006). In this regard, HSP70 and 90 are suggested to chaperone tumor antigens (refs 13,14 in 5320 (Menoret and Chandawarkar 1998; Milani et al. 2002). In an MHT study in which tumors were implanted bilaterally in the lumbar region of rats and MNPs were inoculated into one tumor followed by A/C MF, HSP expression reached its maximum 24 hours after treatment and increased to its maximum at 48 hours; interestingly, tumors on both sides were significantly reduced (Ito et al. 2001). When immature dendritic cells were injected into mouse subcutaneous T-lymphomas that had been previously treated by direct inoculation of MNPs followed by AMF, complete regression of tumors occurred in 75% of the mice (Tanaka et al. 2005).

Tumors were implanted bilaterally in rat femurs, and one tumor only received intratumoral MNPs followed by A/C MF exposure in a report in 1998 (Yanase et al. 1998). Intriguingly, 28 days later both tumors were gone, and subsequent studies revealed that $CD4^+$ and $CD8^+$ lymphocytes had invaded. The rat immune systems had recognized tumor antigens exposed by the hyperthermia.

Small amounts can be effective: Our group has used core shell/Fe_3O_4 MNPs injected intratumorally into mouse subcutaneous melanomas to show that small amounts followed by three ten minute A/C magnetic exposures could slow tumor growth (Balivada et al. 2010); see Fig. 3. The inorganic center of these MNPs was protected against rapid biocorrosion by organic dopamine-oligoethylene glycol ligands. TCPP (4-tetracarboxypehnyl porphyrin) units were attached to the dopamine-oligoethylene glycol ligands. The temperature increase of the cancer tissue upon exposure of the A/C magnetic field (5kA/m) was directly measured with a fiber optic probe. Interestingly, the amount of MNP used in these experiments is much less than what was previously thought to be an effective amount of MNP in magnetic hyperthermia (12 $\mu g/cm^3$ compared to 1 mg/cm^3). This could be the result of iron (0)-MNPs, which are a more effective heater than Fe_2O_3 or Fe_3O_4 MNPs (Habib et al. 2008), but other

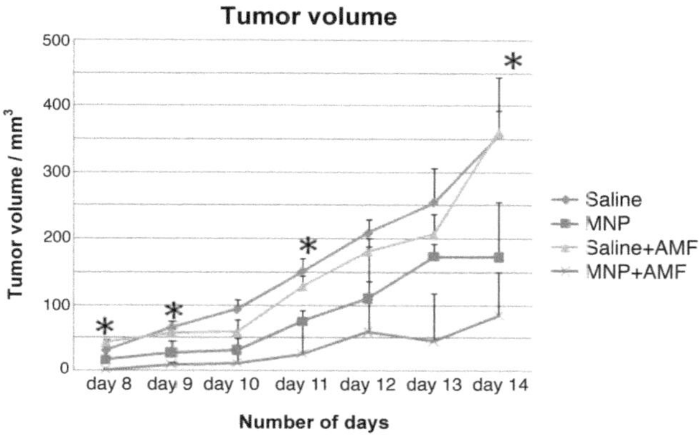

Fig. 3. Attenuation of melanoma with small amounts of MNPs and A/C MF. MNPs were administered intratumorally (final concentration: 12µg/cm_3) (Balivada et al., BMC Cancer, permission not required for reproduction).

factors cannot be excluded such as biocorrosion and enhanced oxidative damage (Winterbourn 1999) or disruption of cellular membranes due to mechanical rotation of the MNPs that is caused by Brownian relaxation. Initially, Brownian rotation will be slow in many biological systems. However, in supraparamagnetic MNPs, there is always a Néel component. Hergt and coworkers have shown in a series of model calculations that the nanosphere around a MNP can heat up by several degrees C in a viscous environment (Hergt et al. 1998).This local heating effect can decrease the microfluidity in biological membranes and may help to switch on Brownian rotators (Glöckl et al. 2006).

Clinical Cancer and Magnetic Hyperthermia

An advantage of hyperthermia is that it allows frequent repeated treatment; however, these treatments should not be given within 48 hours of each other to preclude thermotolerance caused by upregulated Hsp activity (Ito et al. 2006). Complex changes have been noted during and after hyperthermia regarding nutrient, oxygen and blood supply to tumors, signal transduction, immune-modulation and pharmacologic effects (Hildebrandt et al. 2002). Improved response and survival rates have been observed in patients treated with whole body hyperthermia and radiotherapy compared to radiotherapy alone, in some cases with a very significant clinical benefit (Hildebrandt et al. 2002). In Germany, a magnetic field of 100kHz and variable field strength of 0–18 kA/m is generated in the applicator MFH®300F (MagForce Nanotechnologies AG, Berlin, Germany). Due to its universal design, malignancies in several places in the body can be treated (Thiesen and Jordan 2008). Also in Germany, recent Phase II clinical trials utilizing magnetic fluid hyperthermia have been evaluated histopathologically (van Landeghem et al. 2009). The results underscored the need for multiple trajectories of inoculation.

Targeting Strategies

Attempted tumor specific targeting can be either passive or active. Passive targeting is achieved by the smaller MNPs via enhanced permeability and retention (Nie et al. 2007; Bossmann 2009). Tumor vasculature is hyperpermeable compared to healthy tissue accumulate in tumor interstitial spaces (Nie et al. 2007). The hydrophilic surfactant nature of MNPs also helps in passive targeting by allowing MNPs vasculature, which helps smaller MNPs to avoid plasma protein absorption (Shinkai 2002). Active targeting can be achieved by attaching tumor vasculature-specific and tumor cell-specific molecules onto the MNPs (see below).

Some drugs such as lidocaine or amphotericin B are cytotoxic only at elevated temperatures (Hildebrandt et al. 2002). This raises the possibility of using such drugs as anticancer 'heat acitivated' prodrugs.

Ligand attachment: A major roadblock to the use of hyperthermia as a resource for patients and their physicians is the limiting targeting ability of currently available MNPs (Latorre and Rinaldi 2009). This cripples the ability to concentrate MNPs within a tumor, negatively affecting homogeneous distribution of heat. This limitation is usually circumvented by direct injection of MNPs into the tumor. However, delivery of MNPs to tumors via intratumoral injection is feasible only when tumors are accessible (Gazeau et al. 2008). This approach is unfortunately not applicable to deep or metastatic tumors. One possible approach toward the goal of accessing these tumors is to attach ligands to the MNPs that allow them to bind to vasculature or tumor cells after they have gained entry via the enhanced permeability and retention (EPR) effect which takes advantage of 'leaky' tumor neoangiogenic vessels. In some cases, antibodies have been attached (Shinkai et al. 1995); in other cases, peptides have been attached (Montet et al. 2006). With regard to peptides, new and much more powerful peptides that allow dramatic increases within the tumor have recently been described. Ruoslahti's group has shown that co-administration of a peptide known as 'iRGD' enhances, in a tumor-specific manner, vascular and tissue permeability allowing drugs or NPs or magnetic nanoworms co-administered with it to penetrate deeply into extravascular tumor tissue (Sugahara et al. 2010). This peptide initially tracks to tumors by binding αv integrins specifically expressed on tumor perivasculature. It is then cleaved in the tumor to produce CRGDK/R, and the resultant peptide binds to neruopilin-1 (NRP-1), which triggers tissue penetration. Earlier work had used a conjugated iRGD peptide (Sugahara et al. 2009), but the ability to co-inject it with various therapeutic molecules or nanomaterials such as MNPs is even more impressive.

Thermolabile materials: Another targeting strategy involves the use of thermosensitive liposomes (Tang et al. 2010). Folate receptor-targeted, thermosensitive magnetic liposomes loaded with doxorubicin (MagFolDox) resulted in enhanced tumor cytotoxicity indicated the multidrug resistance could be overcome using this approach (Pradhan et al. 2010). Doxycycline-loaded magnetite PNIPAM composite MNPs were synthesized in which the lower critical solution temperature (LCST) allowed much greater release of drug at elevated temperatures (Purushotham and Ramanujan 2010; Purushotham et al. 2009). Thermosensitive liposomes have been developed for temperature-dependent release of drugs with hyperthermia(Needham et al. 2000).

Use of magnets to target: External magnetic fields have been used to target drug/carrier complexes in the form of ferrofluid (magnetic drug targeting; MDT). The MNPs are given to the patient systemically, after which high-gradient magnetic fields are used to concentrate the complex within tumors. When the MNP-drug complex localizes to the tumor, the drug can be released by tumor-specific conditions such as pH or enzymes (Alexiou et al. 2000).

Cell based strategies: Ikehara et al. injected oligomannose-coated liposomes encapsulating Fe_3O_4 and 5-fluorouracil (5-FU) into the peritoneal cavity with subsequent uptake by peritoneal macrophages and delivery to peritoneal gastric tumors in a mouse model (Ikehara et al. 2006). After AMF exposure, they observed significant attenuation of the peritoneal tumors. Interestingly, the localized hyperthermia caused release of the 5-FU from the macrophages by induced exocytosis.

Some groups have utilized SPION-loaded stem cells and MRI to detect tumors. For example, Loebinger and colleagues labeled human bone marrow mesenchymal stem cells (MSCs) into mice and showed that as few as 1,000 of the cells could be detected by MRI on month after their co-injection with breast cancer cells that formed subcutaneous tumors. They then used MRI to track SPION loaded mesenchymal stem cells to metastatic lung tumors in mice (Loebinger et al. 2009). In a similar approach, MSCs from Fischer 344 rats were co-labeled with SPION and GFP and given IV 3 days after transplanting glioma cells. MRI imaging of cells in tumors were followed by histological evidence of Prussian blue and GFP positive cells (Wu et al. 2008).

Our group has previously used stem cells we discovered in the umbilical cord (Weiss et al. 2006) for targeted gene therapy in a mouse model of human cancer (Rachakatla et al. 2007; Ganta et al. 2009; Rachakatla and Troyer 2009); see Fig. 4). We have shown that it is possible to pre-load delivery stem cells (in this case, neural progenitor cells) with the MNPs we have developed (described above), allow the stem cells to ferry them to the tumors after IV injections, and subject mouse melanomas to A/C MF exposure (Rachakatla et al. 2010); see Fig. 5. We developed aminosiloxane-protected Fe/Fe_3O_4 MNPs that possess tethered porphyrins groups as "bait" for cancer cells. NPCs were efficiently loaded with core/shell Fe/Fe_3O_4 MNPs with minimal cytotoxicity; the MNPs accumulated as aggregates in the cytosol (Fig. 6). The MNPs with attached porphyrins are shown in Fig. 7. The NPCs loaded with MNPs could travel to subcutaneous melanomas, and after A/C magnetic field (AMF) exposure, attenuate mouse melanomas. To our knowledge, this is the first report showing that tumor-tropic stem cells loaded with MNPs *ex vivo* and administered intravenously can result in regression of preclinical tumors after A/C magnetic field exposure.

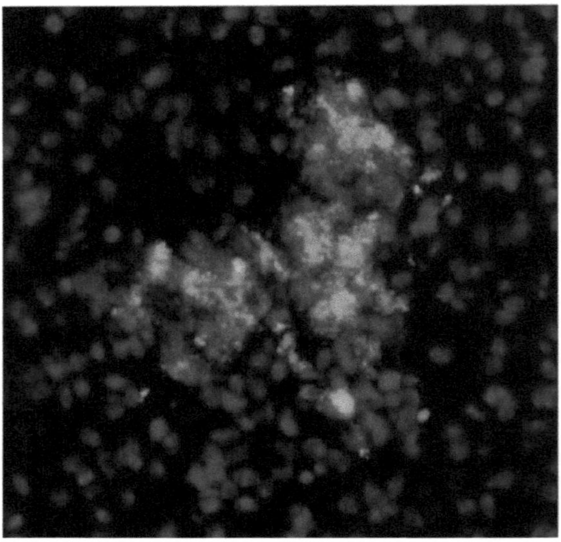

Fig. 4. Umbilical cord matrix stem cells (UCMSC) home to melanoma and deliver therapeutic proteins. Dye loaded human UCMSC (red or orange), engineered to express β-interferon, home to metastatic human breast cancer tumors in SCID mouse lungs (green) and attenuate the tumors (unpublished).

Color image of this figure appears in the color plate section at the end of the book.

Other uses of MNPs in Cancer Therapy

Magnetic separation: Magnetic bead separation of some cell populations from others has been available for many years. In cancer biomedical

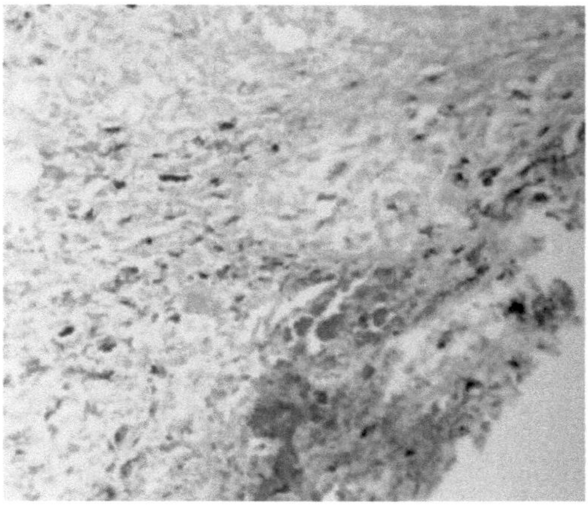

Fig. 5. Murine Neural Progenitor Cells (NPCs) (blue) home to melanomas. Image was captured four days after intravenous administration of stem cells into mice bearing subcutaneous B16F10 melanoma. (Reproduced with permission from (Rachakatla et al. 2010) Copyright: American Chemical Society 2010).

Color image of this figure appears in the color plate section at the end of the book.

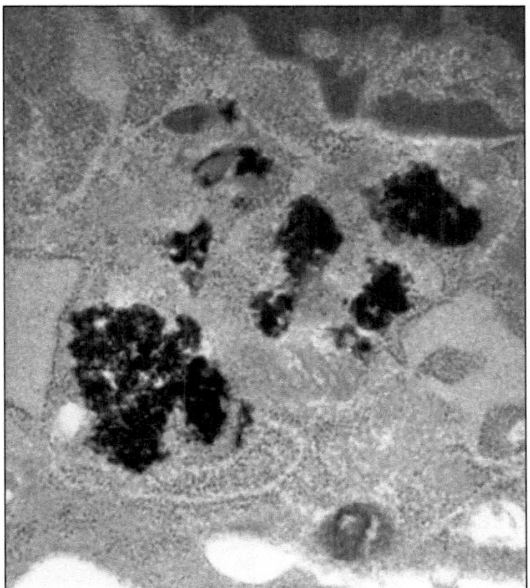

Fig. 6. Stem cell loaded with MNPs. Note the presence of dense aggregates in the cytosol of the delivery cell. (Magnification: 30,000x).

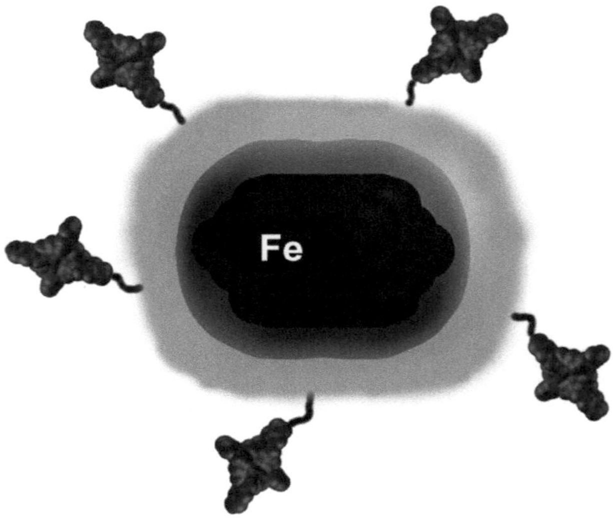

Fig. 7. Aminosiloxane (blue layer)-stabilized Fe/Fe_3O_4 composite core/shell nanoparticles featuring a dopamine-anchored organic stealth-layer (green layer) and tetrakis-carboxyphenyl-porphyrin (TCPP) labels for the enhanced uptake of the nanoparticle by the LDL (low density lipid) receptor, which is overexpressed in numerous cancer cell lines (Reproduced with permission from (Rachakatla et al. 2010) Copyright: American Chemical Society 2010).

Color image of this figure appears in the color plate section at the end of the book.

applications, this approach could be used to enrich for cancer stem cells but could be further improved. It has been used and is being further developed to detect circulating tumor cells (CTCs). It can also be used to identify tumor associated cells, or immune cells within tumors.

Gene delivery: Polyethylenimine-coated MNP showed a marked increase in transfection efficiency for both viral and non viral vectors compared to standard vectors alone (Scherer et al. 2002). This is termed 'magnetofection' and appears to offer advantages to standard *in vitro* transfection protocols (Shubayev et al. 2009).

Applications to Health and Disease

Magnetic nanoparticles are relevant to health due to their MRI imaging potential for detection of a variety of diseases, including cancer in otherwise healthy individuals. Sensitive detection of circulating tumor cells may enable early detection or evaluation of therapy. Other detection methods include tumor cell-specific, protease-sensitive cleavage of otherwise quenched fluorescent molecules. Magnetic nanoparticles have been successfully used to create elevated temperatures in tumors (magnetic hyperthermia) after exposure to an external alternating magnetic field. Most applications of MHT have involved direct injection of milligram amounts of MNPs into accessible tumors. Targeting of the MNPs to tumors is becoming increasingly realistic by attaching peptides to them or by cell-mediated delivery. Magnetic drug delivery utilizing magnetic nanoparticles and an external magnet have also been used.

Summary Points

- Magnetic nanoparticles are 'theranostic agents'; i.e., they are useful both for imaging and treatment of cancer.
- Magnetic hyperthermia involves the accumulation of MNPs in tumors followed by application of an external alternating magnetic field.
- Magnetic hyperthermia exerts potent stimulatory effects on the innate and acquired immune systems to enhance tumor attenuation or elimination.
- Magnetic nanoparticles can be administered intratumorally or systemically if they are tagged with tumor-specific peptides or other ligands, or carried by tumor-tropic cells.
- Depending on the size and nature of the MNPs, and the viscosity of the medium, heat is generated by Neel or Brownian relaxation.
- There are many reports in the literature of successful tumor attenuation in animal models and human patients.
- Magnetic hyperthermia is most effective when used as adjuvant therapy along with radiotherapy or chemotherapy.
- Apart from MHT, MNPs have different uses in biomedicine such as drug delivery, gene delivery, cell tracking, biosensing, cell isolation, and cellular proteomics.

Definitions

Alternating current magnetic field: In electromagnetism, a changing electric field generates a magnetic field and a changing magnetic field produces a magnetic field; both are capable of inducing each other. The field strength of the magnetic field is maximal if the field strength of the electric field is zero, and vice versa. Both fields change their direction periodically.

Brownian relaxation: One of two main contributing mechanisms to the Specific Absorption Rate (SAR) in single domain magnetic nanoparticles. In Brownian relaxation, the rotation of the entire nanoparticle is observed. Therefore, Brownian relaxation is very dependent on the viscosity of the surrounding medium. It is greatly diminished if the nanoparticles are bound to biological structures.

Coercivity: Intensity of the applied magnetic field required to reduce the magnetization of a magnetic material to zero after the magnetization of the sample has been driven to saturation. Coercivity (also called the coercive field or coercive force) is measured A m^{-1}.

Coil: An electromagnetic coil is formed when a conducting wire is wound around a rod-like core. Coils are inductors or, to use a more common terms, electromagnets.

Hyperthermia: Elevation of temperature above normal body temperature.

Hysteresis loss: When an external magnetic field is applied to a magnetic material, its atomic structure will become aligned with the field. As a consequence, the relationship between the magnetic field strength (H) and magnetic flux density (B) is not linear: at some strength of H, B will reach a maximum (magnetic saturation). If H is reduced to zero, a certain flux density will remains, because of prior alignment with H. This phenomenon is called remanence. The coercive field (coercivity) is the strength of the applied magnetic field required to reduce the magnetization to zero after the magnetization of the sample has been driven to saturation. These phenomena occur every time when B follows the alternating changes of H in an A/C magnetic field. As a consequence, magnetic energy is converted to heat. In multi-domain particles the dominant heating mechanism is hysteresis loss due to the movement of domain walls. In single domain magnetic nanoparticles, either Néel or Brownian relaxation is observed.

Magnetic hyperthermia: Hyperthermia due to generation of heat by magnetic nanoparticles in response to an external alternating magnetic field.

Néel relaxation: The second main contributing mechanism to the Specific Absorption Rate (SAR) in single domain magnetic nanoparticles. In Néel relaxation, random flipping of the spin (without rotation of the particle) relaxations is observed. In smaller magnetic nanoparticles ($d < 10$–20 nm, depending on material and AMF frequency), Néel relaxation is the main contribution to A/C-magnetic heating in tissue.

Magnetic Field strength (H): H (unit: A m^{-1}; A: Ampere) is caused only by macroscopic currents.

Magnetic flux density (B): B (unit: T (Tesla), $1T = 1NsC^{-1}m^{-1}$; N:Newton, C:Coulomb) is also called magnetic induction. B depends on macroscopic and nanoscopic currents.

Specific Absorption Rate (SAR): Determined by $SAR = C^* \Delta T / \Delta t$, where C is the specific heat capacity of the sample (or surrounding tissue) and T and t are the temperature and time, respectively. The SAR (W g^{-1}) is a measure of the ability of a nanoparticle to heat their surrounding medium.

Specific Loss Power (SLP): Identical with specific absorption rate.

Superparamagnetic: Particles that are magnetic only in a magnetic field but are otherwise not magnetic. Superparamagnetic materials have a low remanence (see hysteresis loss).

Abbreviations

A/C MF or: AMF		alternating current magnetic field
ADAM	:	a disntegrin and metalloproteinase
CLIO	:	Cross-linked iron oxide magnetic nanoparticles
CD	:	cluster of differentiation
CT	:	computed tomography
CTCs	:	circulating tumor cells
DC	:	direct current
EPR	:	enhanced permeability and retention
Fas	:	apoptosis stimulating fragment
5-FU	:	5-fluorouracil
TNF-R	:	tumor necrosis factor receptor
Grp96	:	glucose-regulated protein 96
HCC	:	hepatocellular carcinoma
HIF-1α	:	Hypoxia inducible factor—1 alpha
Hsp70	:	Heat shock protein 70
Hsp90	:	heat shock protein 90
HT	:	hyperthermia
kHz	:	kilohertz
LCST	:	lower critical solution temperature
LH-RH	:	leutinizing hormone-releasing hormone
MDR	:	multidrug resistance
MDT	:	magnetic drug targeting
MHC	:	major histocompatability antigen
MHT	:	magnetic hyperthermia
MMP	:	matrix metalloproteinase
MNPs	:	magnetic nanoparticles
MSCs	:	mesenchymal stem cells
MRI	:	Magnetic resonance imaging
NPCs	:	neural progenitor cells
PEG	:	polyethylene glycol
SAR	:	specific absorption rate
SPION	:	superparamagnetic iron oxide nanoparticles
TNFα	:	Tumor necrosis factor alpha
Wnt	:	wingless

References

Alexiou, C., W. Arnold, R.J. Klein, F.G. Parak, P. Hulin, C. Bergemann, W. Erhardt, S. Wagenpfeil and A.S. Lubbe. 2000. Locoregional cancer treatment with magnetic drug targeting. Cancer Res. 60: 6641–8.

Balivada, S., R.S. Rachakatla, H. Wang, T.N. Samarakoon, R.K. Dani, M. Pyle, F.O. Kroh, B. Walker, X. Leaym, O.B. Koper, M. Tamura, V. Chikan, S.H. Bossmann and D.L. Troyer. 2010. A/C magnetic hyperthermia of melanoma mediated by iron (0)/iron oxide core/shell magnetic nanoparticles: a mouse study. BMC Cancer 10: 119.

Bossmann, S.H. 2009. Nanoparticles for hyperthermia treatment of cancer. Fabrication and Bio-Application of Functionalized Nanomaterials (ahead to print).

Chen, M., S. Yamamuro, D. Farrell and S.A. Majeticha. 2003. Gold-coated iron nanoparticles for biomedical applications. Journal of Applied Physics 93.

Fortin, J.P., C. Wilhelm, J. Servais, C. Menager, J.C. Bacri and F. Gazeau. 2007. Size-sorted anionic iron oxide nanomagnets as colloidal mediators for magnetic hyperthermia. J. Am. Chem. Soc. 129: 2628–35.

Ganta, C., D. Chiyo, R. Ayuzawa, R. Rachakatla, M. Pyle, G. Andrews, M. Weiss, M. Tamura and D. Troyer. 2009. Rat umbilical cord stem cells completely abolish rat mammary carcinomas with no evidence of metastasis or recurrence 100 days post-tumor cell inoculation. Cancer Res. 69: 1815–20.

Gazeau, F., M. Levy and C. Wilhelm. 2008. Optimizing magnetic nanoparticle design for nanothermotherapy. Nanomedicine (Lond) 3: 831–44.

Gilchrist, R.K., R. Medal, W.D. Shorey, R.C. Hanselman, J.C. Parrott and C.B. Taylor. 1957. Selective inductive heating of lymph nodes. Ann. Surg. 146: 596–606.

Glöckl, G., R. Hergt, M. Zeisberger, S. Dutz, S. Nagel and W. Weitschies. 2006. The effect of field parameters, nanoparticle properties and immobilization on the specific heating power in magnetic particle hyperthermia Journal of Physics and Condensed Matter 18.

Gordon, R.T., J.R. Hines and D. Gordon. 1979. Intracellular hyperthermia. A biophysical approach to cancer treatment via intracellular temperature and biophysical alterations. Med. Hypotheses 5: 83–102.

Habib, A.H., C.L. Ondeck, P. Chaudhary, M.R. Bockstaller and M.E. McHenry. 2008. "Evaluation of iron-cobalt/ferrite core-shell nanoparticles for cancer thermotherapy." Journal of Applied Physics 103.

Hergt, R., W. Andra, C.G. d'Ambly, I. Hilger, W.A. Kaiser, U. Richter and H.G. Schmidt. 1998. Physical limits of hyperthermia using magnetite fine particles. Magnetics, IEEE Transactions on 34: 3745–3754.

Hildebrandt, B., P. Wust, O. Ahlers, A. Dieing, G. Sreenivasa, T. Kerner, R. Felix and H. Riess. 2002. The cellular and molecular basis of hyperthermia. Crit. Rev. Oncol. Hematol. 43: 33–56.

Hilger, I., W. Andra, R. Hergt, R. Hiergeist, H. Schubert and W.A. Kaiser. 2001. Electromagnetic heating of breast tumors in interventional radiology: *in vitro* and *in vivo* studies in human cadavers and mice. Radiology 218: 570–5.

Huber, D.L. 2005. Synthesis, properties, and applications of iron nanoparticles. Small 1: 482–501.

Ikehara, Y., T. Niwa, L. Biao, S.K. Ikehara, N. Ohashi, T. Kobayashi, Y. Shimizu, N. Kojima and H. Nakanishi. 2006. A carbohydrate recognition-based drug delivery and controlled release system using intraperitoneal macrophages as a cellular vehicle. Cancer Res. 66: 8740–8.

Ito, A., H. Honda and T. Kobayashi. 2006. Cancer immunotherapy based on intracellular hyperthermia using magnetite nanoparticles: a novel concept of "heat-controlled necrosis" with heat shock protein expression. Cancer Immunol. Immunother. 55: 320–8.

Ito, A., M. Shinkai, H. Honda and T. Kobayashi. 2001. Heat-inducible TNF-alpha gene therapy combined with hyperthermia using magnetic nanoparticles as a novel tumor-targeted therapy. Cancer Gene Ther. 8: 649–54.

Ito, A., M. Shinkai, H. Honda, K. Yoshikawa, S. Saga, T. Wakabayashi, J. Yoshida and T. Kobayashi. 2003. Heat shock protein 70 expression induces antitumor immunity during intracellular hyperthermia using magnetite nanoparticles. Cancer Immunol. Immunother. 52: 80–8.

Johannsen, M., B. Thiesen, A. Jordan, K. Taymoorian, U. Gneveckow, N. Waldofner, R. Scholz, M. Koch, M. Lein, K. Jung and S. A. Loening. 2005. Magnetic fluid hyperthermia (MFH) reduces prostate cancer growth in the orthotopic Dunning R3327 rat model. Prostate 64: 283–92.

Jordan, A., R. Scholz, K. Maier-Hauff, M. Johannsen, P. Wust, J. Nadobny, H. Schirra, H. Schmidt, S. Deger, S. Loening, W. Lanksch and R. Felix. 2001. Presentation of a new magnetic field therapy system for the treatment of human solid tumors with magnetic fluid hyperthermia. Journal of Magnetism and Magnetic Materials 225: 118–126.

Jordan, A., R. Scholz, K. Maier-Hauff, F.K. van Landeghem, N. Waldoefner, U. Teichgraeber, J. Pinkernelle, H. Bruhn, F. Neumann, B. Thiesen, A. von Deimling and R. Felix. 2006. The effect of thermotherapy using magnetic nanoparticles on rat malignant glioma. J. Neurooncol. 78: 7–14.

Jordan, A., R. Scholz, P. Wust, H. Fahling, J. Krause, W. Wlodarczyk, B. Sander, T. Vogl and R. Felix. 1997. Effects of magnetic fluid hyperthermia (MFH) on C3H mammary carcinoma *in vivo*. Int. J. Hyperthermia 13: 587–605.

Jordan, A., R. Scholz, P. Wust, H. Schirra, S. Thomas, H. Schmidt and R. Felix. 1999. Endocytosis of dextran and silan-coated magnetite nanoparticles and the effect of intracellular hyperthermia on human mammary carcinoma cells *in vitro*. Journal of Magnetism and Magnetic Materials 194: 185–196.

Jordan, A., P. Wust, H. Fahling, W. John, A. Hinz and R. Felix. 1993. Inductive heating of ferrimagnetic particles and magnetic fluids: physical evaluation of their potential for hyperthermia. Int. J. Hyperthermia 9: 51–68.

Kawai, N., M. Futakuchi, T. Yoshida, A. Ito, S. Sato, T. Naiki, H. Honda, T. Shirai and K. Kohri. 2008. Effect of heat therapy using magnetic nanoparticles conjugated with cationic liposomes on prostate tumor in bone. Prostate 68: 784–92.

Kawai, N., A. Ito, Y. Nakahara, H. Honda, T. Kobayashi, M. Futakuchi, T. Shirai, K. Tozawa and K. Kohri. 2006. Complete regression of experimental prostate cancer in nude mice by repeated hyperthermia using magnetite cationic liposomes and a newly developed solenoid containing a ferrite core. Prostate 66: 718–27.

Kotitz, R., P.C. Fannin and L. Trahms. 1995. Time-Domain Study of Brownian and Neel Relaxation in Ferrofluids. Journal of Magnetism and Magnetic Materials 149: 42–46.
Krishnamachari, Y., S.M. Geary, C.D. Lemke and A.K. Salem. 2011. Nanoparticle Delivery Systems in Cancer Vaccines. Pharm Res. 28: 215–236.
Latorre, M. and C. Rinaldi. 2009. Applications of magnetic nanoparticles in medicine: magnetic fluid hyperthermia. P. R. Health Sci. J. 28: 227–38.
Le, B., M. Shinkai, T. Kitade, H. Honda, J. Yoshida, T. Wakabayashi and T. Kobayashi. 2001. Preparation of tumor-specific magnetoliposomes and their application for hyperthermia. Journal of Chemical Engineering of Japan 34: 66–72.
Loebinger, M.R., P.G. Kyrtatos, M. Turmaine, A.N. Price, Q. Pankhurst, M.F. Lythgoe and S.M. Janes. 2009. Magnetic resonance imaging of mesenchymal stem cells homing to pulmonary metastases using biocompatible magnetic nanoparticles. Cancer Res. 69: 8862–7.
Menoret, A., and R. Chandawarkar. 1998. Heat-shock protein-based anticancer immunotherapy: an idea whose time has come. Semin. Oncol. 25: 654–60.
Milani, V., E. Noessner, S. Ghose, M. Kuppner, B. Ahrens, A. Scharner, R. Gastpar and R.D. Issels. 2002. Heat shock protein 70: role in antigen presentation and immune stimulation. Int. J. Hyperthermia 18: 563–75.
Montet, X., K. Montet-Abou, F. Reynolds, R. Weissleder and L. Josephson. 2006. Nanoparticle imaging of integrins on tumor cells. Neoplasia 8: 214–22.
Mornet, S., S. Vasseur, F. Grasset and E. Duguet. 2004. Magnetic nanoparticle design for medical diagnosis and therapy. J. Mat. Chem. 14: 2161–2175.
Needham, D., G. Anyarambhatla, G. Kong and M.W. Dewhirst. 2000. A new temperature-sensitive liposome for use with mild hyperthermia: characterization and testing in a human tumor xenograft model. Cancer Res. 60: 1197–201.
Nie, S., Y. Xing, G.J. Kim and J.W. Simons. 2007. Nanotechnology applications in cancer." Annu. Rev. Biomed. Eng. 9: 257–88.
Ohno, T., T. Wakabayashi, A. Takemura, J. Yoshida, A. Ito, M. Shinkai, H. Honda and T. Kobayashi. 2002. Effective solitary hyperthermia treatment of malignant glioma using stick type CMC-magnetite. *In vivo* study. J. Neurooncol. 56: 233–9.
Pakhomov, A.B., Y.P. Bao and K.M. Krishnan. 2005. Effects of surfactant friction on Brownian magnetic relaxation in nanoparticle ferrofluids. J. Appl. Phys. 97: 10Q305/1-10Q305/3.
Pradhan, P., J. Giri, F. Rieken, C. Koch, O. Mykhaylyk, M. Doblinger, R. Banerjee, D. Bahadur and C. Plank. 2010. Targeted temperature sensitive magnetic liposomes for thermo-chemotherapy. J. Control Release 142: 108–21.
Purushotham, S., P.E. Chang, H. Rumpel, I.H. Kee, R.T. Ng, P.K. Chow, C.K. Tan and R.V. Ramanujan. 2009. Thermoresponsive core-shell magnetic nanoparticles for combined modalities of cancer therapy. Nanotechnology 20: 305101.
Purushotham, S. and R.V. Ramanujan. 2010. Thermoresponsive magnetic composite nanomaterials for multimodal cancer therapy. Acta Biomater 6: 502–10.
Rachakatla, R.S., S. Balivada, G.M. Seo, C.B. Myers, H. Wang, T.N. Samarakoon, R. Dani, M. Pyle, F.O. Kroh, B. Walker, X. Leaym, O.B. Koper, V. Chikan, S.H. Bossmann, M. Tamura and D.L. Troyer. 2010. Attenuation of Mouse

Melanoma by A/C Magnetic Field after Delivery of Bi-Magnetic Nanoparticles by Neural Progenitor Cells. ACS Nano.

Rachakatla, R.S., F. Marini, M.L. Weiss, M. Tamura and D. Troyer. 2007. Development of human umbilical cord matrix stem cell-based gene therapy for experimental lung tumors. Cancer Gene Ther. 14: 828–35.

Rachakatla, R.S., and D. Troyer. 2009. Wharton's jelly stromal cells as potential delivery vehicles for cancer therapeutics. Future Oncol. 5: 1237–44.

Ravel, B., E.E. Carpenter and V.G. Harris. 2002. Oxidation of iron in iron/gold core/shell nanoparticles. Journal of Applied Physics 91: 8195–8197.

Scherer, F., M. Anton, U. Schillinger, J. Henke, C. Bergemann, A. Kruger, B. Gansbacher and C. Plank. 2002. Magnetofection: enhancing and targeting gene delivery by magnetic force *in vitro* and *in vivo*. Gene Ther. 9: 102–9.

Shapiro, M.G., T. Atanasijevic, H. Faas, G.G. Westmeyer and A. Jasanoff. 2006. Dynamic imaging with MRI contrast agents: quantitative considerations. Magn. Reson. Imaging 24: 449–62.

Shinkai, M. 2002. Functional magnetic particles for medical application. J. Biosci. Bioeng. 94: 606–13.

Shinkai, M., M. Suzuki, S. Iijima and T. Kobayashi. 1995. Antibody-conjugated magnetoliposomes for targeting cancer cells and their application in hyperthermia. Biotechnol. Appl. Biochem. 21 (Pt 2): 125–37.

Shinkai, M., M. Yanase, M. Suzuki, H. Hiroyuki, T. Wakabayashi, J. Yoshida and T. Kobayashi. 1999. Intracellular hyperthermia for cancer using magnetite cationic liposomes. Journal of Magnetism and Magnetic Materials 194: 176–184.

Shubayev, V.I., T.R. Pisanic and S. Jin. 2009. Magnetic nanoparticles for theragnostics. Adv. Drug Deliv. Rev. 61: 467–77.

Sugahara, K.N., T. Teesalu, P.P. Karmali, V.R. Kotamraju, L. Agemy, O.M. Girard, D. Hanahan, R.F. Mattrey and E. Ruoslahti. 2009. Tissue-penetrating delivery of compounds and nanoparticles into tumors. Cancer Cell 16: 510–20.

Sugahara, K.N., T. Teesalu, P.P. Karmali, V.R. Kotamraju, L. Agemy, D.R. Greenwald and E. Ruoslahti. 2010. Coadministration of a tumor-penetrating peptide enhances the efficacy of cancer drugs. Science 328: 1031–5.

Sun, S., C.B. Murray and H. Doyle. 1999. Controlled assembly of monodisperse epsilon -cobalt-based nanocrystals. Advanced Hard and Soft Magnetic Materials, MRS Spring Meeting, San Francisco, CA: 385–398.

Tanaka, K., A. Ito, T. Kobayashi, T. Kawamura, S. Shimada, K. Matsumoto, T. Saida and H. Honda. 2005. Heat immunotherapy using magnetic nanoparticles and dendritic cells for T-lymphoma. J. Biosci. Bioeng. 100: 112–5.

Tang, M.F., L. Lei, S.R. Guo and W.L. Huang. 2010. Recent progress in nanotechnology for cancer therapy. Chin. J. Cancer 29: 775–80.

Thiesen, B., and A. Jordan. 2008. Clinical applications of magnetic nanoparticles for hyperthermia. Int. J. Hyperthermia 24: 467–74.

van Landeghem, F.K., K. Maier-Hauff, A. Jordan, K.T. Hoffmann, U. Gneveckow, R. Scholz, B. Thiesen, W. Bruck and A. von Deimling. 2009. Post-mortem studies in glioblastoma patients treated with thermotherapy using magnetic nanoparticles. Biomaterials 30: 52–7.

Wang, X., R. Zhang, C. Wu, Y. Dai, M. Song, S. Gutmann, F. Gao, G. Lv, J. Li, X. Li, Z. Guan, D. Fu and B. Chen. 2007. The application of Fe3O4 nanoparticles in cancer research: a new strategy to inhibit drug resistance. J. Biomed. Mater. Res. A. 80: 852–60.

Weiss, M.L., S. Medicetty, A.R. Bledsoe, R.S. Rachakatla, M. Choi, S. Merchav, Y. Luo, M.S. Rao, G. Velagaleti and D. Troyer. 2006. Human umbilical cord matrix stem cells: preliminary characterization and effect of transplantation in a rodent model of Parkinson's disease. Stem Cells 24: 781–92.

Wilcoxon, J.P., and B.L. Abrams. 2006. Synthesis, structure and properties of metal nanoclusters. Chem. Soc. Rev. 35: 1162–94.

Winterbourn, C. 1999. Toxicity of iron and hydrogen peroxide: the Fenton reaction Toxicology Letters 83: 969–974.

Wu, X., J. Hu, L. Zhou, Y. Mao, B. Yang, L. Gao, R. Xie, F. Xu, D. Zhang, J. Liu and J. Zhu. 2008. *In vivo* tracking of superparamagnetic iron oxide nanoparticle-labeled mesenchymal stem cell tropism to malignant gliomas using magnetic resonance imaging. Laboratory investigation. J. Neurosurg. 108: 320–9.

Yanase, M., M. Shinkai, H. Honda, T. Wakabayashi, J. Yoshida and T. Kobayashi 1998. Intracellular hyperthermia for cancer using magnetite cationic liposomes: an *in vivo* study. Jpn. J. Cancer Res. 89: 463–9.

Zeng, Q., I. Baker, J.A. Loudis, Y. Liao, P.J. Hoopes and J.B. Weaver. 2007. Fe/Fe oxide nanocomposite particles with large specific absorption rate for hyperthermia. Applied Physics Letters 90: 233112.

Zhao, X.Q., B. X. Liu, Y. Liang and Z.Q. Hu. 1997. Structure and magnetic properties of the oxide layers on iron ultrafine particles. Applied Physics A: Materials Science & amp; Processing 64: 483–486.

6

Nanoparticle Probes and Molecular Imaging in Cancer

Dongfang Liu[2] and *Ning Gu*[1,*]

ABSTRACT

Molecular imaging can be defined as the *in vivo* visualization, characterization, and measurement of biologic processes at the cellular and molecular level. Contrast to "classical" diagnostic imaging, the molecular imaging aims to probe fundamental molecular processes that are the basis of disease rather than to image the end effects of these molecular alterations. Central to the molecular imaging in cancer is the development of high specific imaging probes for effective detection of cancer markers that are expressed differently in tumor cells or tissues. The applications of nanotechnology in molecular imaging field lead to the emergence of nanoprobes. One of the advantages of such imaging probes is their flexibility when conjugated with targeting ligands and imaging moieties. Additionally, different

[1]Jiangsu Laboratory for Biomaterials and Devices, Nanjing, 210009, Suzhou Key Laboratory of Biomedical Materials and Technology, Suzhou, 215123, State Key Laboratory of Bioelectronics, Nanjing, 210096, School of Biological Science and Medical Engineering; Southeast University, P. R. China.
E-mail: guning@seu.edu.cn

[2]Jiangsu Key Laboratory of Molecular and Functional Imaging. Department of Radiology, Zhongda Hospital, Medical School, Southeast University, Nanjing, China;
E-mail: liudf2008@gmail.com

*Corresponding author

List of abbreviations after the text.

imaging modalities can be merged into a single nanoparticle to achieve multimodalities probes for comprehensive lesions imaging. Furthermore, nanoparticles can be readily modified to alter their pharmacokinetics, prolong their plasma half lives, enhance their stability *in vivo*, improve the targeting efficiency and reduce nonspecific binding. These properties of nanoparticles make them promising candidates of high performance probes in molecular imaging. In this chapter, we summarize the innovative nanoparticle-based imaging probes that have been investigated for *in vivo* molecular imaging in cancer diagnosis.

INTRODUCTION

Cancer, the uncontrolled growth and invasion and sometimes metastasis of malignant cells that can affect almost any tissue of the body and people of all ages, is a major public health burden in the world. In the United States, 1 in 4 deaths is caused by cancer (Jemal et al. 2010), which is the leading cause of death among persons younger than 85 years (Jemal et al. 2010). It is a major challenge to diagnose cancer in the early stage for curative treatment. Molecular imaging, which is broadly defined as the *in vivo* visualization, characterization, and measurement of biologic processes at the cellular and molecular level (Weissleder and Mahmood 2001), is emerging as a promising technology to meet this challenge. The molecular imaging aims to probe molecular processes that are the basis of disease rather than to image the end effects of these molecular alterations, which enabled much earlier detection of disease as well as real-time monitoring of therapeutic responses. Central to the molecular imaging in cancer is the development of high specific imaging probes for effective detection of cancer markers that are expressed differently in tumor cells or tissues. An ideal imaging probe used for *in vivo* imaging should generally meet several key criteria: (a) high specificity and affinity to the target tissues or cells; (b) *in vivo* safety; and (c) high target-to-background ratio.

Nanoparticles have shown great potentials as imaging probes due to their unique physical and biological properties and special interactions with biomolecules. One of the advantages of such imaging probes is their flexibility when conjugated with targeting ligands and imaging moieties. The large surface areas of nanoparticles make it possible to bind multiple targeting ligands and imaging moieties to a single nanoparticle, leading to significantly improved binding affinity via a polyvalent effect and amplified signals at the target regions. Alternatively, different imaging modalities can be merged into a single nanoparticle to achieve multimodalities probes for comprehensive lesions imaging. Furthermore, nanoparticles can be readily modified to alter their pharmacokinetics, prolong their plasma half lives, enhance their stability *in vivo*, improve the targeting efficiency and reduce nonspecific binding. These properties of nanoparticles make them promising candidates of high performance probes in molecular imaging. In the past decades, there have been significant developments in the field of nanoparticle-based target molecular imaging. Several review articles have summarized these recent advantages and have discussed their unique design and applications (Lee et al. 2010; Qiao et al. 2009; He et al. 2010). In this chapter, we will overview the innovative nanoparticle-based imaging probes, which are often called the nanoprobes, that have been investigated for *in vivo* molecular imaging in cancer diagnosis.

DESIGN OF NANOPARTICLE PROBES

To meet diverse imaging applications, nanoparticles should be properly surface modified and functionalized. Although the surface modification strategies vary for different nanoparticles, the purpose of the surface decoration is always to yield nanoparticles with high physiological stability, facile labeling sites for targeting ligands and long plasma half-time that is enough for nanoparticles to accumulate in tumors.

For active tumor targeting, nanoparticles should be decorated with specific targeting ligands, including monoclonal antibodies (Tada et al. 2007), small molecule ligands (Rossin et al. 2004), peptides (Hood et al. 2002), aptamers (Missailidis and Perkins 2007), and polyaccharides (Mehvar 2003), to active targeting tumor tissues. In most cases, targeting ligands are coupled to the nanoparticles by bio-conjugation techniques. Some of commonly used techniques can be categorized as follows (Fig. 1). Antibodies or amine terminated ligands are normally conjugated with carboxylated nanoparticles via amide linkage with EDC/sulfo-NHS

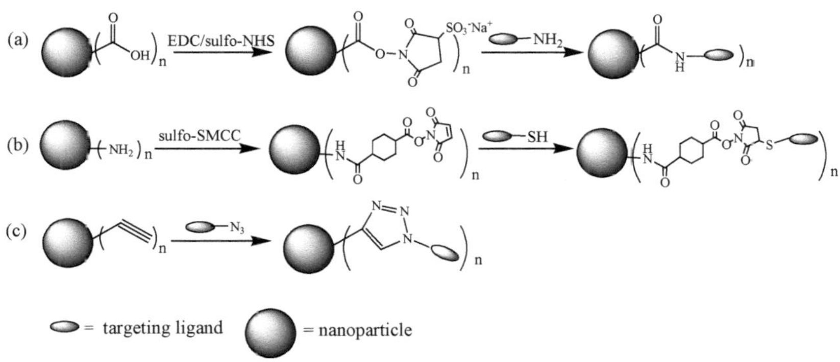

Fig. 1. Nanoparticle-labeled targeting ligands to be the nanoprobes (unpublished figure).

as catalysts (Pan et al. 2004). Similarly, nanoparticles with amines on the surface are coupled to the carboxyl (Chandrasekar et al. 2007) or aldehyde (Fischer-Durand et al. 2007) terminated ligands via amide or imine linkage. Alternatively, the nanoparticles presenting amino groups can be first reacted with sulfosuccinimidyl-4-[N-maleimidomethyl]cyclohexane-1-carboxylate (sulfo-SMCC), and then conjugated with thiolated targeting ligands (Jun et al. 2005).

NANOPARTICLE PROBES FOR MR IMAGING

Magnetic resonance imaging (MRI) is one of the prominent non-invasive diagnostic techniques based on computer-assisted imaging of relaxation signals of nuclei spins within tissues excited by radiofrequency waves in a strong magnetic field. Although there are many medically relevant nuclei, the proton has been the most widely studied in clinical practice due to its high gyromagnetic ratio and high abundance in nature and biological tissues. In addition to proton, ^{19}F has a similar high gyromagnetic ratio as proton and a very high (100%) natural occurrence, which make it an interesting choice for MR imaging (Longmaid et al. 1985).

Dendritic Probes for T1 MR Imaging

For the molecular MRI, a main challenge for the T1 probes is represented by low accumulation and sensitivity of monovalent probes when the local concentration of biological receptors is commonly very low. The insufficient accumulation can be compensated by attaching multiple MRI labels to a single scaffold. Nanotechnologies provided unique platforms for the incorporation of multiple MR imaging moieties. Dendrimers, the highly branched macromolecules with nanoscopic dimensions, high end functionality and precisely defined structure, have been successfully employed as multivalent scaffolds carrying multiple imaging moieties (Langereis et al. 2007). In addition to the multiplication of the MRI signals, the sensitivity in terms of relaxivity (r_1) of the Cd(III)DTPA-based contrast agents can be also enhanced by conjugation with dendrimers in magnetic fields of 0.5–1.5 T (Wiener et al. 1994; Kobayashi et al. 2003; Rudovský et al. 2005; Langereis et al. 2004). For example, Wiener and coworkers (Wiener et al. 1994) have demonstrated that the relaxivity of the Gd(III)DTPA complex can be enhanced by six times when conjugated to six generation of PAMAM dendrimer. The strong increase in r_1 was ascribed to the lower molecular tumbling rate of the Gd(III)DTPA complex at the periphery of the dendrimer (Wiener et al. 1996). Due to the higher generations of dendritic MRI contrast agents possess relative long blood circulation time, in contrast to low molecular Cd(III) chelate, they are usually referred to as blood pool agents, suitable for *in vivo* MR angiography, lymphography, as biometric nanoprobes to detect vascular permeability, and blood pool imaging (Kobayashi et al. 2003; Dong et al. 1998). For effectively tumor imaging, target-specific dendritic MRI probes were also described. Very recently, León-Rodríguez and coworkers reported a peptoid-$(Cd)_8$-polylysine dendron for MRI detection of the vascular endothelial growth factor receptor 2 (VEGFR2) (León-Rodríguez et al. 2010), which is an important target for tumor metastasis. A dimeric form of a 9-residue peptoid sequence (GU40C4), with high affinity to VEGFR2, was conjugated to the poly (DOTA-lysine) dendron by reaction of thiol group of a Cys residue of the peptoid and the maleimide moiety of the dendron. The peptoid-Gd_8-dendron probe was formed after chelating of Gd^{3+} by DOTA. Mice bearing VEGFR2-positive MDA-MB-231 tumor xenografts were imaged before and after administration of the dendron probe (Fig. 2). T1-weighted images show that tumors in mice were maximally enhanced at ~4h postinjection, while the image intensity in tumors after injection of the control peptoid Gd8-dendron conjugate returned to baseline levels by 4 h, indicating that the dendron-GU40C4 peptoid conjugation can be

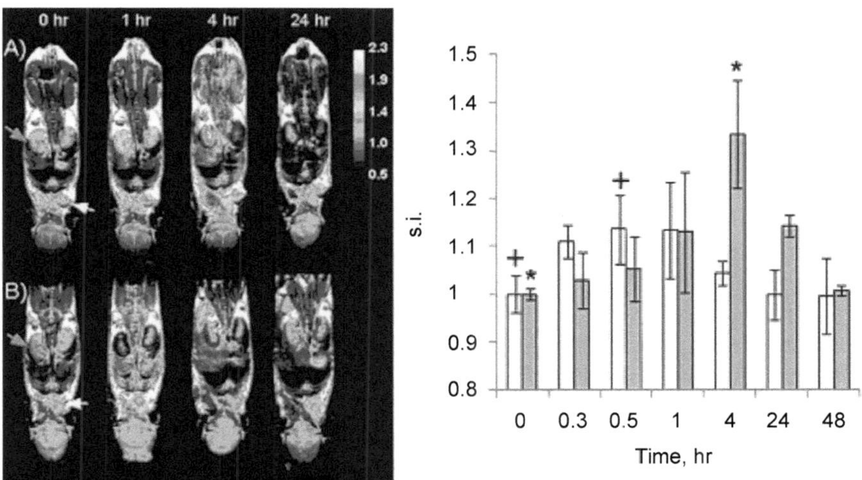

Fig. 2. 9.4 T MR T_1-weighted coronal images of nude mice with subcutaneous cell tumor MDA-MB-231 xenografts at various time points following the intravenous (i.v.) tail injection of Gd_8-dendron peptoid conjugates. (A) images of a tumor-bearing mouse before and after injection of Gd_8-dendron peptoid probe. (B) Images of another tumor-bearing mouse before and after injection of an indentical Gd_8-dendron conjugated to a control scrambled peptoid (with permission from León-Rodríguez et al. 2010), arrows indicate tumors.

used as high performance T1 relaxicity probes for molecuarl imaging of VEGFR2 *in vivo* by MRI. The Gd(III) complexes are currently the most widely applied contrast agents for MRI due to their predominant positive signal enhancement. However, slow excretion of Gd(III)-based contrast agents may cause serious toxic side effects. Mn(II)-based targeted contrast agent has relatively lower side effects because Mn(II) ion is an essential component of cells and a cofactor for enzymes and receptors (Crossgrove and Zheng 2004). Tan and coworkers reported a targeted nanoglobular Mn(II) chelate conjugate for MR cancer imaging (Tan et al. 2010). The nanoglobular probes was achieved by the conjugation of a G3 nanoglobule-(Mn-DOTA)$_{44}$-(PEG-azido)$_4$ with CLT1 peptide via click chemistry between the reactive propargyl group of the peptide and azide group of G3 nanoglobule. CLT1 peptide binds to fibrin-fibronectin complexes or oncofetal fibronectin in the stroma of solid tumors (Pilch et al. 2006). *In vivo* T1 weighted MRI in female nude mice bearing MDA-MB-231 human breast carcinoma xenografts resulted in a significant signal enhancement in tumor tissue. The CLT1 conjugated nanoblogular probes were found to target the fibronectin positive MDA-MB-231 tumor cells with minimal nonspecific uptake in the liver of tumor bearing mice. Other reports on *in vivo* cancer imaging using dendric probes include the accumulations of conjugates of folate with PEG-G3-Gd(III)-DTPA dendrimer (Chen et al. 2010) or PAMAM (Konda et al. 2001) in xenografted ovarian tumors, and the conjugates of CLT1 peptide and Gd(III)-DOTA with a G3 nanoglobules in breast xenografted tumors (Tan et al. 2010).

Supermagnetic Magnetic Nanoparticle Probes for T2 MR Imaging

Supermagnetic iron oxide nanoparticles (SPIONs) are some of the most studied nanoparticles in the field of molecular imaging, mainly due to their ability to enhance the T2-weighted MRI signal, biocompatibility and ease preparation and modification. Upon conjugation with appropriated targeting molecules, SPIONs can be utilized for the active detection of cancer. However, the effectiveness of promising nanoparticles is usually compromised by insufficient accumulation of these contrast agents within tumors because of the nonspecific recognition and clearance of the nanoparticles by the reticuloendothelial system (RES) prior to reaching target tissue (Pankhurst et al. 2003). The fate of these nanoparticles upon intravenous administration is highly dependent on their hydrodynamic size and surface chemical structures (Weissleder et al. 1995; Gupta and Gupta 2005). One strategy to overcome this limitation is to prepare SPIONs with small hydrodynamic size. Xie et al. reported an ultrasmall RGD coated SPIONs with hydradynamic size of 8.5 nm (Xie et al. 2008). The c(RGDyk) peptides were covalently conjugated to 4-methylcatechol coated SPIONs via a Mannich reaction and rendered as stable NPs in physiological conditions. *In vivo* targeting can be achieved through an RGD-integrin $\alpha_v\beta_3$ interaction. *In vivo* T2 weighted MRI in mice bearing $\alpha_v\beta_3$-positive U87MG tumors resulted in a significant signal reduction in tumor tissue (Fig. 3). The RGD conjugated SPIONs were found to target the integrin-positive tumor vasculature and tumor cells with little to no macrophage uptake.

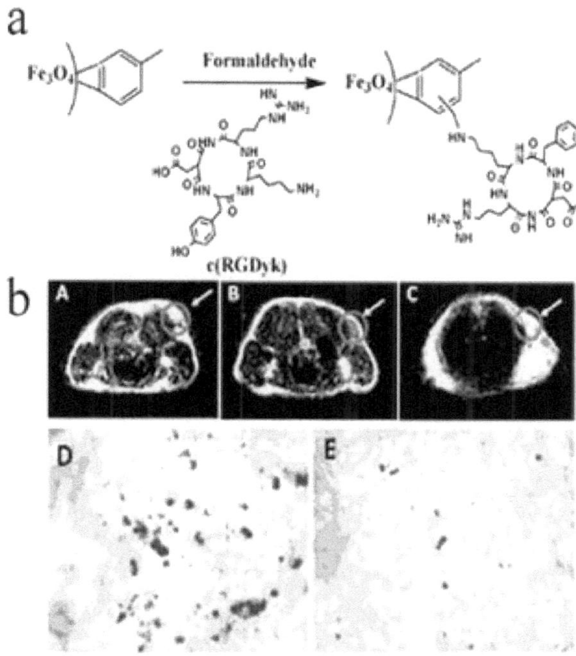

Fig. 3. (a) Schematic diagram of c(RGDyK)-MC-Fe3O4 NPs. (b) MRI of the cross section of the U87MG tumors implanted in mice: (A) without NPs, (B) with the injection of 300 μg of c(RGDyK)-MC-Fe3O4 NPs, and (C) with the injection of c(RGDyK)-MC-Fe3O4 NPs and blocking dose of c(RGDyK); and Prussian blue staining of U87MG tumors in the presence of (D) c(RGDyK)-MC-Fe3O4 NPs and (E) c(RGDyK)-MC-Fe3O4 NPs plus blocking dose of c(RGDyK) (with permission from Xie et al. 2008).

An alternate strategy commonly used for resisting unspecific uptake of contrast agents by RES is to modify the surface of nanoparticles with anti-biofouling polymers (Lee et al. 2006; Amstad et al. 2009; Sun et al. 2008; Hu et al. 2006). In which, PEG is the most useful surface coating to minimize the biofouling of SPIONs in blood circulation. Sun et al. reported a PEGylated SPION nanoprobe composed of an iron oxide core coated with PEG and conjugated with the targeting agent, chlorotoxin (Sun et al. 2008). The nanoprobe targeted gliomas expressing membrane-bound matrix metalloproteinase-2 (MMP-2) with high-level specificity both *in vitro* and *in vivo*. Hu et al. developed a 'one-pot' thermal decomposition method for preparing Fe_3O_4 nanoparticles coated with PEG terminated with carboxylic groups (Hu et al. 2006). Anti-carcinoembryonic antigen monoclonal antibody rich 24 (rch 24 mAb) is conjugated to the PEG-coated Fe_3O_4 nanoparticles via EDC chemistry, forming a MR molecular imaging probe. Both *in vivo* and *in vitro* investigation indicate that the resultant probe can specifically target human colon carcinoma cells and xenograft implanted in nude mice (Hu et al. 2006).

Improved molecular imaging of cancer is possible utilizing $MnFe_2O_4$ probes, which shows a r_2 of 358 $mM^{-1} S^{-1}$, which is more than two-times higher than the values of conventional SPIONs related MR contrast agents (Lee et al. 2007). The increased MR imaging capability of $MnFe_2O_4$ is advantageous in cancer molecular imaging. Lee et al. reported the *in vivo* imaging of breast cancer using the $MnFe_2O_4$-herceptin nanparticle probe (Lee et al. 2007). Effective R2 increase of ~25% (1 h) and ~34% (4 h), much higher than the values achieved using CLIO-Herceptin (~5%), are achieved from post-injection T2-weighted MR images which are acquired with a 1.5 T clinical MR instrument. By using the $MnFe_2O_4$-herceptin probe, tumors as small as ~50 mg can be detected.

NANOPARTICLE PROBES FOR OPTICAL MOLECULAR IMAGING

Optical imaging is one of the most widely used imaging modalities with the advantages of high sensitivity, use of nonradioactive materials, easy mode of operation and comparatively low cost. Optical imaging provides the potential for noninvasive study of molecular imaging inside the body of living animal with appropriate optical probes.

Due to their novel electronic and size-dependent fluorescence properties, semiconductor quantum dots (QDs) are emerging as a new class of molecular probes. In comparison with organic dyes, the advantages of QDs involve bright photoluminescent, high photochemical endurance, and narrow and more symmetric emission spectra, which makes QDs suitable for *in vivo* imaging applications. In 2002, Akerman and coworkers first introduced the *in vivo* applications of QDs (Akerman et al. 2002). They conjugate QDs to peptides with affinity for breast tumor cells and their vasculatures. After intravenous injection of these probes into tumor bearing mice, microscopic fluorescence imaging of tissue sections demonstrated the specific accumulation of QDs in the tumor vasculature. Since the pioneer work of Akerman and coworker introduced above, conjugates of antibodies and peptides with QDs receive much attention as potential probes for *in vivo* imaging of various cancers (Gao et al. 2004).

Although optical imaging with QDs as probes has been used in living animal models, it is still limited by the poor transmission of visible light through biological tissues. To address this issue, QDs with emission in the near-infrared fluorescence (NIRF) spectra window (650–900 nm), where biological tissues display low absorption, have received considerable attention. Cai et al. successfully demonstrated a RGD-labeled NIRF QD for *in vivo* tumor vasculature imaging (Fig. 4) (Cai et al. 2006). Conjugation of thiolated RGD to PEG-coated QD 705 was achieved via 4-maleimidobutyric acid N-hydroxysuccinimide ester as a heterobifunctional linker. The QD probe can specifically target $\alpha_v\beta_3$-integrin *in vitro, ex vivo,* and in living mice bearing U87MG tumor. Tumor vasculature fluorescence reached maximum at 6h postinjection of the QD probes with good contrast using nude mice bearing subcutaneous U87MG human glioblastoma. In addition, this probe was further developed as a dual modality probe for both optical and PET imaging by binding [^{64}Cu-DOTA] on the surface of the QD-peptide (Cai et al. 2007), which will be introduced in detail in the latter section. Smith and coworkers reported the targeting and imaging of newly formed tumor neovasculature expressing $\alpha_v\beta_3$ integrin in living subjects using RGD-QD800 and intravital microscope (Smith et al. 2008). Diagaradjane and coworkers developed epidermal growth factor (EGF)

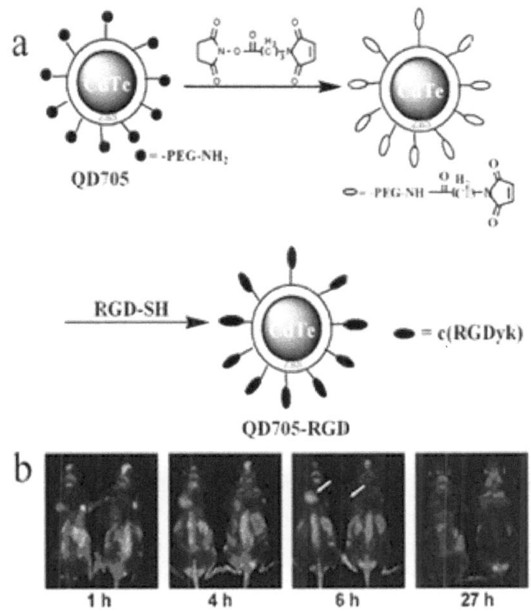

Fig. 4. (a) Schematic diagram of QD705-RGD. (b) *In vivo* NIR fluorescence imaging of U87MG tumor-bearing mice at various time points after intravenous injection of QD705-RGD (left) or QD705 (right) (with permission from Cai et al. 2006). Arrows indicate tumors.

conjugated QDs800 as probes to imaging the EGF receptor expression in human colon cancer xenografts (Diagaradjane et al. 2008). Their results indicate that the QD-EGF conjugates can be used for the imaging of EGF receptor that is expressed in tumor.

The above investigations display the potentials of QDs-based nanoprobes for *in vivo* imaging of tumors and tumor vasculatures. Yet, a major concern has been raised over the use of QDs for clinical applications due to their toxicity, which are mainly raised from their metallic components, such as Cd^{2+}, Hg^{2+}, Pb^{2+}, and As^{3+}, and their nanodimensions. Presently the most commonly used QDs contain divalent cadmium, a potential carcinogen that accumulates in kidney, liver, and many other tissues inducing DNA and protein damages. Although the toxicity of QDs from cadmium can be temporary avoided by coating protecting layers such as ZnS shells, it has been demonstrated that those shells can gradually deteriorate on the *in vivo* environment. It is still unclear if these toxic ions will impact the use of QDs as clinical contrast agents.

Surface-enhanced Raman scattering (SERS) is a nondestructive analytical technique providing great potential for non-toxic imaging of cells. Recently, Qian and coworkers reported surface-enhanced Raman reporter Au nanoparticles for *in vivo* tumor targeting (Qian et al. 2008). Au NPs are bifunctionalized with malachite as a SERs reporter and a ScFv antibody as a targeting ligand for recognizing human EGF receptor. In the *in vivo* experiments, the Au probes were systemically administered in nude mice bearing EGF receptor-positive human head and neck tumor. SERS results indicated ScFv-conjugated Au NPs were able to specifically target the EGF receptor-positive tumors. The Au nanoparticles were above 200 times brighter than NIR emitting QDs, allowing spectroscopic detection of tumors as small as 0.03 cm^3 at a penetration depth of 1–2 cm (Qian et al. 2008).

MULTIMODAL NANOPROBES FOR MOLECULAR IMAGING

For successfully diagnosis various diseases, various imaging modalities, including positron emission tomography (PET), single photon emission computed tomography (SPECT), MRI, optical imaging, computed X-ray tomography (CT), and ultrasound have been developed. However, each imaging modality has its own merits and deficiencies (Massoud and Gambhir 2003). For example, MRI and CT have the advantages of being noninvasive and high spatial resolution but low target sensitivity. Although the radioactive imaging techniques like PET have very high target sensitivities, they are limited by poor spatial resolution. Other optical imaging methods such as fluorescence have relatively good sensitivity and spatial resolution but suffer from low tissue penetration depths. Accordingly, no single imaging modality is sufficient to provide formation on all aspects of structure and function. Therefore, synergistic combination of different modalities into a single system, multimodal imaging methods have promising applications for complicated tumor diagnosis. Due to these combined advantages, nanoparticle probes with multimodalities are very attractive for biomedical applications and clinics.

One of the well-studied examples is the nanoparticle probes in which magnetic nanoparticles are chemically conjugated with fluorophores for the MRI-optical dual-modality imaging of tumor *in vivo*. Moore and coworkers developed a novel imaging probe consisting of crosslinked iron oxide nanoparicles (CLIO)-Cy5.5 conjugated to peptide EPPT (Moore et al. 2004), which can specifically recognize underglycosylated mucin-1 antigen (uMUC-1) on various tumor cells. The synthesized probe displayed high specificity toward variety of human adenocarcinomas overexpressing uMUC-1 *in vitro*. Additionally, *in vivo* MR and NIRF imaging showed specific accumulation of the probe on uMUC-1-positive tumors whereas no signal in control tumors. The magnetic nanoparticles can also be coupled with radionuclide labels for dual-mode MRI-radionuclide imaging (Lee et al. 2008). MRI-PET has the potential for providing better spatial resolution with anatomical information as well as improved signal sensitivity. The use of ^{64}Cu labeled iron oxide nanoparticle probes for *in vivo* detection of the U87MG tumor is a good example (Lee et al. 2008). In this work, the polyaspartic acid (PASA) coated iron oxide (PASA-IO) nanoparticles were prepared using a coprecipitation method, and DOTA was coupled via NHS chemistry for PET after labeling with ^{64}Cu. For active targeting the $\alpha_v\beta_3$-integrin positive tumors, RGD was also coupled to the PASA-IO nanoparticles by another typical conjugation method. Both small animal PET and T_2-weighted MRI results show integrin specific delivery of the nanoparticle probe to tumor sites and prominent RES uptake. Recently, our group has developed a novel SPION-embedded microbubble (EMB) which can be applied as MRI/ultrasound dual-modality contrast agent (Yang et al. 2009). The EMBs were prepared by using a double emulsion solvent evaporation interfacial deposition (water-in-oil-in-water emulsion) process in the presence of hydrophobic Fe_3O_4 nanoparticles. The *in vitro* and *in vivo* MR experiments demonstrate that the microbubble susceptibility effect can be significantly enhanced through the incorporation of SPIONs in the shell. At the meantime, the resultant EMB can still maintain adequate echogenicity and can be used as US contrast agents. The SPION-inclusion microbubble can be manipulated instantaneously and locally *in vivo* via cavitation by the moderate US irradiation, which it might be possible to monitor *in vivo* contrast changes with MR imaging. Further work on conjugation of the SPION-embedded with targeting agents for MR/ultrasound dual-modality cancer imaging is underway.

Similarly, QD-based dual-modality probes for both PET and optical imaging have been reported via conjugation of ^{64}Cu-DOTA on the surface of QDs. Cai et al. prepared a nanoparticle probe of ^{64}Cu-labeled QDs conjugated with RGD tumor PET/NIRF dual-modality imaging (Cai et al. 2007). To prepare the QD based dual-modality probes, QD705 with amine groups on their surface were double functionalized with RGD peptides for integrin $\alpha_v\beta_3$-targeting and 1,2,7,10-tetraazacyclodocecane-N, N′, N, N′-tetraacetic acid (DOTA) for chelating of ^{64}Cu in one step. The quantification ability and ultrahigh sensitivity of PET imaging enabled the quantitative analysis of the biodistribution and targeting efficacy of this dual-modality imaging probe. *In vivo* imaging was performed in mice bearing $\alpha_v\beta_3$-positive U87MG tumors. The dual functional PET/NIRF probe were shown to specifically target tumor and can render sufficient tumor contrast at a considerably low concentration compared to that required for *in vivo* NIFR imaging, thereby significantly reducing potential toxicity. Another similar probe, consisting of [^{64}Cu DOTA] modified QDs conjugated with VEGF proteins (Chen et al. 2008), successfully targeted tumor vascular VEGF receptors and allowed quantitatively evaluation of the targeting efficacy *in vivo*. However, the full potential of *in vivo* dual-modality imaging were not realized in these studies, as fluorescence was only used as an *ex vivo* imaging tool to validate the *in vivo* results of PET imaging, mainly because of the lower sensitivity of optical imaging in comparison with PET.

Applications to Areas of Health and Disease

Molecular imaging can be used for early diagnosis of cancer by detection of the early stages of molecular abnormalities that are the basis of the disease. There are five main categories of molecular imaging modalities: nuclear medicine techniques, optical imaging, MRI, ultrasound and CT. Each imaging modality has its own merits and deficiencies. The choice of the imaging modality is determined by the sensitivity and special resolution of the imaging system, depth of the biological process and the availability of suitable molecular probes for target imaging.

Table 1. Key Facts of nanoparticle probes.

1. A nanoparticle-based molecular imaging system typically consists of targeting ligands, a nanoparticle platform and imaging modalities (Lee et al. 2010).
2. Optical techniques include two major classes: fluorescence and bioluminescence imaging. The depth resolution of optical imaging techniques is influenced by photon wavelength. Currently, near-infrared photons provide the greatest wavelengths (650 nm to 900 nm) and the best depth of penetration (>1 cm) (Bonekamp et al. 2010).
3. The MRI contrast agents principally work by shortening the T1 or T2 relaxation times of protons located nearby. Elements with unpaired electron spin such as gadolinium, manganese and iron, can effectively reduce the T1 relaxation time, which gives rise to positive contrast. Supermagnetic nanoparticles such as γ-Fe_2O_3 and Fe_3O_4 nanocrystals can shorten the T2, and such changes result in the darkening of the corresponding area in T2- and $T2^*$-weighted images (Qiao et al. 2009).

This table lists the key facts of nanoparticle probes including the explanation of nanoparticle probes, optical techniques and MR imaging.

Definitions

Molecular imaging: the *in vivo* visualization, characterization, and measurement of biologic processes at the cellular and molecular level.

Relaxivity: the ability of magnetic compounds to increase the relaxation rates of the surrounding water proton spins.

Reticuloendothelial system: part of the immune system, consists of the phagocytic cells located in reticular connective tissue, primarily monocytes and macrophages.

Quantum dot: is a semiconductor whose excitons are confined in all three spatial dimensions.

Fluorescence: is the emission of light by a substance that has absorbed light or other electromagnetic radiation of a different wavelength.

Summary

- The potential applications of molecular imaging for early diagnosis of cancer are tremendous.
- These powerful techniques will offer valuable opportunities to study and imaging complex cancer-related biological processes at the molecular level *in vivo*.
- The development of the molecular imaging technologies is not only dependent on the progress of imaging system, but also on the exploitation of imaging probes.

- In this chapter we have provided an overview on the developments of nanoparticle-based probes for the *in vivo* cancer imaging with MR, optical, and synergetic multiple imaging modalities.
- Nanoparticles provide a nanoplatform for the multivalent binding of targeting and imaging moieties, producing molecular imaging probes with high specificity and sensitivity, which provides new opportunities for early diagnosis of cancer.

Abbreviations

QDs	:	Quantum dots
EDC	:	N-(3-dimethylaminopropyl)-N´-ethylcarbodiimide hydrochloride
NHS	:	N-hydroxysuccinimide
sulfo-SMCC	:	Sulfosuccinimidyl-4-[N-maleimidomethyl]cyclohexane-1-carboxylate
MRI	:	Magnetic resonance imaging
DTPA	:	Diethylenetriaminepentaacetic acid
PAMAM	:	Poly(amido amine)
VEGFR2	:	Vascular endothelial growth factor receptor 2
DOTA	:	Tetraazacyclododecanetetraacetic acid
Cys	:	Cysteine
PEG	:	Poly(ethylene glycol)
SPIONs	:	Supermagnetic iron oxide nanoparticles
RES	:	Reticuloendothelial system
RGD	:	Arginine-glucine-aspartate
NPs	:	Nanoparticles
DNA	:	Deoxyribo nucleic acid
rch 24 mAb	:	Monoclonal antibody rich 24
NIRF	:	Near-infrared fluorescence
EGF	:	Epidermal growth factor
SERS	:	Surface-enhanced Raman scattering
PET	:	Positron emission tomography
SPECT	:	Single photon emission computed tomography
CT	:	Computed X-ray tomography
CLIO	:	Crosslinked iron oxide nanoparicles
uMUC-1	:	Underglycosylated mucin-1 antigen
PASA-IO	:	Polyaspartic acid coated iron oxide
EMB	:	microbubble

References

Akerman, M.E., W.C. Chan, P. Laakkonen, S.N. Bhatia and E. Ruoslahti. 2002. Nanocrystal targeting *in vivo*. Proc. Natl. Acad. Sci. USA 99: 12617–12621.

Amstad, E., S. Zurcher, A. Mashaghi, J.Y. Wong, M. Textor and E. Reimhult. 2009. Surface Functionalization of Single Superparamagnetic Iron Oxide Nanoparticles (SPIONs) for Targeted Magnetic Resonance Imaging (MRI). Small 5: 1334–1342.

Bonekamp, D., D.A. Hammoud and M.G. Pomper. 2010. Molecular Imaging: Techniques and Current Clinical Applications. Appl. Radiol. 39: 10–21.

Cai, W., D.W. Shin, K. Chen, O. Gheysens, Q.Z. Cao, S.X. Wang, S.S. Gambhir and X. Chen. 2006. Peptide-Labeled Near-Infrared Quantum Dots for Imaging Tumor Vasculature in Living Subjects. Nano Lett. 6: 669–676.

Cai, W., K. Chen, Z.-B. Li, S.S. Gambhir and C. Xiao. 2007. Dual-function probe for PET and nearinfrared fluorescence imaging of tumor vasculature. J. Nucl. Med. 48: 1862–1870.

Chandrasekar, D., R. Sistla, F.J. Ahmad, R.K. Khar and P.V. Diwan. 2007. The development of folate- PAMAM dendrimer conjugates for targeted delivery of anti-arthritic drugs and their pharmacokinetics and biodistribution in arthritic rats. Biomaterials. 28: 504–512.

Chen, K., Z. –B. Li, H. Wang, W. Cai and X. Chen. 2008. Dual-modality optical and positron emission tomography imaging of vascular endothelial growth factor receptor on tumor vasculature using quantum dots. Eur. J. Nucl. Med. Mol. imaging. 35: 2235–2244.

Chen, W.T., D. Thirumalai, T.T.F. Shih, R.C. Chen, S.Y. Tu, C. Lin and P.C. Yang. 2010. Dynamic contrast-enhanced folate-receptor-targeted MR imaging using a Gd-loaded PEG-dendrimer-folate conjugate in a mouse xenograft tumor model. Mol. imaging biol. 12: 145–154.

Crossgrove, J., and W. Zheng. 2004. Manganese toxicity upon overexposure. NMR biomed. 17: 544–553.

Diagaradjane, P., J.M. Orenstein-Cardona, N.E. Colon-Casasnovas, A. Deorukhkar, S. Shentu, N. Kuno, D.L. Schwartz, J.G. Gelovani and S. Krishnan. 2008. Imaging epidermal growth factor receptor expression *in vivo*: pharmacokinetic and biodistribution characterization of a bioconjugated quantum dot nanoprobe. Clin. Cancer Res. 14: 731–741.

Dong, Q., D.R. Hurst, H.J. Weinmann, T.L. Chenevert, F.J. Londy and M.R. Prince 1998. MR angiography with gadomer: a study in canines. Invest. Radiol. 33: 699–708.

Fischer-Durand, N., M. Salmain, B. Rudolf, L. Jugé, V. Guérineau, O. Laprévote, A. Vessières and G. Jaouen. 2007. Design of a New Multifunctionalized PAMAM Dendrimer with Hydrazide-Terminated Spacer Arm Suitable for Metal–Carbonyl Multilabeling of Aldehyde-Containing Molecules. Macromolecules 40: 8568–7855.

Gao, X.H., Y.Y. Cui, R.M. Levenson, L.W.K. Chung and S.M. Nie. 2004. *In vivo* cancer targeting and imaging with semiconductor quantum dots. Nat. Biotechnol. 22: 969–976.

Gupta, A.K., and M. Gupta. 2005. Synthesis and Surface Engineering of Iron oxide Nanoparticles for Biomedical Applications. Biomaterials. 26: 3995–4021.

He, X., K. Wang and Z. Cheng. 2010. *In vivo* near-infrared fluorescence imaging of cancer with nanoparticle-based probes. Wiley Interdiscipl. Rev. Nanomed. Nanobiotechnol. 2: 349–366.

Hood, J.D., M. Bednarski, R. Fraustoand S. Guccione, R.A. Reisfeld, R. Xiang and D. Cheresh. 2002. Tumor regression by targeted gene delivery to the neovasculature. Science 296: 2404–2407.

Hu, F., L. Wei, Z. Zhou, Y. Ran, Z. Li and M. Gao. 2006. Preparation of Biocompatible Magnetite Nanocrystals for *In Vivo* Magnetic Resonance Detection of Cancer. Adv. Mater. 18: 2553–2556.

Jemal, A., R. Siegel, J. Xu and E. Ward. 2010. Cancer Statistics. CA Cancer J. Clin. 60: 277–300.

Jun, Y.W., Y.M. Huh, J.S. Choi, J.H. Lee, H.T. Song, S. Kim, S. Yoon, K.S. Kim, J.S. Shin, J.S. Suh and J. Cheon. 2005. Nanoscale Size Effect of Magnetic Nanocrystals and Their Utilization for Cancer Diagnosis via Magnetic Resonance Imaging. J. Am. Chem. Soc. 127: 5732–5733.

Kobayashi, H., and M.W. Brechbiel. 2003. Dendrimer-based macromolecular MRI contrast agents: characteristics and application. Mol. Imaging 2: 1–10.

Kobayashi, H., S. Kawamoto, S.-K. Jo, H.L. Bryant Jr., M.W. Brechbiel, Jr. and R.A. Star. 2003. Macromolecular MRI contrast agents with small dendrimers: pharmacokinetic differences between sizes and cores. Bioconjugate Chem. 14: 388–394.

Konda, M.D., M. Aref, S. Wang, M. Brechbiel and E.C. Wiener. 2001. Specific targeting of folate-dendrimer MRI contrast agents to the high affinity folate receptor expressed in ovarian tumor xenografts. MAGMA 12: 104–113.

Langereis, S., A. Dirksen, T.M. Hackeng, M.H.P. Genderen and E.W. Meijer. 2007. Dendrimers and magnetic resonance imaging. New J. Chem. 31: 1152–1160.

Langereis, S., Q. de Lussane, M.H.P. van Genderen, W.H. Backes and E.W. Meijer. 2004. Terminated Poly(propylene imine) Dendrimers for Magnetic Resonance Imaging. Macromolecules. 37: 3084–3091.

Lee, H.-Y., Z. Li, K. Chen, A.R. Hsu, C. Cu, J. Xie, S. Sun and X. Chen. 2008. PET/MRI Dual-Modality Tumor Imaging Using Arginine-Glycine-Aspartic (RGD)–Conjugated Radiolabeled Iron Oxide Nanoparticles. J. Nucl. Med. 49: 1371–1379.

Lee, H., E. Lee, D.K. Kim, N.K. Jang, Y.Y. Jeong and S. Jon. 2006. Oxide Nanoparticles as Potential Magnetic Resonance Contrast Agents for *In Vivo* Cancer Imaging. J. Am. Chem. Soc. 128: 7383–7389.

Lee, J.-H., Y.-M. Huh, Y.-W. Jun, J.-W. Seo, J.-T. Jang, H.-T. Song, S. Kim, E.-J. Cho, H.-G. Yoon, J.-S. Suh and J. Cheon. 2007. Artificially engineered magnetic nanoparticles for ultra-sensitive molecular imaging. Nat. Med. 13: 95–99.

Lee, S., J. Xie, and X. Chen. 2010. Peptides and Peptide Hormones for Molecular Imaging and Disease Diagnosis. Chem. rev. 110: 3087–3111.

León-Rodríguez, D.L.M., A. Lubag, D.G. Udugamasooriya, B. Proneth, R.A. Brekken, X. Sun, T. Kodadek and A.D. Sherry. 2010. MRI Detection of VEGFR2 *in Vivo* Using a Low Molecular Weight Peptoid–(Gd)8-Dendron for Targeting. J. Am. Chem. Soc. 132: 12829–12831.

Longmaid III, H.E., D.F. Adams, R.D. Neirinckx, C.G. Harrison, P. Brunner, S.E. Seltzer, M.A. Davis, L. Neuringer and R.P. Geyer. 1985. *In vivo* 19F NMR imaging of liver, tumor, and abscess in rats. Preliminary results. Invest. Radiol. 20: 141–145.

Massoud, T.F., and S.S. Gambhir. 2003. Molecular imaging in living subjects: Seeing fundamental biological processes in a new light. Genes Dev. 17: 545–580.

Mehvar, R. 2003. Recent trends in the use of polysaccharides for improved delivery of therapeutic agents: pharmacokinetic and pharmacodynamic perspectives. Curr. Pharm. Biotechnol. 4: 283–302.

Missailidis, S., and A. Perkins. 2007. Aptamers as Novel Radiopharmaceuticals: Their Applications and Future Prospects in Diagnosis and Therapy. Cancer Biother. Radiopharm. 22: 453–468.

Moore, A., Z. Medarova, A. Potthast and G.P. Dai. 2004. *In Vivo* Targeting of Underglycosylated MUC-1 Tumor Antigen Using a Multimodal Imaging Probe. Cancer Res. 64: 1821–1827.

Pankhurst, A.Q., J. Connolly, S.K. Jones and J. Dobson. 2003. Applications of magnetic nanoparticles in biomedicine. J. Phys. D. Appl. Phys. 36: R167–R181.

Pilch, J., D.M. Brown, M. Komatsu, T.A.H. Jarvinen, M. Yang, D. Peters, R.M. Hoffman and E. Ruoslahti. 2006. Peptides selected for binding to clotted plasma accumulate in tumor stroma and wounds. Proc. Natl. Acad. Sci. USA 103: 2800–2804.

Qian, X., X.-H. Peng, D.O. Ansari, Q.Y. Yin-Goen, G.Z. Chen, D.M. Shin, L. Yang, A.N. Young, M.D. Wang and S. Nie. 2008. *In vivo* tumor targeting and spectroscopic detection with surface-enhanced Raman nanoparticle tags. Nat. Biotech. 26: 83–90.

Qiao, R., C. Yang, and M. Gao. 2009. Superparamagnetic iron oxide nanoparticles: from preparations to *in vivo* MRI applications. J. Mater. chem. 19: 6274–6293.

Rossin, R., D. Pan, K. Qi, J.L. Turner, X. Sun, K.L. Wooley and M.J. Welch. 2004. 64Cu-labeled folate conjugated shell cross-linked nanoparticles for tumor imaging and radiotherapy: synthesis, radiolabeling, and biologic evaluation. J. Nucl. Med. 46: 1210–1218.

Rudovský, J., P. Hermann, M. Botta, S. Aime and I. Lukeš. 2005. Dendrimeric Gd(III) complex of a monophosphinated DOTA analogue: optimizing relaxivity by reducing internal motion. Chem. Commun. 41: 2390–2392.

Schulke, N., O.A. Varlamova, G.P. Donovan, D. Ma, J.P. Gardner, D.M. Morrissey, R.R. Arrigale, C. Zhan, A.J. Chodera, K.G. Surowitz, P.J. Maddon, W.D.W. Heston and W.C. Olson. 2003. The homodimer of prostate-specific membrane antigen is a functional target for cancer therapy. Natl. Acad. Sci. USA 100: 12590–12595.

Smith, B.R., Z. Cheng, A. De, A.L. Koh, R. Sinclair and S.S. Gambhir. 2008. Real-time intravital imaging of RGD-Quantum dot binding to luminal endothelium in mouse tumor neovasculature. Nano Lett. 8: 2599–2606.

Sun, C., O. Veiseh, J. Gunn, C. Fang, S. Hansen, D. Lee, R. Sze, R.G. Ellenbogen, J. Olson and M. Zhang. 2008. *In Vivo* MRI Detection of Gliomas by Chlorotoxin-Conjugated Superparamagnetic Nanoprobes. Small 4: 372–379.

Tada, H., H. Higuchi, T.M. Wanatabe and N. Ohuchi. 2007. *In vivo* Real-time Tracking of Single Quantum Dots Conjugated with Monoclonal Anti-HER2 Antibody in Tumors of Mice. Cancer Res. 67: 1138–1144.

Tan, M., X. Wu, E. Jeong, Q. Chen and Z. Lu. 2010. Peptide-Targeted Nanoglobular Gd-DOTA Monoamide Conjugates for Magnetic Resonance Cancer Molecular Imaging. Biomacromolecules. 11: 754–761.

Tan, M., X. Wu, E.-K. Jeong, Q. Chen, D.L. Parker and Z. -R. Lu. 2010. An Effective Targeted Nanoglobular Manganese(II) Chelate Conjugate for Magnetic Resonance Molecular Imaging of Tumor Extracellular Matrix. Mol. Pharm. 7: 936–943.

Weissleder, R., A. Bogdanov, E.A. Neuwelt and M. Papisov. 1995. Long-circulating iron oxides for MR imaging. Adv. Drug Deliv. Rev. 16: 321–334.

Weissleder, R., and U. Mahmood. 2001. Molecular imaging. Radiology. 219: 316–333.

Wiener, E.C., F.P. Auteri, J.W. Chen, M.W. Brechbiel, O.A. Gansow, D.S. Schneider, R.L. Belford, R.B. Clarkson and P.C. Lauterbur. 1996. Molecular Dynamics of Ion-Chelate. Complexes Attached to Dendrimers. J. Am. Chem. Soc. 118: 7774–7782.

Wiener, E.C., M.W. Brechbiel, H. Brothers, R.L. Magin, O.A. Gansow, D.A. Tomalia and P.C. Lauterbur. 1994. Dendrimer-based metal chelates: a new class of magnetic resonance imaging contrast agents. Magn. Reson. Med. 31: 1–8.

Xie, J., K. Chen, H. –Y. Lee, C. Xu, A.R. Hsu, S. Peng, X. Chen and S. Sun. 2008. Ultrasmall c(RGDyK)-Coated Fe3O4 Nanoparticles and Their Specific Targeting to Integrin αvβ3-Rich Tumor Cells. J. Am. Chem. Soc. 130: 7542–7543

Yang, F., Y. Li, Z. Chen, Y. Zhang, J. Wu and N. Gu. 2009. Superparamagnetic iron oxide nanoparticle-embedded encapsulated microbubbles as dual contrast agents of magnetic resonance and ultrasound imaging. Biomaterials. 30: 3882–3890.

7

Iron and Ethylcellulose Nanoplatforms for Cancer Targeting

José L. Arias,[1,a,*] *Margarita López-Viota*[1,b] and *Mª Adolfina Ruiz*[1,c]

ABSTRACT

Current chemotherapy has been frequently demonstrated to display little therapeutic (antitumor) activity against cancer, along with severe toxicity due to the clinical use of very high drug doses. As a very promising alternative, drug delivery nanoplatforms aims to revolutionize cancer treatment. It has been suggested that biocompatible nanoparticulate systems can determine an adequate biodistribution of anticancer molecules upon administration. Subsequently, the antitumor activity of chemotherapy agents will be optimized, and the incidence of systemic side effects will be kept to a very minimum. Further

[1]Department of Pharmacy and Pharmaceutical Technology, Faculty of Pharmacy, University of Granada, Campus Universitario de Cartuja, s/n, 18071 Granada, Spain.
[a]E-mail: jlarias@ugr.es
[b]E-mail: mlvg@ugr.es
[c]E-mail: adolfina@ugr.es
*Corresponding author

List of abbreviations after the text.

functionalization of these nanomedicines on the basis of passive and/or active targeting strategies will definitively improve drug accumulation into cancer site.

In this way, the combined use of inorganic substances (e.g., iron) and/or organic materials (e.g., ethylcellulose) could offer important possibilities in the formulation of drug delivery systems. In this way, the development of inorganic/organic (core/shell) nanocomposites (i.e., iron/ethylcellulose nanoparticles) has been described as a cancer targeting approach that could lead to optimum chemotherapy. It is expected that the magnetic (inorganic: iron) core within such composite nanoparticles will be responsible for the magnetic responsive behavior of the nanocarrier, while the organic (organic/polymeric: ethylcellulose) shell will allow the vehiculización of appropriate amounts of antitumor drugs within the nanoplatform. As a result, the whole drug dose will be concentrated within tumor interstitium with the help of a magnetic gradient. This chapter is focused on the analysis of the current state of the art in the use of iron-, ethylcellulose-, and iron/ethylcellulose nanostructures to develop nanodrugs against cancer.

INTRODUCTION

Despite the efficient *in vivo* antitumor activity of chemotherapy agents, their real possibilities against cancer are markedly conditioned by suffering of several weaknesses, i.e., unfavorable pharmacokinetic profiles short plasma half life ($t_{1/2}$, a consequence of the rapid drug metabolism and/or plasma clearance), non-uniform oral absorption (with very significant patient-to-patient bioavailability variations), and/or stimulation of drug resistant mechanisms in tumor cells. Thus, the need of high doses to beat cancer which concurrently leads to severe dose-limiting toxicity (e.g., myelosupression, cardiotoxicity, haematological side effects, etc.). In order to circumvent such difficulties, nanotechnology has been introduced in the development of more advanced treatments against cancer. Concretely, nanoparticulate drug delivery systems have been formulated to specifically accumulate anticancer drugs into malignant tissues, and to obtain a prolonged and deep contact between the chemotherapy molecule and the tumor cell (Arias 2011).

Numerous research reports have highlighted additional possibilities offered by appropriate engineering approaches in the formulation of such nanomedicines against cancer (Arias 2011). From a basic point of view, polymers and lipids are mainly used to prepare these nanoparticulate systems to concentrate anticancer molecules into tumor tissues. However, the potential use of more complex and multifunctional nanostructures to that objective, e.g., copolymers, magnetic composites, etc. (Reddy 2005; Arias 2011) has been recently described. Such drug delivery nanoplatforms are engineered on the basis of passive and active drug targeting strategies to cancer. The former approach is based on the enhanced permeability and retention (EPR) effect; while the second strategy is related to nanoparticles engineered to be sensitive to external stimuli, and/or surface decorated with biomolecules for ligand-mediated targeting of tumor cells. As a result, the biological fate of chemotherapy agents will be controlled perfectly (Reddy 2005; Arias 2011).

Stimuli-sensitive nanoparticles can alter their physical properties under exposure to an external stimuli (e.g., light, pH, ultrasounds, magnetic gradients, etc.) which can be used to trigger drug release selectively and/or to concentrate the drug dose into the tumor tissue (e.g., magnetically responsive drug nanocarriers). Therefore, drug biodistribution will be minimized (and, subsequently, the undesired toxicity) and drug accumulation into tumor interstitium is maximized (and thus, the anticancer effect) (Arias 2011).

The introduction of magnetic particles in the formulation of drug-loaded nanoplatforms could result in the design of more complex and efficient treatment schedules against cancer (Durán et al. 2008). Magnetically responsive nanoparticles are capable of carrying anticancer molecules or genes to targeted tumors under the guidance of magnetic gradients, while protecting the drug from *in vivo* metabolization and elimination. It has been described that the organic shell of the nanoplatform (e.g., a polymeric matrix) will be responsible for transporting the drug to the tumor tissue (controlling its release), and that meanwhile the magnetic core will allow the magnetic guidance of the nanomedicine towards cancer cells (Arias 2011). Even more, such multifunctional nanoplatforms can simultaneously induce the death of malignant cells by magnetic hyperthermia, and/or (chemo)embolization. These nanosystems have been further described to exhibit combined chemotherapy and diagnostic imaging capabilities (nanotheragnosis) (Arias 2010).

The use of ethylcellulose and iron in the formulation of such nanoplatforms could offer an interesting move towards drug targeting to cancer if it is kept in mind (Arias et al. 2007, 2010): *i*) the extensive experience on the use of ethylcellulose in pharmaceutical technology to prepare oral dosage forms, drug-loaded microparticles, and drug-loaded nanoparticles, with sustained drug release properties, enhanced pharmacotherapy efficacy, and negligible toxicity (Ubrich et al. 2004; He et al. 2008a; Vaghani et al. 2010); *ii*) the excellent magnetic properties of iron nanoparticles (Arias et al. 2007; Arias 2010); and, *iii*) the very promising therapeutic opportunities coming from magnetic nanomedicines (Arias 2010).

This chapter is devoted to the analysis of the biomedical challenges and opportunities coming from the use of ethylcellulose-, iron-, and iron/ethylcellulose-based nanoplatforms in drug delivery to cancer. Emphasis will be given to reproducible techniques leading to the formulation of drug-loaded iron/ethylcellulose (core/shell) nanoparticles. Physicochemical aspects of the preparation conditions and properties of the nanocarriers, along with special engineering and synthesis strategies to incorporate chemotherapy molecules to the nanocomposites will be discussed. *In vitro* and *in vivo* behavior of the drug-loaded nanoplatforms will be also analyzed.

ETHYLCELLULOSE AND CANCER TARGETING

Ethylcellulose (ethylated cellulose, 2-[4,5-diethoxy-2-(ethoxymethyl)-6-methoxyoxan-3-yl]oxy-6-(hydroxymethyl)-5-methoxyoxane-3,4-diol], $C_{20}H_{38}O_{11}$) is obtained upon chemical conversion of the hydroxyl groups of cellulose into ethyl ether groups. This hydrophobic cellulose derivative (molecular weight ≈ 454.51 g/mol, Fig. 1) is classically used as a thin-film coating and as a food additive (i.e., emulsifying agent).

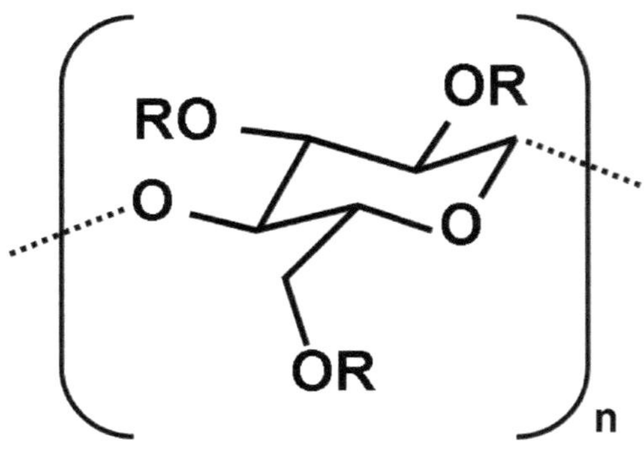

Fig. 1. Chemical structure of ethylcellulose. The number of ethyl groups is determined by the manufacture procedure. In the structure, R = H, or CH_2CH_3.

Ethylcellulose is chemically stable under storage and it is characterized by a significant biocompatibility, great tolerability, and low toxicity (DeMerlis et al. 2005). As a consequence, this macromolecule has been introduced in the formulation of medicines (e.g., preparation of solid dosage forms with controlled drug release properties) (Schmidt and Bodmeier 1999; Rujivipat and Bodmeier 2010; Vaghani et al. 2010), and dietary supplement products (i.e., taste masking agents) (DeMerlis et al. 2005). As an example of the former application, pectin/ethylcellulose film-coated pellets loaded with 5-fluorouracil (5-FU) have been prepared to treat colon cancer. It was described that compared to uncoated pellets, greater drug concentrations into the tumor site with prolonged exposure time were obtained when the coated formulation was given orally to rats (dose equivalent to 15 mg/Kg of 5-FU). Thus, the mean maximum plasma concentration (C_{max}) and area under the time-concentration curve (AUC) values in the coated pellets group were significantly lower (≈ 3.6 ± 2.3 μg/mL and ≈ 9.1 ± 1.2 μg·hr/mL, respectively) than those corresponding to the control group (≈ 23.5 ± 2.9 μg/mL and ≈ 49.1 ± 3.1 μg·hr/mL, respectively). This was the consequence of a different 5-FU release profile from the formulations: drug release from uncoated pellets mainly occurred in the upper gastrointestinal tract, while 5-FU release from coated pellets mainly took place in the cecum and colon. Thus, the later formulation was postulated to have a great potential in the enhancement of the antitumor efficacy of 5-FU with very low systemic toxicity (He et al. 2008a). Ethylcellulose aqueous dispersions have been proposed as oral cisplatin sustained release medicines to treat lung cancer patients (Nakano et al. 1997). The preparation was formulated by combining cisplatin with the water-insoluble polymer and with stearic acid in the ratio 1:10:5. The cisplatin formulation (drug dose: 20 mg/Kg) was administered orally to rats, and the resulting curve of serum drug levels against time was compared with that obtained after intravenous infusion of the same drug dose. It was observed that the oral formulation did not exhibit any toxicity on the gastrointestinal tract, and the resulting oral drug bioavailability was determined to be ≈ 30%. More interestingly, the mean residence time (MRT) of the cisplatin dispersion was ≈ 6.1 ± 0.4 hr, whereas the MRT of the intravenously injected anticancer drug was ≈ 3.9 ± 0.1 hr.

Ethylcellulose-based Micromedicines Against Cancer

The synthesis method frequently followed for the preparation of ethylcellulose microparticles to be used for drug delivery applications is the emulsion solvent evaporation technique (Vanderhoff et al. 1979). A wide range of anticancer molecules have been incorporated into these microplatforms which can further exhibit embolization capability (Table 1). This interesting property results *in vivo* from the occlusion of the tumor blood vessels by the micrometer sized drug delivery system. Such chemoembolization micromedicines have been further postulated to be able to prevent the rapid revascularization of the embolized area (Kato et al. 1981, 1996).

Table 1. Ethylcellulose-based micromedicines against cancer.

Antitumor drug	Microplatform / *In vivo* activity	Type of cancer	Reference
Carboplatinum	Microcapsules (size ≈ 200 μm)/ Chemoembolization	Mid-tongue carcinoma	He et al. 2008b
Mitomycin C, cisplatin, peplomycin	Microcapsules/ Chemoembolization	Liver, kidney, prostate, urinary bladder, uterus, sigmoid colon, Douglas' pouch, lung, head and neck	Okamoto et al. 1985; Kato et al. 1981, 1996
Cisplatin	Microspheres/ Chemoembolization	Oral and maxillofacial region	Yang et al. 1995
5-FU	Microspheres/*in vitro* study	–	Vaghani et al. 2010

Preclinical and clinical investigations have described an enhanced and sustained antitumor activity (with minimized systemic toxicity) when the chemotherapy molecule is loaded to such microparticulate systems. The significantly enhanced therapeutic effect could result in the prolongation of patient's survival.

Other drugs and active agents have been formulated into micro-(nano-)carriers made of ethylcellulose and/or copolymers. For instance, acetaminophen, carbamazepine, chlorpheniramine maleate, diclofenac sodium, ibuprofen, ketoprofen, loratadine, morphine, nimesulide, propranolol, tritorelin acetate, and the sunscreen agent *trans*-2-ethylhexyl-*p*-methoxycinnamate (*trans*-EHMC), to cite just a few (Perugini et al. 2002; Arias et al. 2007; Hasan et al. 2007; Arias et al. 2009; Rujivipat and Bodmeier 2010). Interestingly, it has been also proposed to use ethylcellulose in the formulation of optical nanochemosensors (Borisov et al. 2009).

Ethylcellulose-based Nanomedicines Against Cancer

More recently, ethylcellulose has been introduced in nanotechnology of drug delivery systems, i.e., development of active cancer targeting nanoplatforms, and solid dosage forms with controlled drug release properties (Schmidt and Bodmeier 1999; Arias et al. 2007). It has described the formulation of ethylcellulose nanoparticles by solvent evaporation of nano-emulsions. The later nanosystem (a water/polyoxyethylene 4 sorbitan monolaurate/ethylcellulose oil-in-water nano-emulsion) being formulated by a phase inversion composition method (Calderó et al. 2011). However, easier techniques can be used in the preparation of spherical ethylcellulose nanoparticles (mean diameter ≈ 350 nm, Fig. 2), e.g., a slightly modified emulsion solvent evaporation technique (Arias et al. 2007). Briefly, ethylcellulose (0.2 g) is dissolved in a mixture of organic solvents (i.e., 1 g benzene and 0.2 g ethanol). After 24 hours at room temperature, n-decane (0.1 g, the stabilizer of

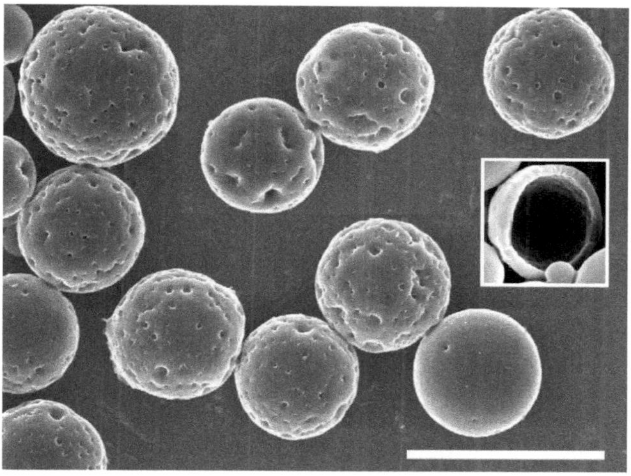

Fig. 2. Scanning electron microscope photograph of ethylcellulose nanoparticles. Inset: detailed picture of an ethylcellulose nanocapsule. Bar length: 500 nm. The rugous surface of the nanoparticles could determine a very important specific surface area for drug adsorption purposes. Ethylcellulose nanoparticles are also characterized by a hydrophobic character, and a negative surface electrical charge which can be adjusted by controlling the pH, and/or ionic strength of the aqueous dispersion. As it is well-known, both properties are key parameters in the formulation of nanoparticle suspensions.

the final emulsion) is added to the polymeric phase. Thereafter, this phase is added to 4 g of a 10^{-3} N nitric acid (HNO_3) aqueous solution [containing 0.125% (w/v) sodium dodecyl sulphate, and 0.375% (w/v) polyethylene glycol 4000]. Before mixing, both phases are heated up to ≈ 70 °C, and the incorporation of the aqueous phase to the polymeric one is gradually done under mechanical stirring. Finally, the organic solvents are then completely evaporated under vacuum to obtain pure ethylcellulose nanoparticles. Finally, it can also introduce a cleaning procedure involving repeated cycles of centrifugation and re-dispersion in deionized and filtered water.

Two procedures have been proposed to load drugs to ethylcellulose nanoparticles (Arias et al. 2009, 2010). The first methodology involves single surface adsorption of the active agent onto preformed nanoparticles. For example, a suspension of polymeric nanoparticles is incubated for 24 hours at room temperature with increasing amounts of drug. The second method (entrapment procedure) followed for drug absorption into the nanopolymer is similar to that above described for the synthesis of ethylcellulose nanoparticles, except that the aqueous phase further contains appropriate amounts of drug molecules. Such methodologies have reported a significant 5-FU loading to these nanoplatforms (Arias et al. 2010). Very interestingly, compared to the surface adsorption procedure, both the entrapment efficiency (%) and the drug loading (%) are significantly enhanced when the chemotherapy agent is incorporated (absorbed) into the polymeric matrix whatever the initial drug concentration fixed. As an example, these parameters can rise from ≈ 20% and ≈ 1.5% (after 5-FU adsorption onto the surface) to ≈ 50% and ≈ 3.5% (when 5-FU is absorbed into the polymer structure), respectively when the drug concentration used is 0.01 M. The most significant parameter influencing 5-FU loading to the nanoparticles resulted to be the drug concentration fixed in the incubation/absorption medium. On the opposite, no significant effect on drug loading occurred when the amount of polymer emulsified into the aqueous phase was enhanced. Concerning the effect of the surfactants employed in the preparation procedure, it has been observed that only when they are used it is obtained a homogeneous distribution of ethylcellulose nanoparticles with reduced size and great uniformity, but without influence on the drug loading efficiency.

Finally, 5-FU *in vitro* release from ethylcellulose nanoparticles have been also described to be extremely dependent on the procedure followed to formulate the drug-loaded nanoparticles, and on the amount of drug incorporated to the nanoplatform (Arias et al. 2010). It was defined that the *in vitro* release of 5-FU adsorbed onto the nanosystem was almost complete after ≈ 1 hour, due to a single and rapid desorption process. On the opposite, the *in vitro* release of 5-FU absorbed into ethylcellulose nanoparticles was a biphasic process with an early rapid release within 1 hour (up to ≈ 40%), the remaining drug being slowly liberated during the next 24 hours. The rapid first phase likely corresponded to the leakage of the surface-associated and/or poorly entrapped chemotherapy molecule. 5-FU release during the slower release phase would result, on the contrary, from drug diffusion through the ethylcellulose matrix. Furthermore, an increase in the amount of drug loaded to the polymeric nanoparticles also enhanced the cumulative 5-FU release.

IRON AND CANCER TARGETING

The high initial magnetic susceptibility and magnetic saturation of iron nanoparticles permits controlling their biological fate once parenterally administered. Iron nanoparticles are biodegradable and characterized by a very low toxicity [lethal dose 50% (LD_{50}): 50 g/Kg] (Chua-anusorn et al. 1999; Whittaker et al. 2002). It is hypothesized that, upon intravenously injected, the nanoparticles will deliver anticancer drugs to targeted tumors under the guidance of externally (and/or internally) applied magnetic gradients. The magnetic force will even focus the iron nanoplatform into the tumor interstitium (or tumor neovasculature) until the drug is completely released (Arias 2010, 2011).

Pure iron nanoparticles are frequently presented in the form of carbonyl iron, a unique form of elemental iron thanks to its small particle size. The synthesis procedure followed to prepare such magnetic nanoparticles involves treating iron with carbon monoxide (CO) under heat and pressure. The resulting pentacarbonyl iron [$Fe(CO)_5$] is then decomposed under controlled conditions, and yields CO and extremely pure iron powder (Whittaker et al. 2002). An easier methodology (a gravitational separation procedure) to obtain nanometer-sized iron nanoparticles has been recently proposed (Arias et al. 2006). This method implies the sonication of 0.3% (w/v) (commercially available) carbonyl iron aqueous suspensions during 5 min. After settling under gravity during 60 min, the upper 10 mm of supernatant are taken. As a result, spherical iron nanoparticles can be obtained with a not very broad size distribution (average diameter ≈ 500 ± 150 nm, Fig. 3).

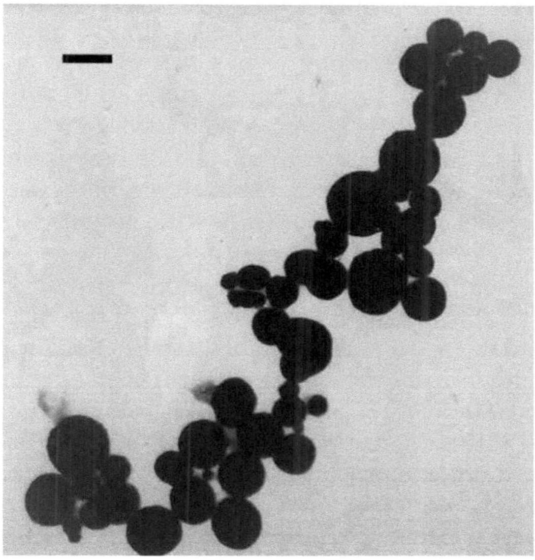

Fig. 3. Transmission electron microscope photograph of iron nanoparticles. Bar length: 500 nm. Iron nanoparticles are characterized by a hydrophilic character, and a by surface electrical charge which can be adjusted by controlling the pH of the aqueous dispersion (Arias et al. 2006). As previously indicated, both properties are key parameters in the formulation of nanoparticle suspensions. Thanks to the small size, it is suggested that most iron nanoparticles will remain in the body, being at last eliminated mainly by renal filtration (Chua-anusorn et al. 1999).

Iron-based Nanoplatforms Against Cancer

Drug loading to iron nanoparticles can be only achieved by surface adsorption (Arias et al. 2009, 2010). Briefly, the methodology involves the incubation of a suspension of iron nanoparticles for 24 hours at room temperature with increasing amounts of drug. Despite an enhancement in the drug concentration fixed in the incubation medium has been reported to influence positively the adsorption process, drug vehiculization could be considered negligible. For instance, maximum entrapment efficiency (and drug loading) obtained for diclofenac sodium and 5-FU has been described to be ≈ 7% (and ≈ 1.2%) and ≈ 6% (and ≈ 0.9%), respectively. Additionally, drug release could be considered a very fast process (completed in less than 1 hour).

Due to this very poor drug vehiculization capability (low drug loading capacity and uncontrollable drug release), iron nanoparticles must be embedded into a biocompatible shell (mainly a polymer or a lipid vesicle) to improve drug delivery to cancer (Arias et al. 2006, 2008). For instance, carbonyl iron/poly(butylcyanoacrylate) (core/shell) nanoparticles (mean diameter ≈ 500 ± 200 nm) have been formulated for the transport of 5-FU and ftorafur to tumors. Both drugs were satisfactorily loaded to the nanocomposites: maximum ftorafur and 5-FU absorption densities ≈ 60 and 40%, respectively. Interestingly, drug release from the nanoparticles was found to be a biphasic process, thus, being significantly prolonged up to 10 and 20 hours, respectively (Arias et al. 2008).

More complex nanostructures based on iron have been engineered for an efficient drug delivery to cancer. For example, magnetically responsive iron/gold (core/shell) nanocomposites have been synthesized by a reverse micelle method. Doxorubicin loading was achieved by surface chemical adsorption: the amine group of the chemotherapy agent binds to the gold shell. Very promisingly, only up to ≈ 25% of the adsorbed drug was released in 80 hours (Kayal and Ramanujan 2010). Finally, iron/magnetite (core/shell) nanoparticles can be prepared for magnetic hyperthermia combined with co-delivery of anticancer drugs, or photodynamic therapy (Balivada et al. 2010). Interestingly, iron/magnetite nanoparticles caused a significant antitumor effect on murine B16-F10 melanoma with three (short) ten-minute alternating magnetic field (AMF) exposures. It was further observed that there was a decrease in tumor size after intravenous administration of the nanocomposites followed by three consecutive days of AMF exposure.

IRON/ETHYLCELLULOSE (CORE/SHELL) COMPOSITES AND CANCER TARGETING

As previously indicated, iron nanoparticles do not exhibit suitable properties for an efficient drug delivery despite their very high magnetic susceptibility. As a consequence, it has been proposed to use polymeric or liposomal matrices in combination with iron nanoparticles to improve drug delivery to cancer. Preceding preclinical and clinical investigations have demonstrated that magnetic particles embedded into polymeric nanomaterials do not significantly affect the toxicity of the latter, and that such nanocomposites present low toxicity (Lübbe et al. 1996a,b). In those nanoplatforms, the shell will transport the chemotherapy agent to the tumor interstitium or vasculature (controlling its release), meanwhile the iron cores will permit the magnetic guidance of the nanomedicine to the targeted site (leading to very high drug concentrations) (Arias 2010).

The formulation of ethylcellulose-based magnetic nanoplatforms was firstly described by Kato et al. (1984). The anticancer drug mitomycin C was loaded to ferromagnetic ethylcellulose microcapsules which were prepared by a phase separation methodology. Interestingly, the magnetic nanoparticles were located either on the microplatform surface or into the microcarrier core. Anyhow, both formulations presented sustained drug release characteristics and a sensitive response to magnetic gradients. *In vivo* studies demonstrated that VX2 tumors in the hind limb and urinary bladder of the rabbit were successfully treated with magnetic control of the microdrug. This was considered to be the consequence of an enhanced drug accumulation into the tumor mass for a prolonged period of time. Unfortunately, the micrometer size of the drug delivery system does not assure an extensive biodistribution into the organism and, thus, the accumulation of the drug into tumors located deep in the body. Additionally, it has been hypothesized that (upon intravenous administration) microparticulate drug carriers exhibit a rapid recognition by the reticuloendothelial system (and plasma clearance), compared to nanocarriers (Arias 2010). Hence, recent investigations have been focussed on the development of magnetic nanoparticulate ethylcellulose-based systems for the efficient delivery of chemotherapy agents to cancer (Arias et al. 2010).

Magnetic nanoparticles consisting of a magnetic core (iron) and a biocompatible polymeric shell (ethylcellulose) have been formulated by an emulsion solvent evaporation process (Fig. 4) (Arias et al. 2007). Briefly, the synthesis method is similar to that described for ethylcellulose

Fig. 4. (a) Dark-field high resolution transmission electron microphotograph of iron/ethylcellulose (core/shell) nanoparticles (bar length: 500 nm), and (b) optical microscope photograph (magnification 63X) of a nanocomposite aqueous suspension under the influence of an external magnetic field (B = 200 mT) in the direction of the arrow. Particle chaining in the magnetic gradient direction is an indication of strong magnetic dipolar interactions between the composite nanoparticles. Such interactions very efficiently overcome double layer repulsions between magnetic nanocomposites which would allow their accumulation into the targeted site during the time required. Adapted with permission from Ref. (Arias et al. 2010). Copyright Elsevier (2010).

nanoparticles, except that the aqueous phase is a 3.75% (w/v) iron suspension in a 10^{-3} N HNO_3 solution. Cleaning of the nanocomposites is possible by repeated magnetic separation and re-dispersion in deionized and filtered water, until the supernatant is transparent and its conductivity is ≤ 10 μS/cm. Regarding the biological fate of the magnetic nanocarrier, taking into account the mean particle size (≈ 500 nm) we could hypothesize that it will be removed by the cells of the reticuloendothelial system and mechanically filtrated by the spleen. Therefore, it is expected that after removal of the magnetic field from the tumor site, the whole magnetic nanomedicine will be eliminated by renal filtration (Okon et al. 1994; Gupta and Gupta 2005).

5-FU loading to iron/ethylcellulose nanocomposites has been investigated by surface adsorption onto preformed nanocomposites, and by incorporation/entrapment (absorption) into the magnetic particle (Arias et al. 2010). As it is generally the case when drug molecules are loaded to nanocarriers, 5-FU absorption into the polymeric matrix yielded higher drug loading values and a slower drug release profile as compared with drug adsorption (Fig. 5). For instance when 5-FU concentration was 0.01 M, the entrapment efficiency (%) and the drug loading (%) was increased from ≈ 20 and ≈ 1.4% (when 5-FU was surface adsorbed)

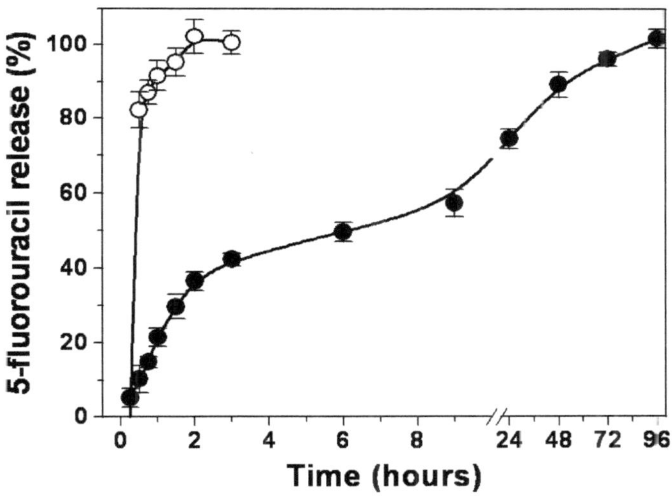

Fig. 5. Cumulative release of adsorbed (○) and entrapped (•) 5-FU from iron/ethylcellulose (core/shell) nanoparticles as a function of the incubation time in PBS (pH = 7.4 ± 0.1). The lines are guides to the eye. The release of 5-FU adsorbed onto the core/shell nanoparticles is almost complete after 1 hour, thanks to a rapid desorption process. On the contrary, the release of 5-FU absorbed into the nanocomposites is a biphasic process with an early rapid release phase (up to ≈ 40% within 2 hours), the remaining drug being slowly liberated during the next 94 hours. The first phase likely corresponds to the leakage of the surface-associated and/or poorly entrapped 5-FU. Drug release during the slower phase may result from 5-FU diffusion through the ethylcellulose matrix. Such a biphasic profile, suggests that the major fraction of the chemotherapy agent is entrapped into the polymeric network. The drug entrapped at the iron/polymer interface can determine a complex 5-FU diffusion through the polymeric shell. Adapted with permission from Ref. (Arias et al. 2010). Copyright Elsevier (2010).

to ≈ 60 and ≈ 2.3% (after drug absorption into the polymeric matrix), respectively. These preliminary results suggested the potential of this stimuli-sensitive drug carrier for cancer targeting.

Conclusions

Active drug targeting strategies to cancer are expected to revolutionize current pharmacotherapy. In this way, iron/ethylcellulose (core/shell) nanoparticles can offer very promising possibilities due to their magnetic responsiveness and their excellent antitumor drug vehiculization capabilities. However, advanced engineering approaches are still needed in the formulation of these nanoplatforms for an efficient drug delivery into cancer, i.e., long-circulating functionalities, surface decoration with specific ligands to receptors overexpressed onto cancer cell surface, and the exclusive disruption inside the malignant cell. Very importantly, it must be *in vivo* optimized relationships between nanoparticle structure and biological fate, anticancer effect, and nanotoxicity. Regarding the latter, greater progress in the formulation and nanoengineering of such nanomedicines would allow overcoming the problem of nanoparticle biodegradation, particularly in long-treatment schedules to avoid overloading the body with foreign material.

Summary Points

- Antitumor drugs are characterized by a little specificity for cancer cells which induces severe side effects.
- Active drug targeting approaches based on the use of magnetically responsive nanoparticles can very efficiently enhance drug accumulation into malignant cells.
- Iron nanoparticles are characterized by excellent magnetic properties (i.e., very high initial magnetic susceptibility and magnetic saturation values) which could be used to control their biological fate externally. Unluckily, the very poor drug vehiculization capacity of such nanomaterials severely limited their clinical use in chemotherapy.
- Very interestingly for drug delivery applications, ethylcellulose nanomatrices offer higher anticancer drug loading values and a slower (biphasic) drug release profile.
- Iron/ethylcellulose (core/shell) nanocomposites can take advantage on the characteristics of both materials for tumor targeting. The shell will transport the anticancer molecule to the tumor site (triggering its release), while the iron cores will allow the magnetic guidance of the nanomedicine to the targeted cancer cells.

- Greater nanoengineering developments on *in vivo* studies are required to definitively allow the introduction of iron and ethylcellulose nanoplatforms in cancer arena.

Dictionary

Chemoembolization: combined therapeutic effect through tumor (micro) infarction (occlusion of the blood vessels feeding the cancer cells) and prolonged local drug (release) activity into the tumor tissue.

Magnetic hyperthermia: heat production due to magnetic hysteresis loss when magnetic nanoparticles are under the influence of an alternating magnetic gradient of high frequency (≈ 1 MHz). This temperature increment can even be cytotoxic to tumor cells. Additionally, heat can induce drug desorption from the nanocarrier. Thus, drug release can be triggered by hyperthermia which also induces tumor permeability to the chemotherapy agent.

Median lethal dose, lethal dose 50% (LD_{50}): amount of the substance required (generally per body weight) to kill 50% of the test population. Dose that kills half (50%) of the animals tested.

Micro-(nano-)sphere: micro-(nano-)meter-sized matrix system with the drug molecules dispersed along a continuous particle core.

Micro-(nano-)capsule: micro-(nano-)meter-sized vesicular system with the molecules of active agent trapped into an oily or aqueous cavity surrounded by a lipidic or polymeric shell.

Photodynamic therapy: technique that involves the administration of a photosensitizer and its activation with localized light irradiation. It is used to obliterate premalignant sites and cancers in early stages, and to reduce tumor size in end-stage cancers.

Key Facts

- Key facts of the EPR effect. It is based on the "leaky" vasculature of tumor tissues which permits the selective extravasation and accumulation of nanomedicines into cancer.
- Key facts of ligand- or receptor-mediated cancer targeting. Selective delivery of the nanomedicine to cancer occurs thanks to specific molecular recognition processes. Chemical conjugation of the drug-loaded nanoparticle to targeting (tissue-, or cell-specific) ligands generally leads to receptor-mediated cell internalization of the nanomedicine (e.g., endocytosis).
- Key facts of long-circulating drug nanocarriers. Surface functionalization of the nanoparticle with hydrophilic chains (i.e.,

polyethylene glycol) will slow down the recognition of the nanocarrier by the reticuloendothelial system and plasma clearance. An extended $t_{1/2}$ of the drug delivery system will assure higher intratumor drug levels.

- Key facts of the introduction of ethylcellulose in the formulation of micro- and nano-medicines against cancer. Since 1980s, an intense research has led to the development of numerous ethylcellulose-based microparticles and nanoparticles (e.g., spheres and capsules) loaded with chemotherapy agents.
- Key facts of the biological fate of iron, ethylcellulose, and iron/ethylcellulose nanomedicines. Nanoparticle geometry, surface thermodynamics, and surface charge are the major aspects in the engineering of these nanomedicines for drug delivery purposes.
- Key facts of nanotoxicity in drug delivery to cancer. Nanoparticle biocompatibility is determined by both the biodegradability and toxicity of the nanocarrier and breakdown products. Biodegradation must takes place, and the degradation products must be cleared out of the body in the shortest period of time.

Applications to Areas of Health and Disease

Iron and ethylcellulose nanoplatforms could offer significant possibilities in the delivery of drug molecules to severe diseases, e.g., cancer, infectious diseases, inflammatory diseases, etc. More promisingly, the combination of both materials into the same nanoparticle has been recognized as a very interesting active cancer drug targeting approach which can improve chemotherapy while keeping to a very minimum the systemic toxicity.

It could be further expected that more resistant cancer cells to antitumor molecules will be very sensitive to such nanomedicines. However, more advanced functionalization approaches are needed to prolong the *in vivo* residence time of the nanoplatforms, and to bypass the blood brain barrier to reach brain tumors. Even more, iron/ethylcellulose nanoparticles could be used to combine chemotherapy with hyperthermia and/or photodynamic therapy against cancer, and to increase the specificity and sensitivity of diagnostic tools.

Abbreviations

AMF	:	alternating magnetic field
AUC	:	area under the time-concentration curve
C_{max}	:	maximum plasma concentration
CO	:	carbon monoxide
EPR effect	:	enhanced permeability and retention effect
$Fe(CO)_5$	:	pentacarbonyl iron

HNO_3	:	nitric acid
LD_{50}	:	lethal dose 50%, median lethal dose
MRT	:	mean residence time
PBS	:	phosphate buffered saline
$t_{1/2}$	:	plasma half life
trans-EHMC	:	*trans*-2-ethylhexyl-*p*-methoxycinnamate
5-FU	:	5-fluorouracil

References

Arias, J.L., V. Gallardo, F. Linares-Molinero and A.V. Delgado. 2006. Preparation and characterization of carbonyl iron/poly(butylcyanoacrylate) core/shell nanoparticles. J. Colloid Interface Sci. 299: 599–607.

Arias, J.L., M. López-Viota, M.A. Ruiz, J. López-Viota and A.V. Delgado. 2007. Development of carbonyl iron/ethylcellulose core/shell nanoparticles for biomedical applications. Int. J. Pharm. 339: 237–245.

Arias, J.L., F. Linares-Molinero, V. Gallardo and A.V. Delgado. 2008. Study of carbonyl iron/poly(butylcyanoacrylate) (core/shell) particles as anticancer drug delivery systems. Loading and release properties. Eur. J. Pharm. Sci. 33: 252–261.

Arias, J.L., M. López-Viota, J. López-Viota and A.V. Delgado. 2009. Development of iron/ethylcellulose (core/shell) nanoparticles loaded with diclofenac sodium for arthritis treatment. Int. J. Pharm. 382: 270–276.

Arias, J.L. 2010. Drug targeting by magnetically responsive colloids. Nova Science Publishers Inc., New York, USA.

Arias, J.L., M. López-Viota, A.V. Delgado and M.A. Ruiz. 2010. Iron/ethylcellulose (core/shell) nanoplatform loaded with 5-fluorouracil for cancer targeting. Colloids Surf. B. Biointerfaces 77: 111–116.

Arias, J.L. 2011. Drug targeting strategies in cancer treatment: An overview. Mini-Rev. Med. Chem. 11: 1–17.

Balivada, S., R.S. Rachakatla, H. Wang, T.N. Samarakoon, R.K. Dani, M. Pyle, F.O. Kroh, B. Walker, X. Leaym, O.B. Koper, M. Tamura, V. Chikan, S.H. Bossmann and D.L. Troyer. 2010. A/C magnetic hyperthermia of melanoma mediated by iron(0)/iron oxide core/shell magnetic nanoparticles: a mouse study. BMC Cancer 30: 119.

Borisov, S.M., T. Mayr, G. Mistlberger, K. Waich, K. Koren, P. Chojnacki and I. Klimant. 2009. Precipitation as a simple and versatile method for preparation of optical nanochemosensors. Talanta 79: 1322–1330.

Calderó, G., M.J. García-Celma and C. Solans. 2011. Formation of polymeric nano-emulsions by a low-energy method and their use for nanoparticle preparation. J. Colloid Interface Sci. 353: 406–411.

Chua-anusorn, W., D.J. Macey, J. Webb, P. de la Motte Hall and T.G. St. Pierre. 1999. Effects of prolonged iron loading in the rat using both parenteral and dietary routes. Biometals 12: 103–113.

DeMerlis, C.C., D.R. Schoneker and J.F. Borzelleca. 2005. A subchronic toxicity study in rats and genotoxicity tests with an aqueous ethylcellulose dispersion. Food Chem. Toxicol. 43: 1355–1364.

Durán, J.D.G., J.L. Arias, V. Gallardo and A.V. Delgado. 2008. Magnetic colloids as drug vehicles. J. Pharm. Sci. 97: 2948–2983.

Gupta, A.K., and M. Gupta. 2005. Synthesis and surface engineering of iron oxide nanoparticles for biomedical applications. Biomaterials 26: 3995–4021.

Hasan, A.S., M. Socha, A. Lamprecht, F.E. Ghazouani, A. Sapin, M. Hoffman, P. Maincent and N. Ubrich. 2007. Effect of the microencapsulation of nanoparticles on the reduction of burst release. Int. J. Pharm. 344: 53–61.

He, W., Q. Du, D.Y. Cao, B. Xiang and L.F. Fan. 2008a. Study on colon-specific pectin/ethylcellulose film-coated 5-fluorouracil pellets in rats. Int. J. Pharm. 348: 35–45.

He, H., F. Ping, G. Chen and S. Zhang. 2008b. Chemoembolization of tongue carcinoma with ethylcellulose microcapsuled carboplatinum and its basic study. Artif. Cells Blood Substit. Immobil. Biotechnol. 36: 114–122.

Kato, T., R. Nemoto, H. Mori, M. Takahashi and M. Harada. 1981. Arterial chemoembolization with mitomycin C microcapsules in the treatment of primary or secondary carcinoma of the kidney, liver, bone and intrapelvic organs. Cancer 48: 674–680.

Kato, T., R. Nemoto, H. Mori, R. Abe, K. Unno, A. Goto, H. Murota, M. Harada and M. Homma. 1984. Magnetic microcapsules for targeted delivery of anticancer drugs. Appl. Biochem. Biotechnol. 10: 199–211.

Kato, T., K. Sato, R. Sasaki, H. Kakinuma and M. Moriyama. 1996. Targeted cancer chemotherapy with arterial microcapsule chemoembolization: Review of 1013 patients. Cancer Chemother. Pharmacol. 37: 289–296.

Kayal, S., and R.V. Ramanujan. 2010. Anti-cancer drug loaded iron-gold core-shell nanoparticles (Fe@Au) for magnetic drug targeting. J. Nanosci. Nanotechnol. 10: 5527–5539.

Lübbe, A.S., C. Bergemann, H. Riess, F. Schriever, P. Reichardt, K. Possinger, M. Matthias, B. Doerken, F. Herrmann, R. Guertler, P. Hohenberger, N. Haas, R. Sohr, B. Sander, A.J. Lemke, D. Ohlendorf, W. Huhnt and D. Huhn. 1996a. Clinical experiences with magnetic drug targeting: A phase I study with 4'-epidoxorubicin in 14 patients with advanced solid tumors. Cancer Res. 56: 4686–4693.

Lübbe, A.S., C. Bergemann, W. Huhnt, T. Fricke, H. Riess, J.W. Brock and D. Huhn. 1996b. Preclinical experiences with magnetic drug targeting: Tolerance and efficacy. Cancer Res. 56: 4694–4701.

Nakano, K., O. Ike, H. Wada, S. Hitomi, Y. Amano, I. Ogita, N. Nakai and K. Takada. 1997. Oral sustained-release cisplatin preparation for rats and mice. J. Pharm. Pharmacol. 49: 485–490.

Okamoto, Y., A. Konno, K. Togawa, T. Kato and Y. Amano. 1985. Microcapsule chemoembolization for head and neck cancer. Arch. Otorhinolaryngol. 242: 105–111.

Okon, E., D. Pouliquen, P. Okon, Z.V. Kovaleva, T.P. Stepanova, S.G. Lavit, B.N. Kudryavtsev and P. Jallet. 1994. Biodegradation of magnetite dextran nanoparticles in the rat. A histologic and biophysical study. Lab. Invest. 71: 895–903.

Perugini, P., S. Simeoni, S. Scalia, I. Genta, T. Modena, B. Conti and F. Pavanetto. 2002. Effect of nanoparticle encapsulation on the photostability of the sunscreen agent, 2-ethylhexyl-p-methoxycinnamate. Int. J. Pharm. 246: 37–45.
Reddy, L.H. 2005. Drug delivery to tumors: Recent strategies. J. Pharm. Pharmacol. 57: 1231–1242.
Rujivipat, S., and R. Bodmeier. 2010. Improved drug delivery to the lower intestinal tract with tablets compression-coated with enteric/nonenteric polymer powder blends. Eur. J. Pharm. Biopharm. 76: 486–492.
Schmidt, C., and R. Bodmeier. 1999. Incorporation of polymeric nanoparticles into solid dosage forms. J. Control. Release 57: 115–125.
Ubrich, N., P. Boulliot, C. Pellerin, M. Hoffman and P. Maincet. 2004. Preparation and characterization of propranolol hydrochloride nanoparticles: A comparative study. J. Control. Release 97: 291–300.
Vaghani, S., S. Vasanti, K. Chaturvedi, C.S. Satish and N.P. Jivani. 2010. Stomach-specific drug delivery of 5-fluorouracil using ethylcellulose floating microspheres. Pharm. Dev. Technol. 15: 154–161.
Vanderhoff, J.W., M.S. El-Aasser and J. Ugelstad. 1979. Polymer Emulsification Process. U.S. Patent No. 4,177,177, December 4, 1979.
Whittaker, P., S.F. Ali, S.F. Imam and V.C. Dunkel. 2002. Acute toxicity of carbonyl iron and sodium iron EDTA compared with ferrous sulfate in young rats. Regul. Toxicol. Pharmacol. 36: 280–286.
Yang, J., X.C. Ma, Z.J. Zou and S.L. Wei. 1995. Experimental maxillofacial arterial chemoembolization with encased-cisplatin ethylcellulose microspheres. AJNR Am. J. Neuroradiol. 16: 1037–1041.

8

Matrix Scaffolds in Cancer Cell Biology; Physiological Representation and Nanoscale Characterisation

Hsin-ya Chien,[1] *Sandra Fok,*[1] *Vanisri Raviraj,*[1] *Jiwon Lee,*[1] *Louise Cole,*[2] *J. Guy Lyons*[3] and *Lilian Soon*[1,a,*]

ABSTRACT

In vitro cultures such as 2-dimensional (D) cell culture, semi-3D, 3D, and *in vivo* mouse models represent sub-approximations to native conditions that had led to the development of successful translational applications. However, it is also becoming clear that not all *in vivo* situations are adequately represented and for a subset of patients, their disease conditions and corresponding therapeutic measures may not have been properly addressed. It is essential to match experimental approaches with *in vivo* environmental

[1]Australian Centre for Microscopy and Microanalysis (ACMM), Australian Microscopy & Microanalysis Research Facility (AMMRF); The University of Sydney, Australia.
[a]E-mail: lilian.soon@sydney.edu.au
[2]Advanced Microscopy Facility, Bosch Institute, The University of Sydney.
[3]Sydney Head & Neck Cancer Institute, Royal Prince Alfred Hospital, Sydney Cancer Centre, Sydney Medical School.
*Corresponding author

List of abbreviations after the text.

conditions for cellular studies. Some of these include the use of clustered cells called spheroids and single tumour cells embedded in matrices with differing densities and architecture. These configurations mimic, for example, desmoplastic tumours in breast cancers. Concomitantly, appropriate analytical tools that can accommodate samples in their native conditions and/or enable nanostructural studies are critical to evaluate the adherence of novel *in vitro* models to physiological conditions.

INTRODUCTION

Cancer has become more prevalent and diverse as human lifespans have increased. There is almost no natural recourse, making cancer the primary cause of death worldwide (WHO). Yet, we struggle to understand the most life-threatening aspects—metastasis, cancer recurrence and associated tumour dormancy. The vast majority of human cancers are carcinomas, which by definition derive from epithelial tissues. In tumour formation, normal epithelial cells acquire genetic and/or epigenetic changes that bypass normal check-points for cell division resulting in excessive cell proliferation. There may be further changes that mimic a developmental process called epithelial to mesenchymal transition (EMT) that allow tumour cells to migrate away from the primary tumour (Lafleur et al. 2006; Soon 2009). Metastasis is a late stage disease where secondary tumours grow at distant sites away from the primary tumour such as the lung and bones. This spread of cancer requires several steps including invasion into the stroma, intravasation, extravasation and growth at secondary sites (Soon et al. 2003; Liotta and Kohn 2001). There is also evidence to suggest that cell invasion occurs at an early stage in cancer well before metastasis sets in (Pantel et al. 2003).

In order to derive mechanisms of cell migration, a key process in cancer, it is imperative to have understanding of cell to substratum or matrix interactions. Escape from the primary tumour requires movement of tumour cells from the tumour mass and breaching of the basement membrane (BM) that borders the epithelium and stroma. The BM contains nanoscale structures that include a conglomeration of ridges and pores that represent crests and low-points in the surface topography. Present also are fibrils of ~80 nm that cells are able interact with using fine filopodia which act as sensing and traction instruments. Similarly, once inside the stroma and free from the constraints of the BM, tumour cells encounter nanoscale structures and topographical features that can dictate cell behaviour including migration and matrix remodelling.

This article discusses the *in vitro* and *in vivo* models of breast carcinoma (BrCa), the recent improvements made to simulate physiological conditions and corresponding new information derived and the nanoscale characterisation techniques that can be used to quantify the physical aspects of the models.

LIMITATIONS OF ANIMAL CELL AND TUMOUR MODELS

Animal disease models have been developed in order to evaluate mechanisms of gene function and drug targets. Syngenic mice are important tools that are being used to verify targets in preclinical trials. However, there are a number of differences between rodents and humans that can lead to misinterpretation of data when extrapolating from one species to another. These include the life spans, sizes and biology of the species. For example, tumour progression in mice is more rapid, where the accumulation of mutations occurs within eighteen months. Subversion of normal genetic programming by introducing oncogenes or carcinogens appears to be easier in rodents (Balmain and Harris 2000; Kim 2005). Whilst these differences are due to intrinsic factors of the species and are unavoidable, other differences are attributable to the design and technical attributes of the models and these can be controlled.

MAMMOGRAPHIC DENSITY, DESMOPLASIA AND BrCa DEVELOPMENT

A clear difference between animal tumour models and human cancers is the scant matrix material in the former. This is pertinent to tumours generated from the injection of tumour cells alone into a representative area such as in the mammary gland. These models are simple to generate but may not address one of the highest risk factors in the development of breast cancer—women within the highest classification of 60–75% overall dense tissues have up to a 6-fold increase BrCa risk compared to women in the lowest category with fatty tissue as the main breast constituent (Boyd et al. 2006). Dense tissues in early, pre-diagnosis mammograms are predisposed to developing ductal carcinomas *in situ* (DCIS) (Ursin et al. 2005). The predominant component of high-MD tissue is collagen (Alowami et al. 2003) and there is a stronger correlation between breast cancer and increased connective tissue density than with epithelia (Li et al. 2005).

A notable 3D model developed by Amatangelo et al. 2005, utilised fibroblasts from various stages of carcinogenesis to reconstitute matrices. These cell-free stromal fibroblast-derived matrices contained the necessary molecular and structural stimulants to induce desmoplastic differentiation of normal stromal cells (Amatangelo et al. 2005). A transgenic model of high-MD was also developed in mice with a type-1 collagen mutation in the $\alpha1$(I) chain that prevented collagen proteolysis (Provenzano et al. 2008). The mammary glands of the mice exhibited increased collagen deposition with signs of hyperplasia such as irregular epithelium boundaries. This elegant study confirms the importance of high-MD as an indicator of higher cancer risks. Furthermore, high matrix content causes increased matrix stiffness that can, in turn, stimulate the proliferation of normal and transformed cells leading to tumour progression (Paszek and Weaver 2004; Wozniak et al. 2003). Mammographically dense breast tissue contains elevated levels of stromal collagen whereby cancerous breast tissues are significantly stiffer than healthy tissue (Paszek and Weaver 2004). Young's modulus is used to measure the stiffness or resistance to deformation of collagen matrices, which are dependent on fibrillar cross-links and fibre densities (Lian et al. 1973; Wolf et al. 2009; Zaman et al. 2006). High matrix density and accompanying nanoscale parameters of decreased pore size and networked fibrillar organisation are important aspects of microenvironmental regulation of cell signalling and biology.

IN VITRO MODELS OF BrCa

Cell migration mechanisms. Tumour cell migration in three-dimensional (3D) matrices can be classified into broad categories known as collective, amoeboid and mesenchymal. These migration modes are dependent on environmental conditions and intrinsic cell properties such as polarity, adhesion and cytoskeletal functions. Tumour cells are able to adapt to variations in these factors and change migratory modes accordingly. In the lower spectrum of matrix densities, amoeboid-like tumour cells exhibit non-proteolytic migration regulated by Rho-associated coiled-coil forming kinase (ROCK)-regulated contractility or protrusion-led mechanisms (Lammermann et al. 2008; Sahai and Marshall 2003; Tchou-Wong et al. 2006). Mesenchymal migration incorporates matrix proteolysis and the cells adhere to the matrix via focal adhesions at the protrusions and rear of the cell (Friedl and Wolf 2003). Amoeboid- and mesenchymal-like cells might utilise proteolysis for migration depending on the density (Zaman et al. 2006), the presence of cross-links (Sabeh et al. 2004), and the fibrillar or non-fibrillar nature of the matrices (Wolf et al. 2007). Collective migration is the migration of cohort tumour cells that are adhered to one another via E-cadherin or occludin (Ilina and Friedl 2009). In this migration mode, the cells at the periphery of the cohort that are leading the migration tend to be polarised whilst the cells trailing behind maintain the appearance of epithelial cells (Ilina and Friedl 2009).

Cell invasion studies are mainly performed using 1–2 mg/ml matrix concentrations that represent the content of BMs that line most endothelia and epithelia. There are other physiological and pathophysiological extracellular matrices that constitute thickened and dense basement membranes such as those underlying the renal glomeruli and the airway epithelium of asthmatics (Davies et al. 1978; Roche et al. 1989) or dense material in fibrotic states (Thannickal and Loyd 2008), desmoplasia (Walker 2001) and high mammographically dense (HMD) tissues, about which little is known. How tumour cells respond to dense matrices is an area of study that requires more investigations.

Spheroids in Matrices and Epithelial-mesenchymal Transition

Multicellular tumour spheroids (MCTS) are spherical, heterogeneous aggregates of tumour cells composed of cells with different phenotypes, such as proliferating, quiescent, and necrotic cells (Kim 2005). The centre of spheroids is cut off from nutrient and oxygen supply and therefore has a tendency to become necrotic. A proportion of cells in the outer layers are thus proliferative whilst those in the core areas are quiescent or necrotic (Santini and Rainaldi 1999; Sutherland 1988).

Invasive carcinomas in human are complex 3D tissues with extensive cell-cell and cell-extracellular matrix (ECM) interactions. In conventional, monolayer or suspension cultures, the extent to which cell-cell and cell-extracellular matrix interactions can occur is limited whereas 3D spherical models resemble *in vivo* solid tumours more closely, as they preserve specific morphological and functional characteristics (Santini and Rainaldi 1999). In addition, MCTS also represent a more realistic model for proliferation dynamics. MCTS have been characterised with two growth phases: early exponential phase followed by a period of retarded growth (Santini and Rainaldi 1999). The advantages of 3D spheroid culture as a realistic model for *in vivo* solid tumours are clear. We propose that to mimic desmoplastic conditions accurately, the spheroids could be further embedded within dense matrices. This experimental model will be highly applicable for the evaluation of EMT involving the migration of tumour cells away from the primary tumour into the surrounding matrix.

Biophysical and Nanoscale Aspects of Dense Matrices

Spheroids resemble small tumour cell masses suitable for the examination of tumour biology *in vitro*. However, most studies observe free-floating spheroids in solution. In other studies where spheroids are embedded in gels, the matrix milieu usually represents a specific aspect of tumour biology, which is the basement membrane. Tumours comprise of cells and matrix material that present as rigid forms in palpitation examinations. Rigidity or stiffness indicates the degree by which tumours resist deformation. Conversely, elasticity or compliance dictates the degree by which tumours can be deformed. The elastic modulus of tumours was measured to be 4049 ± 938 Pa compared to 167 ± 31 Pa for normal mammary gland, and 175 ± 37 Pa for reconstituted basement membrane (Paszek and Weaver 2004). The latter is similar to the elastic modulus of collagen at 1–2 mg/ml, a concentration generally used to produce *in vitro* matrices. Studies performed using these low concentration matrices therefore recapitulate the breaching of basement membranes that line the epithelium as tumour cells cross the boundary into the extracellular matrix (ECM). However, once in the ECM, the cells may encounter conditions that are significantly more rigid but there are as yet few studies that explore cellular behaviour in dense matrices. (Paszek and Weaver 2004), determined that changes in matrix stiffness in small increments between 170 to 1200 Pa caused an increase in the size of tumour colonies, and produced other tumorigenic phenotypes such as loss of lumen, tissue disorganisation, translocation of β-catenin from adherens junctions to the cytoplasm and concomitant loss of polarity.

Although we are beginning to understand the complexities of cell signalling in response to biophysical traits associated with dense matrices, there is no conciliation with physical properties such as nanoscale topographic features. There is evidence that eutherian cells react to nanoscale features that affect biological processes such as cell adhesion that in turn, can regulate proliferation, migration and differentiation. Nanoscale ridges and grooves can increase the interfacial or contact area allowing cells to spread and divide more (Lim et al. 2005). Similarly, reducing the interfacial area can reduce cell size and proliferation rates (Lenhert et al. 2005).

NANOSCALE IMAGING OF NATIVE BREAST CANCER TISSUE

In order to understand the relationship between cancer cells and the stroma, light and electron microscopy studies were conducted using high mammographically dense (HMD) tissue. In the tissue, cohorts of tumour cells are surrounded by dense collagen fibrils as observed from second harmonic imaging and transmission electron microscopy (TEM) (Fig. 1). Under the multiphoton, collagen fibrils have a wavy appearance suggestive of a conformation that confers tensile strength to the structure. However, in cancer tissue, the ordered appearance of collagen is frequently disturbed (Fig. 1A). Furthermore at the light resolution, only large ~1 μm collagen fibrils are visible where visualisation of smaller fibrils is limited by diffraction (Fig. 1A). However, under the TEM, fibrils in the nanoscale range are easily distinguishable. The orientation of the fibrils can also be observed whereby cross-sectional views of fibrils are evident at the periphery of tumour cells (Fig. 1B).

Masson's trichrome staining of HMD matrices demonstrated infiltration of ductal carcinoma *in situ* (DCIS) cells mainly in regions consisting of fibrillar collagen (Fig. 2A). Several migration modes, as

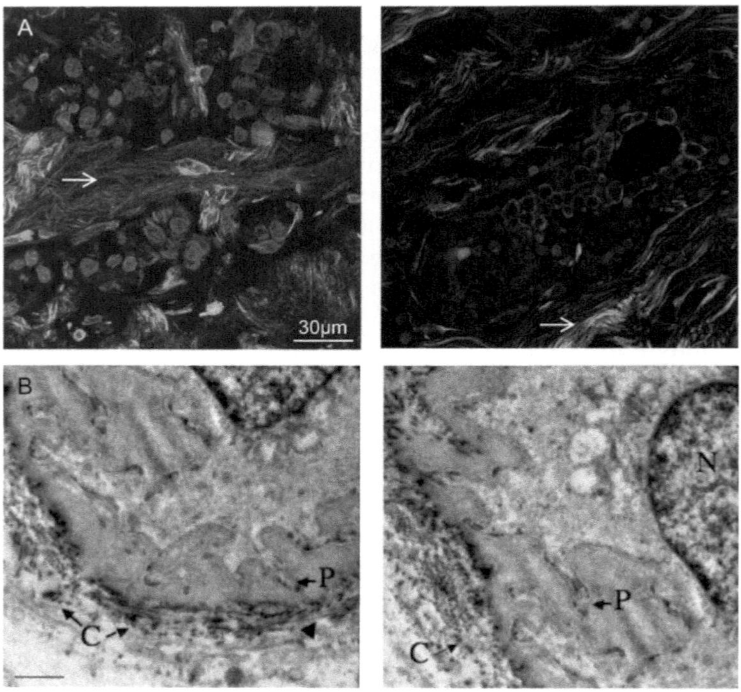

Fig. 1. Multiphoton, second harmonic and TEM imaging of tumour cells and collagen. A. Cohorts of tumour cells from the acini of the breast are stained by cytokeratin-19 antibodies (red). The tumour mass is surrounded by dense collagen areas imaged by second harmonic (cyan) (arrows). B. The protrusions (P) of tumour cells extend into the extracellular matrix and appear to be surrounded by collagen fibrils (C) that are mostly imaged in cross-section.

Color image of this figure appears in the color plate section at the end of the book.

deduced from cellular morphology and organisation were observed in the histological sections. Collective migration included cells migrating in a cohort fashion in isolated clusters with each cluster containing at least 6 cells (Friedl and Wolf 2003). Collective migrating cells generally occurred in regions less than 50 μm away from the primary tumour mass. The leading cells were elongated and polarised towards the dense collagen matrix, while the trailing cells were spherical in shape (Fig. 2A). Scanning electron microscopy (SEM) of HMD tissue showed that there were two organisations of collagen fibrils; mesh-like and large fibrillar bundles. The former consisted of networks of collagen fibrils, while the latter were composed of five or more intertwined collagen fibrils (Fig. 2B). Larger bundles measure from 200 nm to 500 nm in diameter.

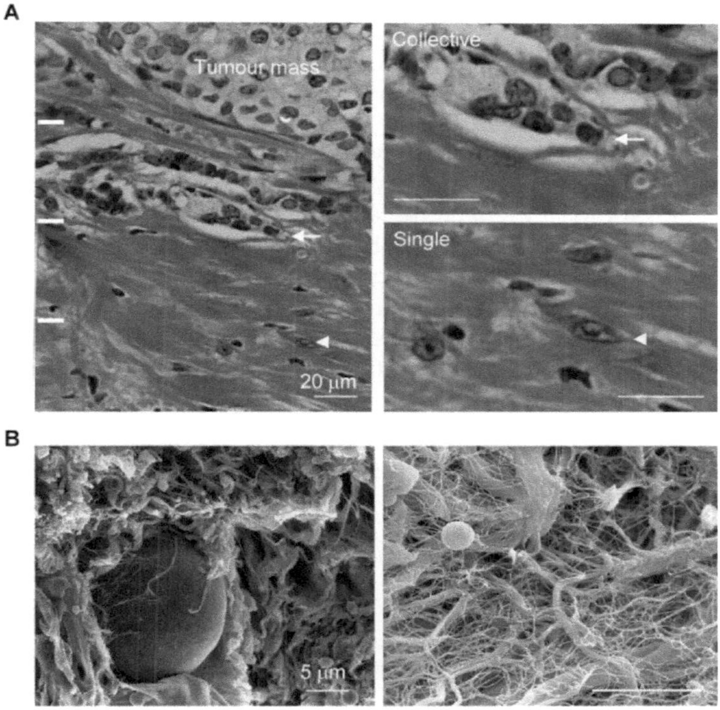

Fig. 2. Collagen fibrils surrounding invasive tumours in HMD matrices. A. Masson's Trichrome staining revealed infiltration of DCIS cells in HMD matrices occurring in regions abundant with fibrillar collagen (blue). Collective migratory cells occur close to the primary tumour mass (arrow). Distance between white markers on left = 50 μm. B. SEM demonstrating close apposition of fibrillar collagen and an epithelial cell. Mesh-like organisation of fibrillar collagen (blue arrow) and bundles of larger collagen fibrils are present throughout the HMD matrix.

Color image of this figure appears in the color plate section at the end of the book.

NANOSCALE CHARACTERISATION OF RECONSTITUTED HIGH-DENSITY MATRICES

We sought to derive high-density collagen matrix *in vitro* that match the density in HMD tissue. In mammograms, the brightest, confluent areas represent dense matrix/tumour regions. These dense fibroglandular HMD regions are macrodissected and assayed for collagen content. The high collagen concentration of HMD tissues is matched *in vitro*. We demonstrate that amoeboid-like tumour cells are able to migrate for long periods in matrices with collagen concentrations raised to 20 mg/ml, similar to that of HMD tissues. Cell behaviour and morphology have also been studied by live-cell imaging and multiphoton microscopy (unpublished data). The collagen structure and similarities to mammary gland matrices have been evaluated using environmental SEM (ESEM) and field emission scanning electron microscopy (FESEM) (see Nanoscale Characterisation) (Figs. 3–5). Experiments are underway to understand the plasticity in behaviour of tumour cells in dense matrices.

In cancer cell biology, the study of live processes is most frequently performed at the light microscopy level. In terms of resolution, light microscopy is approaching nanometre resolutions, however, superresolution techniques are currently more suitable for 2D and fixed sample imaging. On the other hand, electron microscopy has superior resolution but have certain limitations associated with sample preparation and viewing conditions. However, there have been developments to overcome these limitations such as the use of cryofixation and cryoimaging to preserve

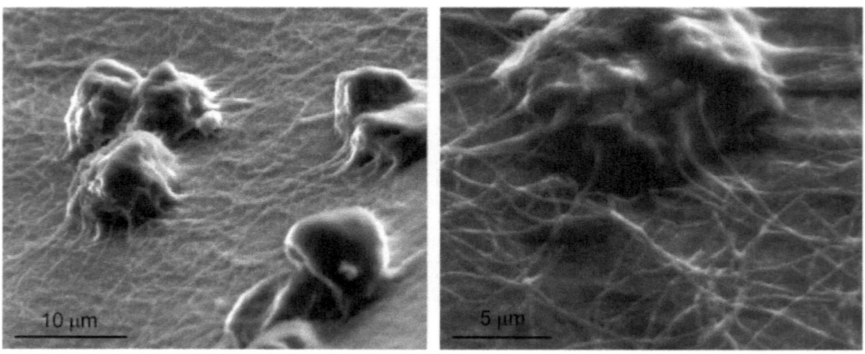

Fig. 3. Imaging tumour cells in collagen matrix using ESEM. Breast carcinoma cells were allowed to migrate into dense matrices for 24 hours prior to wet imaging. Scanning electron images were taken with FEI Quanta 200 3D (Fei Company, Eindhoven, Netherlands) collecting secondary electrons with ESEM mode and gaseous secondary electron detector (GSED) using 8 kV, current 0.21 nA, spot size 3, working distance 6.5 or 6.6 mm and tilting 29°.

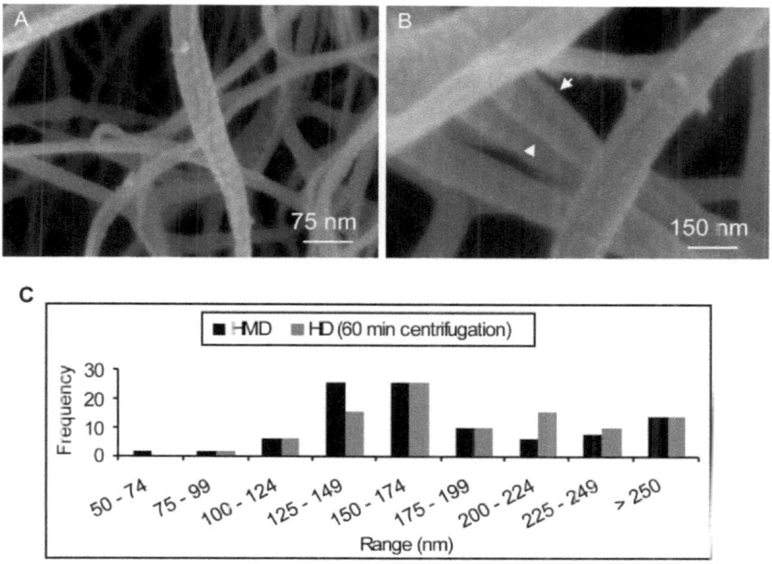

Fig. 4. A, B. FESEM images illustrate the presence of 65 nm D spacings characteristic of collagen fibrils that are normally visible only under the transmission electron microscope. The D-spacings are aligned in fibrils that are intertwined into larger fibrils in right-handed helical twists. **C.** Measurements of collagen fibril diameter show that the majority of high-density collagen fibrils (HD) (60 min centrifugation) and high mammographically dense (HMD) matrices lie between 125–225 nm. HMD mean: 156.98 ± 61.69 nm; HD mean: 167.84 ± 60.02 nm; n = 100.

native states. Here, we will review technological advancements associated with scanning electron microscopy (SEM) that now allow imaging under wet and low vacuum conditions, and resolution that approaches that of transmission electron microscopy.

In SEM, a focused electron beam is guided across the sample in a raster pattern. The electrons interact with atoms near the surface of the sample, imparting a small amount of energy that ionises one or more electrons in an atom. The emerging electrons are the secondary electrons that form the image of the sample. The wavelength of electrons is significantly smaller than that of light, which enables greater resolutions at routine nanoscale levels to be obtained during imaging. Light microscopy operates at a resolution limit of about 200 nm whilst a beam of electrons with a wavelength of 1 nm can be used to resolve ultrastructural details.

SEM traditionally involves imaging under high vacuum using dehydrated and metal-coated samples. Sample preparation is decidedly harsh involving the use of aldehyde-based, cross-linking fixatives, alcohols or acetone and sputter-coating with a layer of metallic substance that might obscure small surface features. High-vacuum conditions in

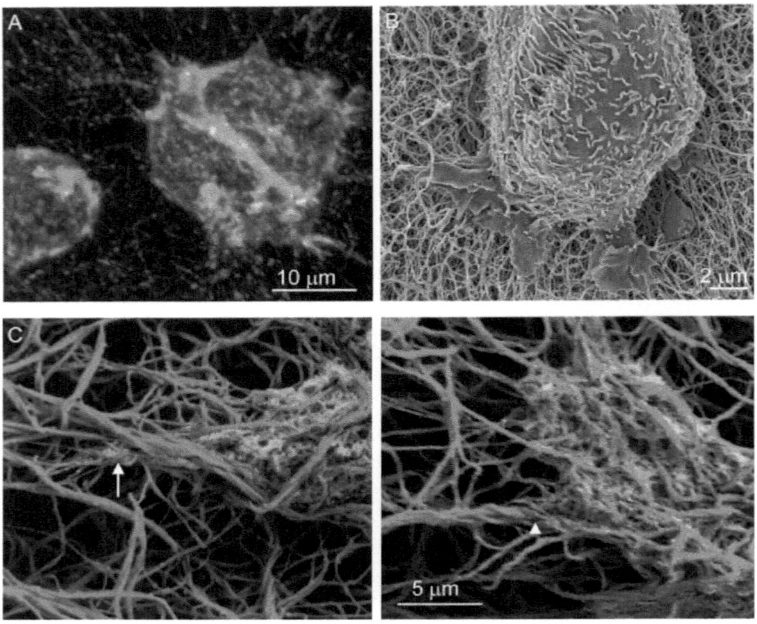

Fig. 5. Confocal and reflection microscopies, and FESEM, illustrating cell-matrix interactions. MTLn3 breast carcinoma cells are seeded onto dense collagen matrices for 48 hours prior to fixation and treatment. **A.** Tumour cells staining for phalloidin actin-Alexa 488 (green) have migrated into the matrices (red). Tumour cell fluorescence was captured by confocal microscopy and the matrix fibrils were imaged using reflection microscopy. **B.** A low resolution FESEM image illustrating a tumour cell extending protrusions into collagen matrix. **C.** Higher resolution FESEM images demonstrate the interactions between the protrusions with matrix fibrils. Arrow shows how the protrusion is enmeshed within tight fibrillar networks. Images have been pseudocoloured using Photoshop according to differences in the intensity of objects.

Color image of this figure appears in the color plate section at the end of the book.

the path between the source of electrons and the sample serve to prevent the scatter of electrons. Fixation and dehydration are necessary whereby fixation stabilises the samples and dehydration prevents the escape of water molecules that might contaminate the column. The application of metal coats to non-conductive, biological samples serves several purposes. It prevents the accumulation of charge on the sample in the insulating vacuum conditions of the microscope. In addition, metal coatings further stabilise the samples against the electron beam and increases contrast in the images. Therefore, conventional SEM imaging has limitations imposed by the necessity to render samples tolerant of vacuum conditions and for the column to accommodate organic-based materials that are normally hydrated, fragile and non-conductive. The greatest limitation for the biologist is the possible introduction of artefacts such as shrinkage and structural changes to the samples including reducing resolutions from the coatings.

Environmental SEM (ESEM) and imaging under wet conditions. ESEM is a technique which introduces low vacuum, gaseous conditions for imaging (Stokes 2003). This technology entails the creation of different vacuum zones whereby the high vacuum column (10^{-5} to 10^{-9} Torr) is more or less isolated from the low-pressure gaseous sample chamber (~50 Torr). This is achieved by introducing at least two closely spaced pressure limiting apertures (Staple et al. 2009) to create an intermediate vacuum zone between the column and the sample chamber. The apertures are small enough to allow electrons to pass from the column into the sample chamber but restrict the escape of gases into the column. Furthermore, gas in the intermediate zone is pumped away to significantly reduce any gas or vacuum leaks between the column and the sample chamber. There is a separate pump that introduces water vapour into the sample chamber. The secondary electrons released from the sample collide with the water molecules that in turn release more secondary electrons that react with other water molecules. The result is a cascade of amplified signals that are collected by the gaseous secondary electron detector and transmitted directly to an electronic amplifier (Danilatos 1991). The gaseous secondary electron detector has a positive bias of a few hundred volts that attracts the secondary electrons and repels the now, positively charged water vapour molecules, having lost secondary electrons. The water vapour ions propagate towards the sample, neutralising the negatively charged sample upon impact (Stokes et al. 2000). Thus, there is an added advantage to gaseous secondary electron detection in the avoidance of charging effects or the accumulation of negative ions commonly observed in conventional SEM imaging.

The ability to image in wet conditions means that it is no longer necessary to alter non-conductive samples to make them conductive, eliminating the need for desiccation and coating samples. This allows imaging at ultraresolutions under near native conditions. ESEM has been used to observe cellular organisation and differentiation status in tissue engineering studies at low resolutions (Baguneid et al. 2004). However, for higher resolution imaging, biological samples pose intrinsic difficulties whereby the environmental water vapour pressures need to be compatible with hydrated samples. Water loss at particular temperatures and osmotic pressures of the samples need to be considered when optimising external conditions for imaging (Muscariello et al. 2005; Stokes et al. 2003). Stokes et al. (2003) showed empirically that eukaryotic cells tolerate sub-saturation conditions at an equilibrium humidity of up to 25% below saturation relative to pure water. The imaging conditions prescribed utilise chamber pressures of 2.8–2.9 torr of water vapour, at 3°C, 50% humidity and intermediate voltage of 10 keV. This has allowed the visualisation of unfixed cells demonstrating the presence of nuclei, nucleolus and plasma membranes. ESEM was also utilised to perform qualitative analyses of the pore arrangements in collagen matrices (Madaghiele et al. 2008).

In high-density matrices, ESEM was useful to demonstrate interactions between cell protrusions and the substratum in hydrated conditions (Fig. 3) and for comparisons with more traditional SEM preparation and imaging conditions (Figs. 2, 3). Under ESEM, the cancer cells appear less collapsed compared to dehydrated samples and the contours of the cells are smoother with less demarcation between the cell body and protrusions. The similarities between SEM and ESEM images are the types of interactions observed between the protrusions and collagen fibrils including intertwining of fibrils with protrusions and draping of fibrils over the cell body (Fig. 3). While resolving to a higher degree than light microscopy and allowing imaging under hydrated conditions, the resolution of the ESEM falls short of other SEM techniques that allow true nanoscale imaging of fine structural features.

Field Emission SEM (FESEM) and nanoscale imaging. The FESEM was developed to improve the resolution of conventional SEMs which now reaches that of transmission electron microscopy (Figs. 4, 5). Signals collected in SEM images are masked by noise so that there is a certain amount of error in the relative brightness of each pixel. This means that contrast or differences in intensities between pixels is compromised (Pawley 1997). The large interaction volumes occurring in low-density, uncoated or thinly-coated biological samples exacerbates this condition. Signal to noise ratios can be improved by lowering the kV to restrict the interaction volume but this comes at the expense of brightness and beam current. This is because thermal tungsten electron sources in conventional SEMs have low brightness under low kV conditions; lowering the voltage from 20 to 1.5 kV to achieve greater contrast is often not feasible due to the significant drop in brightness (Pawley 1997). Low voltages also increase the spread of electron energies, which dominates the probe size and limits the image resolution.

Therefore, low voltage imaging needs to be accompanied by improvements in electron sources (Hefter 1987; Mullerova and Frank 2004). These include the development of the cold-FE and hot- or Schottky-FE electron sources (Orloff 1985). The latter however, provide greater beam stability, is less prone to tip noise and has higher emission currents compared with cold-FE sources. Electrons are extracted directly by an extremely high electrical field. A FE source is a fine needle creating a high-energy field at the tip. The emission area is therefore small and can easily be controlled. These sources are designed to have high current capabilities and low energy spread enabling bright and highly resolved images to be taken.

Other improvements include the design of an immersion lens that allows placement of the specimen in the lens field, reducing the focal length and increasing resolution. In the immersion lens system, a beam booster accelerates electrons in the column. At the end of the column, a retarding field decelerates the electrons. This field serves not only as the electromagnetic lens for the incident beam but also collects secondary electrons, accelerating these to the detector (Frosien et al. 1989; Plies et al. 1999). Furthermore, the in-lens detector located inside the column improves collection of electrons over out of column detectors (Homma et al. 2004). FESEM is now widely used in a number of biomaterials applications (Fang et al. 2010; Zhang et al. 2010) and is a promising technology for application to studying cells and ECM in experimental models of cancer.

Reconstituting dense matrices that simulated the collagen concentrations of HMD tissue was a novel undertaking. It was important to know whether at 20 mg/cm^3 matrix density, the collagen nanoscale structure was similar to those of native tissue. Using FESEM, we were able to visualise the ~60 nm D spacings of individual collagen fibrils. Furthermore, these dark and light intercalating bands are aligned with respect to those present in adjacent fibrils where several fibrils are intertwined together in a right-handed helical twist, generating a thicker fibril (Figs. 4A, B). Normally, D-spacings are only visible at the TEM level. The ability of the FESEM to reach nanoscale imaging of TEMs is an advantage because of topographic details that are intrinsically available to the scanning technology. Other quantitative analyses are also possible. FESEM images allowed quantitation of collagen fibril sizes showing close alignment between those measured from HMD tissue with those from *in vitro* prepared gels, where most fibrils fall between the range of 150–174 nm in size (Fig. 4C). Finally at the cellular level, FESEM illustrated interactions between cell protrusions and collagen fibrils where protrusions are observed to insert into networks of fine collagen fibrils of nanometres in diameter (Fig. 5).

Summary

The bulk of our knowledge on cell signalling and complex cell behaviours such as sensing the external environment leading to directional cell migration and cell differentiation, proliferation and apoptosis are derived from 2D culture studies (DesMarais et al. 2009; Lapidus et al. 2007; Mouneimne et al. 2004; Soon et al. 2005; Soon 2007; Yen and Soon

2009). Further verification might be performed using 3D culture models and animal studies with variable success. The primary reason is that the models may not be directly relatable to one another. Another concern is how representative these are to physiological and pathophysiological conditions. We discuss models that best represent desmoplastic breast cancer tumours such as the study of spheroid and single tumour cells in dense collagen matrices. Finally, imaging techniques are described that can be utilised to evaluate the fidelity of the *in vitro* scaffolds to *in vivo* dense matrices and cell-scaffold interactions at nanometre resolutions and at the nanoscale range.

Acknowledgements

This work was supported by the NHMRC (#402510) and ARC (#DP0881012). The authors also acknowledge the facilities as well as scientific and technical assistance from staff, particularly Mr Steve Moody, Dr Patrick Trimby and Mr Dennis Dwarte in the AMMRF (Australian Microscopy & Microanalysis Research Facility) at the Australian Centre for Microscopy & Microanalysis, The University of Sydney.

References

Alowami, S., S. Troup, S. Al-Haddad, I. Kirkpatrick and P.H. Watson. 2003. Mammographic density is related to stroma and stromal proteoglycan expression. Breast Cancer Research. v. 5: R129–R135.

Amatangelo, M.D., D.E. Bassi, A.J.P. Klein-Szanto and E. Cukierman. 2005. Stroma-derived three-dimensional matrices are necessary and sufficient to promote desmoplastic differentiation of normal fibroblasts. Am. J. Pathol. v. 167: 475–488.

Baguneid, M., D. Murray, H.J. Salacinski, B. Fuller, G. Hamilton, M. Walker and A.M. Seifalian. 2004. Shear-stress preconditioning and tissue-engineering-based paradigms for generating arterial substitutes. Biotechnol. Appl. Biochem. v. 39: 151–157.

Balmain, A., and C.C. Harris. 2000. Carcinogenesis in mouse and human cells: parallels and paradoxes. Carcinogenesis. v. 21: 371–377.

Danilatos, G.D. 1991. Review and outline of environmental SEM at present. Journal of Microscopy-Oxford. v. 162: 391–402.

Davies, M., A.J. Barrett, J. Travis, E. Sanders and G.A. Coles. 1978. Degradation of human glomerular basement-membrane with purified lysosomal proteinases—evidence for pathogenic role of polymorphonuclear leukocyte in glomerulonephritis. Clin. Sci. Mol. Med. v. 54: 233–240.

DesMarais, V., H. Yamaguchi, M. Oser, L. Soon, G. Mouneimne, C. Sarmiento, R. Eddy and J. Condeelis. 2009. N-WASP and Cortactin Are Involved In Invadopodium-Dependent Chemotaxis to EGF In Breast Tumor Cells. Cell Motil. Cytoskeleton. v. 66: 303–316.

Fang, R., E.W. Zhang, L. Xu and S.C. Wei. 2010. Electrospun PCL/PLA/HA Based Nanofibers as Scaffold for Osteoblast-Like Cells. J. Nanosci. Nanotechnol. v. 10: 7747–7751.

Friedl, P., and K. Wolf. 2003. Tumour-cell invasion and migration: diversity and escape mechanisms. Nat. Rev. Cancer v. 3: 362–74.

Frosien, J., E. Plies and K. Anger. 1989. Compound magnetic and electrostatic lenses for low-voltage applications. J. Vac. Sci. Technol. B. v. 7: 1874–1877.

Hefter, J. 1987. Morphological characterization of materials using low-voltage scanning electron-microscopy (LVSEM). Scanning Microscopy. v. 1: 13–21.

Homma, Y., S. Suzuki, Y. Kobayashi, M. Nagase and D. Takagi. 2004. Mechanism of bright selective imaging of single-walled carbon nanotubes on insulators by scanning electron microscopy. Appl. Phys. Lett. v. 84: 1750–1752.

Ilina, O., and P. Friedl. 2009. Mechanisms of collective cell migration at a glance. J. Cell Sci. v. 122: 3203–3208.

Kim, J.B. 2005. Three-dimensional tissue culture models in cancer biology. Semin. Cancer Biol. v. 15: 365–377.

Lafleur, M.A., F.A. Mercuri, N. Ruangpanit, M. Seiki, H. Sato and E.W. Thompson. 2006. Type I collagen abrogates the clathrin-mediated internalization of membrane type 1 matrix metalloproteinase (MT1-MMP) via the MT1-MMP hemopexin domain. J. Biol. Chem. v. 281: 6826–40.

Lammermann, T., B.L. Bader, S.J. Monkley, T. Worbs, R. Wedlich-Soldner, K. Hirsch, M. Keller, R. Forster, D.R. Critchley, R. Fassler and M. Sixt. 2008. Rapid leukocyte migration by integrin-independent flowing and squeezing. Nature. v. 453: 51–55.

Lapidus, K., J. Wyckoff, G. Mouneimne, M. Lorenz, L. Soon, J.S. Condeelis and R.H. Singer. 2007. ZBP1 enhances cell polarity and reduces chemotaxis. J. Cell Sci. v. 120: 3173–3178.

Lenhert, S., M.B. Meier, U. Meyer, L. Chi and H.P. Wiesmann. 2005. Osteoblast alignment, elongation and migration on grooved polystyrene surfaces patterned by Langmuir-Blodgett lithography. Biomaterials. v. 26: 563–70.

Li, T., L. Sun, N. Miller, T. Nicklee, J. Woo, L. Hulse-Smith, M.-S. Tsao, R. Khokha, L. Martin and N. Boyd. 2005. The association of measured breast tissue characteristics with mammographic density and other risk factors for breast cancer. Cancer Epidemiol. Biomarkers Prev. v. 14: 343–9.

Lian, J.B., S. Morris, B. Faris, J. Albright and C. Franzblau. 1973. The effects of acetic acid and pepsin on the crosslinkages and ultrastructure of corneal collagen. Biochimica et Biophysica Acta (BBA)—Protein Structure. v. 328: 193–204.

Liotta, L.A., and E.C. Kohn. 2001. The microenvironment of the tumour-host interface. Nature. v. 411: 375–379.

Madaghiele, M., A. Sannino, I.V. Yannas and M. Spector. 2008. Collagen-based matrices with axially oriented pores. J. Biomed. Mater. Res. A. v. 85A: 757–767.

Mouneimne, G., L. Soon, V. DesMarais, M. Sidani, X. Song, S.-C. Yip, M. Ghosh, R. Eddy, J.M. Backer and J. Condeelis. 2004. Phospholipase C and cofilin are required for carcinoma cell directionality in response to EGF stimulation. J. Cell Biol. v. 166: 697–708.

Mullerova, I., and L. Frank. 2004. Contrast at very low energies of the gold/carbon specimen for resolution testing. Scanning. v. 26: 18–24.

Muscariello, L., F. Rosso, G. Marino, A. Giordano, M. Barbarisi, G. Cafiero and A. Barbarisi. 2005. A critical overview of ESEM applications in the biological field. J. Cell Physiol. v. 205: 328–334.
Orloff, J. 1985. Thermal field-emission for low-voltage scanning electron-microscopy. J. Microsc.. v. 140: 303–311.
Pantel, K., V. Muller, M. Auer, N. Nusser, N. Harbeck and S. Braun. 2003. Detection and clinical implications of early systemic tumor cell dissemination in breast cancer. Clin. Cancer Res. v. 9: 6326–6334.
Paszek, M.J., and V.M. Weaver. 2004. The tension mounts: Mechanics meets morphogenesis and malignancy. J. Mammary Gland Biol. Neoplasia. v. 9: 325–342.
Pawley, J. 1997. The development of field-emission scanning electron microscopy for imaging biological surfaces. Scanning. v. 19: 324–336.
Plies, E., B. Degel, A. Hayn, G. Knell, J. Neumann and B. Schiebel. 1999. Experimental results using a "low-voltage booster" in a conventional SEM. Nuclear Instruments & Methods in Physics Research Section a-Accelerators Spectrometers Detectors and Associated Equipment. v. 427: 126–130.
Provenzano, P.P., D.R. Inman, K.W. Eliceiri, J.G. Knittel, L. Yan, C.T. Rueden, J.G. White and P.J. Keely. 2008. Collagen density promotes mammary tumor initiation and progression. BMC. Med. v. 6: 11.
Roche, W.R., R. Beasley, J.H. Williams and S.T. Holgate. 1989. Subepithelial fibrosis in the bronchi of asthmatics. Lancet. v. 1: 520–524.
Sabeh, F., I. Ota, K. Holmbeck, H. Birkedal-Hansen, P. Soloway, M. Balbin, C. Lopez-Otin, S. Shapiro, M. Inada, S. Krane, E. Allen, D. Chung and S.J. Weiss. 2004. Tumor cell traffic through the extracellular matrix is controlled by the membrane-anchored collagenase MT1-MMP. J. Cell Biol. v. 167: 769–781.
Sahai, E., and C.J. Marshall. 2003. Differing modes of tumour cell invasion have distinct requirements for Rho/ROCK signalling and extracellular proteolysis. Nat. Cell Biol. v. 5: 711–9.
Santini, M.T.M., and G.G. Rainaldi. 1999. Three-dimensional spheroid model in tumor biology. . Pathobiology: Journal of Immunopathology, Mol. Cell. Biol. v. 67: 148–157.
Soon, L., G. Mouneimne, J. Segall, J. Wyckoff and J. Condeelis. 2005. Description and characterization of a chamber for viewing and quantifying cancer cell chemotaxis. Cell Motil. Cytoskeleton. v. 62: 27–34.
Soon, L., A. Tachtsidis, S. Fok, E.D. Williams, D.F. Newgreen and E.W. Thompson . The continuum of epithelial mesenchymal transition—implication of hybrid states for migration and survival in development and cancer. pp. 117–130. In: Lyden, D., D.R. Welch and B. Psaila [eds.]. 2009. Cancer Metastasis: Biologic Basis and Therapeutics. Cambridge Press, New York. USA.
Soon, L.L. 2007. A discourse on cancer cell chemotaxis: where to from here? IUBMB Life. v. 59: 60–67.
Soon, L.L., T.A. Yie, A. Shvarts, A.J. Levine, F. Su and K.M. Tchou-Wong. 2003. Overexpression of WISP-1 down-regulated motility and invasion of lung cancer cells through inhibition of Rac activation. J. Biol. Chem. v. 278: 11465–11470.
Staple, D.B., M. Loparic, H.J. Kreuzer and L. Kreplak. 2009. Stretching, unfolding, and deforming protein filaments adsorbed at solid-liquid interfaces using the tip of an atomic-force microscope. Phys. Rev. Lett. v. 102: 128302.

Stokes, D.J. 2003. Recent advances in electron imaging, image interpretation and applications: environmental scanning electron microscopy. Philosophical Transactions of the Royal Society of London Series a-Mathematical Physical and Engineering Sciences. v. 361: 2771–2787.

Stokes, D.J., S.M. Rea, S.M. Best and W. Bonfield. 2003. Electron microscopy of mammalian cells in the absence of fixing, freezing, dehydration, or specimen coating. Scanning. v. 25: 181–184.

Stokes, D.J., B.L. Thiel and A.M. Donald. 2000. Dynamic secondary electron contrast effects in liquid systems studied by environmental scanning electron microscopy. Scanning. v. 22: 357–365.

Sutherland, R. 1988. Cell and environment interactions in tumor microregions: the multicell spheroid model. Science. v. 240: 177–184.

Tchou-Wong, K.M., S.Y.Y. Fok, J.S. Rubin, F. Pixley, J. Condeelis, F. Braet, W. Rom and L.L. Soon. 2006. Rapid chemokinetic movement and the invasive potential of lung cancer cells; a functional molecular study. BMC Cancer. v. 6: 1–12.

Thannickal, V.J., and J.E. Loyd. 2008. Idiopathic pulmonary fibrosis - A disorder of lung regeneration? American Journal of Respiratory and Critical Care Medicine. v. 178: 663–665.

Ursin, G., L. Hovanessian-Larsen, Y.R. Parisky, M.C. Pike and A.H. Wu. 2005. Greatly increased occurrence of breast cancers in areas of mammographically dense tissue. Breast Cancer Research. v. 7: R605–R608.

Walker, R.A. 2001. The complexities of breast cancer desmoplasia. Breast Cancer Research. v. 3: 143–145.

Wolf, K., S. Alexander, V. Schacht, L.M. Coussens, U.H. von Andrian, J. van Rheenen, E. Deryugina and P. Friedl. 2009. Collagen-based cell migration models *in vitro* and *in vivo*. Semin. Cell Dev. Bio. v. 20: 931–941.

Wolf, K., Y.I. Wu, Y. Liu, J. Geiger, E. Tam, C. Overall, M.S. Stack and P. Friedl. 2007. Multi-step pericellular proteolysis controls the transition from individual to collective cancer cell invasion. Nat. Cell Biol. v. 9: 893–904.

Wozniak, M.A., R. Desai, P.A. Solski, C.J. Der and P.J. Keely. 2003. ROCK-generated contractility regulates breast epithelial cell differentiation in response to the physical properties of a three-dimensional collagen matrix. J. Cell Biol. v. 163: 583–595.

Yen, C.C., and L.L.L. Soon. 2009. Simulating Sharp Gradients for Short-term, Ca2+ Transients and Long-term Chemotaxis in Cancer Cells. Tech. Canc. Res. Treat. v. 8: 241–247.

Zaman, M.H., L.M. Trapani, A.L. Sieminski, A. Siemeski, D. Mackellar, H. Gong, R.D. Kamm, A. Wells, D.A. Lauffenburger and P. Matsudaira. 2006. Migration of tumor cells in 3D matrices is governed by matrix stiffness along with cell-matrix adhesion and proteolysis. Proc. Natl. Acad. Sci. USA. v. 103: 10889–10894.

Zhang, Y.Z., V.J. Reddy, S.Y. Wong, X. Li, B. Su, S. Ramakrishna and C.T. Lim. 2010. Enhanced Biomineralization in Osteoblasts on a Novel Electrospun Biocomposite Nanofibrous Substrate of Hydroxyapatite/Collagen/Chitosan. Tissue Engineering Part A. v. 16: 1949–1960.

9

Fluorescence Nanoprobe Imaging Tumor by Sensing the Acidic Microenvironment

Lu Wang[1,a] and *Cong Li*[1,b,*]

ABSTRACT

Visualizing tumors in their early stages is crucial to increase the survival rate of cancer patients. Optical imaging, as a developing imaging modality, proves to be promising in visualizing tumors non-invasively *in vivo* due to its high sensitivity, no radioactive irradiation and low running cost. The development of fluorescence probes with high tumor targeting specificity will accelerate the translation of optical imaging from the bench to bedside. Compared to the small molecular imaging probes, the tumor targeting nanoprobes demonstrate the advantages, such as the tunable circulation lifetime, the up-regulated tumor accumulation *via* the enhanced permeability and retention (EPR) effect, and improved targeting specificity/sensitivity by labeling multiple targeting/imaging domains on a single nanoparticle. Tumor microenvironment (TME) indicators such as acidosis,

[1]Key Laboratory of Smart Drug Delivery, Ministry of Education & PLA, School of Pharmacy, Fudan University, 826 Zhangheng Rd., Shanghai 201203, China.
[a]E-mail: wldhr1988@gmail.com
[b]E-mail: congli@fudan.edu.cn
*Corresponding author

List of abbreviations after the text.

hypoxia and high interstitial pressure are universal phenomena of solid tumors. Therefore, fluorescent nanoprobes that respond to TME, especially the physiologically acidic pH environment, are promising in visualizing tumor and metastases regardless of the tumor types or even developmental stages. The signal of these pH "responsive" nanoprobes stays "silent" at the normal tissues, but it enhances significantly in the tumor acidic microenvironment. Therefore the signal ratio between the tumor and surrounding normal tissues can increase remarkably, which benefits the visualization of early stage/small volume tumors with high sensitivity. Here, we have summarized the recent advances in the development of fluorescence nanoprobes for tumor visualization by sensing the tumor acidic microenvironment.

INTRODUCTION

More than 70 years ago, Otto Warburg found that while approximately 10% of adenosine triphosphates (ATPs) in the normal cells are generated from glycolysis, this number increases to above 50% in cancer cells (Warburg 1956). Compared to the mitochondria involved aerobic metabolic pathway in which 36 ATPs are generated per glucose, the glycolysis pathway only produces two ATPs per glucose. The reliance of cancer cells on low efficient glycolysis to obtain energy can be explained by the poorly formed tumor vasculatures that lead the large part of a tumor under a hypoxic condition (low levels of O_2) (Denko 2008). To survive in such a hostile physiological environment, tumor cells adapt to the glycolysis pathway for energy production and therefore much more glucose is consumed than normal cells because of the low efficiency of glycolysis. The main products of the glycolysis—lactate and proton—are pumped out of the cytoplasm and lead to the acidification of the tumor extracellular space (pH 6.2–6.9) compared to the normal tissues (pH 7.4) (Fig. 1). Therefore, the acidic extracellular pH (pHe) is the universal characteristic of solid tumors regardless of tumor types or even development stages (Zhang et al. 2010).

Currently, most imaging probes determine the targeting specificity by recognizing the tumor-associated receptors over-expressed in the cancer cells. However, due to the unpredictable expression levels of these receptors in different tumor subtypes, the tumor specificity of these probes *in vivo* is much less satisfactory than its performance *in vitro*. Compared to the limited tumor-associated receptors on the cancer cell, the acidic tumor interstitium is the universal phenomenon of solid tumors (Zhang et al. 2010) and is much easier to access. Therefore, imaging probes that are responsive to the tumor acidic microenvironment are promising for

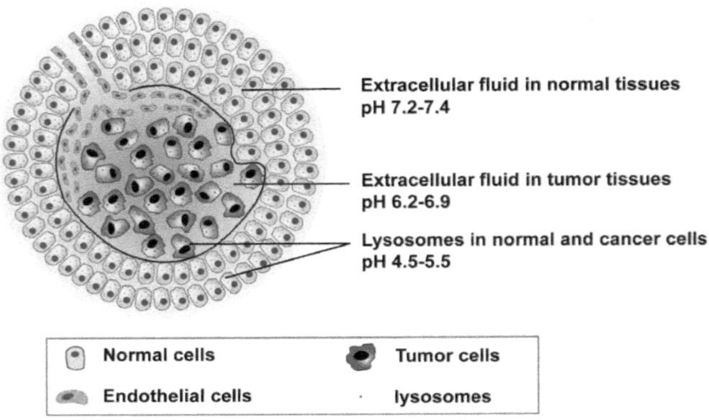

Fig. 1. A schematic of the acidic compartment distribution in normal and tumor tissues. The acidic compartments include the extracellular fluid in the tumor (pH 6.2–6.9), and the lysosomal luments (pH 4.5–5.5) in the normal as well as cancer cells. The pH of the blood, extracellular fluid of the normal tissues and cytosol in both normal and cancer cells stay in the neutral range (7.2–7.4).

Color image of this figure appears in the color plate section at the end of the book.

the visualization of tumors and metastases regardless of tumor types or development stages. Furthermore, the signal output of the conventional receptor-targeting probes usually remains at a high level, which results in a high background signal in normal tissue due to the non-specific binding. In contrast, the signal of the pH responsive probes stays at a low level in the normal tissues, but it increases significantly in the tumor acidic micro-environment, which greatly increases the signal ratio between the tumor and normal tissues (T/N ratio) and helps to detect the small volume tumor/metastases with improved sensitivity (Fig. 2).

Aside from the tumor extracellular fluid, lysosomes as cell organelles provide an acidic environment (pH 4.5–5.5) to help the proteases, such as cathepthins to degrade excess or worn out organelles, food particles, and engulfed viruses or bacteria. Previous works showed that a lot of receptor targeting probes were internalized into the cytoplasm *via* the receptor-mediated endocytosis following the delivery into lysosomes (Le and Wrana 2005). Due to the high expression level of the tumor associated receptors on the cancer cells, the uptake and lysosomal delivery of the probes into cancer cells are more efficient than that in normal cells. Therefore, imaging the acidic lysosomal lumen of cancer cells provides an additional opportunity to visualize the tumors in their early stages (Fig. 3).

Compared to magnetic resonance imaging (MRI) and position emission tomography (PET), optical imaging as a developing imaging modality shows advantages including high sensitivity, no radioactive irradiation,

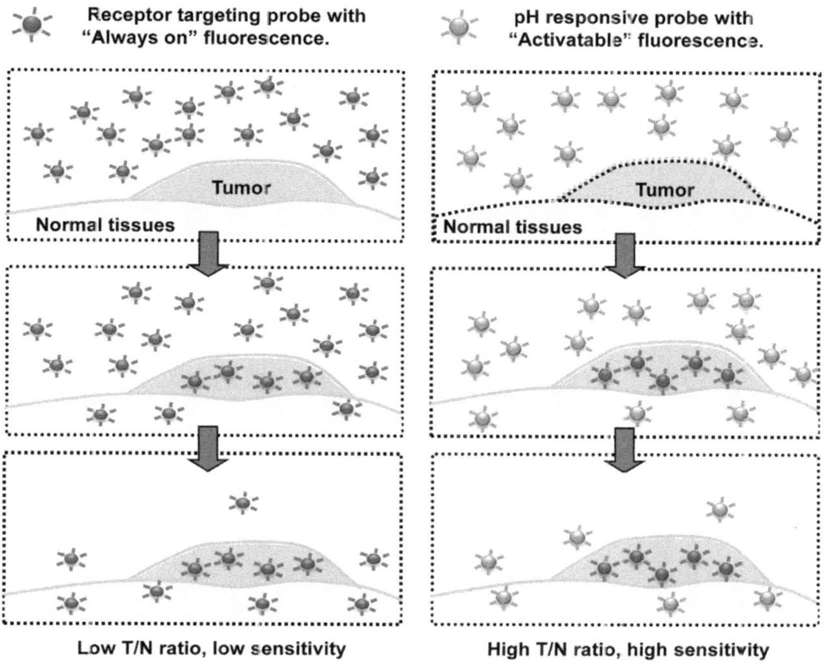

Fig. 2. pH activatable probe significantly increasing the sensitivity of tumor imaging. (A) Schematics of conventional receptor-associated tumor imaging strategy. The probe signal remains at the "On" level during the whole imaging process. (B) Schematics of pH activatable tumor imaging strategy. In the normal tissues, the signal of the probe stays at the "Off" level, but it increases significantly in the tumor acidic microenvironment.

Color image of this figure appears in the color plate section at the end of the book.

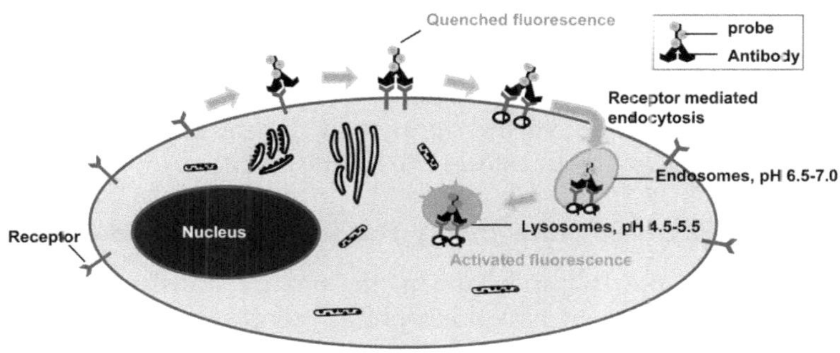

Fig. 3. Probes imaging tumor by sensing the acidic endosome/lysosome lument. The probes labeled in the antibody are quenched in neutral pH. After internalization *via* the receptor-mediated endocytosis, the probes are delivered into acidic endosomes followed by lysosomes, where the probes are activated concomitantly with significant signal enhancement. The picture is reproduced with permission from Ref. 18 © the Nature Publishing Group.

Color image of this figure appears in the color plate section at the end of the book.

low running cost and potential application in image-guided surgery. Another important property of optical imaging is the "switchable" signal output, enabling the development of "responsive" or "smart" probes that only switch on at the intended target (Kobayashi et al. 2010). For example, the fluorescence intensity of a "pH responsive" probe usually stays at the quenched state at the normal physiological pH, but it increases significantly in the tumor acidic microenvironment. Therefore, the "smart" probes demonstrate a higher T/N ratio compared to the conventional receptor-targeting probes, which will aid in the visualization of the small volume tumors or metastases that are hardly detectable by conventional imaging probes.

MECHANISMS OF pH ACTIVATABLE FLUORESCENCE

According to the previous work, there are mainly three mechanisms leading to pH-activatable fluorescence in the physiological pH range (pH 4.5–7.4). They include photon-induced electron transfer (PeT) effect (deSilva et al. 1997), fluorescence resonance energy transfer (FRET) effect (Huebsch and Mooney 2007) and self-aggregation associated energy transfer (SAET) effect (Khairutdinov and Serpone 1997).

The Photon-induced Electron Transfer (PeT) Effect

The pH responding fluorescence induced by PeT can be briefly described in Fig. 4A. When a fluorophore is excited, the electron jumps up from the highest occupied molecular orbital (HOMO) to the lowest unoccupied molecular orbital (LUMO). A photon is generated with the excited electron jumps back from LUMO to HOMO. However, for the fluorophores with PeT effect, after the excitation, the HOMO is occupied immediately by the electron from intramolecular donor, which results in the fluorescence quenching. In the acidic environment, the protonation of fluorophore reduces the energy level of the donor and the intramolecular electron transfer is inhibited, which restores the fluorescence.

The Fluorescence Resonance Energy Transfer (FRET) Effect

Juxtaposed with the PeT mechanism, the energy transfer in the FRET probes happens between two fluorophores conjugated *via* a pH liable bond. After excitation, a donor fluorophore transfers the energy to an acceptor fluorophore (the spatial distance < 10 nm) through a nonradiative dipole–dipole coupling and the fluorescence from the donor is fully quenched. However, in the acidic environment, the cleavage of the pH liable bond between the fluorophores stops the energy transfer and results in emission enhancement from the donor fluorophore (Fig. 4B).

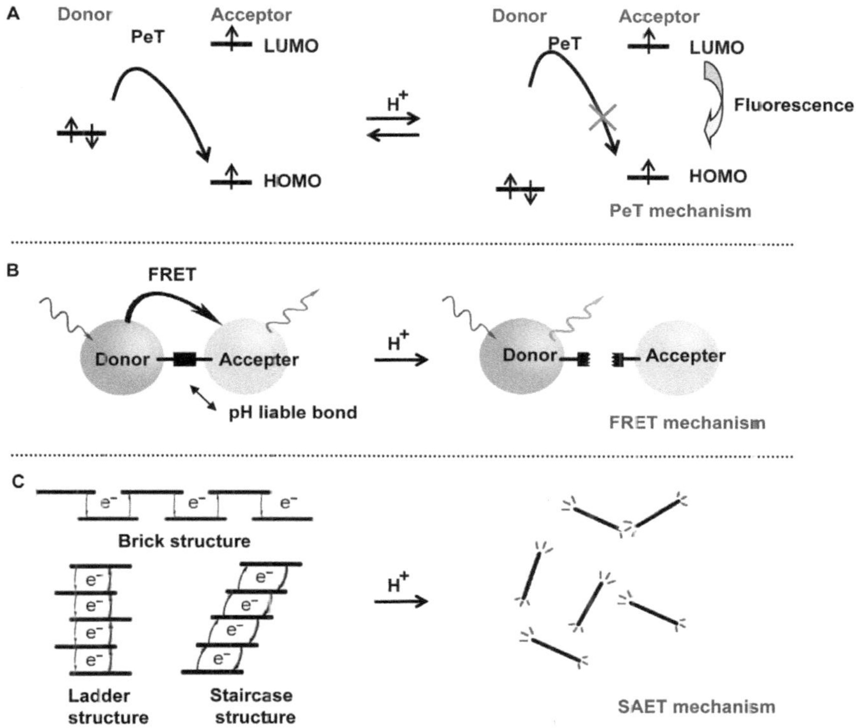

Fig. 4. Schematics of the mechanisms inducing pH responsive fluorescence. (A) The pH activated photon-induced electron transfer (PeT) effect. (B) The pH activated fluorescence resonance energy transfer (FERT) effect. (C) The pH activated self-aggregation associated energy transfer (SAET) effect.

Color image of this figure appears in the color plate section at the end of the book.

The Self-Aggregation Associated Energy Transfer (SAET) Effect

The fluorophores, such as carbocynine derivatives with big-π conjugated system, intend to aggregate homogeneously and form the brick work, a ladder or a staircase-like arrangement in aqueous solution *via* the strong intermolecular van der Waals interaction forces (Fig. 4C). The non-radiative energy transfer between the fluorophores quenches the fluorescence efficiently. According to the above properties, the fluorophores are conjugated into the macromolecular platform with high payload through the pH liable bonds. The fluorescence is quenched in the normal physiological environment due to the self-aggregation. The cleavage of the pH liable bonds in the acidic microenvironment results in the spatial separation of the fluorophores and the concomitant recovery of the fluorescence.

WHY CHOOSE A NANOPROBE TO VISUALIZE TUMORS?

The nanoprobes demonstrate advantages in tumor imaging, including: 1. the controllable particle size and morphology leading to the anticipated circulation lifetime and pharmacokinetics *in vivo*; 2. the passive tumor specificity determined by the enhanced permeability and retention (EPR) effect; 3. the enhanced tumor targeting specificity and imaging sensitivity determined by labeling multiple targeting or imaging groups on the nanoparticles. Therefore the nanoprobe is promising in detecting the tumor or metastases that are hardly visible by small molecular probes.

Nanoprobe Showing the Size-Dependent Pharmacokinetics *in vivo*

The first generation of nanoprobes is made up of the linear polymers such as polylysine (PL) (Fig. 5A). However, their wide molecular weight distribution pattern, non-reproducible chemical composition and low repeatability of the experimental results preventing their successful translation into clinical studies (Liechty et al. 2010). To overcome these problems, second-generation platforms, such as dendrimer, were discovered. Dendrimers are a class of highly branched spherical macromolecules with multiple primary amines on the surface of particle. Unlike the first-generation platform with a random coil structure, nanoprobes based on dendrimer maintain the spherical shape in aqueous solution due to minimal conformational changes (Barrett et al. 2009). Therefore, these nanoprobes demonstrate consistent pharmacokinetic properties. Additionally the bio-distribution of the nanoprobes can be anticipated by choosing the dendrimer with the suitable size (Fig. 5B). For example, small-sized dendrimer-based paramagnetic nanoprobes (diameter < 3 nm) rapidly perfuse throughout the body and excrete

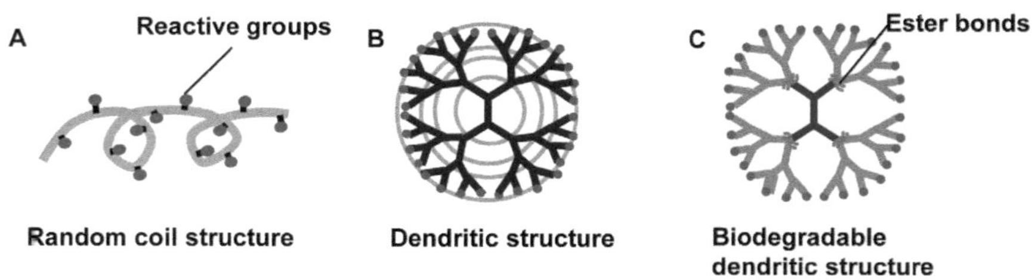

Fig. 5. Structures of the polymers used for the nanoprobe platform. (A) The linear or branched polymers with a random coil-like structure. (B) The dendrimer-like polymer with a unique molecular weight, spherical shape and well-defined topology of the peripheral reactive groups on the particle surface. (C) The biodegradable dendrimer-like polymer possesses the advantages of a dendrimer but with low systemic toxicity.

Color image of this figure appears in the color plate section at the end of the book.

through kidney filtration. Nanoprobes with 5–12 nm in diameter prefer to retain in circulation and are better suited for blood pool contrast agents. The nanoprobes with a diameter larger than 15 nm are quickly trapped in the reticuloendothelial system (liver and spleen) (Kobayashi and Brechbiel 2005). Even though the dendrimer-based nanoprobes showed promising properties, their translation into clinic was still hindered by their non-degradable properties *in vivo*, which leads to the potential systemic toxicity. To address this issue, the third generation of nanoprobe platform consisting of a biodegradable dendritic scaffold was reported (Almutairi et al. 2008). This platform not only inherits the advantages of the conventional dendrimers, but also can be fully degraded by the endogenous esterases *in vivo* (Fig. 5C). Therefore the potential toxicity resulting from the long resident lifetime of the nanoprobes can be minimized.

Nanoprobe Passively Accumulating in Tumor *via* the EPR Effect

The structural differentiations between the normal and tumor vasculatures provide opportunities for the nanoparticles to achieve tumor specific targeting. In contrast to normal vessels, the tumor vasculatures are immature, leaky, tortuous and have large pores between the endothelial cells (Carmeliet and Jain 2000). The nanoprobe preferentially accumulates in solid tumors by passive convective transportation through leaky tumor vasculature, but cannot penetrate normal vessels (Fig. 6). Considerable evidence shows that nanoparticles with a molecular size between 5 nm

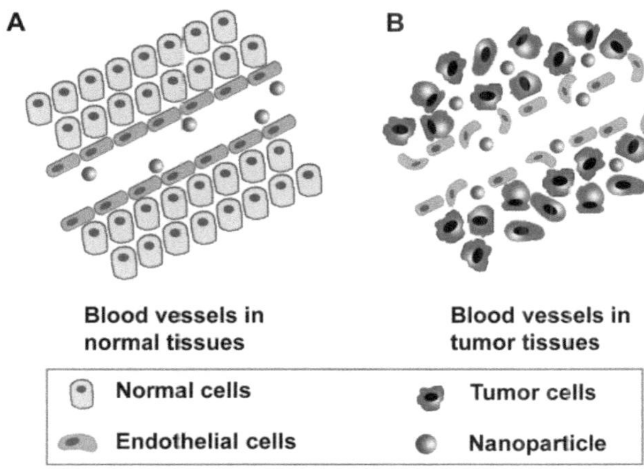

Fig. 6. A schematic of the enhanced permeable and retention (EPR) effect. (A) The nanoparticles are restricted in the normal vasculatures. (B) Due to the high permeability of tumor vasculatures, the nanoparticles penetrate the vessel walls and enter the tumor interstitium.

Color image of this figure appears in the color plate section at the end of the book.

to 1 μm demonstrate the EPR effect in most experimental solid tumors (Maeda 2010). The resident time of nanoparticles in a tumor can last as long as days after administration (Maeda 2010). The long resident lifetime of nanoprobes in a tumor not only provides an extended time-window to dynamically visualize the tumor, but also substantially increases the T/N ratio due to the faster excretion rate of the nanoprobe in normal tissues.

Nanoprobe Shows Improved Tumor Specificity and Sensitivity *via* Multivalent Effect

Multivalent effect is a unique advantage of nanoparticles. The nanoprobes modified with multiple targeting groups usually demonstrate much higher tumor targeting specificity compared to the free targeting damain (Fig. 7). Baker et al. prepared a series of G5 dendrimer-based nanoparticles functionalized with different numbers of folic acids (Hong et al. 2007). Studies showed that the binding affinity of the normalized binding affinity per folic acid in nanoprobe was measured as 15,200 times higher compared to that of free folic acid. Josephson et al. also reported that the cyclic [RGDfK] peptide modified nanoparticles demonstrate obvious multivalent effect to the $\alpha_v\beta_3$ integrin over-expressed in the cells. The binding affinity of the nanoparticle modified with 20 cyclic peptides was 71 times higher compared to that of the free peptide (Montet et al. 2006). Furthermore, the nanoprobes provide the opportunity to modify different types of imaging reporters on a single nanoparticle. In the imaging modalities used in the clinic, MRI possesses the highest spatial resolution but suffers from less than optimal sensitivity. On the other hand, optical imaging is very sensitive but its resolution is relatively low. Therefore, the nanoparticle-based multimodal imaging probes, such as MRI/optical imaging probes, can complement their strengths while minimize their weaknesses (Louie 2010).

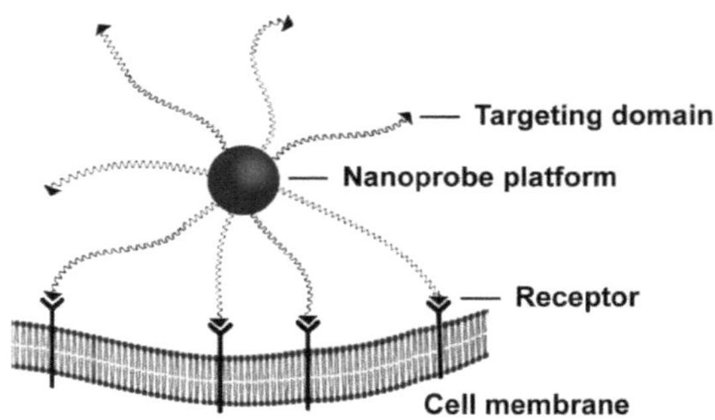

Fig. 7. A schematic of the nanoparticle associated with multivalent effects. The multiple targeting domains labeled in the nanoparticle provide the opportunity to bind receptors on the cell membrane with high affinity.

In brief, the nanoprobes show great promise in visualizing tumors with higher sensitivity than the small molecular probes. At the same time, due to biodegradable properties, the potential toxicity induced by the non-specific accumulation in normal tissues can be minimized. With the commercialization of fused imaging instruments, such as the PET/CT and PET/MR, the multimodal imaging nanoprobes provide the unprecedented opportunity to visualize the tumor or metastases with high sensitivity and spatial resolution.

MATERIALS USED FOR pH RESPONSIVE NANOPROBES

pH Responsive Nanoprobes Based on Biomaterials

The bio-molecules, such as peptides, proteins and antibodies, are chosen as the nanoprobe platform because of its good biocompatibility, low immunogenecity and bio-degradability. Kobayashi et al. reported an antibody-based nanoprobe that successfully visualized the metastastic tumor models *via* the PeT effect (Urano et al. 2009). In aiming for the nanoprobe DiEtN-BDP–trastuzumab, the fluorophore DiEtN-BDP with pH dominated PeT effect was conjugated to the antibody trastuzumab that specifically targets the Her2 receptors over-expressed in many types of breast cancer cells (Fig. 8A). For comparison, a control nanoprobe PhEtN-BDP–trastuzumab without the pH sensitivity was also prepared. The pH

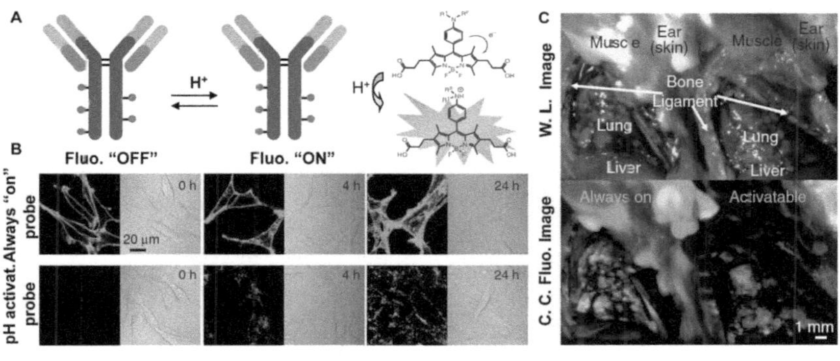

Fig. 8. Nanoprobe imaging tumors *via* the pH induced PeT effect. (A) The schematic of the antibody-based nanoprobe, in which the pH activatable fluorophores were labeled to the HER2 antibody. (B) Confocal fluorescence microscope images of NIH3T3 HER2$^+$ cells at a selected time after the treatment of "always-on" or the pH activatable nanoprobe. (C) White light (top) and color coded fluorescence image (bottom) of the lung metastases after injection of "always-on" (left) or pH-activatable (right) nanoprobes. The activatable probe (bottom right) readily distinguishes between HER2$^+$ (green) and RFP$^+$HER2$^-$ (red) tumors. In contrast, the "always on" nanoprobe cannot distinguish HER2$^+$ or HER2$^-$ tumors. Pictures in panel A–C are reproduced with permission from Ref. 18 © the Nature Publishing Group.

Color image of this figure appears in the color plate section at the end of the book.

responsive fluorescence of the nanoprobe was verified in the live cells. At 24 h post incubation, evidence of fluorescence was only found in the acidic lysosomes in the cytoplasm (Fig. 8B). In contrast, strong fluorescence was shown everywhere in the cells treated with control nanoprobe. pH Responsive nanoprobe DiEtN-BDP–trastuzumab demonstrated higher imaging sensitivity than the control nanoprobe in the metastatic NIH3T3 lung tumor xenograft over-expressed Her2 receptors. The fluorescence intensity ratio between the tumor and surrounding normal tissues (T/N ratio) of this nanoprobe was measured at 22 times higher than the control nanoprobe. Meanwhile only the tumors with the high expression level of Her2 receptors were visualized with high T/N ratio (Fig. 8C).

Engelman et al. recently reported a peptide based nanoprobe demonstrating the high T/N ratio by increasing the cellular uptake *via* a pH-dependent topological change in the tumor acidic microenvironment (Reshetnyak et al. 2006). In this work, a pH (low) insertion peptide (pHLIP) derived from the bacteriorhodopsin C helix was developed. At neutral pH (pH = 7.4), this nanoprobe bound weakly to the cell membrane as a random coil structure. Whereas at acidic pH (pH < 7.0), the nanoprobe changed to a rigid transmembrane helix structure due to the increase of hydrophobicity that resulted from the protonation of Asp residues at low pH. The helix conformation greatly increased the translocation capability of the nanoprobe into the lipid bilayer (Fig. 9A). *In vitro* microscopic investigations showed that the intracellular fluorescence of the HeLa cells increased with the acidification of the cell culture media, and increased 5 times when the media was acidified from pH 7.4 to pH 5.5 (Fig. 9B).

Even above pH activatable nanoprobes successfully visualizing the acidic compartments in cancer cell culture, the limited penetration depth of the green fluorescence hardly contributes to the non-invasive tumor imaging. Recently the fluorescent probes that emit in the near-infrared (NIR) region (650–900 nm) are highly desirable because tissue absorption and autofluorescence of endogenous molecules in this regime are low, which allows the light to penetrate into tissues for distances of centimeters and gives the images with high sensitivity and spatial resolution (He et al. 2010). A few pH responsive fluorescence NIR probes have been developed, but their applications *in vivo* are limited due to the hydrophobic properties, short circulation lifetime, unmatched responsive pH range and low target-to-background signal ratios. To address these challenges, Reshetnyak et al. recently labeled a 36 amino acid pHLIP peptide with a NIR fluorophore Alexa Fluo 750 (Andreev et al. 2007). This nanoprobe showed high sensitivity to breast adenocarcinoma xenografts. A small tumor undetectable to the naked eye was clearly observed by nanoprobe pHLIP-Alexa Fluo 750 as a fluorescent spot in the NIR fluorescence image

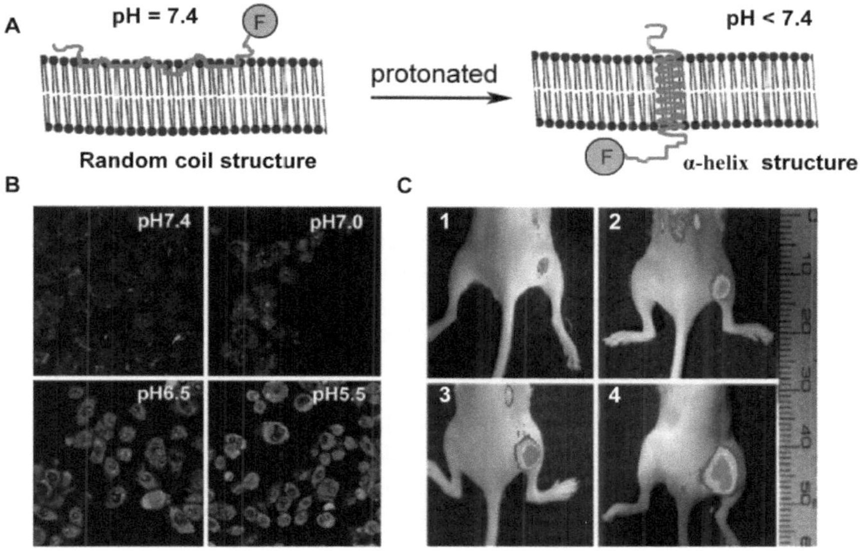

Fig. 9. Nanoprobes imaging tumor by changing the topological structure in the acidic environment. (A) Schematic of the up-regulated nanoprobe uptake induced by its pH responsive topological structure variation. (B) Microscopic fluorescence images of HeLa cells incubated with the pH responsive nanoprobe at pH 5.5, 6.5, 7.0, and 7.4 for 15 min. (C) *In vivo* overlapped white light and NIR fluorescent images of mice bearing tumors in right flanks. Panel 1–4 show a tumor undetected by the naked eye, 3–4 mm, 5–6 mm, and 8–9 mm. Pictures in panel A–C are reproduced with permission from Ref. 21 © the National Academy of Sciences, USA.

Color image of this figure appears in the color plate section at the end of the book.

(Fig. 9C). *Ex vivo* fluorescence imaging studies demonstrated that the high NIR fluorescence was not only observed in the tumor, but also in the kidney. The acidic environment (pH ≈ 6.4) of the renal medulla in healthy mice/rats is well-documented due to the high pCO_2 in the vasa recta blood. Therefore the strong fluorescence in kidneys supports the pH activatable fluorescence of the nanoprobe *in vivo*.

Above all, the biomaterial-based pH responsive nanoprobes are promising in detecting tumors by sensing the acidic compartment in tumors. However, the high cost of preparing these bio-materials limits their wide application. Therefore, the materials that not only possess the advantages of biomaterials, but can also be prepared economically with large quantities are highly desirable.

pH Responsive Nanoprobes Based on Synthetic Polymer

Synthetic polymers attract consistent interest as materials used in a drug delivery vector because of their good processability, low cost and the availability to be prepared in large amounts. However, as a material used for medicinal purposes, several criteria need to be met: 1) the synthetic polymer has to be water-soluble, low-toxic and low-immunogenic; 2) the material has to have low accumulation in normal tissues and be excreted safely. To meet the aforementioned criteria, the polymer needs to be either below the renal excretion threshold (< 6 nm), or it has to be degradable in the physiological surroundings to minimize its long-term toxicity.

Dendrimers, as highly branched synthetic polymer, demonstrate unique molecular weight, spherical shape, well-defined peripheral reactive groups, reliable synthetic procedure and adjustable particle size (Kobayashi and Brechbiel 2005). However, the systemic toxicity of dendrimers induced by their highly positive charges and non-degradable nature prevents their translation into clinical studies (Longmire et al. 2008). To overcome these shortcomings, Frechet et al. first reported a biodegradable dendrimer-based NIR fluorescent nanoprobe to sense the physiologically acidic environment (Almutairi et al. 2008). As shown in Fig. 10, NIR fluorophore cypates were labeled into the multilayer dendrimer/polyethylene glycols (PEGs) copolymer *via* pH liable hydrazone bonds. In the neutral pH, the cypate dyes formed H-type homoaggregates *via* face-to-face stacking and the fluorescence was suppressed through SAET effect. Whereas in the acidic environment, the cleavage of the hydrazone bond drifted the fluorophore apart and the fluorescence intensity amplified simultaneously. The polyester bonds that connect the different layers of the nanoprobe can be further cleaved by the endogenous esterases to accelerate the excretion of the nanoprobe (Fig. 10). This nanoprobe was 8–10 nm in diameter and its NIR fluorescence increased by 6 times after incubation in pH 4.5 buffer solution for 24 h.

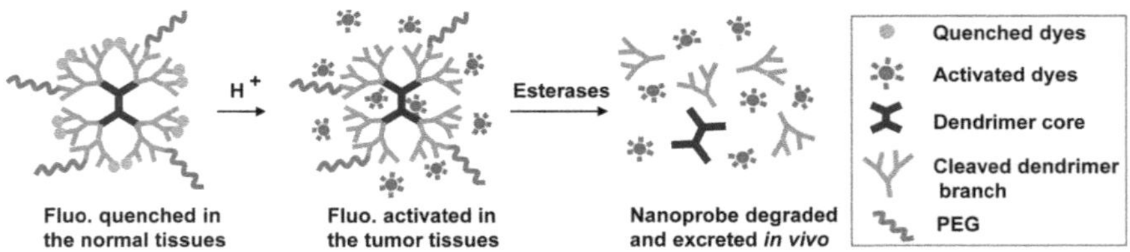

Fig. 10. Schematic of the pH responsive biodegradable nanoprobe. In the physiological pH, the NIR fluorophores labeled on the nanoprobe quench with each other *via* the SAET effect. In the tumor acidic environment, the cleavage of the pH liable bonds results in the pH activated fluorescence. The nanoprobe can be further degraded by endogenous esterases following the excretion *in vivo*.

Color image of this figure appears in the color plate section at the end of the book.

Recently Li et al. reported a dextran based NIR fluorescent nanoprobe **NP1** that demonstrated the feasibility to image tumors non-invasively by sensitizing the tumor acidic microenvironment *in vivo* (Fig. 11A) (Li et al. 2010). In this work, NIR fluorophore IR783 was conjugated into the dextran/polylysine copolymer *via* the pH liable hydrazone bonds. Due to the spatial proximity of the conjugated IR783 in the dextran, the self-quenching between the fluorophores leads to a low signal background in normal tissues. The cleavage of hydrazone bonds in the acidic microenvironment with the concomitantly diminishing of the SAET effect resulted in the activation of NIR fluorescence in the tumor. Kinetic studies demonstrated that the fluorescence of the nanoprobe increased by 6.8 and 4.2 times respectively at pH 5.5 and 6.5 after incubation for 24 h. Confocal microscopic images showed the lysosomal delivery of **NP1** in the U87MG cells (Li et al. 2010) and the average NIR fluorescence in the cells increased by 4.7 times after 24 h of post-incubation.

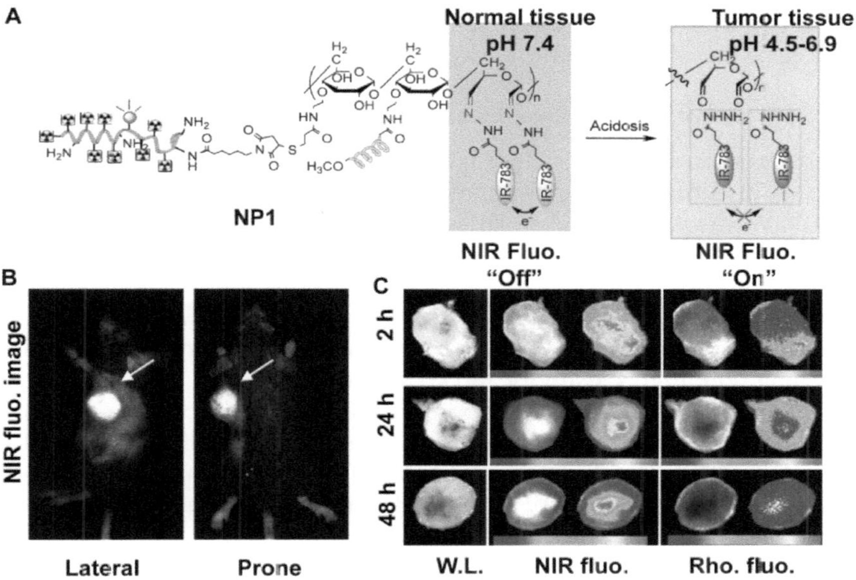

Fig. 11. Nanoprobe imaging tumor *via* the pH induced SAET effect *in vivo*. (A) The schematic of the dextran based nanoprobe, in which the NIR fluorophores are labeled into dextran *via* the pH liable hydrazone bonds. (B) Representative *in vivo* NIR fluorescence images of a mouse bearing a U87MG tumor xenograft at 48 h post-injection of **NP1** *via i.v.* The arrow points to the position of the tumor. (C) Representative white light, NIR and rhodamine fluorescence images of *ex vivo* U87MG tumor sections at 2, 24 and 48 h post-injection of **NP1**. WL: white light; CC: color coded.

Color image of this figure appears in the color plate section at the end of the book.

More importantly, **NP1** determined the tumor targeting specificity by sensing the tumor acidic environment *in vivo*. Dynamical optical imaging studies showed the NIR fluorescence on the U87MG tumor xenograft increased with time after the intravenous (*i.v.*) administration, and the fluorescence T/N ratio reached its maximum value at 48 h post nanoprobe injection (Fig. 11B). *Ex vivo* optical imaging showed that the NIR fluorescence was predominately located in the central area of fresh tumor sections, but the rhodamine fluorescence that indicates the position of nanoprobe itself was mainly present in the peripheral area **NP1** (Fig. 11C). The author explained this phenomenon as the departure of the IR783 dyes from the nanoprobe in the tumor acidic microenvironment and diffusion of the tumor core area due to the high diffusion coefficient of the small molecular fluorophores. Interestingly, bio-distribution studies showed that while nanoprobe uptakes in the liver and kidneys were much higher than that in tumors, the normalized NIR fluorescence in *ex vivo* liver and kidney tissues was lower than that in tumors. This result highlights the important role that pH activatable fluorescence played to achieve the high T/N ratio by minimizing the background signal from normal tissues.

pH Responsive Nanoprobes Based on Inorganic Materials

In comparison to organic fluorophores, inorganic fluorescent materials such as quantum dots (QDs) show excellent optical properties such as broad absorbance spectra, narrow and symmetric emission band, size and composition-tunable fluorescence emission, high quantum yield, as well as high photostability (Resch-Genger et al. 2008). Due to the advantages of the QDs, the modified QDs have been applied for *in vitro* cell imaging, single cell trafficking, FRET-based sensing, and sentinel lymph-node mapping with high sensitivity and spatial resolution.

Chan et al. reported a pH responsive QD, in which a peptide-polymer hybrid polymer was modified in the surface of a core-shell CdSe-ZnS QDs (Wu et al. 2010). The fluorescence of this QD-based nanoprobe increased by about 3 times when the pH increased from 5 to 9. The author explained the pH dependent fluorescence to the conformational rearrangements of the hybrid polymer, which influenced the degree of surface passivation of the hybrid polymer (Fig. 12). In the work of Liu et al., a core/shell/shell QD capped with mercaptoacetic acid (MAA) was reported (Liu et al. 2007). The CdSe/ZnSe/ZnS nanocrystal structure not only makes the QDs robust against photobleaching, but also determines its sensitivity to the intracellular pH variation. The integrated fluorescence of the nanoprobe increased above 5 times from pH 4 to 10 during basification in the buffered solution. The pH dependent fluorescence was explained by the de-association of the MAA coating during the acidification. In the

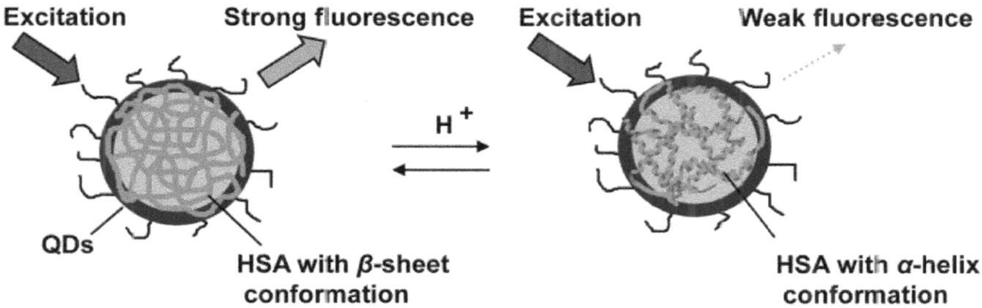

Fig. 12. Quantum dot based pH responsive fluorescence nanoprobe. In the acidic pH, the denatured HSA protein coated on the surface of QDs forms an α-helix conformation, but it turns into the β-sheet conformation in the basic environment. The pH dependent conformation of the protein coating leads to a fluorescence intensity change of the QD.

Color image of this figure appears in the color plate section at the end of the book.

acidic environment, the increased exposure of the nanocrystal surface in the aqueous solution, which is a reservoir to the excited electrons from the nanocrystal, leads to a higher nonradiative relaxation rate and reduced fluorescence intensity.

Even QD-based nanoprobes are promising to image the biologic target or processes—their long-term toxicity induced by the heavy metals that are released, such as cadmium and selenium *in vivo* that prevents their translation into clinic. Nevertheless, due to their outstanding optical properties, QD-based nanoprobes are promising in developing the tumor diagnostic kit for an *in vitro* blood sample screening strategy.

RATIOMETRIC FLUORESCENCE NANOPROBES DETERMINING PHYSIOLOGICAL pH

Because the acidification induced by the anaerobic glycolysis of cancer cells under hypoxic condition is perhaps the most pervasive phenomenon of solid tumors, recent therapeutic approaches have been designed to target the tumor acidic microenvironment either through pH activation of pro-drugs, pH activation of drug release from macromolecular cargoes, or drugs that raise the pH of acidic tumors. Thus, measuring the pH of targeted tissue with high precision and spatial/temporal resolution is not only important in terms of designing the drug with a pH response capability, but also helpful in evaluating the therapeutic response during the pH neutralization treatment.

Advantages of the Ratiometic Fluorescenec Nanoprobe

Due to the heterogeneous nature of tumor vasculatures, it is difficult to determine the local pH by simply measuring the fluorescence intensity of a pH sensitive probe. The advantage of ratiometric probes is that the excitation source fluctuations and probe concentration do not affect the ratio between the fluorescence intensities of the indicator fluorophore and the reference fluorophore. Therefore, the ratiometric probe is a viable option to determine the intratumoral pH accurately with minimized artificial interference.

Even though a series of small molecular ratiometric probes have been employed to determine the pH in buffered solution, few of them are suitable for *in vitro* and *in vivo* pH determinations due to the strong tissue background, fast excretion rate, or inappropriate fluorogenic p*K*a values. To solve these problems, several ratiometric nanoprobes have been developed. The common characteristic of these nanoprobes is that both the pH sensitive indicator fluorophore and the pH insensitive reference fluorophore are physically embedded (Fig. 13A) or covalently labeled (Fig. 13B) in a nanoparticle with multiple copies. In contrast to the reference fluorophore, in which its fluorescence intensity remains constant, the signal of the indicator fluorophore increases significantly with the acidification. Therefore, the pH can be determined by fitting with the calibration curve that presents the pH dependent fluorescence intensity ratio between the indicator and reference fluorophores.

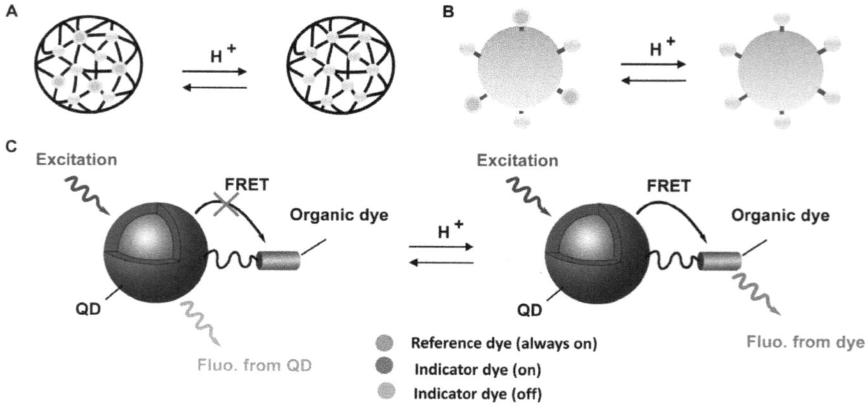

Fig. 13. Schematics of pH responsive ratiometric fluorescence nanoprobes. (A) The nanoprobe in which both reference and indicator dyes are embedded inside the polymer. (B) The nanoprobe in which reference and indicator dyes are labeled covalently on the surface. (C) The inorganic nanoprobe measures the pH *via* the intramolecular FRET effect between the quantum dot and the organic dyes modified on its surface.

Color image of this figure appears in the color plate section at the end of the book.

Synthetic Polymer Based Ratiometic Fluorescence Nanoprobe

Sun et al. reported a ratiometric nanoprobe, in which the rhodamine as the reference dye and *N*-fluoresceinylacrylamide as the pH sensitive indicator dye, were covalently labeled a polyacrylamide matrix by microemulsion copolymerization (Sun et al. 2006). The nanoprobe demonstrated a narrow size distribution with an average hydrodynamic diameter of 50 nm. On average 5.7 indicator dyes and 75 reference dyes were labeled in one nanoprobe. This ratiometric nanoprobe demonstrated the reversible pH sensitivity, fast responding time (< 0.4 s) and a lower photobleaching rate compared to the free fluorophore. Importantly, a pKa value of 6.4 of this nanoprobe was determined, which helps to determine the accurate pH measurement in the physiological pH range.

Bacteriophage Based Ratiometic Fluorescence Nanoprobe

Weissleder et al. developed a bacteriophage-based NIR fluorescent nanoprobe for the pH determination in live cancer cells (Hilderbrand et al. 2008). In this work, hundreds of pH responsive NIR fluorophore HcyC-646 and pH insensitive NIR fluorophore Cy7 were labeled into the M13 bacteriophage. In the acidic environment, HCyC-646 exists primarily in its protonated form with fluorescence emission at 670 nm. However, in neutral or basic environments, the dye is deprotonated and the absorbance shifted hypsochromically, which results in the diminishment of the NIR fluorescence. This nanoprobe with a pKa value of 6.2 demonstrated the feasibility to measure the intracellular pH in RAW cells.

Quantum Dot Based Ratiometric Fluorescence Nanoprobe

Recently a QD based ratiometric fluorescence nanoprobe was reported. In this nanoprobe, the CdSe-ZnS core was encapsulated with poly (acrylic acid) coating that was further labeled with pH responsive squaraine dyes (Fig. 13C) (Snee et al. 2006). The pH sensibility of the QD-squaraine nanoprobe was imparted by modulation of the FRET efficiency between the pH-sensitive squaraine dye absorption spectrum and the pH insensitive quantum dot emission spectrum. Owing to the pH dependent absorbance of squaraine, the spectral overlap between the squaraine absorption and the QD emission changed according to the pH variation. At the basic condition, FRET was inefficient as the spectral overlapping is small. Hence, the emission spectrum was dominated by fluorescence from the QD at 613 nm. With the acidification, the spectrum overlapping increased and the squaraine fluorescence at 650 nm increased continuously. Interestingly the fluorescence at the isosbestic point of 640 nm stayed independent of the pH variation. Therefore the pH can be read out precisely by taking the ratio of the emission of either the QD or the squaraine to the intensity at the isosbestic point, which functions as an internal reference. This QD-squaraine nanoprobe precisely showed the pH determination ability with an error below 5%.

THE APPLICATION OF FLUORESCENCE NANOPROBES TO OTHER HEALTH OR DISEASE ISSUES

Besides the tumor, the physiological acidification is also observed in inflammation-related diseases such as arthritis, asthma and pulmonary tuberculosis. More evidence shows that the phagocytose activity serves as a host response of cells against heterogeneous bacteria leading to the acidification of the inflammative tissues. The extracellular pH of inflammative tissues is measured in the range of 6.2–7.0, which provides the opportunity to visualize the inflammation-related diseases by using the pH responsive nanoprobe with high imaging sensitivity. Recently the pH responsive NIR fluorescence nanoprobe pHLIP-Cy5.5 that successfully visualized acidic tumor microenvironment demonstrated the feasibility to image anthritis in animal models. The fluorescence intensity in the anthritis model was 5–6 times higher than that in the normal knee joint of the same mouse (Almutairi et al. 2008).

Key Factor of Tumor Microenvironment

- Cancer is the most complex and lethal disease that threatens human health and the tumor microenvironment (TME) plays crucial role in the evolution of the malignant phenotype and resistance to therapeutic treatments.
- TME is used to describe the infrastructure that surrounds and supports cancer cells, which includes the extracellular matrix (ECM), cancer associated fibroblasts (CAFs), pericytes, tumor associated macrophages (TAMs), and vascular and lymphatic endothelial cells.
- The characteristics of TME include hypoxia, acidic pH, and high interstitial pressure, which are induced by the poor perfusion and high metabolic rate of cancer cells.
- Cancer cells demonstrate the formidable capacity to adapt and survive in a hostile TME that would be ordinarily lethal to normal cells.
- TME provides a sanctuary for cancer cells to shield harmful effects from therapy.
- For cancer diagnosis, TME can be visualized instead of cancer cells because it is easier to be accessed.

Definitions

Glycolysis: The metabolic pathway that converts glucose into pyruvate. The free energy released in this process is used to form the high-energy compounds such as ATP (adenosine triphosphate).

Hypoxia: A pathological condition in which the body as a whole or a region of the body is deprived of adequate oxygen supply.

Near-infrared fluorescence: The fluorescence with a wavelength range between 650 to 900 nm.

Quantum dot: A semiconductor whose excitons are confined in all three spatial dimensions.

Dendrimer: A repeatedly branched, roughly spherical polymer.

Summary Points

- Developing different media to non-invasively visualize tumors and metastases in their early stages is crucial to achieving a satisfactory therapeutic response in clinic.
- Compared to conventional receptor targeting probes, the probes that sense TME, especially the acidic pH, provide an additional or even better solution for solid tumor diagnosis because the acidic environment is easier to access.
- Acidification is a universal phenomenon of all solid tumors, therefore the pH responsive probe can visualize tumors without the limitation of tumor types or even development stages.
- Fluorescence nanoprobes provide an important niche to image tumor acidic microenvironment due to their high switchable signal output, adjustable pharmacokinetic properties, and unique EPR and multivalent effect.
- With the rapid development of nano- and imaging -technologies, the pH responsive nanoprobes that are currently being explored at the bench are promising in translating to the bedside in the near future.

Acknowledgements

This work was supported by the National Basic Research Program of China (973 Program, No. 2011CB910404), and the National Natural Science Foundation of China (No. 81171384).

Abbreviations

ATP	:	adenosine triphosphates
MRI	:	magnetic resonance imaging
CT	:	computed tomography
PET	:	position emission tomography
PeT	:	photon-induced electron transfer
FRET	:	fluorescence resonance energy transfer
SAET	:	self-aggregation associated energy transfer
HOMO	:	highest occupied molecular orbital
LUMO	:	lowest unoccupied molecular orbital
EPR	:	enhanced permeability and retention
FOV	:	field of view
PL	:	polylysine
PEO	:	polyethylene oxide
T/N ratio	:	tumor/normal tissue signal ratio
NIR	:	near-infrared
PEG	:	polyethylene glycol
i.v.	:	intravenous
QD	:	quantum dots
TME	:	tumor microenvironment
MAA	:	mercaptoacetic acid
TAM	:	tumor associated macrophages
pHLIP	:	pH (low) insertion peptide

References

Almutairi, A., S.J. Guillaudeu, M.Y. Berezin, S. Achilefu and J.M.J. Frechet. 2008. Biodegradable pH-sensing dendritic nanoprobes for near-infrared fluorescence lifetime and intensity imaging. J. Am. Chem. Soc. 130: 444–445.

Andreev, O.A., A.D. Dupuy, M. Segala, S. Sandugu, D.A. Serra, C.O. Chichester, D.M. Engelman and Y.K. Reshetnyak. 2007. Mechanism and uses of a membrane peptide that targets tumors and other acidic tissues in vivo. Proc. Natl. Acad. Sci. USA 104: 7893–7898.

Barrett, T., G. Ravizzini, P.L. Choyke and H. Kobayashi. 2009. Dendrimers in Medical Nanotechnology Application of Dendrimer Molecules in Bioimaging and Cancer Treatment. Ieee. Eng. Med. Biol. 28: 12–22.

Carmeliet, P., and R.K. Jain. 2000. Angiogenesis in cancer and other diseases. Nature 407: 249–257.

Denko, N.C. 2008. Hypoxia, HIF1 and glucose metabolism in the solid tumour. Nat. Rev. Cancer 8: 705–713.

deSilva, A.P., H.Q.N. Gunaratne, T. Gunnlaugsson, A.J.M. Huxley, C.P. McCoy, J.T. Rademacher and T.E. Rice. 1997. Signaling recognition events with fluorescent sensors and switches. Chem. Rev. 97: 1515–1566.

He, X., J. Gao, S.S. Gambhir and Z. Cheng. 2010. Near-infrared fluorescent nanoprobes for cancer molecular imaging: status and challenges. Trends Mol. Med. In press.

Hilderbrand, S.A., K.A. Kelly, M. Niedre and R. Weissleder. 2008. Near infrared fluorescence-based bacteriophage particles for ratiometric pH imaging. Bioconjug. Chem. 19: 1635–1639.

Hong, S., P.R. Leroueil, I.J. Majoros, B.G. Orr, J.R. Baker and M.M. Banaszalk Holl. 2007. The binding avidity of a nanoparticle-based multivalent targeted drug delivery platform. Chem. Biol. 14: 107–115.

Huebsch, N.D., and D.J. Mooney. 2007. Fluorescent resonance energy transfer: A tool for probing molecular cell-biomaterial interactions in three dimensions. Biomaterials 28: 2424–2437.

Khairutdinov, R.F., and N. Serpone. 1997. Photophysics of cyanine dyes: Subnanosecond relaxation dynamics in monomers, dimers, and H- and J-aggregates in solution. J. Phys. Chem. B 101: 2602–2610.

Kobayashi, H., and M.W. Brechbiel. 2005. Nano-sized MRI contrast agents with dendrimer cores. Adv. Drug Deliver Rev. 57: 2271–2286.

Kobayashi, H., M. Ogawa, R. Alford, P.L. Choyke and Y. Urano. 2010. New strategies for fluorescent probe design in medical diagnostic imaging. Chem. Rev. 110: 2620–2640.

Le Roy, C., and J.L. Wrana. 2005. Clathrin- and non-clathrin-mediated endocytic regulation of cell signalling. Nat. Rev. Mol. Cell Bio. 6: 112–126.

Li, C., J.A. Xia, X.B. Wei, H.H. Yan, Z. Si and S.H. Ju. 2010. pH-Activated Near-Infrared Fluorescence Nanoprobe Imaging Tumors by Sensing the Acidic Microenvironment. Adv. Funct. Mater. 20: 2222–2230.

Li, C., K. Li, H. Yan, G. Li, J. Xia and X. Wei. 2010. Dextran based pH-activated near-infrared fluorescence nanoprobe imaging the acidic compartments in cancer cells. Chem. Commun. 46: 1326–1328.

Liechty, W.B., D.R. Kryscio, B.V. Slaughter and N.A. Peppas. 2010. Polymers for Drug Delivery Systems. Annu. Rev. Chem. Biomol. 1: 149–173.

Liu, Y.S., Y. Sun, P.T. Vernier, C.H. Liang, S.Y. Chong and M.A. Gundersen. 2007. pH-sensitive photoluminescence of CdSe/ZnSe/ZnS quantum dots in human ovarian cancer cells. J. Phys. Chem. C 111: 2872–2878.

Longmire, M., P.L. Choyke and H. Kobayashi. 2008. Clearance properties of nano-sized particles and molecules as imaging agents: considerations and caveats. Nanomedicine 3: 703–717.

Louie, A. 2010. Multimodality Imaging Probes: Design and Challenges. Chem. Rev. 110: 3146–3195.

Maeda, H. 2010. Tumor-Selective Delivery of Macromolecular Drugs via the EPR Effect: Background and Future Prospects. Bioconj. Chem. 21: 797–802 .

Montet, X., M. Funovics, K. Montet-Abou, R. Weissleder and L. Josephson. 2006. Multivalent effects of RGD peptides obtained by nanoparticle display. J. Med. Chem. 49: 6087–6093.

Resch-Genger, U., M. Grabolle, S. Cavaliere-Jaricot, R. Nitschke and T. Nann. 2008. Quantum dots versus organic dyes as fluorescent labels. Nat. Methods 5: 763–775.

Reshetnyak, Y.K., O.A. Andreev, U. Lehnert and D.M. Engelman. 2006. Translocation of molecules into cells by pH-dependent insertion of a transmembrane helix. Proc. Natl. Acad. Sci. USA 103: 6460–6465.
Snee, P.T., R.C. Somers, G. Nair, J.P. Zimmer, M.G. Bawendi and D.G. Nocera. 2006. A ratiometric CdSe/ZnS nanocrystal pH sensor. J. Am. Chem. Soc. 128: 13320–13321.
Sun, H., A.M. Scharff-Poulsen, H. Gu and K. Almdal. 2006. Synthesis and characterization of ratiometric, pH sensing nanoparticles with covalently attached fluorescent dyes. Chem. Mater. 18: 3381–3384.
Urano, Y., D. Asanuma, Y. Hama, Y. Koyama, T. Barrett, M. Kamiya, T. Nagano, T. Watanabe, A. Hasegawa, P.L. Choyke and H. Kobayashi. 2009. Selective molecular imaging of viable cancer cells with pH-activatable fluorescence probes. Nat. Med. 15: 104–109.
Warburg, O. 1956. On respiratory impairment in cancer cells. Science 124: 269–270.
Wu, Y., S. Chakrabortty, R.A. Gropeanu, J. Wilhelmi, Y. Xu, K.S. Er, S.L. Kuan, K. Koynov, Y. Chan and T. Weil. 2010. pH-Responsive quantum dots via an albumin polymer surface coating. J. Am. Chem. Soc. 132: 5012–5014.
Zhang, X., Y. Lin and R.J. Gillies. 2010. Tumor pH and its measurement. J. Nucl. Med. 51: 1167–1170.

10

Magnetic Nanorings and Nanotubes for Cancer Detection and Therapy

Hai Ming Fan[1,*] and *Malini Olivo*[2]

ABSTRACT

Magnetic nanorings and nanotubes, which have a unique hollow structure and more tuneable magnetic properties when compared to their superparamagnetic solid nanoparticles and bulk counterparts, offer new opportunities for designing novel contrast and therapeutic agents for cancer detection and therapy. In this chapter, we review recent developments in the synthesis and cancer-related biomedical applications of magnetic nanorings/nanotubes. The classification of magnetic hollow structures based on their magnetic configurations is discussed. The fabrication method is introduced in detail. Moreover, typical examples in biomedical applications are given for cancer cell

[1]Shaanxi Key Laboratory of Degradable Biomedical Materials, School of Chemical Engineering, Northwest University, Taibai North Road 229, Xi'an, Shaanxi 710069, P.R. China; E-mail: nanofhm@gmail.com

[2]School of Physics, National University of Ireland Galway, University Road, Ireland; E-mail: malini.olivo@nuigalway.ie

*Corresponding author

List of abbreviations after the text.

separation, magnetic resonance imaging, targeted drug delivery/release and magnetic hyperthermia. Superparamagnetic iron oxide nanorings/nanotubes and vortex-state nanorings are more commonly used in biomedical applications compared to various other magnetic materials because of their high biocompatibility and ability to form stable ferrofluids. The reported results clearly suggest that by elaborating design and implementation, the advantages of iron oxides nanorings and nanotubes, such as high chemical and physical stability, high saturation magnetization, hollow structure and unique shape-dependent magnetic configuration, can be utilized to effectively improve cell separation, MRI sensitivity and drug delivery. They are also expected theoretically to be an agent for magnetic hyperthermia treatment of cancer cells or tumours. It is thus believed that the development of magnetic hollow nanoparticles based cancer diagnostic and therapeutic techniques will play a major role in future cancer research and nanomedicine.

INTRODUCTION

The idea of using magnetism for healing and therapy can be traced back to the ancient Greeks and Egyptians, who believed that the magical rock, now known as magnetite, was a "cure" for most common ailments. A similar belief was found in the ancient cultures of India and China, where people also used the power of magnets for healing. Nowadays, nanoscale magnetic materials offer novel size- and shape-dependent chemical and physical properties that can be used as an effective and efficient contrast and therapeutic agent for cancer cell labelling, cell death, separation and detection (Portney and Ozkan 2006; Goya et al. 2008; Sekhon and Kamboj 2010). Thanks to the rapid development of nanomaterials fabrication, people are now able to perform controllable growth of various magnetic nanoparticles including all kinds of ferrites and some metal alloys (Frey et al. 2009). Most of the nanoparticle based cancer work, however, is focused on the biomedical application of ultra-small superparamagnetic nanoparticles (usually spherical iron oxide nanoparticle), in which the small size results in randomized magnetization flip at room temperature, thereby leading to a stable magnetic colloid or ferrofluid (Xie et al. 2009). Though the superparamagnetic nanoparticles have been shown to have many advantages in biomedical applications and some of them are also commercially available, many drawbacks remain that need improvement. For instance, as a contrast agent for magnetic resonance imaging (MRI) applications, superparamagnetic iron oxide nanoparticles (SPIO) with low saturation magnetization as well as low chemical and physical stability induced by its small size lead to poor enhancement in the spin-echo MR effect (Jun et al. 2005). In this chapter, we will focus on novel hollow structure magnetic nanoparticles such as nanorings and nanotubes. Tailoring the size and shape of these nanoparticles together with their unique porous structures enhances tuneability and magnetic properties, thereby leading to high performance practical applications in biomedicine as compared to solid SPIO (Cheng and Sun 2010). The potential use of magnetic nanorings/nanotubes in cancer diagnostics and therapy will be discussed. Because magnetite (Fe_3O_4) has high biocompatibility, it is the most prevailing candidate in biomedicine and therefore, we will focus primarily on the biomedical applications of iron oxide nanorings/ nanotubes in this chapter; other magnetic hollow materials will not be discussed as they may exhibit similar properties. We will also discuss the various approaches to fabricate such hollow structures and their unique magnetic properties. A few examples on the use of these magnetic nanorings and nanotubes in cancer detection and therapy will be provided. We would like to emphasize that the area of magnetic nanorings and nanotubes on nanomedicine has not been well developed and many novel phenomena that occur in such a system require further investigation. The aim of this chapter is to provide an overview on this rapidly developing area and to explore its potential in cancer detection and therapy.

Applications in the Areas of Health and Disease

Like other magnetic nanoparticles, magnetic nanorings and nanotubes have broad applications in the areas of health and disease. These hollow nanoparticles offer novel and unique shape-dependent chemical and physical properties in contrast to the commonly used superparamagnetic nanoparticles, which make them powerful tools for cancer diagnostics and therapy. These nanorings/nanotubes have the potential to improve currently available nanoparticle-based medical techniques including cancer cell separation, visualization of malignant cells using nanoprobes, high-sensitivity magnetic resonance imaging, targeted delivery of large amounts of chemotherapeutic agents to desired cells, magnetic hyperthermia therapy, activation of photosensitizers and production of cytotoxic species for photodynamic therapy etc. The principles developed in magnetic nanoring and nanotube related cancer research can also be applied to the diagnostics and therapy of other conditions such as magnetic immunoassay analysis of infectious diseases or to health related environmental protection such as water treatment.

UNIQUE MAGNETIC NANORINGS AND NANOTUBES

Compared to the traditional 0 dimensional (0D) nanocrystal or 1D nanowire/nanorod, nanorings and nanotubes are special hollow structures. Generally speaking, nanorings and nanotubes are materials that have through-holes in the solid nanoparticles. Such hollow nanostructures may demonstrate unique physical and chemical properties that are far beyond this obvious difference in appearance. To understand their characteristics better, the magnetic hollow structure can be divided into five types based on their magnetic configurations. They are ferromagnetic/ferrimagnetic nanotubes, onion-state nanorings, vortex-state nanorings, paramagnetic nanorings and superparamagnetic nanorings/nanotubes. The curling-state (where the magnetization curls along the nanotube) nanotubes and antiferromagnetic nanorings/nanotubes will not be discussed in this chapter. As stated earlier the differences in magnetic properties of nanorings and nanotubes strongly depend on their shape. Figure 1a demonstrates schematically the shape-dependent magnetic properties of nanorings and nanotubes. By using the LLG Micromagnetics Simulator, the shape-dependent magnetic configurations of single crystalline Fe_3O_4 nanorings and nanotubes can be obtained as shown in Fig. 1b. It should be noted that the Fe_3O_4 nanorings and nanotubes used for this simulation have a ratio of inner/outer diameter of $\beta=0.6$. More direct observation of the magnetic vortex state in a nanoring structure can be seen from an off-axis electron hologram in Fig. 1c and d where the magnetic moment circulates around a magnetite nanoring. In Fig. 1e, we show the clear differences in hysteresis loops of ferrimagnetic magnetite nanotubes, FVIO and SPIO at 300K. One can see that the magnetic hollow nanostructure is indeed a more complex system as compared to superparamagnetic nanoparticles. An appropriate hollow nanostructure should be chosen for a particular biomedical application. Among these magnetic hollow nanostructures, only superparamagnetic hollow structures and vortex-state nanorings can form stable ferrofluids due to the weak magnetic inter-particle forces that are created in the absence of an external field. Therefore, they are more attractive in the applications of cancer detection and therapy.

Key Features of magnetite (Fe_3O_4) nanorings and nanotubes

- Magnetite nanorings and nanotubes are special ferrimagnetic hollow structured nanoparticles. Because of high biocompitability, they are excellent candidates for biomedical applications.
- Magnetite nanorings and nanotubes have rich size and shape-dependent magnetic properties such as vortex state, onion state and superparamagnetism. Among them, vortex state and superparamagnetic nanorings can form stable magnetic colloid.

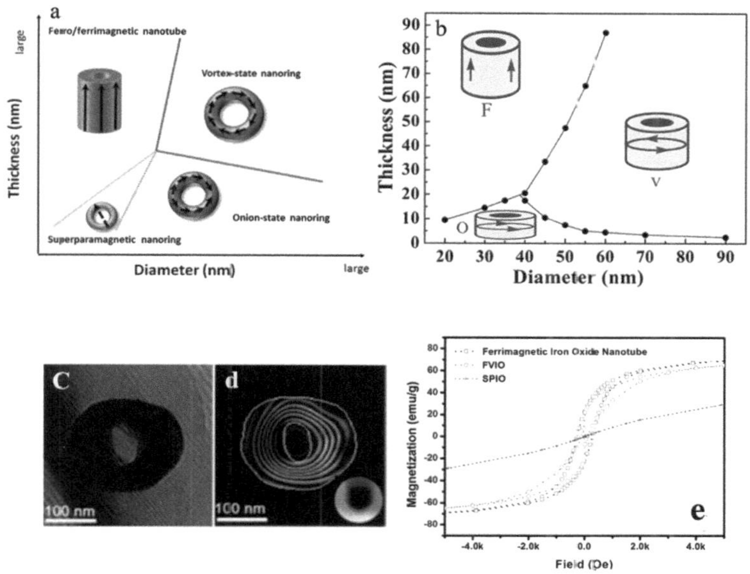

Fig. 1. Unique properties of magnetic nanorings and nanotubes. (a) Schematically demonstrates the shape-dependent magnetic properties of nanorings and nanotubes. (b) The shape-dependent magnetic configurations of single crystalline Fe_3O_4 nanorings and nanotubes obtained from LLG Micromagnetics simulation. (Reprinted with permission from Yang et al. unpublished data, provided by Dr. Jun Ding in Materials Science and Engineering, National University of Singapore) (c) Off-axis electron hologram of a single magnetite ring with an average thickness of 50 nm. In the form of bending interference fringes, the phase changes can be seen. (d) Direction of the magnetic induction under field-free conditions following magnetisation, indicated by colour as shown in the colour wheel in the inset (red = right, yellow= down, green = left, blue = up). (e) The hysteresis loops of a ferrimagnetic magnetite nanotube (370 nm in length and 80 nm in outer diameter), FVIO (50 nm in thickness and 70nm in outer diameter), and SPIO (10 nm) at 300K. (Reprinted with permission from Jia et al. 2008 and Fan et al. 2009, ©2008 and 2009 ACS Publishing Group.)

Color image of this figure appears in the color plate section at the end of the book.

- Magnetite nanorings and nanotubes have large surface and high peroxidase-like catalytic property, which can be used for laboratory immunoassays.
- Magnetite nanorings and nanotubes can offer high T2 (T2*) or T1 MR relaxivities for cellular imaging application.
- Magnetite nanorings and nanotubes with many voids inside the nanoparticle can be developed as nanoscale carriers that have high drug loading for targeted drug delivery.
- Superparamagnetic and vortex-state magnetite nanorings are promising materials as hyperthermia agents for highly efficient magnetic hyperthermia due to their unique ring-like geometry.

FABRICATION OF MAGNETIC NANORINGS AND NANOTUBES

Over the past decades, synthesis of nanomaterials with controllable size and shape has been studied extensively for numerous applications (Goesmann and Feldmann 2010). To date many synthetic protocols, including both conventional top-down nanofabrications and self-assembly based bottom-up methods, have been well-established for various magnetic nanoparticles. The physical and chemical properties of hollow structures strongly correlate to their geometric parameters such as inner/outer diameter and thickness, however, the controlled growth of a uniform 3D nanostructure with a desirable size and shape is much more difficult compared to the formation of a monodispersed nanocrystal. In general, it is favoured to form a solid particle during the crystal growth process due to thermodynamic stability. To achieve a hollow nanostructure, we need to make a hole in a solid particle. Etching a hole by physical or chemical approaches in a solid nanoparticle is particularly difficult because this process occurs in a nanoscale system where traditional chemical or physical reaction mechanisms for bulk materials may not work well. Alternatively, we can also assemble small nanoparticles into a big ring- or tube-shape nanoparticle by a template confinement method (Zhang et al. 2009). Learning from previous work in the controllable fabrication of nanoparticles, researchers have found some ways to generate uniform nanorings and nanotubes using template assisted growth, surfactant assisted wet-chemical methods, nanoscale Kirkendall effect methods, as well as other methods. In this section, we will introduce the approaches for the fabrication of magnetic nanorings and nanotubes.

Template Assisted Nanolithography

The basic principle of the template based strategy for the fabrication of hollow nanostructures is to use a specially designed template as a shape confinement. During this process, the template plays a role as a geometric frame and does not participate in the chemical reaction of the growth of the targeted material. The targeted material is first deposited conformably onto the template by sputtering, atomic layer deposition (ALD) or chemical vapour deposition (CVD). After the conformal growth is complete, the template is removed by chemical etching or physical evaporation. A typical example is a porous anodic aluminium oxide (AAO) template; other materials such as polystyrene (PS) and silica colloid, and semiconductor nanowires are also used. Due to the chemical inactivity of the template, nano-scale materials with annular or tubular structure have been successfully fabricated in a variety of materials. Hobbs et al. (2004) reported a typical process for nanoring fabrication by sputter redeposition using AAO templates, which are fabricated using a two-step anodization process. Figure 2 schematically illustrates the two methods to fabricate nanorings using an AAO template. While a through-hole alumina template is used as a sputter mask in method I for nanoring fabrication, it can also be used as an evaporation shadow mask in method II followed by ion-etching to make an array of nanorings or nanotubes. Figure 2 g and h show the SEM images of a magnetic Ni nanorings array fabricated using methods I and 2, respectively. Another common route is colloidal

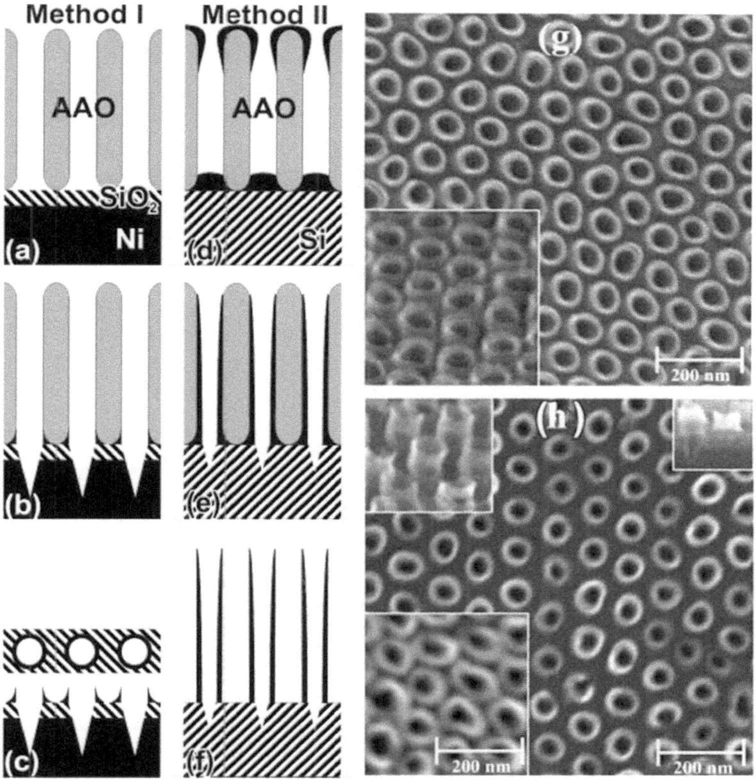

Fig. 2. Fabrications of magnetic nanorings and nanotubes through template assisted Nanolithography Template. Schematic view of methods I and II. For method I: (a) shows the AAO template on 20 nm of SiO_2 on the desired ring material, (b) shows sputter redeposited material around the pore walls after sputter etching, (c) shows the rings after the AAO mask is removed (with a plane-view added to guide the eye). For method II: (d) shows the AAO mask on a silicon substrate after the ring material is evaporated down the mask, (e) shows sputter redeposited material around the pore walls after sputter etching, (f) shows the rings after the AAO mask is removed, (g) and (h) show SEM images of Ni nanorings fabricated by method I and method II, respectively. (Reprinted with permission from Hobbs et al. 2004, ©2004 American Chemical Society.)

lithography (CL), in which colloidal arrays are used as lithographic masks or templates to fabricate nanostructures. Sun et al. (2008) have reported a universal approach to fabricate various nanorings consisting of different polymer and inorganic components using a combined CL and reactive ion etching (RIE) method. First, silica colloidal crystals are prepared from a composite film consisting of Fe_3O_4 nanoparticles and polymer. After thermal moulding at high temperatures, the colloidal crystals are removed. RIE treatment is used for removing redundant polymers that then results in a magnetic nanoring array. Yu and his co-workers (2009) have proposed a modified 3D CL method to fabricate nanorings via geometrically irreversible processing. This idea was inspired by the fact that conformal etching is not the geometric reverse of conformal growth at locations with negative surface curvature. In a 3D colloidal crystal, there are abundant negative curvatures at the contact point of each pair of spheres. After controllable etching of the coating layer, the ring is left at the contact points between the spheres. The template-assisted nanolithography theoretically can be used to prepare any kind of material with controllable geometric parameters by applying a similar principle. However, the elaborated design of experimental process and compatible template are necessary for the successful fabrication of desirable nanoring and nanotube.

Coordination-assisted Dissolution Process

It is well known that molecules can be selectively absorbed onto a specific crystal facet. If the absorbed molecule can react with the materials to form a dissolvable complex in solution at the same time, it is highly possible to make a hole in the solid nanoparticles. As a result, hollow nanostructures can be obtained during the coordination-assisted dissolution process. It was first reported by Jia and colleagues in 2005 that the hematite (Fe_2O_3) nanotubes could be formed through hydrothermal treatment of $FeCl_3$ in the presence of $NH_4H_2PO_4$ (Jia et al. 2005). By varying the concentration of two reactants or the ratio of double anion (phosphate and additional sulphate), the size and shape of hollow hematite nanoparticles can be modulated from the nanotube to nanoring (Fan et al. 2008; Jia et al. 2008). Figure 3 (a)–(d) shows the time-dependent evolution of the morphology of the hematite hollow nanoring during hydrothermal growth. This clearly demonstrates the formation of hollow α-Fe_2O_3 nanoparticles via the coordination-assisted dissolution process where the iron-phosphate complex is involved. The obtained α-Fe_2O_3 nanorings and nanotubes can be further transformed into cubic spinel magnetite Fe_3O_4, maghemite γ-Fe_2O_3 and ferrite MFe_2O_4 (M = Co, Mn, Ni, Cu) using a thermal transformation process without losing their single-crystalline nature as shown in Fig. 3(e).

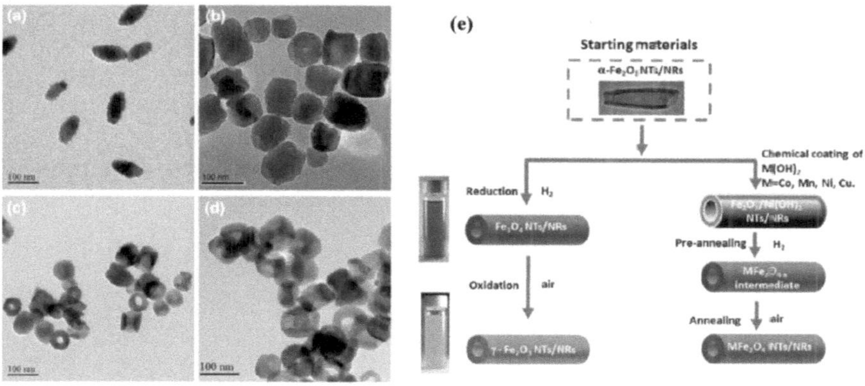

Fig. 3. Fabrications of iron oxide nanorings and nanotubes through coordination-assisted dissolution process. (a)–(d) Morphological evolutions of the α-Fe_2O_3 nanorings with different reaction times of (a) 3 h, (b) 10 h, (c) 24 h, and (d) 48 h, as revealed by TEM images. (e) The overall scheme for the synthesis of spinel MFe_2O_4 (M = Co, Mn, Ni, Cu) nanotubes/nanorings. The insets are the optical images of Fe_3O_4 and γ-Fe_2O_3 nanotubes water dispersion after surface modification by citric acid. (Reprinted with permission from Fan et al. 2008 and Fan et al. 2009, ©2009 American Chemical Society.)

Color image of this figure appears in the color plate section at the end of the book.

Nanoscale Kirkendall Effect

The Kirkendall effect originally describes the movement of the interface between diffusion couples. Such atomic diffusion occurs through vacancy exchange with the different diffusion rate at an elevated temperature where the diffusion is possible. Directional material flows resulting from coupled reaction-diffusion phenomena at solid/gas or solid/liquid interfaces lead to deformation, void formation, or both during the growth of metal oxide. Sun and Peng (2007) have reported a solution phase route to fabricate monodispersed magnetite nanorings by controlled oxidation of amorphous core-shell Fe-Fe_3O_4 nanoparticles through the nanoscale Kirkendall effect. Both Fe and Fe_3O_4 in such a core-shell system are in an amorphous state at the beginning. Controlled oxidation of these core-shell nanoparticles in the presence of the oxygen-transfer reagent Me_3NO gives a Fe_3O_4 shell that contains polycrystalline Fe_3O_4 grains. This controlled oxidation is directed by the nanoscale Kirkendall effect, in which Fe metal diffuses outward faster than oxygen does inward. This process leads Fe_3O_4 collecting at the metal-oxide interface rather than in the interior of the core, and consequently results in a superparamagnetic Fe_3O_4 polycrystalline nanoring. An et al. (2008) have reported that heating the magnetite nanocrystals in technical grade trioctylphosphine oxide (TOPO) can yield hollow nanoparticles by applying the same principle. Impurities such as alkylphosphine oxides, alkylphosphinates, alkylphosphonic acids, and alkylphosphinic acids in commercially available TOPO can react with magnetite and form dissolvable metal-phosphate complexes. This reactive process is also directed by the nanoscale Kirkendall effect, thus leading to the synthesis of magnetite nanorings. However, it should be noted that the nanoring obtained is of amorphous metal phosphate and its magnetisation behaviour is paramagnetic.

Wrap-bake-peel Process

The wrap-bake-peel process is quite simple and is developed for the synthesis of hollow iron oxide with porous shells (Piao et al. 2008). First, a thin layer of silica is coated onto the spindle-shaped akagenite (β-FeOOH) solid nanoparticle in solution. These silica coated spindle nanoparticles are subjected to a thermal treatment at 500 °C. Due to a significant volume change from a loose akagenite structure to a dense haematite phase, the inner iron oxide becomes porous hollow nanocapsule-like structures with a hematite (α-Fe_2O_3) phase. Finally, magnetite nanocapsules with porous shells are obtained by the gas-solid reduction reaction and the selective removal of the silica shell under a strong basic solution. It is worth noting that the resultant nanocapsules without any surfactant coating show excellent water dispersibility. The magnetite nanocapsules obtained by this method are superparamagnetic at room temperature.

CELL SEPARATION AND IMMUNOASSAY TEST

Magnetic separation is a process in which magnetically susceptible material is extracted from a mixture using magnetic force. The attractiveness of magnetic bioseparation arises from its scalability, efficiency, simplicity and mild conditions. The rapid advance in nanoscience and nanotechnology, in particular the development of superparamagnetic nanoparticles, strongly promotes applications such as magnetic separation of cells, DNA and proteins (Safarik and Safarikova 1999). The protocol for cell separation is basically not very different for solid SPIO and hollow nanoparticles. In a typical cell separation protocol, a monoclonal antibody specific for a target cell is immobilized on the surface of the nanoparticle as shown in part 1 of Fig. 4a. The cell mixture is incubated with the antibody-nanoparticles system, allowing the antibody to bind to the corresponding antigen. Magnetic separation then isolates the resulting cell-antibody-nanoparticle complex as shown in part 2 of Fig. 4a. After isolation of the cell is complete, antibodies are removed by degradation or replacement and the cells are released. Commercially available protocols offer cell recoveries that have a yield in the range of 85% to 99%. In some cases, it

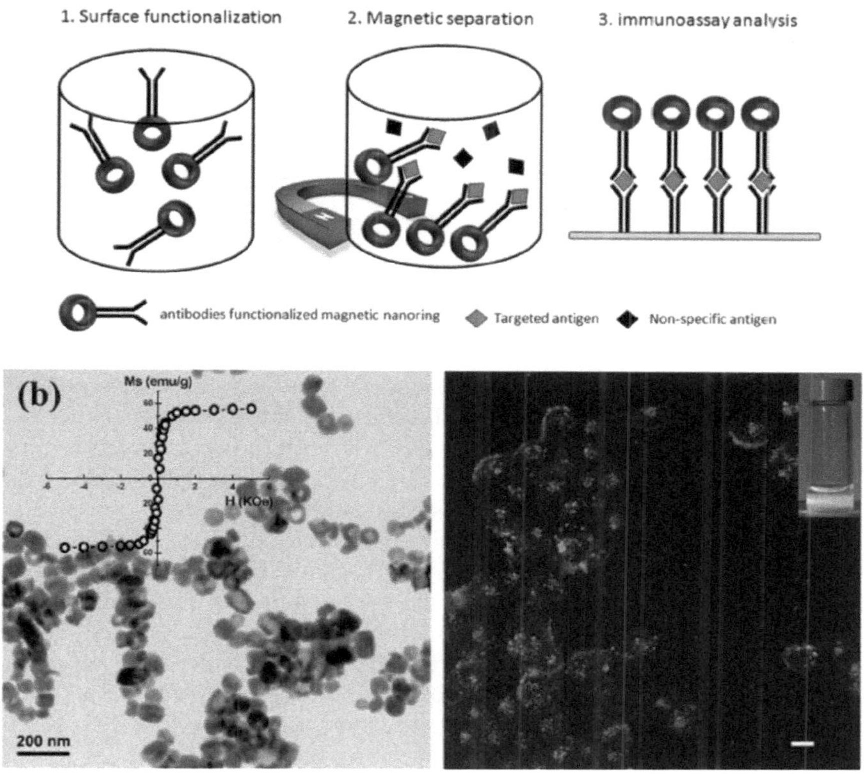

Fig. 4. Magnetic nanoring for cell separation and immunoassay. (a) Schematic illustration of a magnetic cell separation and capture-detection immunoassay. (b) TEM images of luminescent/magnetic nanorings. The insert is the hysteresis loop (luminescent/magnetic) at 300K. (c) Fluorescence images of lung cancer cells labelled with the luminescent/magnetic nanorings. The insert is the luminescent image of luminescent/magnetic nanoring dispersion in water. (Reprinted with permission from Fan et al. 2009, ©2009 American Chemical Society.)

Color image of this figure appears in the color plate section at the end of the book.

will be more convenient to combine cell separation with immunoassay analysis without removing the antibody. As shown in part 3 of Fig. 4a, the antibody-nanoring-antigen complex can be transferred and immobilized onto a plate with the secondary antibody. The plate can then be used to perform immunoassays via a biochemical test or through detection by a magnetic reader (called magnetic immunoassay).

In practical applications, the vortex-state nanorings will be an excellent candidate for cell separation compared to other magnetic hollow structures because they can provide a large magnetic drag force due to high saturation magnetization as compared to superparamagnetic nanorings and at the same time it have low magnetic inter-particle interaction in contrast to ferromagnetic nanotubes, which facilitate the formation of magnetic colloids. Fan et al. (2009) have demonstrated the magnetic separation of A431 lung cancer cells using vortex-state γ-Fe_2O_3 nanorings as shown in Fig. 4b. For the purpose of fluorescence imaging of separated cancer cells, γ-Fe_2O_3 nanorings were conjugated with luminescent quantum dots and anti-EGFR antibody. Since the vortex domain structure in nanocomposites possess a much higher saturation magnetization and can be easily saturated with a relatively small field, the labelled cells can be separated efficiently by applying a small external field (~1000 Oe) using a permanent magnet. The fluorescent images of the separated lung cancer cells are shown in Fig 4b. By contrast, typical superparamagnetic nanocrystals show much lower magnetization (<10 emu/g) under the same external field. Consequently, a high magnetic gradient must be offered. Moreover, an intrinsic peroxidase-like catalytic property, which is widely used for laboratory immunoassays, has been discovered in Fe_3O_4 nanoparticles (Gao et al. 2007). Such a catalytic property has been measured using Fe_3O_4 nanotubes and compared with that of solid nanoparticles of similar size (Fan et al. 2009). The catalytic constant (Kcat) of magnetite nanotubes (370 nm in length) is larger than that of spherical nanocrystals (300 nm in diameter) by an order of magnitude due to its large surface to volume ratio. These results suggest that we can combine the cell separation and immunoassay together using vortex magnetite nanorings for the detection of cancer cells. Such an attempt has yet to be made. However, we believe that the simple but highly sensitive magnetic Fe_3O_4 nanoring-based immunoassay, which combines the cell separation and detection techniques, will be attractive in the future.

MRI APPLICATION

MRI is perhaps the most successful medical imaging technique currently available in radiology to visualize internal structures of the body. This technique is non-invasive, non-destructive, has a short acquisition time and provides good contrast between the different soft tissues of the body without limitation in volume or depth of the analyzed target. These merits make MRI especially useful in cancer diagnosis and therapy as compared to other medical imaging techniques such as CT or X-rays. To improve the sensitivity of MRI, contrast agents are employed (Na et al. 2009). There are two types of contrast agents, positive (T1) contrast agents act primarily to shorten the relaxation time (T1) and generate a bright image while negative (T2 and T2*) contrast agents act mainly to shorten the transverse relaxation time (T2) and lead to signal reduction, that is, a dark image. SPIO is the first nanoparticulate MRI contrast agent, and is commonly used as a negative contrast agent. However, the ultra-small crystal size of SPIOs (typically less than 12 nm) makes it difficult to retain their stoichiometry, size uniformity and magnetism during the complex protocol for water-soluble nanoparticles, which in turn leads to rather poor enhancement of the spin-echo MR signal. In addition, effective tissue-labelling of cellular imaging requires a large number of SPIOs to be loaded into cells, however, both cell division and biodegradation can dilute the label below detectable ranges. Therefore, magnetic iron oxide nanorings/nanotubes with superior shape-induced magnetic property and the tuneable size (20–300 nm in outer diameter) are developed to provide significant enhancement on the MR T2* signal. In particular, the vortex-state iron oxides nanorings are more attractive contrast agents as compared to ferromagnetic nanotubes because of their ability to form stable colloid.

The first MRI test of magnetic vortex nanorings was performed using a quantum dot (CdSe/ZnS) capped, ferrimagnetic vortex-state iron oxide nanoring (QD-FVIO; Fan et al. 2010). The QD-FVIOs displayed varying hydrodynamic diameters of 310 nm (QD-FVIO1) and 155nm (QD-FVIO2), respectively. Figure 5 (up) shows a qualitative comparison of a T2*-weighted spin-echo MRI of VIO and commercial ferucarbotran (SPIO) with respect to the varied echo time (TE). Clearly, QD-FVIOs result in significantly greater signal reduction (darker images) at the designated TE from 10 to 30 ms in contrast to ferucarbotran. The MR relaxation states of QD-FVIOs and ferucarbotran are presented in the Table seen in Fig. 5 (down). The r2* values of QD-FVIOs almost quadruple that of the commercial ferucarbotran while the r2*/r1 ratios are greater by two orders of magnitude. Previous theoretical and experimental studies revealed that the r2* relaxation rate strongly depends on the local field inhomogeneity, which is correlated with the relative volume fraction, magnetic moment of magnetic core and susceptibility difference between the particle and water. Hence, the extremely large r2* relaxation states in QD-FVIOs obviously arise from their ring-like shape and magnetization process from vortex to onion that provide both high relative volume fraction and susceptibility. In addition, the internal field inhomogeneity of QD-FVIOs that originates from the susceptibility difference between the inner and outer surface of the magnetite nanoring may also contribute to the enhancement of the r2* value. Furthermore, these MR relaxation results (r2* >> r2) are not consistent with that of the clustered SPIO of similar size. The relaxation behaviour of FVIO is said to be in a strongly echo-limited regime, indicating that FVIO nanoring systems are of more complex construction than SPIO. The effects of a magnetic vortex core on the T2* relaxation time has not been well characterized. However, these *in vitro* MRI measurements show that FVIO nanorings could achieve significant enhancement of the MRI signal in T2* weighted sequences and have potential for cancer imaging applications.

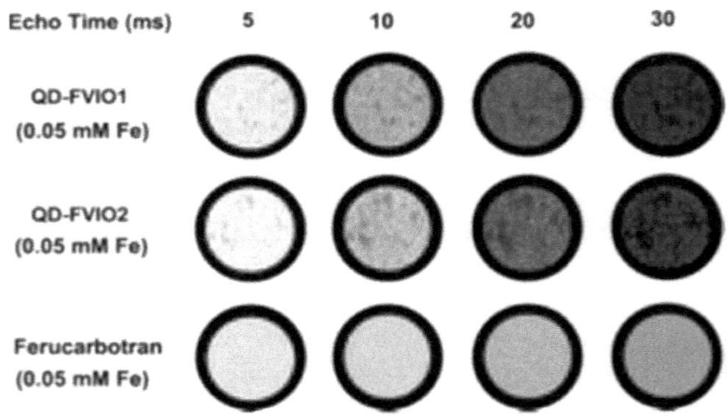

Sample	$r1$ ($s^{-1}mM^{-1}$)	$r2$ ($s^{-1}mM^{-1}$)	$r2/r1$	$r2^*$ ($s^{-1}mM^{-1}$)	$r2^*/r1$
QD-FVIO1	0.44	73.8	168	1079	2450
QD-FVIO2	0.59	55.1	93	976	1654
Ferucarbotran	11.3	225	19.9	254	22.5

Fig. 5. Magnetic vortex-state nanoring for MRI application. (Up) *In vitro* T2* weighted MRI of QD-FVIOs in 2% agarose and commercial ferucarbotran in water. (Down) MR relaxation rates of QD-FVIOs and commercial ferucarbotran at 1.5T are listed in the table. (Reprinted with permission from Fan et al. 2010, ©2010 American Chemical Society.)

TARGETED DRUG DELIVERY

Today, most of the anticancer therapeutic agents developed are non-specific and do not target the malignant cells, so the maximum dose of a given drug is limited due to toxicity related side-effects to the entire body after drug administration. The concept of nanoparticle-based nanovectors or nanocarriers, which have the ability to selectively deliver the therapeutic agents to the desired target and also help in controllable drug release after intravenous administration, serves as a promising alternative for cancer treatment. The mechanisms that drive the nanoparticles towards the target and helps in overcoming the biological barrier are almost the same in all kinds of magnetic nanoparticle candidates. However, unlike the solid SPIO, hollow magnetic nanorings and nanotubes contain many voids that could be used for drug storage, have more surface to be functionalized and also display high saturation magnetization to attain highly sensitive MRI tracking. Hence, it is of importance to develop novel nanocarriers of magnetic hollow nanoparticles, to seek highly efficient mechanism for drug storage and release by using the unique void inside the nanoparticles, and explore their potential application in cancer therapy.

A good example has been demonstrated by Cheng et al. (2009), using porous magnetite nanorings for targeted delivery and controlled release of cisplatin for the treatment of breast cancer. Cisplatin is widely used as a powerful therapeutic agent against numerous solid tumours. However, its therapeutic applications are largely restricted by some disadvantages like chemical instability, non-specificity to tumours, and poor water solubility. Carrier systems, such as hydrophilic polymer or biocompatible SPIO nanoparticles have been developed for the transport, controlled release and cellular distribution of cisplatin. Nevertheless, it still lacks the desired target specificity and therapeutic efficacy and because of this, a porous hollow Fe_3O_4 system has been developed for cisplatin delivery and controlled release. The fabrication of porous Fe_3O_4 hollow nanoparticles (PHNP) has been described earlier in the chapter. The spherical interior cavities are surrounded by discrete polycrystalline Fe_3O_4 domains (seen clearly from high resolution TEM images in Fig. 6 c and d). Interestingly, the porous wall is stable in neutral or basic physiological conditions, however, in low pH (<6) conditions, the wall tends to form wider pore gaps due to acidic etching. This characteristic can be used to control the release of drugs in response to external pH in physiological environments. Figure 6a schematically shows the procedure for simultaneous surfactant exchange and cisplatin loading into a PHNP and functionalization of this PHNP with Herceptin. The maximum quantity of encapsulated cisplatin in the PHNPs is about 25%. It is worth noting that the loading quantity is correlated to the pore size and experimental loading process, so that it can be further developed in the future by the design of nanoparticles and optimizing experimental conditions. The kinetics of drug release indicates that the escape of cisplatin was by diffusion-controlled slow process with $t_{1/2}$=16 h in neutral or basic physiological conditions. However, faster release of cisplatin with $t_{1/2} < 4$ h have been observed in low pH (<6) conditions. PHNPs have been further developed for specific targeting of breast cancer cells by Herceptin (a Humanized IgG1 monoclonal antibody) conjugation. The *in vitro* cytotoxicities of Pt-PHNPs and Her-Pt-PHNPs on SK-BR-3 cells are presented in Figure 6b. It can be seen that Her-Pt-PHNPs exhibit higher cytotoxicity than free cisplatin after 48 h of incubation, indicating that the targeted release and high cytotoxicity of the encapsulated cisplatin in the

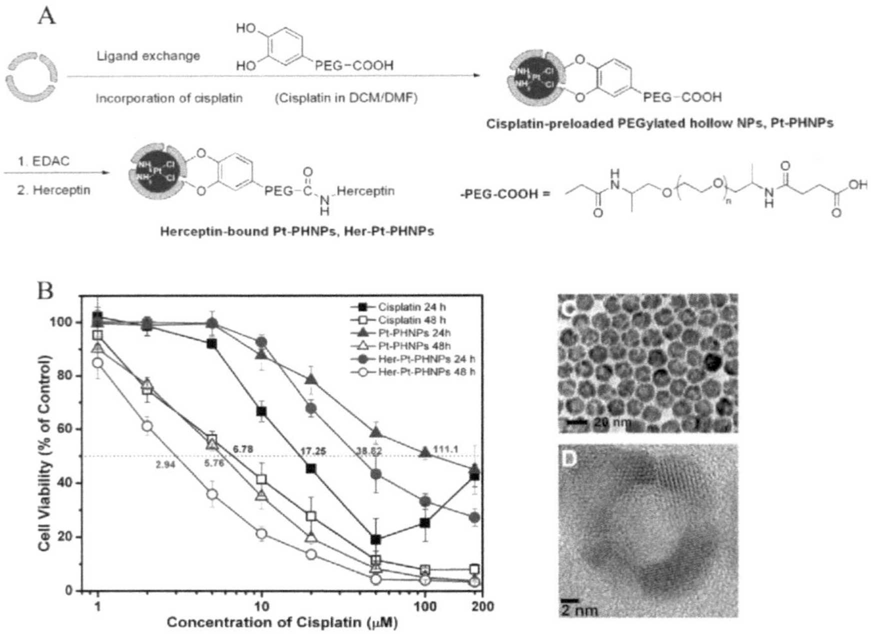

Fig. 6. Magnetic hollow nanoparticles for drug delivery and release. (a) Schematic illustration of simultaneous surfactant exchange and cisplatin loading into a PHNP and functionalisation of this PHNP with Herceptin. (b) Cytotoxicity of cisplatin (black line), cisplatin-loaded Pt-PHNPs (green line, 19.9% of Pt/Fe), and cisplatin-loaded Herceptin-bound Her-Pt-PHNPs (blue line, 19% of Pt/Fe) to SK-BR-3 cells as a function of the cisplatin dose after 24 h and 48 h of incubation, (c) and (d) TEM and high resolution TEM images of PHNP, respectively. (Reprinted with permission from Cheng et al. 2009, ©2009 American Chemical Society.)

cells is triggered by the acidic environment of endosomes and lysosomes. The half maximal inhibitory concentration (IC_{50}) of Her-Pt-PHNPs is 2.9 μM while that of free cisplatin is 6.8 μM. Similar experiments using SK-BR-3 cancer cells have also been reported by Piao et al. (2008), using water-dispersible porous Fe_3O_4 magnetite nanocapsules where doxorubicin was used as a model drug and incorporated into the PEG-coated nanocapsules with a high loading of about 28.9 wt%. These results clearly demonstrate the potential applications of porous magnetic hollow nanoparticles as nanocarriers for anticancer drugs. The existence of nanoscale cavities can not only improve the drug loading but also act as novel "smart" carriers with improved trigger mechanisms that can respond to external stimuli such as pH, temperature, AC magnetic field and light for controlled drug release.

MAGNETIC HYPERTHERMIA

Hyperthermia in oncology therapeutics is a general term for the raising of temperature above the physiologic level (usually in the 40°C–45°C range) over 30min within a targeted tumour without damaging the surrounding healthy tissue. It is well accepted that hyperthermia provokes morphological and physiological changes at the cellular level, which is thought to perturb metabolic pathways preceding cell death. Hyperthermia therapy as a modality in malignant cancer treatment has recently attracted much attention because this modality is expected to significantly reduce a clinical "side effect" and can be used effectively for killing localized or deeply seated tumours. The major issue in current hyperthermia therapy is the low heating temperature resulting from heat loss through a relatively big space gap that is formed between targeted cells and hyperthermia agents, and destruction of healthy cells due to the inability of hyperthermia agents to distinguish normal and tumour cells.

Magnetic nanoparticles based hyperthermia agents (magnetic colloids) offer many biotechnical and physiological advantages that may solve these problems: (a) AC magnetic field induced self-heating (b) can be directly injected into the blood vessels due to their small size, (c) external magnetic field controlled transport through blood vessels to the tumour cell, and (d) possibility for the differentiation of tumour cells from healthy cells by functionalizing the surface of nanoparticles with monoclonal antibodies or ligands to target tumour-related receptors. The basic principle of magnetic induction hyperthermia is that when magnetic fluid is delivered to tumour sites and subjected to an AC magnetic field, the magnetic nanoparticles will produce heat from AC magnetization loss and silence the cancer cells at a controlled temperature. The volumetric power dissipation during magnetic induction hyperthermia can be given by $P = \pi\mu_0 X'' f H_0^2$ (Wang et al. 2010), where $\mu_0 = 4\pi\times10^7$ (Hm^{-1}) is the permeability of free space, H_0 (Am^{-1}) is the magnetic field intensity in the materials, f is the field frequency and χ'' is the imaginary component of complex susceptibility. Therefore, high performance hyperthermia therapy will require well-designed magnetic hyperthermia agents that can form very stable suspensions (magnetic colloids) and have high saturation magnetizations, low demagnetizing fields, and large surface to volume ratios. In this context, superparamagnetic and vortex-state iron oxide nanorings are promising materials as hyperthermia agents for highly efficient magnetic hyperthermia treatment compared to solid SPIO. Unfortunately, there has been no report published on hyperthermia test of magnetic nanorings. Further investigation in this area is on the way.

Summary Points

- Magnetic nanorings and nanotubes, which have high chemical stability, tuneable magnetic properties, unique hollow structures and large surface to volume ratio compared with superparamagnetic nanoparticles, offer new opportunities for designing novel imaging and therapeutic tools for cancer detection and therapy, or to largely improve the performance of currently available nanoparticle-based medicines.
- Among various magnetic hollow nanostructures, superparamagnetic hollow nanostructures and vortex-state nanorings that can form stable ferrofluids are the most promising materials for cancer-related biomedical applications.
- Magnetic nanotubes and nanorings can provide larger magnetic drag forces than SPIO for high efficiency, low energy-consuming cell separation techniques.
- Vortex-state iron oxide nanorings as contrast agents exhibit the largest T2* MR relaxation rates thus far and have shown great potential for MRI applications.
- Superparamagnetic porous nanorings and nanotubes, which have many cavities inside the nanoparticles could be used as nanovehicles for anticancer drug delivery and release. Their unique hollow structure can achieve high drug loading, external stimuli response and drug release.
- Magnetic nanorings with circular shape and large magnetic susceptibility can achieve large damaging heat from AC magnetization loss while minimizing demagnetization effects. Therefore, they can be developed as high performance hyperthermia agents for magnetic hyperthermia treatment of cancer cells or tumours.

Definitions

Superparamagnetic iron oxide nanorings/nanotubes: Like other superparamagnetic iron oxide nanoparticles, nanorings/nanotubes are single-domain ferrimagnetic iron oxide nanoparticles with hollow structures. Due to the small size effect, the energy barrier (KV) of the nanoparticle associated with the magnetization moving from its initial easy axis direction to another easy axis direction is subject to thermal fluctuations where K is the magnetic anisotropy energy and V is the volume of the nanoparticle. Therefore, magnetization can randomly flip direction at room temperature. Superparamagnetic iron oxide nanorings/nanotubes are usually polycrystalline.

Ferrimagnetic vortex-state iron oxide nanoring (FVIO): FVIO is a nanoscale ferrimagnetic iron oxide ring-like particle with a flux-closed magnetic configuration (so called vortex state) where the magnetization runs circumferentially around the ring, either clockwise or counter clockwise.

Ferrimagnetic onion-state iron oxide nanoring (FOIO): In contrast to the FVIO, FOIO is an iron oxide nanoring with an onion state magnetic configuration. An onion state is a bidomain state that consists of two semicircular head to head domains separated by a domain wall.

Ferrofluid: Ferrofluid (also known as magnetic colloid) is a colloidal mixture composed of nanoscale ferromagnetic or ferrimagnetic particles suspended in a carrier fluid that is either organic solvent or water. The ferromagnetic nanoparticles are coated with a surfactant to prevent their agglomeration induced by the inter-particle attraction because of van der Waals and magnetic forces. Ferrofluid can be strongly magnetized in the presence of a magnetic field. Usually, superparamagnetic hollow nanoparticles or ferrimagnetic vortex-state nanorings can form stable ferrofluids.

Surface functionalization: The term of surface functionalization in nanomedicine usually refers to the act of modifying the surface of nanoparticles with a functional polymer, molecule or antibody for a specific application. It can alter a wide range of characteristics of the surface of nanoparticles such as roughness, hydrophilicity, surface charge, surface energy, biocompatibility and reactivity.

Targeted Drug delivery: Targeted drug delivery refers to the method or process of achieving a therapeutic effect in humans or animals by selectively delivering nanoparticles based therapeutic agents to a targeted area of the body (for example, in cancerous tissues), after intravenous administration followed by sustained release of the drug in a controlled manner.

Magnetic Hyperthermia: Magnetic hyperthermia refers to a magnetic nanoparticle based therapeutic method named after an experimental cancer treatment. The basic principle is that when magnetic colloid delivered to tumour sites are subjected to an AC magnetic field at a certain frequency, the magnetic nanoparticles will produce heat from AC magnetization loss and destroy the cancer cells at a controlled temperature.

Abbreviations

MRI	:	magnetic resonance imaging
CT	:	computed tomography scans
PET	:	positron emission tomography
SPIO	:	superparamagnetic iron oxide nanoparticles
FVIO	:	ferrimagnetic vortex-state iron oxide nanoring

FOIO	:	ferrimagnetic onion-state iron oxide nanoring
ALD	:	atomic layer deposition
CVD	:	chemical vapor deposition
AAO	:	anodic aluminium oxide
CL	:	colloidal lithography
RIE	:	reactive ion etching
TOPO	:	trioctylphosphine oxide
QD	:	quantum dot
PHNP	:	porous Fe_3O_4 hollow nanoparticle
IC_{50}	:	the half maximal inhibitory concentration

References

An, K., S.G. Kwon, M. Park, H.B. Na, S. Baik, J.H. Yu, D. Kim, J.S. Son, Y.W. Kim, I.C. Song, W.K. Moon, H.M. Park and T. Hyeon. 2008. Nano Lett. 8: 4252–4258.

Cheng, K., S. Peng, C. Xu and S. Sun. 2009. Porous hollow Fe_3O_4 nanoparticles for targeted delivery and controlled release of cisplatin. J. Am. Chem. Soc. 131: 10637–10644.

Cheng, K., and S. Sun. 2010. Recent advances in syntheses and therapeutic applications of multifunctional porous hollow nanoparticles. Nano Today 5: 183–196.

Fan, H.M., M. Olivo, B. Shuter, J.B. Yi, R. Bhuvaneswari, H.R. Tan, G.C. Xiang, C.T. Ng, L. Liu, S.S. Lucky, B.H. Bay and J. Ding. 2010. Quantum dot capped magnetite nanorings as high performance nanoprobe for multiphoton fluorescence and magnetic resonance imaging. J. Am. Chem. Soc. 132: 14803–14811.

Fan, H.M., J.B. Yi, Y. Yang, K.W. Kho, H.R. Tan, Z.X. Shen, J. Ding, X.W. Sun, M.C. Olivo and Y.P. Feng. 2009. Single-crystalline MFe_2O_4 nanotubes/nanorings synthesized by thermal transformation process for biological applications. ACS Nano 3: 2798–2808.

Fan, H.M., G.J. You, Y. Li, Z. Zheng, H.R. Tan, Z.X. Shen, S.H. Tang and Y.P. Feng. 2008. Shape-controlled synthesis of single-crystalline Fe_2O_3 hollow nanocrystals and their tunable optical properties. J. Phys. Chem. C. 113: 9928–9935.

Frey, N.A., S. Peng, K. Cheng and S. Sun. 2009. Magnetic nanoparticles: synthesis, functionalization, and applications in bioimaging and magnetic energy storage. Chem. Soc. Rev. 38: 2532–2542.

Gao, L.Z., J. Zhuang, L. Nie, J.B. Zhang, Y. Zhang, N. Gu, T.H. Wang, J. Feng, D.L. Yang, S. Perrett and X.Y. Yan. 2007. Intrinsic peroxidase-like activity of ferromagnetic nanoparticles. Nat. Nanotech. 2: 577–583.

Goesmann, H., and C. Feldmann. 2010. Nanoparticulate functional materials. Angew. Chem. Int. Ed. 49: 1362–1395.

Goya, G.F. and V. Grazú, and M.R. Ibarra. 2008. Magnetic nanoparticles for cancer therapy. Curr. Nanosci. 4: 1–16.

Hobbs, K.L., P.R. Larson, G.D. Lian, J.C. Keay and M.B. Johnson. 2004. Fabrication of nanoring arrays by sputter redeposition using porous alumina templates. Nano Lett. 4: 167–171.

Jia, C.J., D.S. Sun, F. Luo, X.D. Han, L.J. Heyderman, Z.G. Yan, C.H. Yan, K. Zheng, Z. Zhang, M. Takano, N. Hayashi, M. Eltschka, M. Klaui, U. Rudiger, T. Kasama,

L. Cervera-Gontard, R.E. Dunin-Borkowski, G. Tzvetkov and J.J. Raabe. 2008. Large-scale synthesis of single-crystalline iron oxide magnetic nanorings. J. Am. Chem. Soc. 130: 16968–16977.
Jia, C.J., L.D. Sun, Z.G. Yan, L.P. You, F. Luo, X.D. Han, Y.C. Pang, Z. Zhang and C.H. Yan. 2005. Single-crystalline iron oxide nanotubes. Angew. Chem. Int. Ed. 44: 4328–4333.
Jun, Y.W., Y.M. Huh, J.S. Choi, J.H. Lee, H.T. Song, S. Kim, S. Yoon, K.S. Kim, J.S. Shin, J.S. Suh and J. Cheon. 2005. Nanoscale size effect of magnetic nanocrystals and their utilization for cancer diagnosis via magnetic resonance imaging. J. Am. Chem. Soc. 127: 5732–5733.
Na, H.B., I.C. Song and T. Hyeon. 2009. Inorganic nanoparticles for MRI contrast agents. Adv. Mater. 21: 2133–2148.
Peng, S., and S. Sun. 2007. Synthesis and characterization of monodisperse hollow Fe_3O_4 nanoparticles. Angew. Chem. Int. Ed. 46: 4155–4158.
Piao, Y., J. Kim, H.B. Na, D. Kim, J.S. Baek, M.K. Ko, J.H. Lee, M. Shokouhimehr and T. Hyeon. 2008. Wrap-bake-peel process for nanostructural transformation from beta-FeOOH nanorods to biocompatible iron oxide nanocapsules. Nat. Mater. 7: 242–247.
Portney, N.G., and M. Ozkan. 2006. Nano-oncology: drug delivery, imaging, and sensing. Anal. Bioanal. Chem. 384: 620–630.
Safarik, I., and M. Safarikova. 1999. Use of magnetic techniques for the isolation of cells. J. Chromatogr. B. 722: 33–53.
Sekhon, B.S., and S.R. Kamboj. 2010. Inorganic nanomedicine—Part 1. Nanomedicine 6: 516–522.
Sun Z., Y. Li, J. Zhang, Y. Li, Z. Zhao, K. Zhang, G. Zhang, J. Guo and B. Yang. 2008. A universal approach to fabricate various nanoring arrays based on a colloidal-crystal-assisted-lithography strategy. Adv. Funct. Mater. 18: 4036–4042.
Wang, X., J. Tang and L. Shi. 2010. Induction heating of magnetic fluids for hyperthermia treatment. IEEE Trans. Magn. 46: 1043–1051.
Xie, J., J. Huang, X. Li, S. Sun and X. Chen. 2009. Iron oxide nanoparticle platform for biomedical applications. Current Medicinal Chem. 16: 1278–1294.
Yang, Y., H.M. Fan and J. Ding. 2010. Stable magnetic vortex nanoring suspension: a LLG micromagnetics simulation. Unpublished.
Yu, X., H. Zhang, J.K. Oliverio and P.V. Braun. 2009. Template-assisted three-dimensional nanolithography via geometrically irreversible processing. Nano Lett. 9: 4424–4427.
Zhang, Q., W. Wanga, J. Goebla and Y. Yin. 2009. Self-templated synthesis of hollow nanostructures. Nano Today 4: 494–507.

11

Silicon Nanowire Arrays and other Silicon Nanostructures and Their Clinical Applications to Cancer

Murugan Veerapandian[2] and *Min-Ho Lee*[1,*]

ABSTRACT

Integration of nanomaterials and biomolecules are always of great interest to many researchers due to its wide range of applications in science and technology. Bioanalytical applications are currently the highly discussed area to have number of practical implementations. In particular, novel biosensors derived from nanomaterials have received much devotion. Over the past decades, several research groups have worked on semiconducting silicon nanowire as biosensors. Silicon nanowires (SiNWs) have been explored as building blocks to fabricate highly sensitive electronic device through the bottom up and top down approach. The underlying mechanism for nanowire biosensors

[1]Medical IT Technology, Korea Electronics Technology Institute, #68 Yatap dong, Bundang Gu, Seongnam Si, Gyeonggi Do, Korea 463-816; E-mail: mhlee@keti.re.kr

[2]College of Bionanotechnology, Kyungwon University, San65, Bokjeong dong, Sujeong Gu, Seongnam Si, Gyeonggi Do, Korea 461-701; E-mail: momugan@gmail.com

*Corresponding author

List of abbreviations after the text.

is a field effect transistor. In this chapter, we review the various properties and fabrication techniques of silicon nanoarrays and other silicon nanostructures. Representative examples of electrical detection mechanism and relative reports on cancer markers are briefly discussed. Furthermore, the key principles involved in applications of silicon nanowires and its other structures as drug carriers for cancer therapy are highlighted.

INTRODUCTION

Point-of-care testing solutions are always desired as a viable option for the rapid and sensitive detection and analysis of disease. Cancer persists to be the most discussed global disease in which prostate, lung, breast and colon cancer are listed to be the high cause for the mortality around the world. New and alternative molecular screening methods are being used to study cancer and these are resulting in a better understanding of the disease. Multi-analyte detection based on potential lab-on-a-chip point-of-care devices (POC) are required to overcome the existing issues in cancer diagnosis. Modern research has tried to develop such interesting and facile novel functionalized nanometer-sized materials, controlling and converting their effects in a much estimated manner to address the needs of specific bio-analysis. Metallic or semiconducting particulate materials with a wide range of aspect ratio, cross-sectional diameter of <1 μm and length as several tens of microns are described as nanowires (Rosi and Mirkin 2005). Nanowires are found to be promising in a number of sensing strategies such as optical, electrical, electrochemical and mass-based approaches, reviewed elsewhere (He et al. 2004). Features such as small size, high-surface to volume ratios, and/or electronic, optical and magnetic properties differ from those of bulk materials, which made the nanowires advanced functional materials. Recent studies revealed the possibilities of incorporating large numbers of nanowires into arrays and complex hetero-architectures for high density biosensing, electronics and mechanical applications (Liber et al. 2007). Furthermore, advanced protocols in ultrasensitive detection using single nanoparticle sensing, nano-fluidics, single-molecule detection, and multiplexing are in construction. In specific, biomedical sciences and biotechnology have blessed significantly from technological improvements of nanoscience. Disease diagnosis and treatment are based on an extensive analysis of biochemical processes. Identification of multiple biomolecular targets, such as nucleic acids or proteins, is important for medical diagnostics and treatment. For instance, multiplexed tests for respiratory pathogens and diagnosing cancers are found to be important for patient recovery and health monitoring (Wulfkuhle et al. 2003).

Nanoscale silica materials are extensively studied and have evinced to be a suitable host system for protein due to its high chemical and thermal stability, large surface area, and final suspending ability in aqueous solution and relative eco-friendliness (Tang et al. 2007). Measurement of changes in surface charges of silica nanostructures in presence of molecular species can modulate the conductance of the nanowires (Ajay et al. 2008); the devices are analogous to biologically gated field effect transistors (FET). Because of its electrostatic stabilization silica surface promotes the dispersion of nanostructure and thus it is considered as highly "soluble", which enhances its applications in solution-based bioassays. Further, electronically switchable properties of semiconducting silicon nanowires provide a sensing modality, a direct and label free electrical readout, which exceptionally impressive. The size of the nanowires can be readily tuned to 100nm and even smaller, which can result in high density device. So, it is feasible for the miniaturized devices to multiple samples in a real time. Silica nanomaterials are also having a high surface silanol concentration which facilitates a large range of surface chemical reactions and the binding of biomolecules. There are many biomolecules that are reported to be conjugated to silica nanostrucutres, including biotin-avidin, antigen-antibodies, peptides, proteins and nucleic acids (Zhao et al. 2004). Due to its multifunctionality silica based nanostructures (nanowires) are found to be the suitable probe towards the analyte and hence utilized much in sensing studies.

The current chapter is organized to review the general properties of silicon nanowire arrays and other silicon nanostructures application in cancer featured by fabrication methods, biosensing mechanism: field effect transistors, usage in electrical detection of cancer markers and brief discussion about the application of other silicon nanostructures as drug delivery system for cancer therapy.

APPLICATION TO OTHER AREAS OF HEALTH AND DISEASE

The versatility of silicon nanowire arrays and other silicon nanostructure has been assessed by recent research developments and implemented into various novel biosensing and drug delivery methods. The merits of silicon nanowires and other nanostructures can be recognized in the breakthrough, compared with other conventional nanomaterials, in terms of both selectivity and sensitivity. Effective integration of biomolecules with nanomaterial based devices clutch great promise for the design of compact and highly sensitive systems. Direct coupling of high efficient silicon nanomaterials and biomolecules can be achieved through several approaches as discussed here and highlights the feasibility of fully utilizing biosensor in disease diagnosis. The specificity of simultaneous detection of multiple analyte using silicon and other material functionalized nanoarrays in any bio-components with significant reproducibility will surely be simplifying the existing issues in health care systems. Such efforts will also lead to extraordinary advances in bio-molecular manipulation. The general interest in nano-biotechnology derived commercialization and increasing activity in this field by small scale laboratory and large scale industrial entities will definitely shorten the paths from lab to practical commercial applications of personalized diagnosis system.

FABRICATION OF SILICON NANOARRAYS AND OTHER SILICON NANOSTRUCTURES

Generally, two methods can be followed to fabricate silica-based nanostructures: the Stöber and reverse microemulsion processes. Both are bottom-up approaches, in which the particles are formed by self assembly (Stöber et al. 1968; Yu et al. 2008). One dimensional SiNWs are particularly of great interest due to their amorphous nature, which results in a material with a theoretical modulus in excess of 100 GPa and strength in excess of 25 GPa. These specific properties are not significant as carbon nanotubes, however SiNWs are more easily fabricated on a bulk scale and their mechanical properties are not compromised by lattice defects. These properties therefore built SiNWs as possible candidates for reinforcement in high implementation nanocomposite materials. A key factor in attaining the excellent functionalities of SiNWs is the preparation and proper orientation of very narrow wires (Paul 2010). Several reported bottom-up approaches have explained that high end device integration using nanowires is tough to achieve, due to the high costs and problem in controlling. However, some recent reports (Bunimovich et al. 2006 and Stern et al. 2007) demonstrated that top-down approaches might rectify the intrinsic degradation caused by the Reactive Ion-Etching (RIE) of the silicon layer, and expressed high sensitivity on pH (Stern et al. 2007) and DNA hybridization (Bunimovich et al. 2006), the problem of relocation to desired substrates for integration remains unresolved. In addition, the presence of salt in phosphate buffered saline (PBS) or blood samples dictates the salt screening, in other words the Debye screening effect, which diminishes the efficiency of signal detection (Heitzinger and Klimeck 2007). For high throughput-sensing devices it is important to maintain the easy accessing towards active substrates and controlling of salt levels, but this remains a major issue. In order to address this issue, we introduced a top-down method to fabricate SiNWs, which is similar to the conventional micro-electromechanical system than other SiNWs fabrication approaches. This new method of transferring of SiNWs array is well lined up with the active substrate and has a high yield. Specific oxidation, patterning process, dry etching and KOH derived wet etching followed by oxidation and nanowire formation are key steps involved in the generation of well organized SiNWs array (Fig. 1 and Fig. 2). Intrinsic degradation is prevented from the RIE process. Further it can be suitable for large scale production and much interesting for their use in biological application (Lee, K.N. et al. 2007 and Lee, M.H. et al. 2008). Table 1 describes the different methods of fabrication of nanowires and its special features.

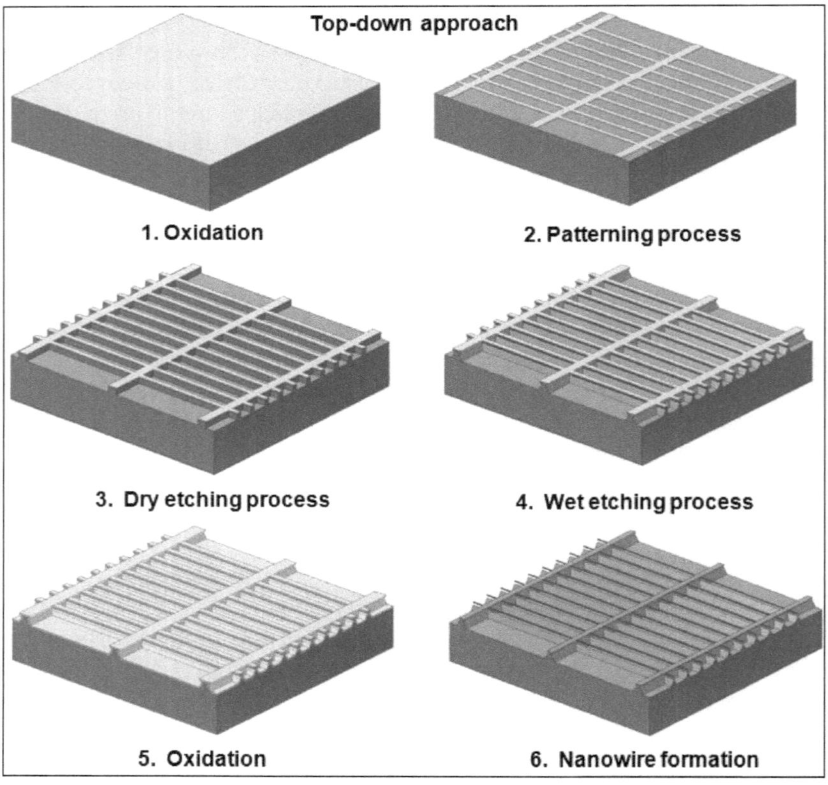

Fig. 1. Steps involved in the fabrication of silicon nanowire arrays (unpublished material of the author).

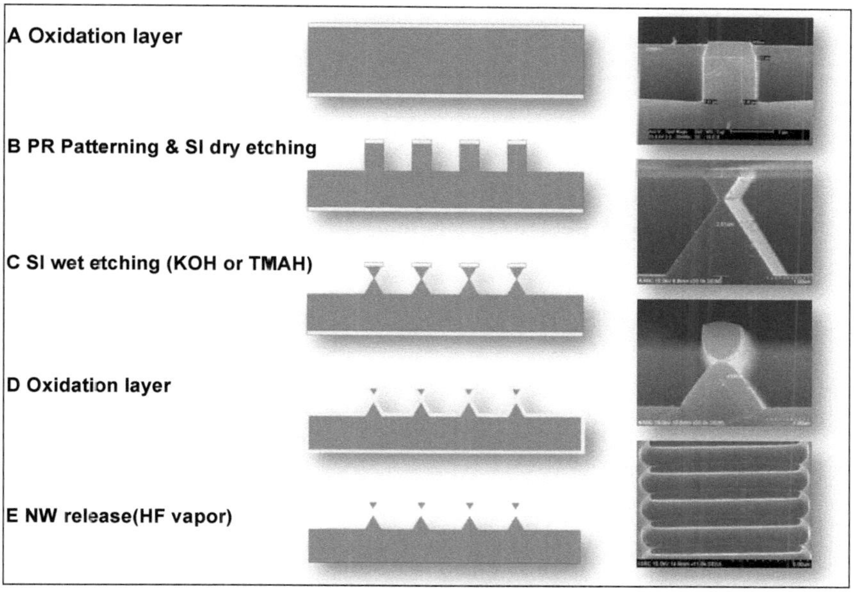

Fig. 2. A schematic of the SiNW array fabrication process and corresponding SEM images of each step. Overall process consists of silicon anisotropic etching, thermal oxidation, photoresist (PR) patterning, silicon wet etching, thermal oxidation, and SiNW release (TMAH, tetramethylammonium hydroxide; HF, hydrofluoric acid). (Lee et al. 2010) (Reprinted with permission from Elseveir).

MECHANISM OF SINWS BIOSENSING: FIELD EFFECT TRANSISTORS (FET)

Recent studies are greatly devoted to the development of SiNWs as ultrasensitive biosensors and chemical sensors (Lin et al. 2009). The ultrahigh sensing ability is attributed to the low-dimensional semiconducting nature of the materials featured by high surface-area-to-volume ratios, enabling local charge transfers to result in current change due to a field-effect, for instance specific response to the nanowire surface binding analytical molecules (Wang et al. 2006). The interface and physicochemical compatibility of Si/SiO_2 made its compatibility with more traditional Si-based technologies. A strong electrical field causes the flow of electricity within a nearby semiconductor, based on the surface modification (analyte molecules) a weak to high electrical signal coming out through an electrode creates an electrical field by the rest of the transistor. The FET was also termed as a "unipolar" transistor due to the fact that current transported by carriers of major polarity, whereas in the traditional bipolar transistor carriers of both polarities (majority

Table 1. Describes the different methods and features utilized for the fabrication of nanowire arrays (unpublished material of the author).

	Yale Univ.	*Caltech.*	*Harvard Univ.*	*KETI*
Device Fabrication	Top-down; Anisotropic etching (100 nm)	Top-down; Super lattice, SNAP (17nm, RIE)	Bottom-up; Flow assembly (20 nm, VLS)	Top-down; NW transfer (50-200 nm)
Features	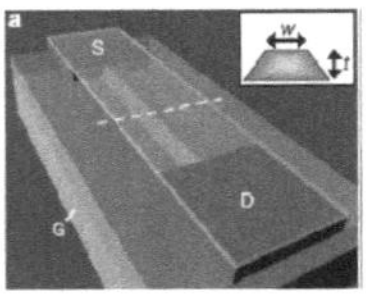	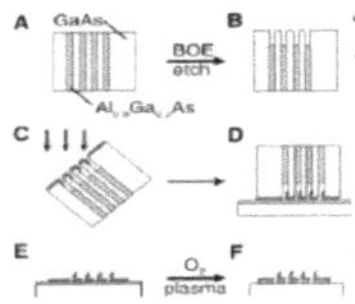	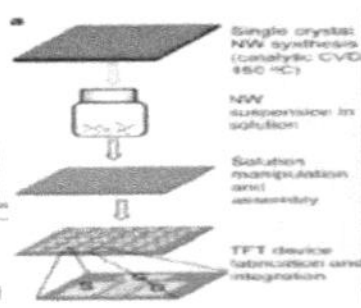	
On-off ration	~ 10^5 (1 NW)	< 10^6 (8 NWs)	10^8 (91 NWs)	>10^5 (1 NW)
S/D length		4 μm (=gate)	5 μm (=gate)	7, 15 μm(gate= 4)
$I_{on\ max}$	~ 1 μA	~ 5 μA	~ 100 μA	~ 10 μA
Subthreshold swing (SS)	Over 500 mV/dec. (Backgate)	83 mV/decade (Top-gate)	600 mV/decade (Bottom-gate)	160 mV/decade (Top-gate)
Contact	Doped Si (Ti/Au)	Annealing (Ti/Au)	Annealing (Ti/Au)	As made (Ti/Au)
Applications	Biosensor	Logics (inverter)	FET transistor, Biosensors	FET transistor, Biosensors
Device substrates	SOI	SOI	Any substrates; flexible,	Any substrates; flexible,
Comments	CMOS compatible	RIE; deterioration	Not compatible with mass production	Mass production compatible

and minority) are involved. FET group may be further classified as Junction FET, Depletion mode Metal-Oxide-Semiconductor (MOS) FET and Enhancement mode MOSFET. Notably electrical energy transfer reactions play a similar vital role in all biological systems by which all the biological cells capture and use energy. Oxidoreductase enzymes, NADH-dependent dehydrogenases or redox proteins, biofuel cells or bioelectro synthesis are some of the significant examples for this bioenergetics (Bard 1994). By utilizing the above principle several research groups worked on SiNWs based biosensors. Recently a simple and low cost method was used to fabricate poly-SiNW FET for biosensing application (Hsiao et al. 2009). In which the poly-SiNW channel was fabricated by employing the poly-silicon side wall spacer technique, the electronic properties which is comparable with current commercial semiconductor process. The electronic property of the poly-SiNW FET is similar to those of single-crystal SiNW FET in aqueous solution. Biotin and avidin/streptavidin sensing system was used to demonstrate the biosensing ability. That result indicated that poly-SiNW FET is quite stable in variety of non-interacting molecules, which is the key factor for biomedical applications (Hsiao et al. 2009). SiNWs are also used as label-free biosensor for other biomacromolecules such as MicroRNAs (miRNA). miRNA are noncoding RNA molecules that regulate the gene expression in both flora and fauna. They play a critical role in developmental and cell biology, including stem cell differentiation and development. Different diseases including cancer and neurological disorders are also associated with miRNA expression. Hence it acts as a molecular diagnostic for disease and new targets in drug development. The traditional methods currently in use for detection of miRNA includes Northern blotting (Valoczi et al. 2004, Ramkissoon et al. 2006), microarray (Liu et al. 2004), polymerase chain reaction (PCR) (Raymond et al. 2005), surface Plasmon resonance (SPR) (Fang et al. 2006), nanoparticles, conducting nanowires-based and bioluminescence-based techniques, reviewed elsewhere (Zhang et al. 2009). Even though there are a wide variety of well reported methods to analyze miRNA, due to their individual limitations such as low sensitivity, difficulty in routine analysis, inability for much smaller primer/promoters, multiple steps and auto-fluorescence (microarray with fluorescent dyes) limits the conventional methods for novel application. Therefore, an *in situ* label-free electrical readout process is required to detect the miRNA. SiNWs are reported to have highly sensitive, sequence-specific and label free nucleic acid sensing ability (Lin et al. 2009). In relation to this, a report was recently demonstrated that a recognition receptor Peptide Nucleic Acid (PNA) is surface immobilized on the surface of SiNW device without further labeling and measured the resistance change before and after hybridization, which correlates the concentration of hybridized target miRNA (Zhang et al. 2009). Detection limit of 1 fM was analyzed using optimized assay, they correlated their identification with the fully matched and mismatched miRNA sequences. In addition to that, SiNW device is believed to have detection capability of miRNA in total RNA extracted from Hela cells. That gives an additional significance of SiNWs as asuitable biosensor for early detection of miRNA as a biomarker in disease (cancer) diagnostics. On the other hand, SiNW is also studied to have high sensitive and rapid detection of reverse-transcription-polymerase chain reaction (RT-PCR) product of Dengue serotype 2 (DEN-2). In that study, the PNA is covalently attached to surface of SiNW and later utilized for the hybridization of complementary fragments of DEN-2. Measurement of difference in the resistance before and after hybridization on PNA functionalized SiNW can detect the binding of RT-PCR product on PNA sequence and believed to have potential point of care diagnostic applications (Zhang et al. 2010). Other silicon nanostructures such as nanoparticles have also been used to immobilize biomolecules such as enzymes, providing a good and simple approach for preparing high quality biosensors. SiO_2 nanoparticles have been introduced into the construction of enzyme field-effect-transistors (ENFETs) to create a glucose sensitive ENFET and providing a biocompatible environment to improve the enzyme activity, preventing leakage as well (Luo et al. 2004). Boron doped SiNWs have been reported (Cui et al. 2001) to create highly sensitive, real time electrically based sensors for biological and chemical species. Biotin was selectively modified on the surface of SiNWs and used to detect the streptavidin down to picomolar concentration. Further, antigen functionalized SiNWs result in real time reversible antibody binding with a manner of concentration dependent detection of metabolic indicator Ca^{2+}. The electrochemical and electrocatalytic properties of hemeproteins such as myoglobin, hemoglobin and horseradish peroxidase in layer-by-layer films assembled with SiO_2 nanoparticles at pyrolytic graphite molecules and the proteins have been demonstrated. The proteins in the films can display reasonable electrocatalytic activities toward various substrates such as oxygen, H_2O_2, tricholoroacetic acid and nitrite, which has showed a potential applicability in fabrication of new kind of biosensor without any additional mediators (He et al. 2004). The reports discussed in the current section regard the basic mechanism of FETs and biosensing application derived from SiNWs shows the viable role of silicon nanoarrays in disease diagnostics.

SILICON NANOWIRE ARRAYS FOR DETECTION OF CANCER MARKERS

Tumor related antigens are a substance generally used as an indicator or biomarker for normal biological processes, pathogenic processes or cancer diagnosis (Ibtisam 2009). These are simply cellular molecules that can be detected from tumor cells and body fluids like blood, urine and observed to be over-expressed due to cancer development. To date several range of biomarkers from different types of cancers are identified. Table 2 describes a list of biomarkers associated with different cancers. Recently our group involved in diagnosis of serum α-fetoprotein, carcinoembryonic antigen and carbohydrate antigen in 83 patients, who had undergone surgery for gastric cancer. The selection of these biomarkers is because of their relation towards stages of diseases. Further we examined the concentration of C-reactive protein (CRP) from sera of gastric cancer patients using our SiNW FETs. The particular reason for sensing CRP is due to its role in inflammation and possible indicator of future progression of the cancer. We designed and fabricated SiNWs to be sensible to CRP. Among the 83 patients, six samples from the patients who have marked elevation of CRP (> 3 to 10 mg/dL, according to clinical measurements) were selected

Table 2. Known biomarker associated with cancer diagnosis and prognosis. (Ibtisam 2009) (Reprinted with permission from Elseveir).

Cancer type disease	Biomarker
Prostate	PSA, PAP
Breast	CA15-3, CA125, CA27.29, CEABRCA1, BRCA2, MUC-1, CEA, NY-BR-1, ING-1
Leukaemia	Chromosomal abnormalities
Testicular	α-Fetoprotein (AFP), β-human chorionic gonadatropin, CAGE-1, ESO-1
Ovarian	CA125, AFP, hCG, p53, CEA
Any solid tumour	Circulating tumour cells in biological fluids, expression of targeted growth factor receptors
Colon and pancreatic	CEA, CA19-9, CA24-2, p53
Lung	NY-ESO-1, CEA, CA19-9, SCC, CYFRA21-1, NSE
Melanoma Tyrosinase	NY-ESO-1
Liver	AFP, CEA
Gastric carcinoma	CA72-4, CEA, CA19-9
Esophagus carcinoma	SCC
Trophoblastic	SCC, hCG
Bladder	BAT, FDP, NMP22, HA-Hase, BLCA-4, CYFRA 21-1

and subjected to measurement with SiNW FETs. Oxygen plasma treated SiNWs surface is chemically reacted with APTES, glutaraldehye and later functionalized with anti-CRP. On the other hand, gold colloidal nanoparticles are prepared by citrate reduction and surface modified with antigen CRP. Polydimethylsiloxane (PDMS) based microfluidics is designed connecting with two micropumps. Illustration of the designed system is shown in Fig. 3A. PBS solution was first injected into the PDMS microchannel; the NWs were then permitted to contact the solution as it flowed through the microchannel (Fig. 3B). After completion of significant washing procedure the waste reservoir will collect the PBS solution and replace it with the target protein solution. The size of the microchannel was 600 μm in length and 400 μm in height. Fig. 4 shows that anti-CRP functionalized SiNWs only responded upon CRP injection (1 ng/mL, PBS), evinced that the current changes was resulted by the appropriate binding between CRP and anti-CRP conjugated SiNWs (Lee et al. 2010). Similarly, another group (Wu et al. 2009) has worked on SiNWs FET for biosensing of BRAF^{V599E} a specific mutation gene, which correlates the occurrence of cancers (melanomas, colorectal cancers, gliomas, lung cancers, sarcomas, ovarian carcinomas, breast cancers, papillary thyroid carcinomas, and liver cancers) as the target DNA sequence (Xu et al. 2003). Upon hybridization of mutation gene with the captured DNA strands on the SiNW, there is a significant increase in the threshold voltage is identified, whereas in the original stage of de-hybridization of the gene the decreased voltage

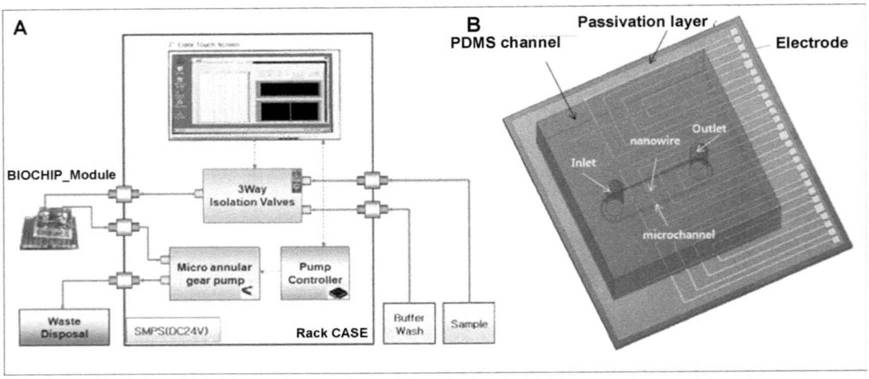

Fig. 3. (A) Schematic of automated fluidic control system. Three-way isolation valves were used—one for buffer, one for target samples, and one for delivery to the biochip module. The waste solution was directly pushed into waste disposal by a microannular gear pump. All pumps and times were carefully controlled by a custom-made lab view–based program. (B) PDMS microfluidic channel through which sample solution flows, and where SiNW FETs sensing took place. (Lee et al. 2010) (Reprinted with permission from Elseveir).

Color image of this figure appears in the color plate section at the end of the book.

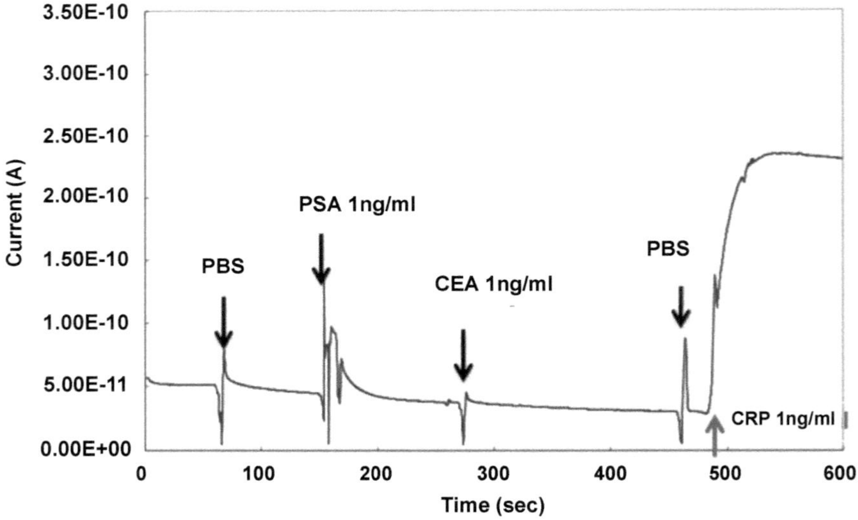

Fig. 4. CRP selectivity test with anti-CRP functionalized SiNWs. Prostate Specific Antigen (PSA) and Carcinoembryonic antigen CEA proteins were injected for the control. Upon injection of CRP, current increase by approximately fourfold was observed. On the contrary, CEA and PSA caused no detectable signal changes. (Reprinted with permission from Elseveir).

is observed. That report further revealed the logarithmic relationship of electrical signal with respect to the concentration of the mutation gene of upto six orders of magnitude, with the detection limit of sub-femto molar level. Also there is a distinguished detection result obtained from the mismatched DNA sequences and complementary DNA gene by SiNWs FET sensor (Wu et al. 2009). By developing such advanced biosensing device, it is possible to increase the survival rate and assist the medical specialist to provide the rationalized therapy for the early recovery of patients suffering from diseases.

SILICA NANOSTRUCTURES AS DRUG CARRIER FOR CANCER THERAPY

Silica is generally accepted as a biocompatible material and has been previously used as surgical implant. It is well studied that controlling of size and morphology are achievable in the synthesis of silica nanoparticles (Jana et al. 2007). Silica nanostructure acts as an excellent platform for facile loading of a wide range of imaging and therapeutic agents, making them a good candidate for theragnosis application. Several reports are studied about the novel application of silica nanoparticles in drug delivery for instance, 2-devinyl-2-(1-hexyloxyethyl)pyropheophorbide (HPPH), a hydrophobic photosensitizing anticancer drug is successfully loaded in silica matrices. The fluorescent property of HPPH is more on its functionalization with silica matrices than in the free form and can lead to efficient killing of cancer cells when irradiated with a laser (Kim et al. 2007). Fabrication of mesoporous structures is important factor for successful loading of drug molecules. To attain such nanoporous structures several methods have been reported, chemically and physically. For instance, n-alkyltrialkoxysilane or other surfactants are mixed with other reacting precursors and are incorporated into the matrices during particle formation. These surface acting agents are later removed from the nanostructures through post-synthesis solvent extraction or calcinations process to obtain the mesoporous structure (Jin et al. 2010; Slowing et al. 2008; Manzano et al. 2009 and Vallet-Regi et al. 2001). Through simple physical interaction large number of pharmacologically active molecules can be loaded on the porous structures. Advantageous of being biocompatible and biodegradable in aqueous solutions made siliceous mesoporous materials for number of drug delivery applications. Pure mesoporous silica material was used as a controlled delivery system (Vallet-Regi et al. 2001) and devoted to tailor the chemical properties, loading and release of guest molecules. In order to provide the sustained release of active moiety from the central core of silica nanoporous surface, sometimes other nanoparticles were surface modified. A photoinduced intracellular controlled release of anticancer drug paclitaxel by gold-capped mesoporous silica nanosphere was studied (Vivero-Escoto et al. 2009). Delivery of drug and self destruction of the same engineered nanoparticles is always desired to avoid the nanotoxicity. Recently a preliminary study demonstrated that luminescent porous silica nanoparticles (LPSiNPs) can have the self-destruction *in vivo* and be renally cleared within a relatively short period of time, therefore reducing the risk of their accumulation and causing toxic effects to normal organs. That LPSiNPs were loaded with doxorubicin and their drug release and cytotoxicity was studied *in vitro* (Park et al. 2009). Functionalization of active ingredients on the surface of nanoparticles will enhance the therapeutic activity of the free form pharmaceuticals. A series of organic N -[3-(3,8-diamino-6-phenyl-5-phenanthridinium)propyl]-3-aminopropyl-functionalized MCM-41 type mesoporous silica nanoparticle materials (PAP-LP-MSN and AP-PAP-MSN) was recently synthesized and characterized with different pore sizes (5.7 nm and 2.5 nm) (Vivero-Escoto et al. 2010). Further they decorated the external particle surface of PAP-LP-MSN and the interior pore surface of AP-PAP-MSN with an oligonucleotide intercalating phenanthridinum functionality. The different chemical accessibility of PAP-functionalized mesoporous silica nanoparticles and its interaction with intracellular oligonucleotides and endocytosis efficiencies were demonstrated. Due to the low permeability of phenanthridinum group such as ethidium bromide and propidium iodide are generally exempted from healthy viable cells. But, it has been studied in the literature that these chemicals can pass through damaged plasmic and nuclear membranes of animal cells (Sailer et al. 1996 and Steinkamp et al. 1999). Because of their strong affinity to nuclear DNAs and RNAs via intercalation and thereby influence the gene replication and transcription processes. Vivero-Escoto et al. tried to examine the function of two phenanthridinium-functionalized MSN materials using live human cervical cancer cells (HeLa). Their report suggests that the *in vitro* biocompatibilities and endocytic properties of AP-PAP-MSN seems to be higher than that of PAP-LP-MSN, which is well agreed with their previous observation (Slowing et al. 2006), emphasizing that positively charged MSN particles are at ease and more cell membrane permeable. Whereas the same intercalator group functionalized on the interior surface of AP-PAP-MSN and doesn't have interaction with cytoplasmic oligonucleotides. The possible mechanism is explained based on the behavior of mesoporous silica nanoparticles due to the size discrimination effect which prevents the oligonucleotides from approaching the PAP groups inside the AP-PAP-MSN. Several drug delivery materials have connecting network of porous morphology, such as dendrimers with significant porous structure and liposomes with a large void core and a porous shell. A "perfect" capping is required in order to achieve the desired "zero premature release" of active pharamaceuticals, because the pore encapsulated active molecules can have the chance of leakage through the connecting pores, when some of the pores are not capped well. Whereas, MSN is consists of honeycomb-like, 2D hexagonal porous structure (Fig. 5A and B) with cylindrical pores running from one end of sphere to the other. There is no such inter connecting pores between other channels. This distinct feature limits the "leaking" chance even in the case of improper capping, which means that the individual cylindrical pores can still be suitable for independent reservoir for drug encapsulation and release (Slowing et al. 2008). As discussed by Slowing et al., "zero premature release" is useful in delivering toxic drugs like anti-cancer agents. By using the phenomena of "gate keeping concept" (Lai et al. 2003) these MSN tend to be used as stimuli-responsive controlled release systems. Factors such as photochemical, pH responsive and redox active gatekeepers are applied in the release of various active molecules from MSN. From these studies it is observed that optimal tuning and surface functionalization of biocompatible silica material will lead the new nanodevices for delivery of cancer drugs and other biological applications.

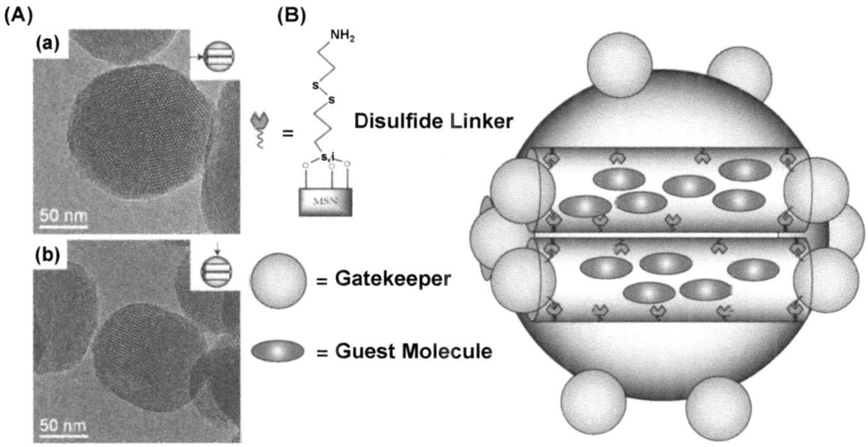

Fig. 5. (A) TEM images of MSNs materials recording from the direction (a) parallel or (b) perpendicular to the long axis of the mesochannels. (B) Representation of an MSN loaded with guest moleculesand end-capped with a general gatekeeper. (Slowing et al. 2008) (Reprinted with permission from Elsevier).

Color image of this figure appears in the color plate section at the end of the book.

OUTLOOK

The value of nanomaterials in medicine is inevitable. This chapter has described a brief introduction about the general properties of silicon nanowire arrays, different methods of fabrication and mechanism of biosensing via field effect transistors. Key references regard the diagnostic application of silicon nanowire in biosensing of biomarkers for cancer and specific application of silica nanoparticles in cancer drug delivery is also highlighted. However, this report gives the initial results for some novel diagnostic application of silicon nanowire arrays. Optimization of the physical, chemical and fabrication parameters is necessary to ensure exploitation of these nanoarrays from small scale laboratory diagnostics to large scale industrial application. We are confident that nanotechnology and its intersection between other fields will make a major contribution to the advancement of early disease diagnosis and treatment by helping drugs to be targeted more efficiently to specific sites. This may also be a means to enrich the research and development to eradicate the existing issues in health care systems.

Summary Points

- Silicon nanowire arrays with a direct sensing modality and label free electrical readout is exceptionally impressive for biomedical sensing.
- Low dimensional semiconducting nature, high surface-area-to-volume ratios enabled local charge transfer results in current change due to field-effect, which are found to be the underlying mechanism behind the silicon nanowire biosensing.
- Conjugation of biological entities (such as antibody or nucleic acids) on the surface of silica nanoparticles/nanowires are very useful in bioanalysis through its analyte recognition and/or signal generation.
- Silica nanoparticle is act as a suitable carrier for facile loading of a wide range of pharmaceutical agents, which making them a good candidate for theragnosis application.
- The large surface area of the pores allows the silica nanoparticles to be filled with active ingredients. Depending on the nature of chemicals functionalized on the outside of the particles, the biological cells will interact and taken up through endocytosis.
- The addition of suitable imaging and therapeutic ligands on the surface of silica nanoparticles that target the tumor cell can enhance its dual theragnostic efficiency.

Key Facts of Silicon Nanoarrays and other Silicon Nanostructures

- The use of nanoscale materials in medicine and disease diagnosis is rapidly increasing due to their unusual physicochemical and biological properties.
- Nanoscale formulation of active ingredients and functionalization of pharmacologically benign molecules on nanomaterials (silica nanoparticles) not only enhance the therapeutic value, but also increase the stability and several other pharmacokinetic profiles.
- Organization of nanoparticles or nanowires into highly ordered two dimensional (2D) arrays (on various active supports) is being increasingly considered for several bioanalytical studies.
- Several efforts (top-down and bottom-up approaches) were made worldwide to design and develop strategies that improve the nanofabrication process.
- Micro Electro Mechanical System (MEMS) based technology has the potential role in the revolution of nanoscale devices. Silicon nanoarrays is one of the typical nanostructures obtained by this technique in large scale, which has a renowned application in biosensing.
- Highly sensitive and specific nanomaterials based biosensors and drug delivery systems will surely simplify greatly the existing biochemical tests, (where rapid, accurate and reliable measurements are needed) and provide a potentially safe and effective therapy.

Definition of Key Terms

Nanowires: It is a type of nanoscale materials with a solid, cylindrical wire like structure usually has the diameter of the order of a nanometer (10^{-9} meters). Many different classes of nanowires exist, including molecular nanowires having repeating units (e.g., DNA), inorganic metallic (e.g., Au, Pt, Ni, etc.), semiconducting (e.g., Si, InP, GaN, etc.), and insulating (e.g., SiO_2, TiO_2).

Nanoarrays: An array of nanoscale objects, that is a form of patterns which have unusual properties for signaling.

Biosensor: It is an analytical device for the detection of an analyte that fuse the biological component with a physicochemical detector component. It is composed of three parts, including a sensitive biological element, transducer and signal processor.

Field Effect Transistor: A field-effect transistor (FET) is a type of transistor generally used for weak-signal amplification (for instance, for amplifying analog or digital signals). It relies on an electric field to control the shape and hence the conductivity of a channel of one type of charge carrier in a semiconductor material. It can also switch direct current or function as an oscillator.

Functionalization: It is a process by which active functional group (for example, drug or chemicals) are introduced on the surface of other material via physicochemical or biological processes. By this way, the material with linking groups on their surfaces can be designed for suitable surface engineering and its derived applications.

C-reactive protein (CRP): It is a protein, which is normally found in the plasma, is synthesized by hepatocytes. In response to injury, infection, or any acute inflammation, CRP concentrations can increase 10^4 -fold. Various clinical studies have demonstrated the use of CRP levels for diagnostic purposes in cardiovascular disease or as prognostic indicator in gastroesophageal cancer.

Acknowledgement

This work was supported by the Ministry of Knowledge and Economy (International Collaborative R&D Project for Semiconductor, Grant no. 10030694)

Abbreviations

$BRAF^{V599E}$	:	B-Raf proto-oncogene serine/threonine-protein kinase, B-Raf proto-oncogene serine/threonine-protein kinase, BRAF1, RAFB1)
CEA	:	Carcinoembryonic antigen
CRP	:	C-reactive protein
DEN-2	:	Dengue serotype 2
ENFETs	:	Enzyme-field-effect-transistors
FETs	:	Field-effect-transistors
HPPH	:	2-devinyl-2-(1-hexyloxyethyl)pyropheophorbide
KOH	:	Potassiun hydroxide
LPSiNPs	:	Luminescent porous silica nanoparticles
miRNA	:	MicroRNAs
MOS	:	Metal-oxide-semiconductor
MSN	:	Mesoporous silica nanoparticles
NADH	:	Nicotinamide Adenine Dinucleotide (reduced form)
(PAP-LP-MSN and AP-PAP-MSN)	:	N-[3-(3,8-diamino-6-phenyl-5-phenanthridinium) propyl]-3-aminopropyl-functionalized MCM-41 type mesoporous silica nanoparticle materials
PBS	:	Phosphate buffered saline
PCR	:	Polymerase chain reaction

PDMS	:	Polydimethylsiloxane
PNA	:	Peptide Nucleic Acid
POC	:	Point-of-care
PSA	:	Prostate Specific Antigen
RIE	:	Reactive-ion-etching
RT-PCR	:	Reverse-transcription-polymerase chain reaction
SiNWs	:	Silicon nanowires
SPR	:	Surface plasmon resonance

References

Ajay, A., K. Buddharaju, I.K. Lao, N. Singh, N. Balasubramanian and D.L. Kwong. 2008. Silicon nanowire sensor array using top-down CMOS technology. Sensor Actuat. A-Phys. 145–146: 207–213.

Bard, A.J. 1994. Integrated chemical systems. A chemical approach to nanotechnology. John Wiley, New York, USA.

Bo, H., T.J. Morrow and C.D. Keating. 2008. Nanowire sensors for multiplexed detection of biomolecules. Curr. Opin. Chem. Biol. 12: 522–528.

Bunimovich, Y.L., Y.S. Shin, W.S. Yeo, M. Amori and G. Kwong. 2006. Quantitative real-time measurements of DNA hybridization with alkylated nonoxidized silicon nanowires in electrolyte solution. J. Am. Chem. Soc. 128: 16323–16331.

Cui, Y., Q. Wei, H. Park and C.M. Lieber. 2001. Nanowire nanosensors for highly sensitive and selective detection of biological and chemical species. Science 293: 1289–1292.

Fang, S., H.J. Lee, A.W. Wark and R.M. Corn. 2006. Attomole Microarray Detection of MicroRNAs by Nanoparticle-Amplified SPR Imaging Measurements of Surface Polyadenylation Reactions. J. Am. Chem. Soc. 128: 14044–14046.

He, P., and N. Hu. 2004. Electrocatalytic Properties of Heme Proteins in Layer-by-Layer Films Assembled with SiO_2 Nanoparticles. Electroanalysis 16: 1122–1131.

Heitzinger, C., and G. Klimeck. 2007. Computational aspects of the three dimensional feature-scale simulation of silicon-nanowire field-effect sensors for DNA detection. J. Comput. Electron. 6: 387–390.

Hsiao, C.Y., C.H. Lin, C.H. Hung, C.J. Su, Y.R. Lo, C.C. Lee, H.C. Lin, F.H. Ko, T.Y. Huang and Y.S. Yang. 2009 Novel poly-silicon nanowire field effect transistor for biosensing application. Biosens. Bioelectron. 24: 1223–1229.

Ibtisam E.T. 2009. Biosensors for cancer markers diagnosis. Semin. Cell Dev. Biol. 20: 55–62.

Jana, N.R., C. Earhart and J.Y. Ying. 2007. Synthesis of water-soluble and functionalized nanoparticles by silica coating. Chem. Mater. 19: 5074–5082.

Jin, X., S. Lee and X. Chen. 2010. Nanoparticle-based theragnostic agents. Adv. Drug Deliv. Rev. (in press).

Kim, S., T.Y. Ohulchanskyy, H.E. Pudavar, R.K. Pandey and P.N. Prasad. 2007. Organically modified silica nanoparticles co-encapsulating photosensitizing drug and aggregation-enhanced two-photon absorbing fluorescent dye aggregates for two-photon photodynamic therapy. J. Am. Chem. Soc. 129: 2669–2675.

Lai, C.Y., B.G. Trewyn, D.M. Jeftinija, K. Jeftinija, S. Xu, S. Jeftinija and V.S.Y. Lin. 2003. A mesoporous silica nanosphere-based carrier system with chemically removable CdS nanoparticle caps for stimuli-responsive controlled release of neurotransmitters and drug molecules. J. Am. Chem. Soc. 125: 4451–4459.

Lee, K.N., S.W. Jung, K.S. Shin, W.H. Kim, M.H. Lee and W.K. Seong. 2007. Fabrication of suspended silicon nanowire arrays. Small 4: 642–8.

Lee, M.H., D.H. Lee, S.W. Jung, K.N. Lee, Y.S. Park and W.K. Seong. 2010. Measurements of serum C-reactive protein levels in patients with gastric cancer and quantification using silicon nanowire arrays. Nanomed. Nanotechnol. 6: 78–83.

Lee, M.H., K.N. Lee, S.W. Jung, W.H. Kim, K.S. Shin and W.K. Seong. 2008. Quantitative measurements of C-reactive protein using silicon nanowire arrays. Int. J. Nanomed. 3: 1-8.

Lieber, C.M., and Z.L. Wang. 2007. Functional nanowires. MRS Bull. 32: 99–108.

Lin, C.H., C.H. Hung, C.Y. Hsiao, H.C. Lin, F.H. Ko and Y.S. Yang. 2009. Poly-silicon nanowire field-effect transistor for ultrasensitive and label-free detection of pathogenic avian influenza DNA. Biosens. Bioelectron. 24: 3019–3024.

Liu, C.G., G.A. Calin, B. Meloon, N. Gamliel, C. Sevignani, M. Ferracin, D.C. Dumitru, M. Shimizu, S. Zupo, M. Dono, H. Alder, F. Bullrich, M. Negrini and C.M. Croce. 2004. An oligonucleotide microchip for genome-wide microRNA profiling in human and mouse tissues Proc. Natl. Acad. Sci. USA 101: 9740–9744.

Luo, X.L., J.J. Xu, W. Zhao and H.Y. Chen. 2004. A glucose biosensor based on ENFET doped with SiO_2 nanoparticles. Sensor. Actuat. B-Chem. 97: 249–255.

Manzano, M. and M. Colilla and M. Vallet-Regi. 2009. Drug delivery from ordered mesoporous matrices. Expert Opin. Drug Deliv. 6: 1383–1400.

Park, J.H., L. Gu, G. von Maltzahn, E. Ruoslahti, S.N. Bhatia and M.J. Sailor. 2009. Biodegradable luminescent porous silicon nanoparticles for in vivo applications. Nat. Mater. 8: 331–336.

Paul, C. 2010. Preparation of Ultrafine Silica Nanowires Using an Iron Based Surfactant. Langmuir 26: 1405–1407.

Ramkissoon, S.H., L.A. Mainwaring, E.M. Sloand, N.S. Young and S. Kajigaya. 2006. Nonisotopic detection of microRNA using digoxigenin labeled RNA probes. Mol. Cell. Probes 20: 1–4.

Raymond, C.K., B.S. Roberts, P. Garrett-Engele, L.P. Lim and J.M. Johnson. 2005. Simple, quantitative primer-extension PCR assay for direct monitoring of microRNAs and short-interfering RNAs. RNA 11: 1737–1744.

Rosi, N.L., and C.A. Mirkin. 2005. Nanostructures in biodiagnostics. Chem. Rev. 105: 1547–1562.

Sailer, B.L., A.J. Nastasi, J.G. Valdez, J.A. Steinkamp and H.A. Crissman. 1996. Interactions of intercalating fluorochromes with DNA analyzed by conventional and fluorescence lifetime flow cytometry utilizing deuterium oxide. Cytometry. 25: 164–72.

Slowing, I., B.G. Trewyn and V.S.Y. Lin. 2006. Effect of surface functionalization of MCM-41-type mesoporous silica nanoparticles on the endocytosis by human cancer cells. J. Am. Chem. Soc. 128: 14792–14793.

Slowing, I.I., J.L. Vivero-Escoto, C.W. Wu and V.S. Lin. 2008. Mesoporous silica nanoparticles as controlled release drug delivery and gene transfection carriers. Adv. Drug Deliv. Rev. 60: 1278–1288.

Steinkamp, J.A., B.E. Lehnert and N.M Lehnert. 1999. Discrimination of damaged/dead cells by propidium iodide uptake in immunofluorescently labeled populations analyzed by phase-sensitive flow cytometry. J. Immunol. Methods 226: 59–70.

Stern, E., J.F. Klemic, D.A. Routenberg, P.N. Wyrembak, D.B. Turner-Evans, A.D. Hamilton, D.A. LaVan and T.M. Fahmy. 2007. Label-free immunodetection with CMOS-compatible semiconducting nanowires. Nature 445: 519–522.

Stöber, W., A. Fink and E. Bohn. 1968. Controlled growth of monodisperse silica spheres in the micron size range. J. Colloid. Interf. Sci. 26: 62–69.

Tang, D., R. Yuan and Y. Chai. 2007. Magnetic Control of an Electrochemical Microfluidic Device with an Arrayed Immunosensor for Simultaneous Multiple Immunoassay. Clin. Chem. 53: 1323–1329.

Vallet-Regi, M., A. Ramila, R.P. Real and J. Perez-Pariente. 2001. A new property of MCM-41: drug delivery system. Chem. Mater. 13: 308–311.

Valoczi, A., C. Hornyik, N. Varga, J. Burgyan, S. Kauppinen and Z. Havelda. 2004. Sensitive and specific detection of microRNAs by northern blot analysis using LNA-modified oligonucleotide probes. Nucleic Acids Res. 32: e175

Vivero-Escoto, J.L., I.I. Slowing, C.W. Wu and V.S.Y. Lin. 2009. Photoinduced Intracellular Controlled Release Drug Delivery in Human Cells by Gold-Capped Mesoporous Silica Nanosphere. J. Am. Chem. Soc. 131: 3462–3463.

Vivero-Escoto, J.L., I.I. Slowing and V.S.Y. Lin. 2010. Tuning the cellular uptake and cytotoxicity properties of oligonucleotide intercalator-functionalized mesoporous silica nanoparticles with human cervical cancer cells HeLa. Biomaterials 31: 1325–1333.

Wang, D., B.A. Sheriff and J.R. Heath. 2006. Complementary symmetry silicon nanowires logic: power-effiecient inverters with gain. Small 2: 1153-1158.

Wu C.C., F.H. Ko, Y.S. Yang, D.L. Hsia, B.S. Lee and T.S. Su. 2009. Label-free biosensing of a gene mutation using a silicon nanowire field-effect transistor. Biosens. Bioelectron. 25: 820–825.

Wulfkuhle, J.D., L.A. Liotta and E.F. Petricoin. 2003. Early detection: proteomic applications for the early detection of cancer. Nat. Rev. Cancer 3: 267-275.

Xu, X. and R.M. Quiros, P. Gattuso, K.B. Ain and R.A. Prinz. 2003. High Prevalence of BRAF Gene Mutation in Papillary Thyroid Carcinomas and Thyroid Tumor Cell Lines. Cancer Res. 63: 4561–4567.

Yu, H., J. Jiang, S.S. Lee and J.Y. Ying. 2008. Reverse Microemulsion-Mediated Synthesis of Silica-Coated Gold and Silver Nanoparticles. Langmuir 24: 5842-5848.

Zhang, G.J., J.H. Chua, R.E. Chee, A. Agarwal and S.M. Wong. 2009. Label-free direct detection of MiRNAs with silicon nanowire biosensors. Biosens. Bioelectron. 24: 2504–2508.

Zhang, G.J., L. Zhang, M.J. Huang, Z.H.H. Luo, G.K. Ignatius Tay, E.J. Andy Lim, T.G. Kang and Y. Chen. 2010. Silicon nanowire biosensor for highly sensitive and rapid detection of Dengue virus. Sensor. Actuat. B-Chem. 146: 138–144.

Zhao, X., L. Hilliard, K. Wang and W. Tan. Bioconjugated silica nanoparticles for bioanalysis, encyclopedia of nanoscience and nanotechnology. pp. 255. In: H.S. Nalwa [ed.] 2004. American Scientific Publishers. Stevenson Ranch, CA.

12

Silver Nanomedicine and Cancer

P.V. Asharani,[1,a] Luxmi Fatimathas,[c] Hui Kheng Lim,[1,b]
Suresh Valiyaveettil[3,d] and M. Prakash Hande[2,*]

ABSTRACT

Silver nanoparticles are the most commercialised nanoparticles in today's market, according to the latest statistics. Nano sized silver has additional advantages over ionic or metallic forms due to its highly reactive nature, the ease at which it can be distributed and its amenability for surface modification. For example; targeting ligands such as antibodies can be immobilized on the surface of silver nanoparticles to control their localization. In addition to the attractive material chemistry these nanoparticles demonstrate, they are also excellent anti-microbial agents with broad spectrum activity against many bacteria and viruses. This places them at

[1]Department of Physiology, Yong Loo Lin School of Medicine, MD9, 2 Medical Drive, National University of Singapore, Singapore, 117597.

[a]E-mail: g0700143@nus.edu.sg
[b]E-mail: g0900019@nus.edu.sg

[2]Associate Professor, Department of Physiology, Yong Loo Lin School of Medicine, MD9, 2 Medical Drive, National University of Singapore, Singapore, 117597;
E-mail: phsmph@nus.edu.sg
[c]Email: luxfatimathas@googlemail.com

[3]Associate Professor, Department of Chemistry, Faculty of Science, National University of Singapore, Singapore, 117543.
[d]E-mail: chmsv@nus.edu.sg

*Corresponding author

List of abbreviations after the text.

the forefront of nanomedicine. Recent findings implicate silver nanoparticles as potential as anti-proliferative agents. The sunny side of silver nanotechnology is however overshadowed by the toxicity profile of these particles. Reports have clearly shown that silver nanoparticles are cytotoxic, genotoxic and toxic to ecosystem. The mechanism of toxicity is proposed to be through mitochondrial dysfunction, reactive oxygen species release and oxidative damage of DNA. In many cells, as well as in *in vivo* systems, silver nanoparticles halted cell proliferation and increased apoptosis. This toxicity can however be used to our advantage and is currently being employed as a therapeutic approach in many existing cancer drugs. The advantage of silver nanoparticles over current chemotherapeutic agents is the excellent targeting achievable through surface conjugation of specific ligands and their high diffusibility. Silver nanoparticles are capable of crossing the blood brain barrier, which is a promising advantage that could be employed in treating brain tumours. The imbalance created between nanomedicine and nanotoxicology is strong enough to impede sustainable growth of this technology at this time. This review analyses the current status of silver nanotechnology from both nanotoxicology and nanomedicine perspectives. The properties of silver nanoparticles that make them potentially detrimental to the human body could at the same time be useful in cancer therapy—a paradox that shall be discussed further.

INTRODUCTION

Silver salts have been in use for thousands of years in jewellery, cooking utensils, photography and medicine. Colloidal silver in particular is well known for its anti-microbial potential. In 1000 BC, the ancient Greeks used silver vessels for cooking and storage to prevent bacterial overgrowth. In the eighth century silver was introduced as a medicine for the treatment of blood-related disorders and heart palpitations. During the seventeenth and eighteenth centuries silver compounds were utilised in the treatment of ulcers (Klasen 2000). By the nineteenth century, the common anti-septics used in surgery were replaced with silver salts (Klasen 2000). However, this growing trend in the medicinal application of silver disappeared following World War II. In the decades that followed the clinical use of silver was limited to the application of silver sulphadiazine to burn infections (Klasen 2000).

The emergence of nanotechnology in the latter half of the twentieth century has seen the re-emergence of silver, not only in medicine but also commercially. This is due to the novel properties of silver nanoparticles, which due to their small size exhibit different properties to their bulk material counterparts. These differences often include changes in optical properties, melting points, surface plasmon resonance, quantum confinement and cellular penetrance.

Silver nanoparticles in particular show an increased ability to combat microbial infections, when compared to bulk material silver. They have been shown to abolish the replication of hepatitis B virus (Lu et al. 2008) and reduce the infectivity of HIV (Elechiguerra et al. 2005). In the broader context, the enhanced anti-microbial activity of silver nanoparticles has been harnessed in commercially available wound dressings.

According to the latest studies, silver nanoparticle-based products are one of the fastest growing product categories in the current market (Chen and Schluesener 2008). Silver nanoparticles are currently not only used in wound dressings but also contraceptives, bone prostheses and catheters (Chen and Schluesener 2008). Non-therapeutic silver nano products include air fresheners, electrical appliances, wall paint, detergents and clothes, especially socks to control odour and bacterial growth.

EXPOSURE RISKS

The ultra small size and unique physiochemical properties of nanoparticles facilitate easy uptake and distribution to vital organs of the body. Silver nanoparticles in particular have been shown to penetrate the blood brain barrier and deposit in the brain, in several vertebrate model systems. Nanoparticles were also found distributed in liver, lung, spleen and kidneys (Kim et al. 2008; Asharani et al. 2009b; Asharani et al. 2008). However, the question still remains as to whether exposure routes influence the intracellular distribution of silver nanoparticles.

The real-life scenario of nanoparticle exposure during large scale synthesis, processing and in usage is another area that requires attention. The concentration and size of silver nanoparticles in the air, medical devices and consumer products is largely elusive. Knowing such statistics is critical to the assessment of exposure risk. Silver is at the forefront of highly commercialised nanoparticles, yet the least assessed for exposure risks. The exposure dynamics of silver nanoparticles in a manufacturing facility has been assessed (Park et al. 2009). The study was conducted in a silver nanoparticle production facility that manufactures 3,000 kg of silver nanoparticles per month. Results showed the concentration and size of silver nanoparticles released into the air, during the liquid phase synthesis process, increased rapidly over time. Particles over 100 nm in size filled the air predominately when the reactor hatch was opened for an hour and a half (Fig. 1). Similar studies were conducted using commercially available silver nanoparticle-impregnated socks, which were shown to release 1.36 mg of silver per gram of socks (Benn and Westerhoff 2008). These studies clearly highlight the exposure risk to humans and calls for attention to the regulation of silver nanotechnology. Such studies should be extended to identify the uptake, distribution and host responses to such particles from real life scenarios, rather than applying arbitrary user defined concentrations as in many current toxicology studies. Unfortunately, there are few studies available from which to kick start exposure risk assessments and control measures. This is therefore an important field for future expansion.

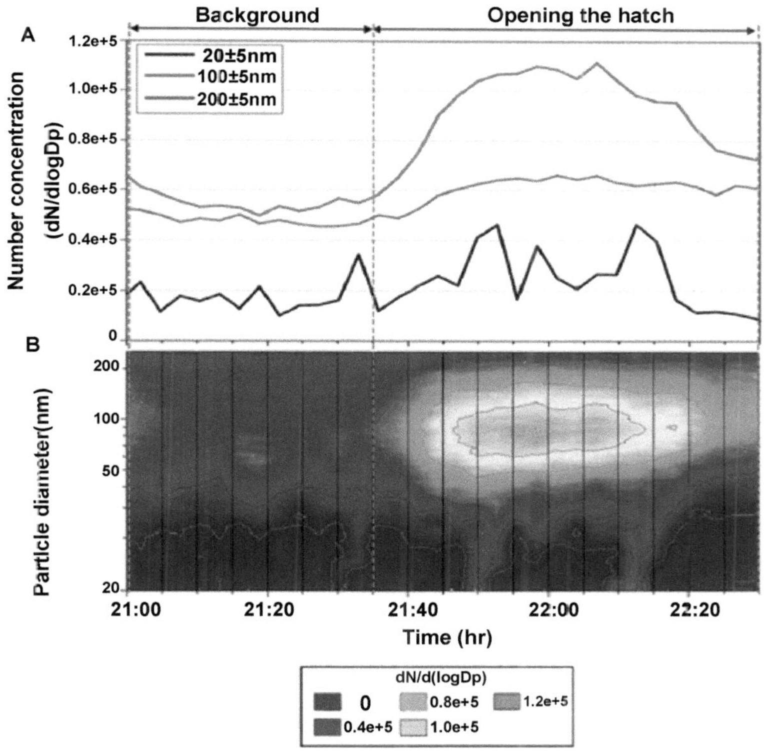

Fig. 1. Time series for the total silver nanoparticle number concentration measurement at a manufacturing facility. The particle concentration increased over time, with 100 nm silver nanoparticles predominating when the reactor hatch was opened for 1.5 hr. (A) Change in the number concentrations of 20 ± 5, 100 ± 5, and 200 ± 5 nm diametre particles and (B) change in particle number concentrations between 20 and 250 nm with time is shown as a filled contour plot (Adapted from Park et al 2009 with permission).

Color image of this figure appears in the color plate section at the end of the book.

Another important aspect that needs to be investigated is the environmental release of silver nano waste. At present there are vague hypotheses to support the safe environmental disposal of silver nanoparticles. Exposure of silver nanoparticles to fish showed a clear indication towards toxicity, suggesting significant risks involved in waste disposal. However it should be noted that statistics on parameters such as the size and stability of nanoparticles in water and the concentration of silver nanoparticles in waste are hard to predict. Silver nanotechnology has progressed to an extent that the public are keen to see convincing reports and reasoning for the use of such technology commercially.

Silver nanoparticles have not only infiltrated the retail sector but are also making significant headway in the medical field, with particular reference to the treatment of wound and burn infections (Klasen 2000). Silver nanoparticle based wound dressings are hitting the market at an increasingly fast pace. The list of commercially available wound dressings is listed in Table 1. Early reports indicate that silver nanoparticles possess anti-proliferative properties, making them ideal candidates for cancer therapy

Table 1. Commercially available nanoparticle based wound dressings.

Product name	**Use**	**Manufacturer**
Aquacel-Ag hydrofibre	Silver nanoparticle impregnated fibres that continuously release Ag+ ions upon hydration	Convatec, Skillman.
Actisorb silver 220	Actisorb pads adsorb bacteria to the charcoal and are killed by nanoparticles	Johnson & Johnson.
Acticoat-7	Mesh coated with silver nanoparticles provide large amount of silver to the wound site	Smith and Newphew
Silverlon	Polymeric fabric containing Silver nanoparticles	Argentum Medical
V.A.C. GranuFoam Silver	Wound dressings containing silver nanoparticles	Kinetic Concepts Inc
Curad® Silver Bandages	Wound dressings: Polyethylene mesh with silver nanoparticles	Medline Industries, Inc.
Hansaplast® med Silver Technology with P.U.R. ® Gel	Polyethylene mesh coated with silver nanoparticles	Beiersdorf AG
Hansaplast® med Silver Technology	Polyethylene mesh coated with silver nanoparticles	Beiersdorf AG
Elastoplast® SilverHealing™	Wound dressing with silver nanoparticles	Beiersdorf AG
Elastoplast® FAST SILVERHEALING™ with P.U.R. ® Gel	Silver nanoparticles in Polyurethane gel pad	Beiersdorf AG
Beiersdorf AG Silver Technology	Wound dressings with silver nanoaprticles	Beiersdorf AG

(Asharani et al. 2009a). Silver nanoparticles have the added advantage of a large surface area that is available for surface functionalisation with targeting ligands such as antibodies or peptides. This has the potential to enhance their innate anti-proliferative effects. Withdrawing silver nanoparticles from the medical market altogether is not a practical way to overcome the toxic effects of the most commercialised nanoparticle. It is rather more realistic to employ a non-cytotoxic dose of highly targeted and functionalised silver nanoparticles for therapeutic applications. This kind of approach will justify the use of these nanoparticles in medicine despite their toxicity.

Key Facts About Silver Nanoparticles

- The first silver nanoproduct, the Lycurgus cup, dates back to Roman era. The Lycurgus cup is a dichroic glass vessel made of silver and gold nanoparticles which gives it two different colours. Red/magenta colour in transmitted light contributed by the gold nanoparticles and greenish brown colour in reflected light due to the presence of silver nanoparticles. Nanoparticles solutions show intense colour due to surface Plasmon resonance (SPR). SPR is the result of oscillations of negatively charged electrons in the nanoparticle core embedded in positively charged lattice. The oscillations occur in the presence of electromagnetic radiation.
- Silver particles can be both medicinal and toxic. The absorption of silver salts through the skin can cause argyria, a bluish discolouration of skin. Conversely, the burn dressing Acticoat contains silver nanoparticles with the specific intent of combating microbes at burn sites and facilitating faster healing. However treatment of a burn victim with Acticoat was shown to cause argyria. This example illustrates the dual and often opposing characteristics of silver nanoparticles.
- Human beings are constantly exposed to silver nanoparticles, whether knowingly or unknowingly. The most common sources are food products, containers, clothes, washing machines, air conditioners and drinking water supplemented with silver nanoparticles for disinfection. The concentration of silver nanoparticle emission from these products is not well studied. The European Parliament, the American Food and Drug Administration (FDA) and the British Standards Institute are all working towards implementing defined controls to impose on companies manufacturing nanoparticles. This is both for the safety of factory employees during the production of nanoparticle-based products and for the safety of consumers purchasing nanoparticle-based products.

- Silver nanomaterials can be synthesised in various sizes and shapes depending on their application. Silver nanospheres are the most common nanomaterial employed. Other shapes include nanorods, prisms and cubes. Controlling the shape and size of silver nanoparticles is more difficult than for gold nanoparticles. This has impeded research on shape-dependant therapeutic applications of silver nanomaterials. In contrast, gold nanorods are already being employed as cancer destruction agents due to their vascular retention effects.

DNA DAMAGE

DNA damage, also termed genotoxicity, can result in irreparable changes to a cell's DNA. This may lead to decreased cell viability and ultimately cell death. The exposure of cells to nanoparticles has been shown to induce such DNA damage. Aqueous solutions of silver nanoparticles have been shown to release silver ions (Navarro et al. 2008). DNA damage may therefore result either from the effects of the silver nanoparticles themselves, or their by-product silver ions. Damage can be induced by both of these species through direct binding of DNA or via downstream effects of their toxicity. Current evidence for the direct binding of silver nanoparticles to DNA has thus far only been shown in bacteria (Yang et al. 2009) and viruses (Lu et al. 2008).

DNA can be damaged in a variety of ways, including alkylation of bases, hydrolysis of bases and bulky adduct formation. However the most widely accepted form of nanoparticle-induced genotoxicity is oxidative damage. Oxidative stress results in the production of excess reactive oxygen species (ROS), such as superoxide (O_2^-) and hydroxyl radicals (·OH). ROS can then go on to oxidise DNA bases and generate both single strand breaks (SSBs) and more lethal double strand breaks (DSBs) (Evans et al. 2004).

Recent studies into silver nanoparticle induced DNA damage have utilised a variety of experimental techniques to evidence these changes. A common approach is to use the comet assay (Kumaravel et al. 2009). Using the alkaline comet assay, silver nanoparticle exposure has been shown to increase levels of SSBs in human cells (Asharani et al. 2009b; Grigg et al. 2009). This increase was dose-dependent and greater in human cancer cells than in normal human cells (Asharani et al. 2009b). Nanoparticle treated cells also underwent cell cycle arrest, possibly as a consequence of DNA damage. Electron microscopy images showed silver nanoparticles present in mitochondria and in the nucleus (Fig. 2). The nuclear localisation of silver nanoparticles suggests they are in close enough proximity to the DNA to inflict damage directly. The mitochondrial localisation of silver

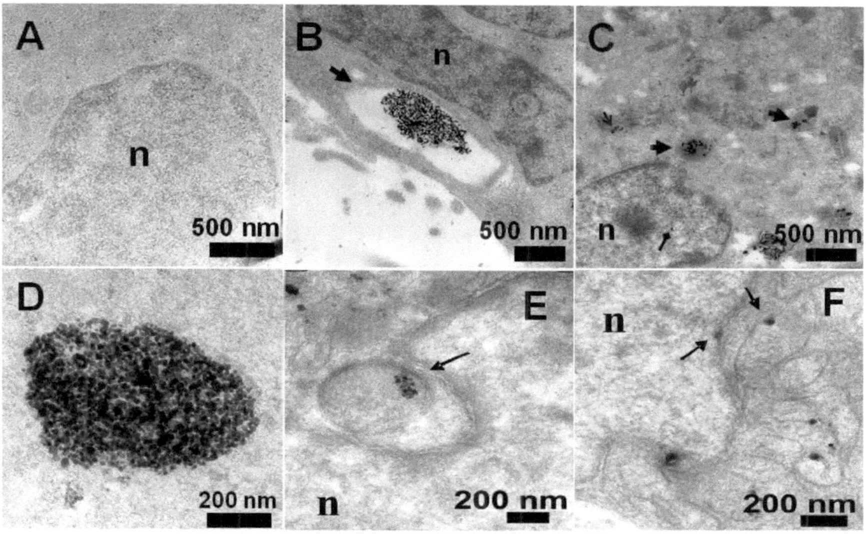

Fig. 2. Intracellular distribution of silver nanoparticles; Silver nanoparticles tend to deposit in the mitochondria and nucleus of cells. (A) Untreated cells, (B) cells treated with silver nanoparticles showing large endosomes near the cell membrane with many nanoparticles inside, (C) electron micrographs showing lysosomes with nanoparticles inside (thick arrows) and scattered in cytoplasm (open arrow). Diamond arrow shows the presence of the nanoparticle in the nucleus, (D) magnified images of nanogroups showed that the cluster is composed of individual nanoparticles rather than clumps, (E) image shows endosomes in cytosol that are lodged in the nuclear membrane invaginations (F) and the presence of nanoparticles in mitochondria and on the nuclear membrane (reproduced from Asharani et al. 2009b with permission from American Chemical Society).

nanoparticles suggests they may also interfere with the mitochondrial electron transport chain, thereby increasing ROS production and indirectly causing DNA damage. The resulting damage may take the form of single or double strand breaks.

SSBs, although deleterious to genomic stability, are more readily repaired than DSBs. Several studies (Ahamed et al. 2008; Kim et al. 2009) have shown silver nanoparticles to induce DSBs in a dose dependent manner. These studies utilised the γH2AX assay which specifically detects DSBs. Following DSBs, the histone family protein H2AX is phosphorylated at Ser-139. This phosphorylation event can be detected by antibodies and then subject to immunostaining or flow cytometry. Silver nanoparticle induced DSBs were prevented by the addition of *N*-acetylcysteine, an antioxidant, suggesting the DNA damage was oxidative (Kim et al. 2009; Fig. 3).

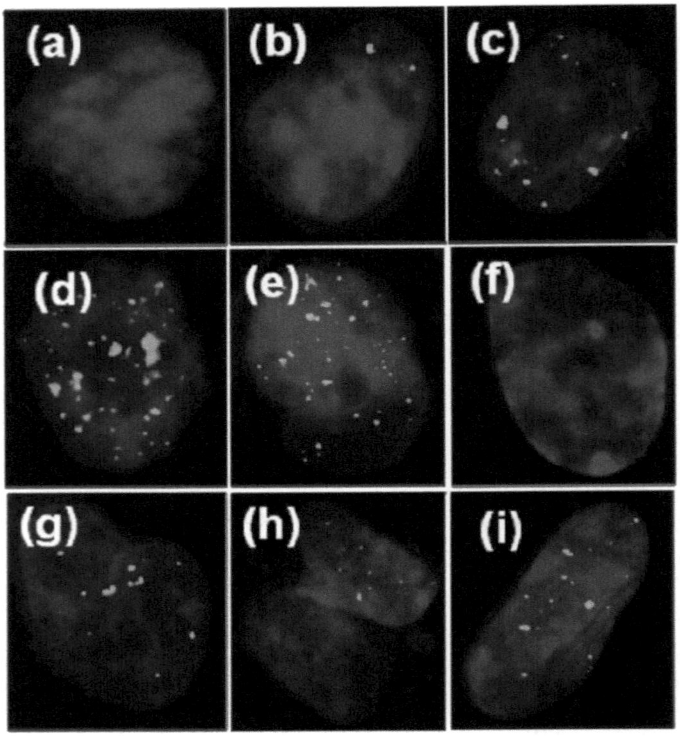

Fig. 3. Gamma H2AX staining; Silver nanoparticles induce DNA damage mainly through oxidative stress. γH2AX staining of untreated human cancer (U251) cells (a) showing minimal foci, (b) cells exposed to 25 µg/mL of Ag-np-3 (c) 50 µg/mL (d) 100 µg/mL and (e) 10 µM H_2O_2, showing multiple foci. (f) Untreated IMR-90 cells shows no foci whereas, treatment with (g) 25 µg/mL of Ag-np-3 (h) 50 µg/mL (i) 100 µg/mL and (K) 10 µM H_2O_2, shows multiple foci.

Color image of this figure appears in the color plate section at the end of the book.

Oxidation is not the only way silver nanoparticles can induce DNA damage. In a study designed for investigating the haemocompatibility of silver nanoparticles, indirect DNA damage was recorded as a consequence of massive haemolysis (Asharani et al. 2010), indicating that the mechanism of action of silver nanoparticles is multifactorial and complex. Apart from exerting DNA damage in cells, silver nanoparticles were shown to be involved in chromosomal aberrations causing aneuploidy, deletions and fragmentation (Wise et al. 2010; Asharani et al. 2009a; Fig. 4).

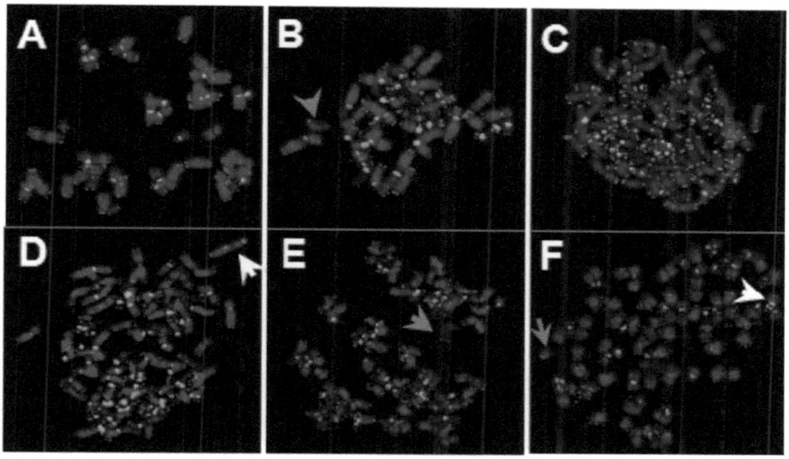

Fig. 4. Chromosomal aberrations in silver nanoparticle treated human cells. Metaphase spreads from the untreated cells show no apparent damage in the chromosomes (A) silver nanoparticle treated human normal fibroblast cells (IMR-90) show acentric and centric fragments (B). Arrow indicates acentric fragments, (C) untreated cancer cells with no aberrations, metaphases show dicentric chromosomes in untreated cells (D) and treated cells. White arrow points to a dicentric chromosome. Cancer cells treated with silver nanoparticles also show acentric fragments (E) and centric fragments, (F) red arrow points to a chromosome fragment (G) summary of the frequency of aberrations observed in silver nanoparticle treated cells. A minimum of 50 metaphases per sample was scored for chromosome analysis (Reproduced from Asharani et al. 2009a).

Color image of this figure appears in the color plate section at the end of the book.

Recently, studies have uncovered the downstream signalling pathways involved in silver nanoparticle induced DNA damage. One such study showed the activation of transcription factors NFKB and NRF2 by silver nanoparticles. This leads to the activation of the mitogen-activated kinase, p38 and subsequent DNA damage, cell cycle arrest and apoptosis (Eom and Choi 2010). Based on our data, a simplistic mechanism of silver nanoparticle toxicity is proposed (Fig. 5). These findings place silver nanoparticles as potential therapeutic agents. Utilising these particles to hijack the apoptotic machinery of infected or cancerous cells could increase their susceptibility to cell death, thereby providing a means to specifically destroy harmful cells.

Taken together these *in vitro* studies illustrate the genotoxic potential of silver nanoparticles. However a recent *in vivo* study in chick embryos has not shown silver nanoparticles to induce oxidative DNA damage (Sawosz et al. 2009). Conversely studies in other model organisms support the *in vitro* data. Recent work in zebrafish has shown the induction of oxidative damage and apoptosis in adult liver (Choi et al. 2010). Further investigation

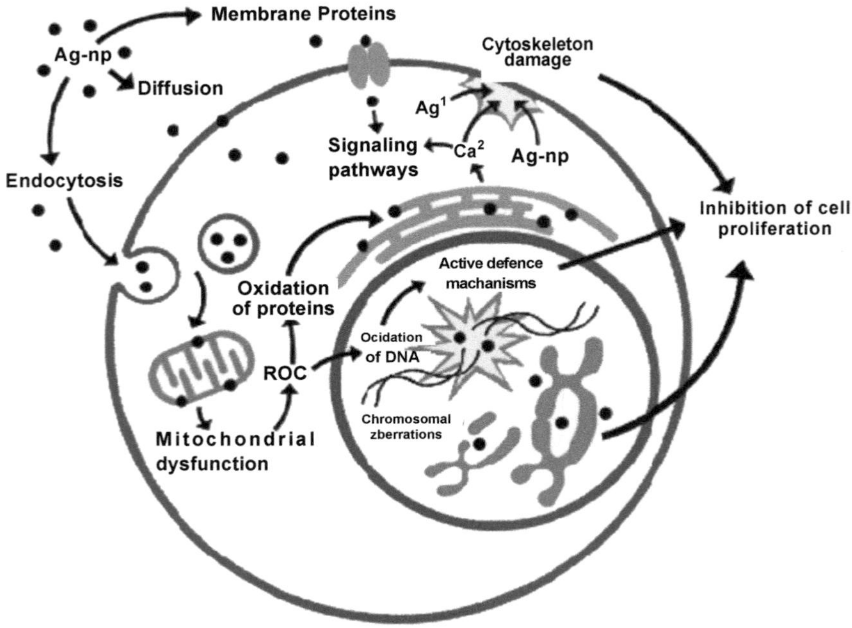

Fig. 5. The proposed mechanism of Ag-np toxicity based on the experimental data obtained from our studies (Reproduced from Asharani et al. 2009a).

Color image of this figure appears in the color plate section at the end of the book.

showed the presence of DSBs, via γH2AX staining, as well as the induction of apoptosis via raised levels of the pro-apoptotic protein p53 and p53-related pro-apoptotic genes. Work carried out in Drosophila further supports the induction of oxidative stress, apoptosis and DNA damage by silver nanoparticles (Ahamed et al. 2010). In response to ingestion of silver nanoparticles Drosophila showed raised levels of antioxidants, increased p53 protein levels and increased levels of p38 protein, which is involved in DNA damage repair.

A substantial body of both *in vitro* and *in vivo* evidence is now building demonstrating the genotoxicity of silver nanoparticles. Further studies are needed to understand the significance of these findings in a human context and how this can be harnessed in a beneficial manner for the treatment of diseases such as cancer.

Key Facts About Nanoparticle Induced DNA Damage

- DNA damage occurs from endogenous or exogenous agents. Endogenous sources are mostly due to reactive oxygen species (ROS). They are natural by-products of metabolism and are highly reactive due to the unpaired valence electrons. ROS can oxidize amino acids in proteins, bases in DNA and fatty acids in lipids, causing extensive damage. They are kept in check by superoxidase dismutase and hydrogen peroxide produced in cells. Exogenous sources of ROS include microbes, radiation and chemicals.

- p38 is a mitogen-activated kinase, also known as RK or CSBP, that can lead to DNA damage. It is activated by a range of stress inducing stimuli such as ultraviolet radiation, cytokines and heat shock. p38 activation requires the conserved TGY (threonine-glycine-tyrosine) motif present in its activation loop. These residues are phosphorylated by the mitogen-activated protein kinase kinases, MKK3 and MKK6, resulting in p38 activation and translocation to the nucleus. Once in the nucleus p38 can activate several transcription factors.
- p53 is a tumour suppressor protein which is mutated in many cancers. It is involved in a wide range of cellular processes including activating DNA repair following damage, inducing apoptosis and growth arrest. p53 can be activated in response to DNA damage, oxidative stress, osmotic shock and deregulation of oncogenes. In normal cells, p53 levels are maintained at low levels by the protein Mdm2. Mdm2 binds to p53, thus inhibiting its actions and sequestering it from its major site of action in the nucleus. Mdm2 also ubiquitinates p53, which targets the protein for degradation. The activation of p53 requires the phosphorylation of its N terminal domain, an increase in its half life and a conformational change. This allows it to go on to act as a transcriptional regulator within the cell nucleus.
- H2AX is a histone 2A family protein required for the formation of histone complexes around which DNA is packaged. Following double strand breaks in the DNA, H2AX becomes phosphorylated and is called γH2AX. This phosphorylation event is carried out by class IV phosphoinositide kinases, in particular Ataxia Telangiectasia Mutated kinase. Phosphorylation of H2AX results in the DNA becoming less condensed. This looser DNA conformation allows room for DNA repair proteins to enter the complex.

CANCER THERAPEUTICS

Cancer remains one of the world's most devastating diseases and current cancer treatments include surgical intervention, radiation and chemotherapeutic drugs. The low molecular weight of these drugs means they are easily excreted, resulting in the use of higher doses and consequently increased toxicity. Their low therapeutic index further confounds the situation, as the effective dose often exceeds therapeutic levels. Furthermore, these drugs when administrated alone, lack specificity and cause significant damage to non-cancerous tissues. The introduction of nanotechnology is now changing the field of anti-cancer pharmacology (Peer et al. 2007).

Silver nanoparticles are among the few metal nanoparticles that possess bactericidal and bacteriostatic properties (Baker et al. 2005). The exact mechanism of interaction is however yet to be established. Many are now exploring the therapeutic promise of these nanoparticles as active components themselves or as physical support for other functional moieties, including current drug treatments.

Recent studies have reported that due to their anti-proliferative activity, nanosilver can be potentially exploited to revolutionise cancer therapy. Silver nanocrystals have been shown to inhibit cancer cell proliferation in a size- and shape-dependent manner (Asharani et al. 2009a; Yang et al. 2008). The rate at which silver nanoparticles inhibited the proliferation of human glioma cells and hepatocellular carcinoma cells was highest amongst silver nanocrystals, followed by silver nanorods, nanowires and lastly bulk silver crystals. Furthermore, the cancer cell proliferation rate was significantly reduced with decreasing size of nanoparticles (Xu et al. 2009). The underlying mechanism of anti-proliferative activity is still not well understood. However experimental evidence suggests that silver nanoparticles damage mitochondria and increase the production of reactive oxygen species (ROS). Pre-treatment of cells with the antioxidant, *N*-acetylcysteine, efficiently prevented genotoxic and cytotoxic changes induced by silver nanoparticles, suggesting the intrinsic toxicity is associated with oxidative damage-dependent pathways (Kim et al. 2009). In addition to inhibiting cancer cell proliferation *in vitro*, silver nanoparticles have reliably inhibited tumour growth *in vivo*. It is believed that the cytostatic effect of silver makes the interaction of silver atoms with DNA and proteins possible, resulting in the inhibition of tumour growth.

The ability of silver nanoparticles to induce apoptosis may also prove to be beneficial in cancer treatment. Evidence shows that apoptosis is the primary mode of cell death induced following silver nanoparticle treatment (Hsin et al. 2008). Most reports indicate that oxidative stress is the main player in silver nanoparticle induced DNA damage, leading to the activation of a mitochondria dependent apoptosis pathway.

Conventional anti-cancer therapies suffer various drawbacks such as a lack of specificity, poor effectiveness in treating advanced stage cancer and extensive toxic side effects. Silver nanoparticles, by themselves or in combination with current treatments, may provide a more efficient and less toxic course of treatment in cancer. This is owing to their anti-proliferative, anti-angiogenic and pro-apoptotic properties.

Key Facts About Cancer Therapeutics

- VEGF is a pro-angiogenic signalling molecule and is therefore an important target in the treatment of cancer. Tumours that over-express VEGF are able to grow and metastasise by inducing the production of new blood vessels to supply the nutrients required for growth. VEGF acts through binding tyrosine kinase receptors on the cells surface. Binding of VEGF to these receptors, results in their dimerisation, transphorylation and subsequent activation of downstream signalling pathways. Anti-VEGF drug treatments are currently available for certain cancers, e.g., Bevacizumab (Avastin) in the treatment of non-small cell lung cancer, metastatic kidney cancer and glioblastoma.
- The combination of nanomedicine and cancer therapy have given new hope to many patients. Several nanoparticle based drugs have passed clinical trials and many more are in the pipeline. Polymeric nano micelles are efficient anti-cancer drug carriers. The first known report on efficient cancer targeting and killing was achieved using magnetite nanomaterials. When magnetite is functionalised with monoclonal antibodies, these nanoparticles became efficient targeting agents and specifically killed tumours in rats.
- The most recent discovery of silver nanocrystal mediated imaging for the study of biological specimens shows promise. Application of this technology will assist in diagnosing clinical conditions such as bone diseases and cancer. The process involves the deposition of silver inside the organ of interest or biological specimen. A specific wavelength of light is then used to excite these silver particles and the subsequent luminescence recorded. The silver nanocrystals act as small mirrors, revealing the transmission properties of the surrounding materials.

CONTROL MEASURES

A major interest of today's commercial industry is to make scientific breakthroughs profitable. However the long-term impact analysis of new materials, such as silver nanoparticles, is often not wholly considered. Consequently, these materials are distributed in the market before fully comprehending their risks to and impacts on the environment. The increasing usage of nanoparticles triggers several potential environmental, health and safety (EHS) concerns, including the following:

- The highly reactive nature of nanoparticles, attributed to their high surface area to volume ratio.
- The lack of adequate risk assessment and toxicity studies.
- High water and air dispersal properties of nanoparticles.

- A high tendency for aggregation and agglomeration of nanoparticles, resulting in the formation of larger particles that may be absorbed by animals or plants and therefore enter the food chain.

It has been shown that nanoparticles are extremely difficult to remove from the environment. For example; in the food chain, as nanoparticles are passed from one organism to the next via ingestion they begin to accumulate and reach more toxic concentrations (Gibson and Pula, 2009). In terms of human exposure, several potential points of contact with nanoparticles are human skin via direct contact, lungs through inhalation and the gastrointestinal tract via accidental ingestion. Once in the body, nanoparticles can travel more freely than their larger bulk mass counterparts. They are able to penetrate cell membranes and cross the blood brain barrier. Early studies found that nanoparticles caused inflammation in the lungs of rodents and killed fish and other organisms in the water and soil, highlighting potential ecological impacts (Gibson and Pula 2009). The unbound and freely dispersed nanoparticles are able to escape into the environment more easily and therefore present an increased threat to the ecosystem (Germano 2008). In the large part these threats have yet to be defined, due to the fact that nanotechnology is still in the early stages of its development.

Specific regulation is only beginning to emerge, due to a greater awareness of the possible threats posed by nanomaterials, coupled with rising levels of human exposure. The European Parliament has called on the relevant parties to review all the applicable legislation within two years in order to ensure safety in all applications of nanomaterials. There has also been an increasing availability of guidelines and codes of conduct for both manufacturers and users of nanoparticles. Nanoscience and nanotechnology research should be conducted with strict adherence to the precautionary principle, anticipating potential EHS impacts and responding with a proportionate level of caution. A commonly held viewpoint is that as long as risk assessment studies on long term safety are not available, the inclusion of nanomaterials into the human system, food items and other commercially available products should be avoided (Seaton et al. 2010).

A report published by the British Standards Institution in 2009 (http://www.bsigroup.com/upload/Standards%20&%20Publications/Nanotechnologies/PD6699-2.pdf) offers detailed guidelines and important instructions on the safe handling of nanomaterials (British Standards Institution 2009). This report covers the proper management of risks associated with exposure to nanoparticles, as well as providing hazard evaluations and underscoring the potential risks associated with a variety of nanomaterials. Related information can also be downloaded from readily available resources on the internet such as SAFENANO (www.

safenano.org), National Nanotechnology Initiative (www.nano.gov/index.html) and NanoAction (www.nanoaction.org). The overall aim of all of these establishments is to help those who are handling nanomaterials to do so in a safe and responsible manner. (Seaton et al. 2010).

Currently the Food and Drug Administration (FDA) is unable to precisely identify or quantify all of the nanotechnology products available and their potential for harm (Wilhelmi 2008). They are therefore predominantly working on a case-by-case and product-by-product basis, without any specific standards to follow, as these have yet to be developed and disseminated. The FDA is currently unable to examine and determine the purity and quality of nanomaterials, whilst standardised procedures for the production of these materials are still lacking (Bowman and Hodge 2007). The FDA can work in collaboration with organisations such as the International Organization for Standarization (ISO), the American National Standards Institute (ANSI) and the Nanotechnology Standards Panel and American Society for Testing and Materials (ASTM), to develop standards and regulatory schemes that will apply across the board to all nanomaterial producers.

Cooperative efforts should be made to develop an integrated scientific database that utilises existing models, such as the Physiologically Based Pharmacokinetic/Pharmacodynamic (PbPk), the Toxicology Data Network (TOXNET) and the Agency for Toxic Substances and Disease Registry (ATSDR), to standardise the format for collection of nanotechnology EHS data. These databases, would serve as a good starting point for the interagency development and utilisation of toxicological data. Due to their wide ranging applications and increasing demand on the market, nanosilver should be prioritised and their potential EHS impacts collated through use of the aforementioned databases.

Key Facts About Control Measures

- Even after many years of research in nanotoxicity, a complete toxicity profile is missing due to the variations in the type of particles used. In response to the outcries of consumers and researchers the Organization for Economic Co-operation and Development (OECD), has recently proposed a common standard for testing silver nanoparticles. The nanoparticles for toxicity studies are distributed by nanoComposix. Silver nanoparticles of different sizes and functionalisation are available along with the acceptable testing concentration ranges. This regulation is expected to harmonise the toxicity research worldwide.
- The main regulatory body monitoring nanotoxicity is the National Nanotechnology Initiative (NNI, established in 2001) which is a group of 25 US based federal agencies. Several well-funded grants

have been distributed by members of the NNI. The National Science Foundation and the Environmental Protection Agency has offering $38 million over a 5 year period. The University of California Center for Environmental Implications of Nanotechnology (UC CEIN) has offered $24 million to develop models to predict the environmental impact of different nanomaterials. The Center for Environmental Implications of Nanotechnology (CEINT) has put forward $14 million to study the fate and transport of natural and manufactured nanomaterials in ecosystems. On a worldwide scale, nanotechnology involves the efforts of chemical, engineering and materials biology scientists and from the U.S., Europe, and Japan. This global community, using funding from their own labs, has announced a new alliance called the International Alliance for NanoEHS Harmonization (IANH). Together the IANH and NNI will develop standard protocols for laboratory tests through a round robin approach.

Summary

- Silver nanoparticles are one of the most promising nanodrugs. Research into this field is as a result growing at an incredible rate. The medical applications of silver nanoparticles include anti-cancer, anti-viral and anti-bacterial therapies. Silver nanoparticles are active against HIV and hepatitis viruses, supporting the role as promising anti-viral drug. They also possess excellent anti-cancer properties both *in vivo* and *in vitro*.
- The use of silver nanoparticles in medicine is challenged by the extreme toxicity of these particles. They have been shown to be toxic in both *in vitro* and *in vivo* models. Animal and fish models such as zebrafish, fathead minnow, mouse and rats showed serious toxicity following nanoparticle exposure. Silver nanoparticles have been implicated in the reduced reproductive capacity of animals. The toxicity spectrum has been shown to spread across a wide variety of species. Until this balance is resolved in favour of nanotherapy over nanotoxicity, the growth of silver nanoparticles in nanomedicine will be hindered.
- Silver nanoparticles are released into the atmosphere during their production, but also from commercialised products in which they are incorporated. The ease with which they can penetrate cells and the blood-brain barrier is of particular concern. They have been shown to aggregate intracellularly and within major bodily organs such as the brain, lungs and kidneys. Clear regulations on the disposal of silver nanoparticle waste have yet to be produced. Further research into silver nanoparticle toxicity is needed to establish more explicit guidelines.

- Silver nanoparticles can cause DNA damage both directly or indirectly. The direct binding of silver nanoparticles to DNA has been shown in lower organisms, such as bacteria and viruses. The deposition of silver nanoparticles within the nuclei of cells, shown in both cell lines and animal models would facilitate the direct action of these particles on DNA. Indirect DNA damage can result from silver ion release and ROS production. Extensive DNA damage as a result of exposure to silver nanoparticles can lead to cell death or growth arrest depending on the cell type.
- Silver nanoparticles share a common mechanism of action with many anti-cancer drugs, whereby they decrease cell proliferation and render cells susceptible to apoptosis through DNA damage. The clear advantage of silver nanoparticles over conventional drugs lies in their ability to cross the blood brain barrier, their high diffusion rate and their long circulation time. A feature unique to these particles, which highlights their promise as a future anti-cancer therapy, is their surface functionalisation with targeting ligands. This will lend enhanced specificity to cancer treatments, resulting in more effective and concentrated attack on cancer cells. However this will also help avoid current off-target effects on healthy cells, which cause the numerous side effects patients endure during chemotherapy.
- Silver nanoparticles can inhibit the proliferation of cancer cells. Silver nanocrystals show the highest anti-cancer potency, followed by silver nanorods and nanowires. *In vivo* studies have shown silver nanoparticles impede the growth of tumours. They can also inhibit the growth of new blood vessels, which may be of benefit in the treatment of vascularised tumours. Furthermore these nanoparticles sensitise cells to apoptosis through oxidative stress, which if harnessed and targeted correctly could promote cancer cell death.
- Control measures are needed to keep in check the highly reactive nature of silver nanoparticles, their ability to readily disperse and their tendency to aggregate and deposit within organisms. The removal of silver nanoparticles from the environment is difficult due to their aforementioned properties. An additional layer of difficulty occurs once these particles are incorporated into the food chain, where they can become concentrated to more toxic levels. The European Parliament, the American Food and Drug Administration (FDA) and the British Standards Institute are all working towards implementing defined controls to impose on companies manufacturing nanoparticles. This is both for the safety of factory employees during the production of nanoparticle-based products and for the safety of consumers purchasing nanoparticle-based products.

Definitions

Antioxidant: A substance that inhibits oxidation, so reducing oxidative damage. For example; glutathione, vitamin E and vitamin C.

Bactericidal: A substance that kills bacteria. For example; disinfectants such as hypochlorites and anti-septics such as iodopovidone.

Bacteriostatic: A substance that inhibits bacterial growth. For example; antibiotics like chloramphenicol and tetracyclines, although at higher doses these can also be bactericidal.

Colloidal silver: A fluid suspension containing fragments of silver between one nanometre and one micrometre in size, with less than 5% silver ion content.

Cytostatic: A substance that inhibits the growth and proliferation of cells.

Metastasis: The spread of cancer cells from the site where they originated (the primary site) to other locations (secondary sites).

Quantum confinement: A phenomenon relating to differences in energy states of particles, which governs their electrical and optical properties.

Surface Functionalisation: The conjugation of an additional moiety to the surface of a structure, such as a nanoparticle. The surface functionalisation of nanoparticles can improve the delivery and targeting of nanoparticle-based therapies, as well as improve nanoparticle-based experimental techniques.

Surface plasmon resonance: The excitation of surface plasmons by light, which forms the basis of many modern biosensor tools. Surface plasmons are oscillations produced from electrons at the surface of an object.

Therapeutic index: A measure of the safety of a drug for use in treatments, calculated by measuring the ratio between the toxic dose and therapeutic dose. A high therapeutic dose meaning the dose required to cause toxicity is well above that required for treatment.

Acknowledgements

This study was supported by grants from the Academic Research Fund, Ministry of Education, Singapore (T206B3108; WBS: 185-000-153-112) to MPH.

Abbreviations

ATSDR	:	Agency for Toxic Substances and Disease Registry
CEINT	:	Center for Environmental Implications of Nanotechnology
DSB	:	Double Strand Break

EHS	:	Environmental, Health and Safety
EPA	:	Environmental Protection Agency
FDA	:	Food and Drug Administration
FGF2	:	Fibroblast Growth Factor 2
H2AX	:	H2A Histone family member X
HIV	:	Human Immunodeficiency Virus
IANH	:	International Alliance for NanoEHS Harmonization
ISO	:	International Organisation for Standarization
NNI	:	National Nanotechnology Initiative
NFκB	:	Nuclear Factor Kappa B
NRF2	:	Nuclear Factor (erythroid-derived 2)-like 2
OECD	:	Organization for Economic Co-operation and Development
p38	:	Protein 38
p53	:	Protein 58
PbPk	:	Physiologically based Pharmacokinetic/ Pharmacodynamic
ROS	:	Reactive Oxygen Species
SSB	:	Single Strand Break
UC CEIN	:	University of California Center for Environmental Implications of Nanotechnology
VEGF	:	Vascular Endothelial Growth Factor

References

Ahamed,M., M. Karns, M. Goodson, J. Rowe, S.M. Hussain, J.J. Schlager and Y. Hong. 2008. DNA damage response to different surface chemistry of silver nanoparticles in mammalian cells. Toxicol. Appl. Pharmacol. 233: 404–410.

Ahamed, M., R. Posgai, T.J. Gorey, N. Nielsen, S.M. Hussain, J.J. Rowe. 2010. Silver nanoparticles induced heat shock protein 70, oxidative stress and apoptosis in Drosophila melanogaster. Toxicol. Appl. Pharmacol. 242: 263–9.

Asharani, P.V., Y.L. Wu, Z. Gong and S. Valiyaveettil. 2008. Toxicity of silver nanoparticles in zebrafish models. Nanotechnology 19: 255102.

Asharani, P.V., M.P. Hande and S. Valiyaveettil. 2009a. Anti-proliferative activity of silver nanoparticles. BMC Cell Biol. 10: 65.

Asharani, P.V., G.K.M. Low, M.P. Hande and S. Valiyaveettil. 2009b. Cytotoxicity and genotoxicity of silver nanoparticles in human cells. ACS Nano 3: 279–90.

Asharani P.V., S. Sethu, S.P. Zhong, C.T. Lim, M.P. Hande and S. Valiyaveettil. 2010. Effects of silver, gold and platinum nanoparticles on normal human erythrocytes. Adv. Funct. Mater. 20: 1233–1242.

Baker, C., A. Pradhan, L. Pakstis, D.J. Pochan and S.I. Shah. 2005. Synthesis and antibacterial properties of silver nanoparticles. J. Nanosci. Nanotechnol. 5: 244–249.

Benn, T.M., and P. Westerhoff. 2008. Nanoparticle Silver Released into Water from Commercially Available Sock Fabrics. Environ. Sci. Technol. 42: 4133–4139.

British Standards Institution 2009 PD 6699-2: 2007 nanotechnologies. Part 2. Guide to safe handling and disposal of manufactured nanomaterials. London, UK: BSI.
Chen, X., and H.J. Schluesener. 2008. Nanosilver: a nanoproduct in medical application. Toxicol. Lett. 176: 1–12.
Choi, J.E., S. Kim, J.H. Ahn, P. Youn, J.S. Kang, K. Park, J. Yi and D.Y. Ryu. 2010. Induction of oxidative stress and apoptosis by silver nanoparticles in the liver of adult zebrafish. Aquat. Toxicol. 100: 151–159.
Bowman, D.M., and G.A. Hodge. 2007. A Small Matter of Regulation: An International Review of Nanotechnology Regulation, Colum. Sci. & Tech. L. Rev. 8: 1–36.
Elechiguerra, J.L., J.L. Burt, J.R. Morones, A. Camacho-Bragado, X. Gao H.H. Lara and M.J. Yacaman. 2005. Interaction of silver nanoparticles with HIV-1. J. Nanobiotechnology 3: 6.
Eom, H.J., and J. Choi. 2010. p38 MAPK Activation, DNA Damage, Cell Cycle Arrest and Apoptosis As Mechanisms of Toxicity of Silver Nanoparticles in Jurkat T Cells. Environ. Sci. Technol. 144: 83337–8342.
Evans, M.D., M. Dizdaroglu and M.S. Cooke. 2004. Oxidative DNA damage and disease: induction, repair and significance. Mutat. Res. 567: 1–61.
Germano, C. 2008. Managing the Emerging Risks of Nanotechnology. ACE Progress Report. Available at *http://www.aceusa.com/news/Pages/News.aspx*
Gibson, K., and D. Pula. 2009. Nanoparticles: Environmental risk and regulation. Environmental Quality Management. 18: 1–7.
Grigg, J., B. Tellabati, S. Rhead, G.M Almeida, J.A. Higgins, K. J. Bowman, G.D. Jones and P.B. Howes. 2009. DNA damage of macrophages at an air-tissue interface induced by metal nanoparticles. Nanotoxicology. 3: 348–354.
Hsin, Y.H., C.F. Chen, S. Huang, T.S. Shih, P.S. Lai and P.J. Chueh. 2008. The apoptotic effect of nanosilver is mediated by a ROS- and JNK-dependent mechanism involving the mitochondrial pathway in NIH3T3 cells. Toxicol. Lett. 179: 130–139.
Kim,Y.S., J.S. Kim, H.S. Cho, D.S. Rha, J.M. Kim, J.D. Park, B.S. Choi., R. Lim, H.K. Chang, Y.H. Chung, I.H. Kwon, J. Jeong, B.S. Han and J.J. Yu. 2008. Twenty-eight-day oral toxicity, genotoxicity, and gender-related tissue distribution of silver nanoparticles in Sprague-Dawley rats. Inhal. Toxicol. 20: 575–583.
Kim, S., J.E. Choi, J. Choi, K.H. Chung, K. Park, J. Yi and D.Y. Ryu. 2009. Oxidative stress-dependent toxicity of silver nanoparticles in human hepatoma cells. Toxicol. *In Vitro*. 23: 1076–1084.
Klasen, H.J. 2000. A historical review of the use of silver in the treatment of burns. II. Renewed interest for silver. Burns 26: 131–138.
Kumaravel, T.S., B. Vilhar, S.P. Faux and A.N. Jha. 2009. Comet assay measurements: a perspective. Cell Biol. Toxicol. 25: 53–64.
Kuzma, J. ed. 2006. The Nanotechnology-Biology Interface: Exploring Models for Oversight. Workshop Report. Available at *http://www.hhh.umn.edu/centres/stpp/nanotechnology.html*
Lu, L., R.W. Sun, R. Chen, C.K. Hui, C.M. Ho., J.M. Luk, G.K. Lau and C.M. Che. 2008. Silver nanoparticles inhibit hepatitis B virus replication. Antivir Ther. 13: 253–262.

Navarro, E., A. Baun, R. Behra, N.B. Hartmann, J. Filser, A.J. Miao, A. Quigg, P.H. Santschi and L. Sigg. 2008. Environmental behavior and ecotoxicity of engineered nanoparticles to algae, plants, and fungi. Ecotoxicology 5: 372–386.

Park, J., B.K. Kwak, E. Bae, J. Lee, Y. Kim, K. Choi and J. Yi. 2009. Characterization of exposure to silver nanoparticles in a manufacturing facility. J. Nanopart. Res. 11: 1705–1712.

Peer, D., J.M. Karp, S. Hong, O.C. Farokhzad, R. Margalit and R. Langer. 2007. Nanocarriers as an emerging platform for cancer therapy. Nat. Nanotechnol. 2: 751–60.

Sawosz, E., M. Grodzik, M. Zielińska, T. Niemiec, B. Olszańska, A. Chwalibog. 2009. Nanoparticles of silver do not affect growth, development and DNA oxidative damage in chicken embryos. Arch. Geflügelk 73: 208–213.

Seaton, A., L. Tran, R. Aitken and K. Donaldson. 2010. Nanoparticles, human health hazard and regulation. J.R.S. Soc. Interface 7: S119–S129.

Wilhelmi, B. 2008. Nanosilver: a test for nanotech regulation. Food Drug Law J. 63: 89–112.

Wise, J.P. Sr., B.C. Goodale, S.S. Wise, G.A. Craig, A.F. Pongan, R.B. Walter, W.D. Thompson, A.K. Ng, A.M. Aboueissa, H. Mitani, M.J. Spalding and M.D. Mason. 2010. Silver nanospheres are cytotoxic and genotoxic to fish cells Aquat. Toxicol. 97: 34–41.

Xu, R., J. Ma, X. Sun, Z. Chen, X. Jiang, Z. Guo, L. Huang, Y. Li, M. Wang, C. Wang, J. Liu, X. Fan, J. Gu, X. Chen, Y. Zhang and N. Gu. 2009. Ag nanoparticles sensitize IR-induced killing of cancer cells. Cell Res. 19: 1031–1034.

Yang, L., H.J. Wang, H.Y. Yang, S.H. Liu, B.F. Zhang, K. Wang, X.M. Ma and Z. Zheng. 2008. Shape-controlled synthesis of protein-conjugated silver sulfide nanocrystals and study on the inhibition of tumor cell viability. Chem. Commun. 26: 2995–2997.

Yang, W., C. Shen, Q. Ji, H. An, J. Wang, Q. Liu and Z. Zhang. 2009. Food storage material silver nanoparticles interfere with DNA replication fidelity and bind with DNA. Nanotechnology. 20: 085102.

13

Nanobubbles and Their Putative Application to Cancer Medicine

Tianyi M. Krupka[1,a] and *Agata A. Exner*[1,b,*]

With ultrasound gaining more ground in cancer diagnosis and therapy, ultrasound contrast agents (UCAs aka bubbles) are receiving unprecedented attention in the scientific community. Other than the conventional application, UCAs also serve as including nuclei for cavitation in ultrasound mediated drug delivery and carriers for drug and/or gene delivery. While *micro*bubbles are widely studied and commercially available for clinical (particularly in Europe and Asia) and experimental use, the utilization of the nanometric version of these bubbles in cancer diagnosis and therapy is still in relatively early stages of development. Despite the understanding that nanobubbles are more desirable for disease targeting, both passively and actively, the lag in nanobubble development in cancer medicine can be largely attributed to challenges with nanobubble echogenicity at clinically relevant ultrasound frequencies (3–20MHz), and lack of commercial availability due to difficulty in cost effective scale up formulation techniques, among others. This chapter provides comprehensive information on topics including the acoustic behavior of bubbles within the ultrasound field, the major types of nanobubbles currently under development, their formulation methods and their working principle in cancer diagnosis and therapy.

[1]Department of Radiology, Case Western Reserve University, 11100 Euclid Avenue, Cleveland, OH 44106, United States.
[a]E-mail: tianyi.krupka@case.edu
[b]E-mail: agata.exner@case.edu
*Corresponding author

List of abbreviations after the text.

Keywords: ultrasound (US), nanobubble, contrast agent, theranostic, cancer diagnostic, cancer therapy, gene/drug delivery

INTRODUCTION

The ultimate goal of cancer drug delivery is to eradicate cancer and save lives, and achieving this goal involves multidisciplinary efforts. In the era of nanotechnology many nanocarrier mediated cancer drug delivery systems (DDS) are either in clinical use or under development. These range from silicon-based structures like nanopores and nanoneedles, to carbon structures such as nanotubes and fullerenes to particle based products such as quantum dots, gold nanoparticles, liposomes and micelles to name but a few. In recent years, nanobubble UCAs have also been investigated and their theranostic (therapeutic and diagnostic) properties examined.

The excitement surrounding nanotechnology can be attributed to the many desirable qualities of the nanocarriers. Their small size makes them excellent for drug delivery to tumors through passive EPR (enhanced permeability and retention) effects that are manifested by leaky vasculature and poorly developed lymphatic drainage system (Jain 1998). Furthermore, many nanocarriers, such as micelles composed of either lipids or polymers, improve the solubility and stability of hydrophobic drugs through the encapsulation of those drugs into the hydrophobic core. Additionally, the hydrophilic corona of these micelles provides stability of these structures in aqueous solutions, making them suitable for parenteral administration. In addition, the increased surface to volume ratio of nanocarrier compared to their micrometer counterparts renders them ideal for ligand conjugation which is necessary in active drug targeting applications. Finally, many nanoparticles can be tailored to serve multiple functions including cancer drug targeting and imaging contrast. All of these tailored features help to improve anticancer drug to tumor selectivity, increase treatment efficacy, and reduce drug distribution in healthy tissue, which ultimately reduces systemic toxicity.

Among the many permutations available for nanoparticle formulation, the unique structure of nanobubbles provides functionality that cannot be achieved with other systems. While they too can be tailored to provide multiple functions including targeting, therapy and imaging simultaneously, their structure also enables the use of external stimulus, low frequency US, to drive payloads into the cells. The following sections provide comprehensive information on topics including the acoustic behaviors of bubbles in US field, the major types of nanobubbles, their formulation methods and their working principle in cancer diagnosis and therapy.

NANOBUBBLE ULTRASOUND CONTRAST AGENTS IN CANCER MEDICINE

Cancer

"Cancer is a disease of malfunctioning cells" (Bergers and Benjamin 2003). According to the American Cancer Society, cancer is the second leading cause of death in the United States. In 2009, there were about 1,500,000 new cancer cases and over 560,000 deaths which was exceeded only by cardiovascular disease (Jemal et al. 2009). The most commonly accepted options for treatment of cancer are surgery, radiation, and chemotherapy. However, once metastatic cancers are diagnosed, systemic chemotherapy often becomes the only treatment option available. Conventional chemotherapy works by taking advantage of the rapid dividing nature of cancerous cells. Inevitably, it also kills normal tissues that have a rapid turnover rate including bone marrow, hair follicles and the cells that line the digestive tracts (Gibson and Stringer 2009), resulting in systemic toxicity. Currently available anticancer drugs may possess curative potency, but the exceedingly high dose required to achieve a complete treatment is nearly impossible with systemic drug administration. As such, one of the major challenges of using chemotherapy to treat cancer is to selectively kill rapidly dividing cells within cancerous lesions, but not normal cells, hence minimizing systemic toxicity. To meet this challenge, the design and implementation of DDS that could direct anticancer agents to the tumor cells but spare the rapidly dividing normal tissue becomes critical. Nanobubble UCAs in combination with US have been shown to hold great potential for this application.

Ultrasound

By definition ultrasound (US, >20 kHz) is sound at frequencies above human hearing (20 Hz–20 kHz). It propagates by transmitting pressure waves through medium such as water, air, and the human body. To acquire an US image, an US pulse is emitted by a transducer. Similarly to electromagnetic waves (light), while some of the waves are transmitted and propagate forward, others will be reflected back by the tissue towards the source of the energy transmission and will be detected. Due to the differences in the ability of tissues to resist sound propagation (acoustic impedance), the amount of sound reflected varies at an interface between two tissue structures. These properties provide the basis for US imaging. Traditional US imaging has primarily served diagnostic purposes. Recently, therapeutic US applications including HIFU (high intensity focused US), lithotripsy, US hemostasis and US assisted thrombolysis have been utilized (Qin et al. 2009). In addition, research over the course of the last decade, in US-activated drug delivery has also been thriving. US drug delivery is based on the principle that US energy produces cavitation events. These cavitation events subsequently perturb cell membrane permeability by creating nonlethal pores (sonoporation) in the cell membrane or the blood vessel walls. This disruption can facilitate gene or drug extravasation from the vasculature and entry into the cells. This approach is attractive from many perspectives—it is cost effective and easily accessible, and the US beam can be directed to deep tissues with predictable energy deposition patterns and millimeter precision. This later feature dictates the treatment to diseased tissues such as cancerous lesions and prevents unnecessary damage of healthy tissue.

Ultrasound Contrast Agents (UCA) and Their Working Principle

UCAs are gas encapsulated bubbles with a typical diameter in the micron range, thus they are generally referred to as 'microbubbles'. The unique ability for microbubbles to scatter sound waves makes them great US contrast enhancers (Fig. 1). There are a handful of microbubble UCAs that are commercially available. While some of them like Albunex®, Imagent, Optison™ and Definity® are FDA approved for specific uses in the United States and other countries, others like Levovist, Sonazoid, Sonovue are currently approved for use only in Europe, Asia and Latin America. In addition, some of these agents were once commercially available and are currently off the shelf, such as Albunex and Imagent; others, such as Cardiosphere and Imagify are going through the process of FDA approval (Kaul 2008). While the manufacturing processes of these micobubbles are similar, with most made by mechanical agitation, the bubble materials vary vastly. The shell materials can be protein, polymer, or lipids; and the gas core can be simple air or high molecular weight, perfluorocarbons with low water solubility.

Within an US field micobubbles exhibit different behaviors (Fig. 2) responding to the change of acoustic pressure, often defined by the mechanical index (MI). At lower acoustic pressure (MI<0.05–0.1), bubbles oscillate linearly. The frequency received is unaltered from the frequency

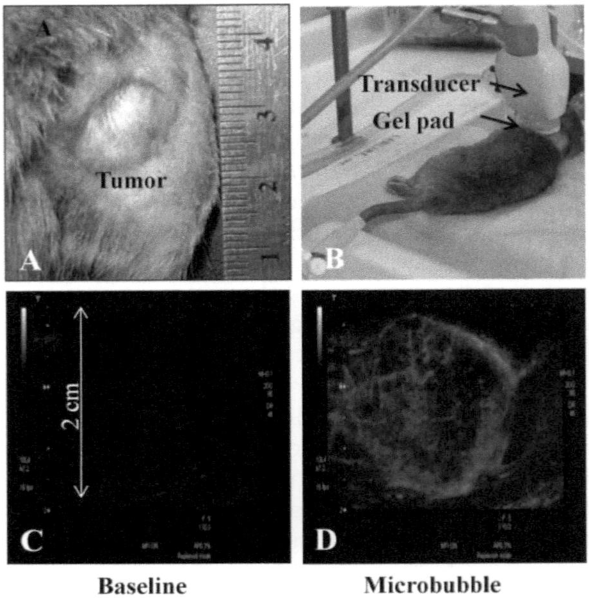

Fig. 1. Contrast enhanced ultrasound imaging of blood flow with microbubbles (A) representative subcutaneous rat tumor; B: tumor imaging setup (C) baseline image of rat tumor without contrast (D) microbubble enhanced ultrasound of tumor vasculature *(unpublished original illustration).*

Color image of this figure appears in the color plate section at the end of the book.

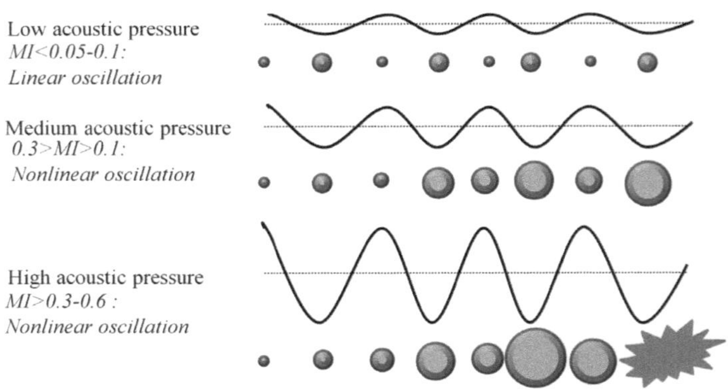

Fig. 2. The behavior of a single bubble under ultrasound irradiation. At lower acoustic pressure, bubbles oscillate linearly, and the frequency received is unaltered from the frequency transmitted by the US transducer. At medium pressure (low power imaging), bubbles oscillate nonlinearly, more resistance to compression than expansion due to gas inward diffusion from the surrounding medium (noninertial cavitation). At high pressure, forced expansion and compression leads to bubble destruction (inertial cavitation) *(unpublished original illustration).*

transmitted by the US transducer, producing no contrast enhancement. At intermediate acoustic pressure (0.3>MI>0.1) non-linear bubble oscillation starts. Here bubbles experience more resistance to compression than expansion (rectified diffusion) leading to the emission of harmonic signals at multiples of the transmitted frequency. The frequency shift amplifies the acoustic impedance mismatch between host tissue and tissues with the presence of bubbles. Consequently, the signal to noise (background) ratio improves. It is the mismatch of impedance between the host tissue and the bubbles that provide the contrast for US imaging (Dijkmans et al. 2004).

In addition to the MI, the matching between the imaging frequency and the natural resonance frequency of the bubbles for bubble detection is also of critical importance. Since the bubbles vibrate most readily when insonated at their resonance frequency, the bubble will produce the highest contrast when imaged at this frequency (Ferrara et al. 2007).

Other than the traditional applications of these bubbles as US signal enhancers in diagnostic imaging, UCA have found relevance in therapeutic applications including targeted gene and drug delivery which is possible due to their unique acoustic properties. While lower and intermediate acoustic pressures produce linear and nonlinear oscillation of bubbles (non-inertial cavitation), at higher acoustic pressures (MI>0.3–0.6), the bubbles undergo forced expansion and compression, and consequently, destruction (inertial cavitation). Under this behavior, the bubble compresses under a higher pressure node and expands at a lower pressure node. However, the degree of expansion is more than that of compression. If these expansion and compression cycles continue, the bubbles will burst and create massive forces that cause the liquid to flow violently, forming shockwaves and micro jets (Dijkmans et al. 2004).

Studies have demonstrated that US alone without the supplementation of contrast bubbles was able to lead to cavitation effects on the cell membranes. However, this occurs in general at very high acoustic pressures which are often invasive. Although under non-inertial cavitation conditions, bubbles can increase membrane fluidity, create transient pores and enhance drug uptake in the cells, it is the destructive behavior of UCAs at higher acoustic pressures which mostly enables the sonoporation process in US mediated gene and drug delivery. It is documented that the supplemented bubbles act as the cavitation nuclei which reduce the threshold of US pressure, thus produce stronger cavitation effects at lower MI. The ability of US imaging to perturb cell membranes or cause cell damage depends on a range of factors which includes but is not limited to the US intensity, irradiation time, and cell type. While the following equation describes the fundamental relationship between UCA properties such as size, shell material, encapsulated gas and their echogenic properties, the details of this complex relationship are described elsewhere (Ferrara et al. 2007).

According to (de Jong et al. 1992), the resonant (natural or fundamental) frequency of an elastic shell stabilized microbubble is related to the bubble diameter and the shell elastic modulus via **Equation 1:**

$$f_0 \approx \frac{1}{2\pi r}\sqrt{\frac{3\gamma}{\rho}\left(P_0 + \frac{\pi}{3\gamma}\frac{S_e}{r}\right)}, \qquad S_e = 8\pi\frac{R_0 - R_i}{1-v}E$$

Equation 1. Bubble resonance frequency. f_0: the resonant frequency of a microbubble, γ: the ideal adiabatic constant of gas, ρ: density of surrounding medium, r: bubble radius, P_0: ambient fluid pressure, R_0–R_i: the thickness of the bubble shell, v: the Poisson's ratio, and E: the Young's modulus.

Based on this equation, the smaller the bubble size, the higher the resonance frequency of the bubble, and the higher the imaging frequency required to adequately visualize the bubbles. Similarly, the stiffer the shell is, the higher the resonance frequency of the bubbles. It should be noted that many of the acoustic theories of bubbles developed are based on those with sizes at the micron scale or larger. However, the same physical principles should be applicable to nanobubble systems.

Current Nanobubble Formulations

Numerous nanobubble formulations have been reported. Broadly, these formulations can be categorized into three groups, 1) single-layer lipid or surfactant nanobubbles, 2) bilayer lipid nanobubbles (the liposome nanobubbles), and 3) other non-lipid bubbles.

Single Layer Lipid or Surfactant Encapsulated Nanobubbles

Post Formulation Manipulations. Single layer lipid or surfactant encapsulated nanobubbles are those gas nanobubbles that are stabilized by one layer of lipids, surfactants, or combination of both. Among these nanobubble formulations, many are obtained through post formulation processing (Fig. 3). Oeffinger and Wheatley (2004) were able to acquire echogenic nanobubbles with a mean diameter of 450–700 nm by first formulating ST68 microbubbles. These microbubbles are comprised of a cocktail containing Span 60, sorbitan monostearate, Tween 80, and polyoxyethylene-sorbitan monooleate as the shell. Perfluoropropane (C_3F_8) is used as the gas core and bubbles are prepared by a probe sonication method. To obtain nanobubbles, the micorbubble formulation was submitted to centrifugation (Oeffinger and Wheatley 2004). This method is based on the principle that larger bubbles have lower density and are more buoyant than the smaller

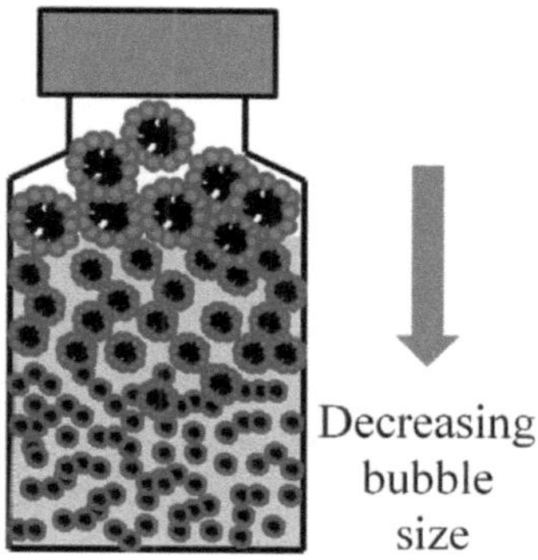

Fig. 3. Bubble size distribution after gradient centrifugation. Larger bubbles are more buoyant rise to the upper layer of the vial while small ones stay in suspension in the lower layers *(unpublished original illustration).*

nanobubbles, hence rise up while smaller bubbles stay in suspension at the lower fluid layers. The authors demonstrated that these nanobubbles provided satisfactory imaging contrast both for *in vitro* test tubes and *in vivo* with rabbits (Wheatley et al. 2006).

Likewise, Xing et al. also used a centrifugation method to acquire nanobubbles (diameter range: 400–600 nm) from parental microbubble formulations made by sonication (Xing et al. 2010). These are C_3F_8 gas bubbles stabilized with a shell consisting of a mixture of the surfactants, Sorbitan monostearate (Span 60) and polyoxyethylene 40. After *in vivo* contrast enhancement studies in the New Zealand white rabbit kidney, the authors concluded that these nanobubbles were appropriate for *in vivo* tumor imaging.

In contrast to the centrifugation fractionation method, Wang Y. et al. collected nanobubbles directly by simply removing the larger bubble layer on top of the parental microbubble formulation. They were able to obtain hexafluoride (SF_6) nanobubbles with mean diameters of 333.1nm. These bubbles are stabilized by soybean lipid (SPC) and Tween 80. By introducing an oil phase in the bubble formulation process, the authors were able to incorporate a model drug coumarin-6 into these nanobubbles. Subsequently, they demonstrated that higher cellular uptake of the model drug was attained even without the presence of US compared to those delivered with liposome and chitosan nanoparticles. The authors speculated that the increased uptake was potentially due to the incorporation of Tween 80 which changed the cell membrane fluidity. The same hypothesis was used to explain why the nanobubbles after systemic administration were echogenic *in vivo* in the liver of nude mice (Wang et al. 2010).

Although effective for selecting nano-sized bubbles, the above methods have drawbacks including: 1) the potential for sample contamination, 2) reducing bubble yield and stability, 3) wasting stock materials, and 4) being labor intensive. In the upcoming section some nanobubbles without post formulation manipulations will be introduced.

Nanobubbles Produced Without Post Formulation Manipulations

While direct nanobubble formulation is not commonly reported, Krupka et al. have recently published an article describing a method using the nonionic triblock copolymer Pluronic as a size modulator to produce echogenic lipid nanobubbles with diameters as small as 180nm (Krupka et al. 2010). These nanobubbles are comprised of a cocktail of 3 types of phospholipids (DPPC, DPPE and DPPA) stabilized by glycerol and Pluronic (Fig. 4). Bubbles were prepared by first dissolving DPPC, DPPA and DPPE in chloroform, followed by evaporation of the solvent and hydration in the presence of glycerol and Pluronic to produce lipid vesicles. After air removal and introduction of C_3F_8, the mixture was agitated to produce the final nanobubble products.

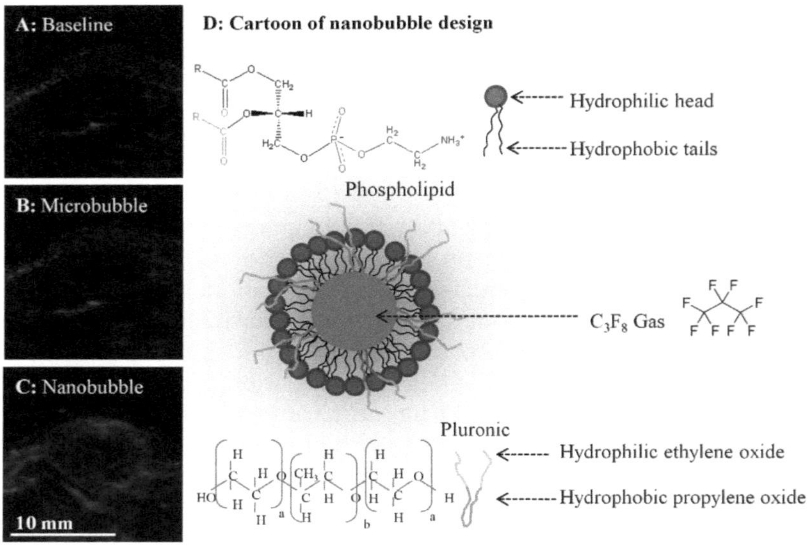

Fig. 4. Ultrasound imaging enhanced with Pluronic nanobubbles. A: Baseline image of subcutaneous rat tumor; B: microbubble enhanced image; C: nanobubble enhanced image; D: Pluronic nanobubbles (C_3F_8 gas stabilized by phospholipid shell and Pluronic, a triblock amphiphilic polymer) (*unpublished data*).

Color image of this figure appears in the color plate section at the end of the book.

Krupka et al. demonstrated that these Pluronic nanobubbles were more stable and more echogenic than the control microbubbles using an *in vitro* gel phantom, and their properties also translated into the *in vivo* settings for tumor contrast enhancement in an allograft, subcutaneous rat colorectal tumor model.

The higher echo produced by the nanobubbles appears to be counterintuitive, since typically the resonant frequency of bubbles is inversely related to bubble diameter. The authors explained that improved contrast produced by Pluronic nanobubbles might be attributable to the lipid shell fluidity, as higher fluidity may be able to compensate the small size resulting in a resonant frequency shift (Krupka et al. 2010).

Both of the *in vitro* and *in vivo* results from the above work are encouraging; however, single layer lipid echogenic nanobubble formulations without post formulation manipulation are still not commonly produced. More extensive research has been published on bilayer lipid (liposomal) nanobubbles.

Bilayer Lipid Nanobubbles

Acoustically Active Liposomes (AALs). AAL nanobubbles have also been called liposomal nanobubbles (or bubble liposomes). These acoustically active liposomes (mean diameter: 150–200 nm) are formulated by sonication of liposome suspensions (composed of 1,2-distearoyl-sn-glycero-phosphatidylcholine (DSPC) and 1,2-distearoyl-sn-glycero-3-phosphatidyl-ethanolamine-methoxypolyethyleneglycol (DSPE-PEG-OMe; NOF)) under the condition of positive pressure with C_3F_8 in sealed glass vials (Fig. 5). After formulation, nanobubbles in the liposome coalesce forming larger microbubbles, a superior contrast enhancer than nanobubbles at clinical relevant frequencies according to the general acoustic theory. These echogenic liposomes are attractive for cancer drug delivery due to: 1) they are potential carriers for various (hydrophilic or lipophilic) genes, drugs, and proteins; 2) payload release at the target site can be facilitated by external stimuli (US) with the entrapped gas serving as a cavitation nuclei; and 3) the entrapped gas serves as an imaging contrast agent providing both pharmacokinetic monitoring potential and tumor localization imaging.

The working principle of the acoustically active liposomes is that drug, which is preloaded into the liposomes, is carried by the liposome bubbles to target tissue. EPR effects will lead to higher tumor tissue accumulation of these drug carrying particles. Upon therapeutic US application, bubbles undergo a cavitation process, membrane perturbation takes place creating transient pores, and cavitation of entrapped gas bubbles release payload from the liposome. All of these processes facilitate drug penetration into the tumor.

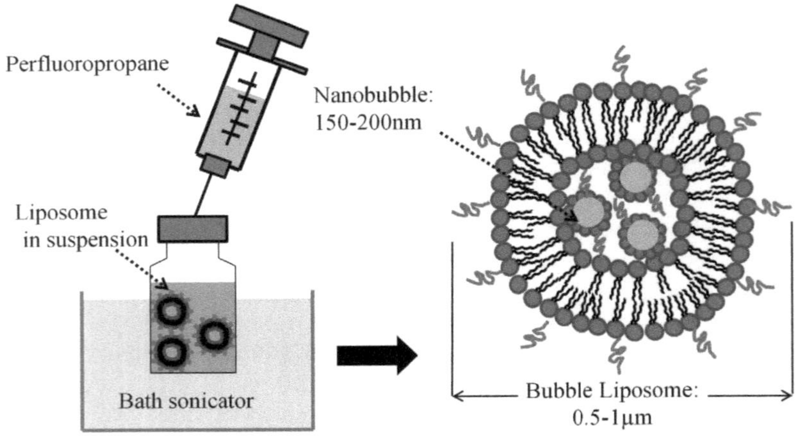

Fig. 5. Schematic of bubble liposome formulation. Sonication of C_3F_8 and suspension of liposomes in a sealed vial lead to the production of nanobubbles, with diameters of 150–200 nm *(unpublished original illustration).*

Color image of this figure appears in the color plate section at the end of the book.

To improve tumor selectivity, liposome nanobubble designs have undergone several iterations of modifications. While their shell construction may vary from one version to another, their makeup and working principles stay mostly unaltered. Between the various versions of these liposome bubbles, they have not only been shown to be suitable US contrast agents, but also been shown to facilitate plasmid DNA transduction into cells both in the *in vitro* and *in vivo* settings (Un et al. 2010a; Un et al. 2010b).

Other acoustically active liposomes have also been developed (Huang and MacDonald 2004; Kodama et al. 2010; Buchanan et al. 2010). While formulation methodology and constituting materials including the shell, drug, and gas of these acoustically active entities varies slightly, similar to the liposome bubbles, all methods involve mechanical agitation of the mixture of liposome and gas.

While the exact configuration of these acoustically active liposomes are not known, three configurations have been proposed: Fig. 6A, inside the liposome, there is a gas/liquid interface; B, in the liposome, there are two compartments, the aqueous one with hydrophilic drugs and the gaseous one encapsulated with single layer of lipids; and C, again there are two separate compartments, single layer nanobubbles suspended in the aqueous core of the liposome. Although, configuration A has been proposed, it is very unlikely due to the hydrophilic and hydrophobic interactions between the aqueous solution and the hydrophobic gas. Hence, it is generally agreed that both configuration B and C are the likely configurations.

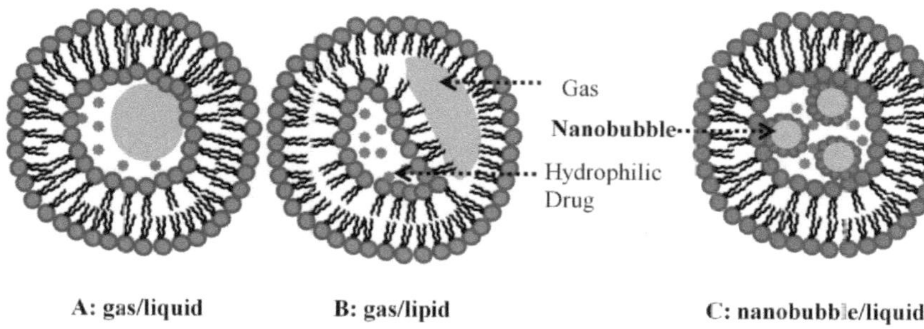

Fig. 6. Potential configurations of acoustically active liposomes. A: gas and hydrophilic drug floating in the aqueous core of the liposome without stabilizing shell; B: gas is encapsulated between the lipid bilayer of the liposome shell; C: bubbles stabilized by a single lipid shell are suspended in the aqueous core of the liposomes *(unpublished original illustration).*

Color image of this figure appears in the color plate section at the end of the book.

Other Non-Lipid Nanobubbles

Several groups have investigated alternatives to the lipid or lipid-surfactant bubble formulations. These include *in vivo* forming nanobubbles such as: plasmonic (Wagner et al. 2010), PEG-PLLA shelled perfluoropentane nanobubbles (Rapoport et al. 2007), and nanobubbles generated from gas generating polymeric nanoparticles (Kang 2009).

Plasmonic Nanobubbles (PNB). PNBs are gold nanoparticle-generated vapor nanobubbles (plasmonic nanobubbles or PNB). Instead of administrating formulations at the gas state, these bubbles are generated on demand at the cell or tissue of interest and exist transiently. These nanobubbles are visualized by a photothermal laser microscope (Wagner et al. 2010). To generate these bubbles, Plasmon/gold nanoparticles (PN) are exposed to laser pulses causing the heating of PN. The heat transportation from the PN to the surrounding aqueous medium leads to the vaporization of a thin layer of liquid, hence the formation of nanobubbles at the surface of the PN (Fig. 7). As the pressure inside the bubble layer grows, it eventually exceeds the surface tension of the surrounding aqueous shell; the bubbles expand and collapse, then cavitate which facilitates drug/gene delivery to cancer cells.

In vitro studies using lung carcinoma cells (A549) have shown that the NPs were internalized through non-specific endocytosis and concentrated in the endosomal compartments. After a short laser exposure (PNB generated), cells were damaged with appropriate laser fluence while cells without laser exposure (no PNBs) had no detectable cell damage. The authors hypothesized that the cell damage was caused by bubble collapse and subsequent cavitation activity that is highly localized.

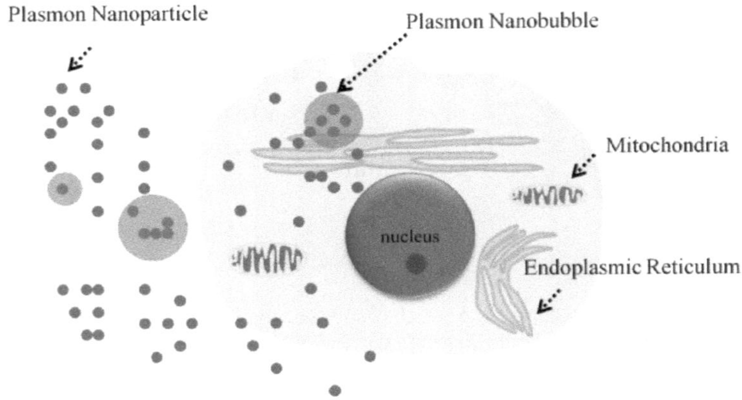

Fig. 7. Plasmon nanobubble formation. Laser pulse at appropriate fluence causes heating of plasmon nanoparticles. Heat accumulation causes evaporation of surrounding medium either around a single particle or clusters of nanoparticles. Evaporated medium forms nanobubbles either in the cell or around the cell and subsequent nanobubbles cavitation leads to cell damage *(unpublished original illustration).*

Color image of this figure appears in the color plate section at the end of the book.

In vivo studies in zebrafish showed that cell damage was detected at laser pulsed areas, while no detectable cell damage was detected at areas without a laser pulse. By tuning the laser fluence, the authors claimed that they were able to detect and destroy single metastatic human prostate cancer cells xenografted in zebrafish (Wagner et al. 2010).

This approach circumvents antibody conjugation and drug incorporation into the nanobubbles, both of which could perturb the stability and size of the bubbles. In addition, prolonged half life is not a requirement, since PNBs are generated on demand. Finally, these bubbles can be imaged acoustically although optical methods to provide better temporal and special resolution. One caveat is that the optical detection of PNBs can be realized in transparent zebrafish but not in disease bearing animals or humans whose bodies are opaque and not suitable for optical imaging (Anderson et al. 2010).

PEG-PLLA (PCL) Nanobubbles. Another type of *in vivo* forming nanobubble formulation has been developed by Rapoport et al. 2007. These nanobubbles are generated by taking advantage of the thermodynamic characteristic of perfluoropentane (PFP). Under the atmosphere pressure, the PFP compound has a boiling point at 29°C indicating a high susceptibility for vaporization upon heating. However, in aqueous solutions, ample boiling point elevation often occurs for small droplets covered with an elastic polymer shell due to Laplace pressure.

Laplace pressure (ΔP, **Equation 2**) is the pressure difference between the interior (P_{in}) and the exterior (P_{out}) of a droplet or bubble; and the larger the interior pressure the higher the boiling point of the compound in the droplet. In addition, the pressure difference is inversely related to the size (radius, r) of the droplet, i.e., the smaller the droplet size, the higher the pressure and the larger the boiling point elevation. σ is the surface tension between the droplet/gas core and the shell materials.

$$\Delta P = P_{in} - P_{out} = \frac{2\sigma}{r}$$

Equation 2. Laplace pressure (ΔP) of droplet or bubbles in solution.

The rationale behind this DDS design is that at lower room temperatures, PFP is in the liquid state and stays as a droplet. After co-administration with drug carrying PEG-PLLA micelles, the droplet maintains the initial state due to Laplace pressure.

The authors theorize that there are three means of droplet to bubble transition, 1) temperature (thermal), 2) sonication (thermal and/or mechanical) and 3) injection through a thin needle (mechanical), among which sonication dominates the process. The sonication induced droplet to bubble transition is called acoustic droplet vaporization, or ADV. Upon administration of PEG-PLLA or PEG-PCL micelles and polymer encapsulated PFP (liquid) droplets, excessive Laplace pressure prevents the PFP droplet to nanobubble phase transition—a favorable property for tumor uptake of these droplets via EPR effects. Upon tumor directed US irradiation, two major processes takes place. First, US triggers droplet to bubble conversion via acoustic droplet vaporization. Second, tumor directed US triggers therapeutic release from drug carriers and unloading cargo to the tumor tissue. Once the phase transition takes place, the nanobubbles coalesce and form larger microbubbles. Compared to nanobubbles, the microbubbles are more echogenic at clinically relevant US frequencies (Fig. 8).

In vivo testing with these DDS using nu/nu mice carrying subcutaneous breast cancer (MDA MB231) or ovarian carcinoma (A2748) xenografts showed that those tumors exposed to nanodroplets combined with US had significantly higher drug uptake than those of micelles combined with US. In addition, US significantly enhanced tumor growth inhibition when systemic nanodroplets encapsulating doxorubicin (DOX) were given and without the presence of US; little tumor growth inhibition was detected. Based on these results, the authors concluded that the bubble system was able to retain DOX tightly, indicating that it was stable *in vivo*. After *in vivo* intratumoral injection, these nanodroplets were able to produce strong US signal lasting for several days. Similarly, 4 hrs after systemic injection of

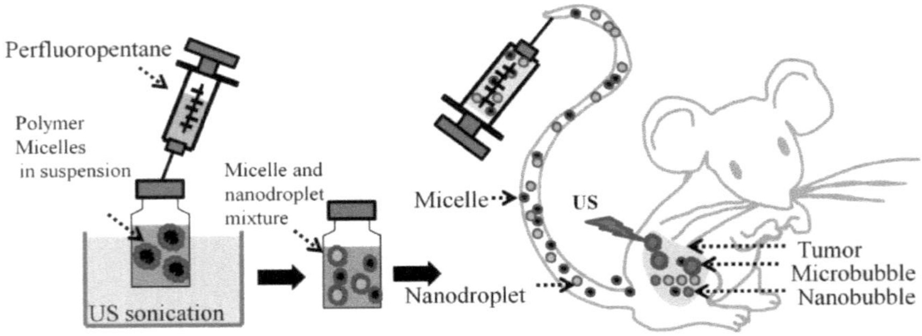

Fig. 8. Schematic of formulation, gene/drug delivery and imaging of PEG-PLLA (PCL) nanobubble system. A mixture of micelle carrying therapeutics and perfluoropentane (PFP) nanodroplets is administered systemically. While PFP is susceptible for vaporization at body temperature, Laplace pressure maintains them in the droplet state. At the tumor site, upon insonation, acoustic droplet vaporization takes place triggering drug release, sonoporation and nanodroplet to bubble phase transition. In the tumor, nanobubbles coalesce forming large microbubbles producing signals that are more suitable for clinically relevant ultrasound imaging *(unpublished original illustration).*

Color image of this figure appears in the color plate section at the end of the book.

these nanodroplets strong US signals were detected at the center of the tumors while no signal was detected in the liver and kidney suggesting strong tumor selectivity.

CO_2 Nanobubbles. CO_2 nanobubble systems were designed using gas generating polymer nanoparticle (GGPNP) which consist of polyesters with carbonate side chains, poly (BL-PO). The working principle of this system is that the inward diffusion of H_2O molecules cleaves the carbonate side chains via hydrolysis. The by-products of this reaction are water, CO_2, cholesterol and ethanol. After intratumoral injection of GGPNP in mice bearing a xenografted subcutaneous squamous cell carcinoma (SCC7), it was shown that US contrast was produced. The proposed mechanism of echo formation is as following. CO_2 accumulation leads to the generation of nanobubbles. Subsequently, the nanobubbles coalesce forming microbubbles providing US contrast (Fig. 9) (Kang 2009).

Other than the above mentioned nanobubbles, Kim, C et al. also developed nanobubbles with PLGA (50-50, MW: 12000 Da) using a modified double-emulsion process (Kim et al. 2010).

Nanobubble Targeting Methods

One of the most desirable features of microbubbles in cancer medicine is that it not only provides imaging contrast, but can also deliver drug/gene to tumor cells. In addition to mechanical stimulation, active microbubble cancer targeting has also been realized either through electrostatic or

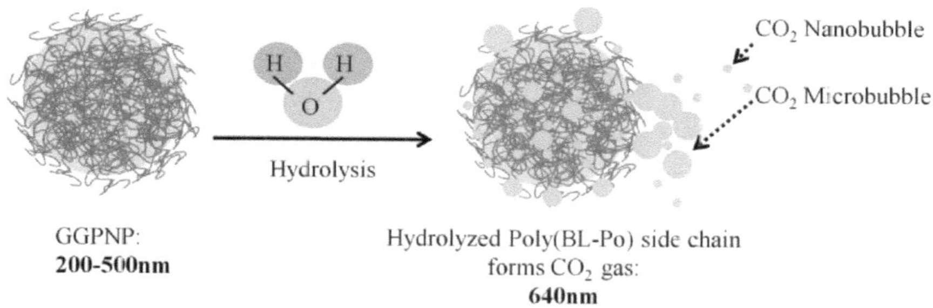

Fig. 9. Schematic of CO_2 nanobubble drug delivery system. Inward diffusion of H_2O molecules into the Poly (BL-PO) nanoparticle cleaves the carbonate side chains via hydrolysis. One product of this chemical process is CO_2. CO_2 accumulation lead to nanobubble formation on the surface of the Poly (BL-PO) nanoparticle and the coalescence of these nanobubbles forms CO_2 microbubbles *(unpublished original illustration).*

Color image of this figure appears in the color plate section at the end of the book.

ligand-receptor interactions. Targeting through electrostatic interaction, i.e., charge-charge interactions, does not provide sufficient specificity and often leads to a lower extent of tumor targeting. In contrast, targeting through ligand (antibodies, peptides and vitamins)-receptor interactions often leads to highly specified tumor targeting. However, currently active tumor targeting can be specific only to tumor endothelial cells (Leong-Poi et al. 2003) since these bubbles are blood pool agents which are too large to extravasate. For example, by targeting α(v)-integrins, Leong-Poi et al. were able to noninvasively assess tumor angiogenesis with US.

Contrary to microbubbles, nanobubble UCAs with diameters smaller than 700 nm should extravasate and reach the cancerous cells (Jain 1998). Since the mechanical and active targeting schemes of microbubbles are all translatable to nanobubbles, cancer targeting of the latter can be achieved through similar means. With the inherent small size, these bubbles can be delivered to tumors via the size-dependent passive targeting. Then, with appropriate ligand conjugation, the nanobubbles can be targeted to cancerous tissue through over expressed cancer cell markers, such as luteinizing hormone releasing hormone (LH-RH) receptors (Safavy et al. 1999) for ovarian, breast, and prostate cancers and human hepatocarcinoma specific antigens expressed on liver cancer cells (Bian et al. 2004) at specific stages of the disease. Finally, since gas bubbles are US responsive, they can be targeted to the tumor by tumor directed therapeutic US (Fig. 10).

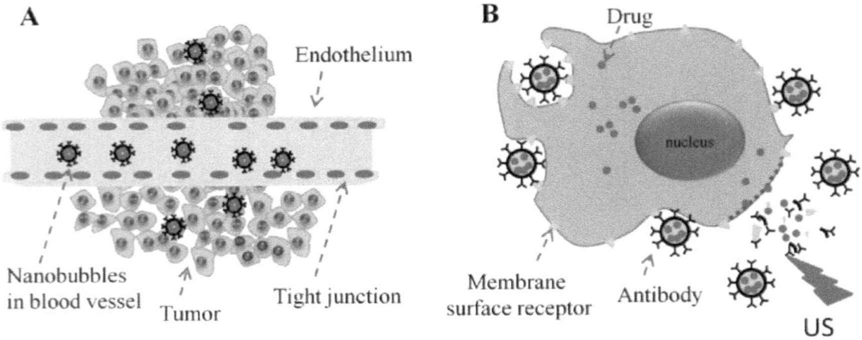

Fig. 10. Nanobubble tumor targeting. (A) Passive tumor targeting: nanoparticles circulating in the blood stream leaks out vasculature through defective fenestra of the vessel wall and remains in tumor tissue due to defective lymphatic drainage (B) Active and mechanical targeting: Nanoparticles are conjugated with antibodies that recognize cancer specific cell surface receptors following extravasation. Ultrasound disrupts nanobubble, increases cell membrane permeability creating transient pores, enhancing drug diffusion into tumor cells *(unpublished original illustration).*

Color image of this figure appears in the color plate section at the end of the book.

Nanobubble Applications to Other Areas of Health and Disease

Other than cancer theranostic applications, nanobubble US contrast agents have shown great potential in areas of health including infectious, cardiovascular and eye diseases, as well as neuromusclular disorders such as Duschenne Muscular dystrophy (Yamashita et al. 2007).

Inflammation has been associated with hypoxia, and studies have shown that dextran nanobubbles are able to deliver oxygen to hypoxic cell and tissues and can be a potential treatment option (Cavalli et al. 2009). In addition, the early stage of atherosclerosis has been manifested with over-expressed intercellular adhesion molecule-1 (ICAM-1), an inflammatory maker. By including ICAM-1 recognizing monoclonal antibodies in the bubble shell, nanobubbles can recognize and adhere to intercellular adhesion molecule-1 (ICAM-1). These types of nanobubble formulations have been used to detect early stages of atherosclerosis, and have also been shown to be effective in detecting acute cardiac transplant rejection (Weller et al. 2003).

Nanobubbles have also been used for treatment of Parkinson's disease. The authors demonstrated that nanobubbles can be used to deliver apomorphine, a particularly beneficial but unstable drug for treating Parkinson's disease, through the blood brain barrier (Hwang et al. 2009).

Although the use of nanobubbles in medicine is in an early development stage, it is possible that in the future, the applications of nanobubbles in medicine will be as far reaching if not more than that of microbubbles whose applications span across the areas of malignant, infectious, cardiovascular and autoimmune diseases.

Current Challenges and Potential Limitations of Nanobubbles in Cancer Theranostic Applications

The present challenges of nanobubble development are both intrinsic and extrinsic. Ideal nanobubble US contrast agents should have a combination of properties including superior tissue specificity, strong echogenecity, and a reasonable degree of tunability for drug incorporation and tumor targeting. In addition, long circulating half life through both inherent stability and ability to resist the actions of the RES are of critical importance. While current nanobubbles are relatively tunable for drug incorporation and tumor targeting, even simple modifications could alter the intrinsic stabilities of the system. Hence, most of the available nanobubble systems suffer from poor stability and poor echogenecity at clinical relevant US frequencies making the goal of uniting diagnostic and therapy an unmet challenge.

In addition, acoustic characteristics of these bubbles are important for therapy and imaging planning. It is noted that for subharmonic imaging, the optimal images are acquired when the transmitted frequency from an US transducer is twice that of the resonance (natural) frequency of the bubbles. However, often even the most rudimentary form of this information is unattainable. While a few versions of mathematical models on the acoustic theory of bubbles do exist, these models are limited to either free bubbles (those that are without shell stabilization) or one or two types of larger bubbles with specific shell materials, hence not applicable for most of the nanobubbles under development. Despite these challenges, there is growing interests in using nanobubbles in cancer medicine. Particularly with the current trend for multidisciplinary approaches to cancer research, we foresee that both nanobubble designs and the theoretical models predicting their behavior will become more sophisticated, and their application in cancer imaging and therapy will continue to advance.

Key Facts about Nanobubbles

- Nanobubbles are gas bubbles stabilized by polymer, lipids or protein shells.
- Nanobubbles can unite diagnosis and therapy.

- The commonly accepted diameter of nanobubbles for cancer targeting is less than 730nm, with the optimal size being less than 200 nm.
- The technological stage of nanobubbles in cancer medicine is still at infancy.
- The developmental hindrance of nanobubbles is the lack of sufficient cost effective scale up manufacturing and characterization techniques.
- The general disadvantages of current existing nanobubbles are low US signal intensity at clinical relevant US frequencies and a short *in vitro* and *in vivo* half life.

Key Terms

Ultrasound contrast agent (UCAs): gas or air bubbles either free or encapsulated in lipid, protein, or polymer shells.

Nanobubble UCAs: nanometric version of the microbubbles for ultrasound signal enhancement.

Sonoporation: the use of sound energy to increase cell membrane permeability by creating transient pores.

Acoustic droplet vaporization (ADV): one means of liquid droplet to gas bubble transition that is caused by sound energy.

Summary Points

- Cancer treatment still relies largely on systemic chemotherapy. While many promising anticancer drugs are available, their effectiveness is often reduced because of their inability to penetrate deep into the solid tumors.
- Combination of US and nanobubble contrast agent-mediated drug/ gene delivery systems can markedly improve the safety and efficacy of cancer therapy. Nanobubbles in combination with US hold great potential in cancer medicine due to its multi-functionality.
- Nanobubbles can be targeted to cancer tissue via 1) passive means taking advantage of the size-dependent EPR effects, 2) active means by modifying the shell materials for ligand targeting and 3) mechanical means by applying tumor-directed US.
- Nanobubbles can be tailored to serve multiple functions simultaneously such as: tumor localization, tumor-directed therapeutic US for targeted delivery of potent anticancer drug, monitoring drug pharmacokinetics, and later for assessing treatment efficacy.
- Microbubble contrast agents which have been commercially available for some time and their clinical applications in areas of cancer diagnosis

and therapy have been well documented; currently no nanobubble formulations are in clinical use.

Abbreviations

AAL	:	acoustically active liposome
ADV	:	acoustic droplet vaporization
C_3F_8	:	perfluoropropane
DDS	:	drug delivery system
DOX	:	doxorubicin
DPPA	:	1,2-dipalmitoyl-*sn*-glycero-3-phosphate
DPPC	:	1,2-dipalmitoyl-*sn*-glycero-3-phosphocholine
DPPE	:	1,2-dipalmitoyl-*sn*-glycero-3-phosphoethanolamine
DSPC	:	1,2-distearroyl-sn-glycero-phosphatidylcholine
EPR	:	enhanced permeability and retention
FDA	:	food and drug administration
GGPNP	:	gas generating polymer nanoparticle
HIFU	:	high intensity focused ultrasound
ICAM-1	:	intercellular adhesion molecule 1
LH-RH	:	luteinizing hormone releasing hormone
MI	:	mechanical index
PBS	:	phosphate buffered saline
PFP	:	perfluoropentane
PN	:	Plasmon nanoparticle
PNB	:	plasmonic nanobubbles
RES	:	reticulo-endothelial system
SF_6	:	hexafluoride
Span 60	:	Sorbitan monostearate
SPC	:	soybean lipid
UCA	:	ultrasound contrast agent
US	:	ultrasound

References

Anderson, L.J., E. Hansen, E.Y. Lukianova-Hleb, J.H. Hafner and D.O. Lapotko. 2010. Optically guided controlled release from liposomes with tunable plasmonic nanobubbles. J. Control Release 144: 151–8.

Bergers, G., and L.E. Benjamin. 2003. Tumorigenesis and the angiogenic switch. Nat. Rev. Cancer 3: 401–10.

Bian, A.N., Y.H. Gao, K.B. Tan, P. Liu, G.J. Zeng, X. Zhang and Z. Liu. 2004. Preparation of human hepatocellular carcinoma-targeted liposome microbubbles and their immunological properties. World J. Gastroenterol. 10: 3424–7.

Buchanan, K.D., S.L. Huang, H. Kim, D.D. McPherson and R.C. MacDonald. 2010. Encapsulation of NF-kappaB decoy oligonucleotides within echogenic liposomes and ultrasound-triggered release. J. Control Release 141: 193–8.

Cavalli, R., A. Bisazza, A. Rolfo, S. Balbis, D. Madonnaripa, I. Caniggia and C. Guiot. 2009. Ultrasound-mediated oxygen delivery from chitosan nanobubbles. Int. J. Pharm. 378: 215–7.
de Jong, N., L. Hoff, T. Skotland and N. Bom. 1992. Absorption and scatter of encapsulated gas filled microspheres: theoretical considerations and some measurements. Ultrasonics 30: 95–103.
Dijkmans, P.A., L.J. Juffermans, R.J. Musters, A. van Wamel, F.J. ten Cate, W. van Gilst, C.A. Visser, N. de Jong and O. Kamp. 2004. Microbubbles and ultrasound: from diagnosis to therapy. Eur. J. Echocardiogr. 5: 245–56.
Ferrara, K., R. Pollard and M. Borden. 2007. Ultrasound microbubble contrast agents: fundamentals and application to gene and drug delivery. Annu. Rev. Biomed. Eng. 9: 415–47.
Gibson, R.J., and A.M. Stringer. 2009. Chemotherapy-induced diarrhoea. Curr. Opin. Support Palliat Care 3: 31–5.
Huang, S.L., and R.C. MacDonald. 2004. Acoustically active liposomes for drug encapsulation and ultrasound-triggered release. Biochim. Biophys. Acta. 1665: 134–41.
Hwang, T.L., Y.K. Lin, C.H. Chi, T.H. Huang and J.Y. Fang. 2009. Development and evaluation of perfluorocarbon nanobubbles for apomorphine delivery. J. Pharm. Sci. 98: 3735–47.
Jain, R.K. 1998. Delivery of molecular and cellular medicine to solid tumors. J. Control Release. 53: 49–67.
Jemal, A., R. Siegel, E. Ward, Y. Hao, J. Xu and M.J. Thun. 2009. Cancer statistics, 2009. CA Cancer J. Clin. 59: 225–49.
Kang, E. 2009. Nanobubbles from Gas-Generating Polymeric Nanoparticles: Ultrasound Imagingof Livingt Subjects. Angew. Chem. Int. Ed. 48: 6.
Kaul, S. 2008. Myocardial contrast echocardiography: a 25-year retrospective. Circulation. 118: 291–308.
Kim, C., R. Qin, J.S. Xu, L.V. Wang and R. Xu. 2010. Multifunctional microbubbles and nanobubbles for photoacoustic and ultrasound imaging. J. Biomed. Opt. 15: 010510.
Kodama, T., N. Tomita, S. Horie, N. Sax, H. Iwasaki, R. Suzuki, K. Maruyama, S. Mori and F. Manabu. 2010. Morphological study of acoustic liposomes using transmission electron microscopy. J. Electron. Microsc. 59: 187–96.
Krupka, T.M., L. Solorio, R.E. Wilson, H. Wu, N. Azar and A.A. Exner. 2010. Formulation and characterization of echogenic lipid-Pluronic nanobubbles. Mol. Pharm. 7: 49–59.
Leong-Poi, H., J. Christiansen, A.L. Klibanov, S. Kaul and J.R. Lindner. 2003. Noninvasive assessment of angiogenesis by ultrasound and microbubbles targeted to alpha(v)-integrins. Circulation 107: 455–60.
Oeffinger, B.E., and M.A. Wheatley. 2004. Development and characterization of a nano-scale contrast agent. Ultrasonics 42: 343–7.
Qin, S., C.F. Caskey and K.W. Ferrara. 2009. Ultrasound contrast microbubbles in imaging and therapy: physical principles and engineering. Phys. Med. Biol. 54: R27–57.

Rapoport, N., Z. Gao and A. Kennedy. 2007. Multifunctional nanoparticles for combining ultrasonic tumor imaging and targeted chemotherapy. J. Natl. Cancer Inst. 99: 1095–106.

Safavy, A., K.P. Raisch, M.B. Khazaeli, D.J. Buchsbaum and J.A. Bonner. 1999. Paclitaxel derivatives for targeted therapy of cancer: toward the development of smart taxanes. J. Med. Chem. 42: 4919–24.

Un, K., S. Kawakami, R. Suzuki, K. Maruyama, F. Yamashita and M. Hashida. 2010a. Development of an ultrasound-responsive and mannose-modified gene carrier for DNA vaccine therapy. Biomaterials 31: 7813–26.

—. 2010b. Enhanced transfection efficiency into macrophages and dendritic cells by a combination method using mannosylated lipoplexes and bubble liposomes with ultrasound exposure. Hum. Gene Ther. 21: 65–74.

Wagner, D.S., N.A. Delk, E.Y. Lukianova-Hleb, J.H. Hafner, M.C. Farach-Carson and D.O. Lapotko. 2010. The *in vivo* performance of plasmonic nanobubbles as cell theranostic agents in zebrafish hosting prostate cancer xenografts. Biomaterials 31: 7567–74.

Wang, Y., X. Li, Y. Zhou, P. Huang and Y. Xu. 2010. Preparation of nanobubbles for ultrasound imaging and intracelluar drug delivery. Int. J. Pharm. 384: 148–53.

Weller, G.E., E. Lu, M.M. Csikari, A.L. Klibanov, D. Fischer, W.R. Wagner and F.S. Villanueva. 2003. Ultrasound imaging of acute cardiac transplant rejection with microbubbles targeted to intercellular adhesion molecule-1. Circulation. 108: 218–24.

Wheatley, M.A., F. Forsberg, N. Dube, M. Patel and B.E. Oeffinger. 2006. Surfactant-stabilized contrast agent on the nanoscale for diagnostic ultrasound imaging. Ultrasound Med. Biol. 32: 83–93.

Xing, Z., J. Wang, H. Ke, B. Zhao, X. Yue, Z. Dai and J. Liu. 2010. The fabrication of novel nanobubble ultrasound contrast agent for potential tumor imaging. Nanotechnology 21: 145607.

Yamashita, T., S. Sonoda, R. Suzuki, N. Arimura, K. Tachibana, K. Maruyama and T. Sakamoto. 2007. A novel bubble liposome and ultrasound-mediated gene transfer to ocular surface: RC-1 cells *in vitro* and conjunctiva *in vivo*. Exp. Eye Res. 85: 741–8.

14

Nano-Strategies in the Delivery of Antitumor Drugs to Cancer

José L. Arias

ABSTRACT

Current chemotherapy against cancer frequently fails due to the little or almost null specificity of antitumor drugs for malignant cells. Even more advanced anticancer molecules can exhibit a very low accumulation into tumor mass, and an extensive biodistribution into healthy tissues. As a consequence, long-term chemotherapy is characterized by significant inefficacy and severe toxicity. Tumor cells can also develop (multiple) resistance mechanisms against chemotherapy molecules which will contribute to treatment failure, even when combined chemotherapy schedules are used at high doses.

In order to overcome the problem, the introduction of nanotechnology into oncology is under intense development. Nano-oncology is considered as a revolutionary concept in cancer arena which primarily involves the formulation of nanoparticulate carriers for antitumor drugs. The use of biodegradable nanoplatforms loaded with anticancer agents

Department of Pharmacy and Pharmaceutical Technology, Faculty of Pharmacy, University of Granada, Campus Universitario de Cartuja s/n, 18071 Granada, Spain;
E-mail: jlarias@ugr.es

List of abbreviations after the text.

can result in improved drug localization into the tumor region with negligible distribution into non-targeted sites. As a result, the anticancer activity of such molecules will be significantly enhanced, and the systemic side effects will be minimized as well. However, recent *in vivo* investigations have shown the limitations on the use of conventional nanocarriers. Interestingly, novel nano-engineering strategies have been introduced in the formulation of nanoparticulate platforms for the delivery of anticancer drugs to malignant tissues. Such approaches are even capable of overcoming multi-drug resistance mechanisms overexpressed by tumor cells. This chapter is devoted to the analysis of the current state of the art in the use of passive and/or active targeting nano-strategies for controlling the biodistribution profiles of chemotherapy agents.

INTRODUCTION

Thanks to the deep investigation into the biology and physiology of cancer, numerous molecular targets have been identified for chemotherapy. Recent investigations have led to the design of novel molecules with greater antitumor activity compared to traditional anticancer agents. On the contrary, the clinical use of such drugs does not offer significant possibilities over the (frequently limited) conventional chemotherapy. This is clearly the consequence of an insufficient accumulation of the drug into the targeted site in optimal quantities and/or an inefficient anticancer activity. Hence, treatment failure can occur in sensitive cancers even if more advanced chemotherapy schedules are used. It has been described that the unfavourable pharmacokinetics and poor physicochemical properties of these molecules (i.e., hydrophobicity) dramatically contribute to the problem, and limit their clinical use. For instance, antitumor drugs are characterized by a short plasma half life ($t_{1/2}$) due to a very rapid plasma clearance and biodegradation. Additionally, poor tumor selectivity and large biodistribution (inducing severe toxicity in healthy tissues) frequently come across. Last but not least, these agents are sensitive to enzymatic systems, and are supposed to induce drug resistance in cancer cells. For that reason, highly toxic doses and rigorous treatment schedules are normally used to obtain the therapeutic effect (Brigger et al. 2002; Arias 2009).

As a revolutionary approach to the problem, nanotechnology has been introduced in the cancer arena. The development of nanoparticulate drug delivery systems have been postulated to result in a specific accumulation of the anticancer molecules into the tumor mass, and in a prolonged and deep contact between the chemotherapy agent and the malignant cell. Supplementary benefits to the vehiculization (and protection) of these drugs into a nanocarrier are the improvement of their pharmacokinetic profile, and the minimization of the associated toxicity (Davis et al. 2008, Arias 2009). Biodegradable materials are principally used in the formulation of nanoplatforms for the delivery of chemotherapy agents to cancer tissues. Concretely, such drug-loaded nanoparticles are made of biodegradable polymers [e.g., chitosan, poly(ε-caprolactone) (PCL), poly(D,L-lactide-*co*-glycolide) (PLGA), and poly(alkylcyanoacrylates)] or lipid-based nanoparticulate systems [mainly, liposomes, niosomes, and solid lipid nanoparticles (SLNs)] (Cho et al. 2008; Arias 2009). Recent investigations have focused special attention on the use of composite materials, i.e., iron oxide nanoparticles associated to polymers, liposomes, or silica (Brigger et al. 2002; Arias 2010, 2011).

Despite the fact that numerous *in vitro* research reports have described the very promising possibilities coming from the use of these drug-loaded nanoparticles (lately named nanomedicines, or nanodrugs), the *in vivo* efficacy of such approach is severely limited. This has been defined to be the consequence of the deep interaction between the nanoplatforms and the cells responsible for plasma clearance of foreign material (i.e., macrophages). As a result, drug nanocarriers have a natural tendency to accumulate into the mononuclear phagocyte system (MPS) ($t_{1/2} \approx$ three minutes) which can only be advantageously used against those cancers exclusively located into MPS organs (i.e., liver, spleen, bone marrow, lungs) (Arias 2009).

In order to beat this very important difficulty to an efficient nano-chemotherapy of tumors located out of the MPS, special engineering strategies are under development (Table 1). Namely, passive targeting tactics, which take advantage of the enhanced permeability and retention (EPR) effect; and active targeting tactics, based on the engineering of nanomaterials to be sensitive to an external stimulus, and/or on the surface functionalization of the nanoplatform with biomolecules for ligand-mediated targeting of cancer cells. Such advanced functionalization of the nanomedicines will allow controlling the biological fate of any given antitumor agent (Arias 2009, 2011).

This chapter is devoted to the analysis of the challenges and opportunities coming from the use of these targeting nano-strategies for drug delivery to cancer. Emphasis will be given to nano-approaches to the problem of multi-drug resistance (MDR), a condition enabling cancer cells to become resistant to chemotherapy.

Table 1. Targeting strategies of nanomedicines to cancer.

Drug targeting strategy	Basic aspects
Passive drug targeting.	Naturally occurs thanks to the EPR effect. Involves the formulation of long-circulating nanoparticles.
Active drug targeting through ligand-receptor interactions.	These interactions determine the intracellular uptake of the nanoparticle mainly by endocytosis. To that aim, the nanoparticles must be surface functionalized with monoclonal antibodies (MAbs), peptides (including integrins), aptamers, folate moieties, and/or transferrin.
Active drug targeting by the use of stimuli-sensitive nanoplatforms.	An external stimulus will induce the complete accumulation and specific release of the anticancer drug within the tumor interstitium. For this purpose, such nanomaterials must be disrupted by acidic environments, enzymatic systems, temperature changes, light, ultrasounds, and/or magnetic gradients.

The most significant approaches to control the biological fate of chemotherapy molecules loaded to a nanocarrier are based on the EPR effect (passive targeting), ligand-mediated endocytosis into cancer cells (active targeting), and/or stimuli-sensitive materials (active targeting).

PASSIVE TARGETING STRATEGIES

Nanomedicines against cancer are commonly intended for the parenteral route. Specifically, they are administered intravenously or intratumoraly. The *in vivo* fate of any given conventional drug nanocarrier rapidly involves a strong interaction with the MPS (essentially, spleen and liver). As a result of an opsonization process which depends strongly on its characteristics, the nanodrug is removed from blood by macrophages. If cancer is exclusively located into the MPS, a controlled accumulation of the nanoparticle within the tumor site will occur. However, the very short $t_{1/2}$ of the drug-loaded nanoplatforms (≈ three minutes) will make targeting other tumor sites unattainable (Brigger et al. 2002; Arias 2009, 2011).

MPS recognition of the nanomedicines can be made difficult by retarding the opsonization process. This will only be possible if the nanocarrier is properly formulated (Table 2) (Arias 2009; Decuzzi et al. 2009). In addition, passive drug targeting to cancer by biodegradable nanoparticles is based on the unique properties of tumor microenvironment: *i*) the structural irregularities of the tumor vasculature (i.e., gap junction between endothelial cells from 100 to 600 nm), whose permeability to macromolecules is higher than that of healthy tissues; and, *ii*) a dysfunctional lymphatic drainage, which results in enhanced fluid retention in the tumor interstitium (Maeda et al. 2009; Arias 2010). The formulation of long-circulating nanomedicines (with prolonged $t_{1/2}$) to exploit the leaky vasculature and locally enhanced capillary permeability of the tumor, not only involves a proper definition of the intrinsic properties of the nanomaterial (Table 2), but also the introduction by physical adsorption or chemical conjugation onto its surface of hydrophilic polymers (Fig. 1). The later approach will provide a shell of hydrophilic and neutral chains able to repel plasma proteins (i.e., opsonins) (Brigger et al. 2002; Arias 2009; Decuzzi et al. 2009; Maeda et al. 2009). Alternatively, long-circulating drug delivery systems can be made of block copolymers with hydrophilic and hydrophobic domains (Cho et al. 2008; Arias 2011).

Table 2. Principal aspects to be considered in the formulation of nanomedicines against cancer.

Characteristics of the nanodrug	***In vivo* fate**
Spherical shape. Size less than 100 nm. Hydrophilicity. Negligible surface electrical charge. Specific drug release into the tumor site. Null drug leakage *in vitro* (under storage) and *in vivo* (in the way to the targeted site).	Directed drug biodistribution and accumulation into the targeted tissue. Enhanced pharmacokinetic profile (i.e., extended $t_{1/2}$). Prolonged contact time between the drug and malignant cells. Reduced drug toxicity.
Physicochemical stability. Biocompatibility, and biodegradability. Vehiculization of adequate drug amounts, without overloading the body with foreign material.	Negligible antigenicity and toxicity.

An efficient antitumor activity strongly relies on a proper engineering of the drug-loaded nanoparticulate platforms.

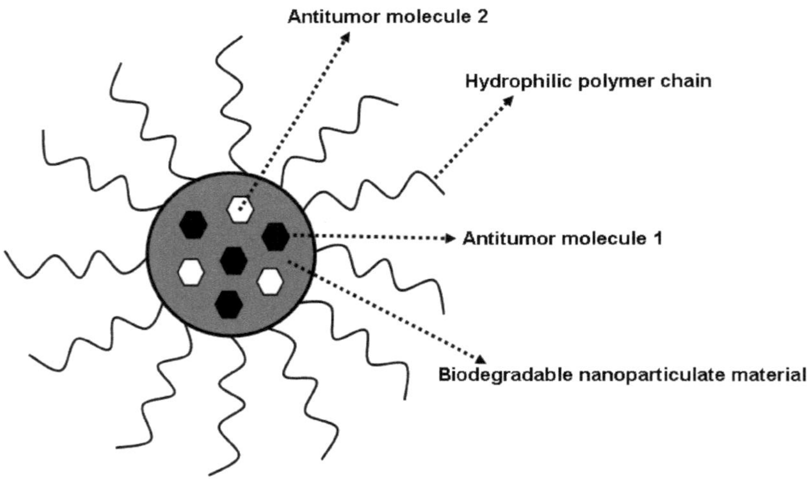

Fig. 1. Long-circulating anticancer nanomedicine for passive tumor targeting. The structure of the nanodrug consists of a biodegradable polymer (or lipid-based material), chemotherapy agent(s), and hydrophilic macromolecules [i.e., polyethylene oxides (such as, polyethylene glycol, PEG), poloxamers, poloxamines, or polysaccharides].

Long-circulating nanomedicines will be capable of targeting tumors located out of the MPS regions, with a selective extravasation and accumulation. This tumor specific disposition is known as the enhanced permeability and retention (EPR) effect (Fig. 2) (Decuzzi et al. 2009; Maeda et al. 2009). Long-circulating polymeric nanoparticles have been formulated for passive drug targeting to cancer. For instance, the surface of PCL nanoplatforms loaded with tamoxifen were surface functionalized with poly(ethylene oxide) chains (i.e., pluronic® F-68 or F-108) by physical adsorption. The intravenous administration of tamoxifen in the form of this nanodrug to mice bearing MDA-MB-231 xenografts, a human breast cancer cell line, resulted in much greater concentrations into the tumor mass (≈ 26% at six hours post injection), compared to the drug administered as an intravenous solution (Shenoy and Amiji 2005). It has been described that liposomes surface modified with hydrophylic chains (e.g., PEG) can also exhibited long-circulating times. As an example, it has been described that these vesicular systems are characterized by ≈ 200-fold decrease in plasma clearance (from 22 to 0.1 L/h), and ≈ 100-fold increase in the area under the time-concentration curve (AUC). Thanks to their almost null interaction with healthy tissues upon administration, long-circulating liposomes are further characterized by ≈ 50-fold decrease in the volume of distribution (from 200 to 4.5 L) (Allen 1994).

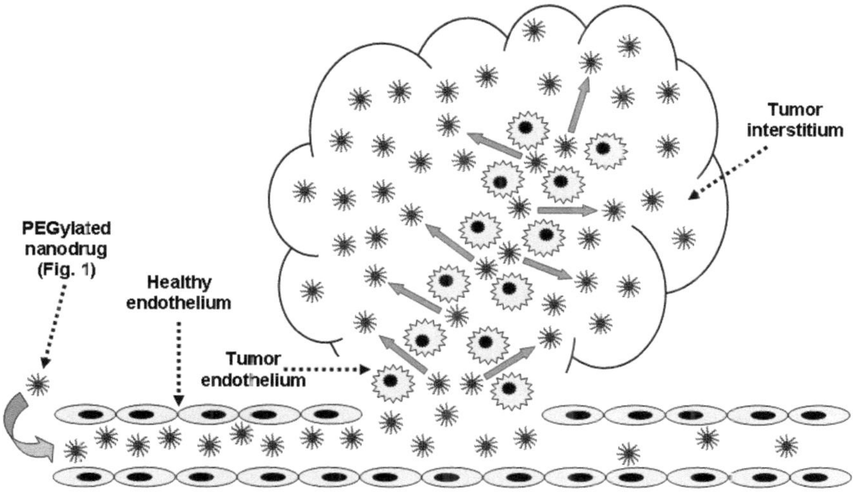

Fig. 2. Specific accumulation of long-circulating anticancer nanomedicines into the tumor interstitium by passive diffusion or convection through the hyperpermeable endothelium. A proper surface functionalization with hydrophilic polymers (e.g., PEGylation) will retard nanoparticle opsonization (and plasma clearance). The EPR effect is based on the "leaky" vasculature of the malignant tissues. Blood recirculation is expected to contribute to a complete accumulation of the PEGylated nanomedicine into the tumor site.

The EPR effect can be advantageously used to treat MDR tumor cells. For instance, PCL nanoparticles surface decorated with PEG moieties have been prepared for the co-administration of paclitaxel and ceramide to a MDR human ovarian cancer cell line. It was proved that this formulation very importantly enhances paclitaxel cytotoxicity in the MDR cancer cell line thanks to the restitution of the apoptotic threshold (van Verklen et al. 2007).

ACTIVE TARGETING STRATEGIES

Even though passive drug targeting nano-strategies to cancer can enhance the antitumor effect of chemotherapy agents, the low specificity of the nanomedicine for the malignant tissue could become a limitation to the clinical success. Advanced investigations on the physiopathology of cancer have revolutionized the application of nanotechnology to chemotherapy. A more selective delivery of the drug nanocarrier has been hypothesized when it is surface functionalized with targeting biomolecules specific to ligands that are unique to tumor cells (ligand- or receptor-mediated targeting). As a (complementary) alternative, the nanomedicine could be engineered with nanomaterials sensitive to external stimuli. This approach is based on the capability of such advanced nanoplatforms to experience alterations in their structure and physical properties under small changes in the environment. The consequence will be a controlled delivery of the nanomedicine to the cancer site and/or triggered drug release into the tumor mass (stimuli-sensitive nanocarriers). It is expected that the real progress in the battle against cancer will come from the concurrent use of these active and passive targeting strategies in the development of nanomedicines (Arias 2011).

Ligand- or Receptor-Mediated Targeting

This concept is based on a selective delivery of the nanomedicine to cancer by specific molecular recognition processes. Ligand-receptor or antibody-antigen interactions between the drug-loaded nanoparticle and the tumor cell are made possible by the chemical conjugation of the former to targeting ligands which are tissue-, or cell-specific (Fig. 3). As a result, selective accumulation of the nanodrug into the tumor site will occur (Arias 2011).

To that aim, it is needed the use of ligands targeting biomolecules overexpressed onto endothelial cells forming the neovasculature of growing tumors (e.g., integrin surface receptors), and/or onto cancer cell membranes (e.g., folate receptors) (Table 3) (Arias 2009). Normally, nanomedicine-cell interactions lead to receptor-mediated internalization of the nanodrug (i.e., by endocytosis) (Fig. 4) (Brigger et al. 2002; Decuzzi et al. 2009).

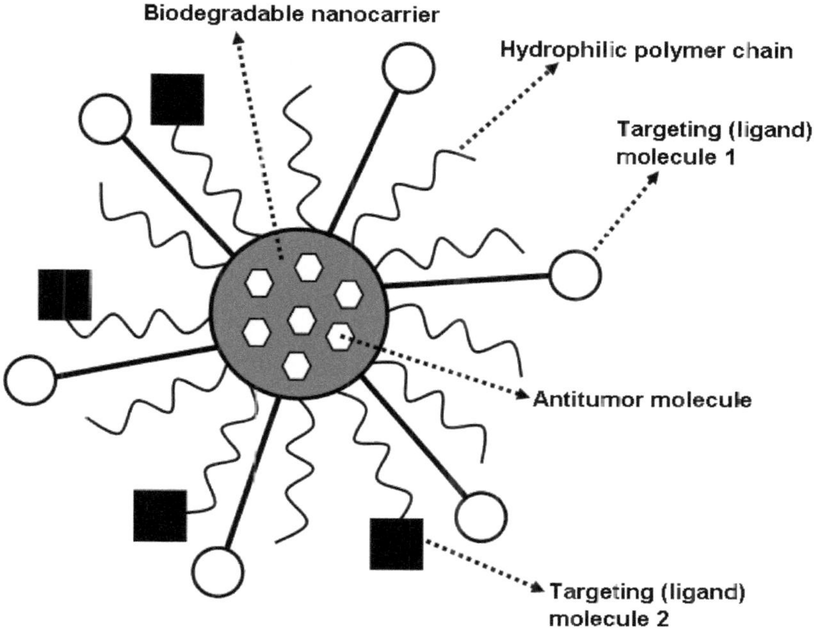

Fig. 3. Nanomedicine surface decorated with targeting moieties for ligand-mediated cancer cell targeting. Surface functionalization typically involves the use of aptamers, folate, transferrin, MAbs, and/or peptides. Chemical conjugation of these biomolecules to the nanodrug can be carried out directly onto the surface and/or through the hydrophilic chains (i.e., PEG) responsible for passive drug targeting mechanism.

Table 3. Examples of targeted cell biomolecules and targeting ligands in the specific delivery of nanodrug to cancer (Arias 2011).

Ligand- or receptor-mediated targeting strategy	**Targeted biomolecule (targeting ligand)**
MAb-mediated targeting.	Human epidermal growth factor receptor-2 (human EGFR2) (anti-human EGFR2 MAb: trastuzumab), mouse transferrin receptor (rat 8D3 MAb), P-glycoprotein (MRK-16). Other targeting MAb: anti-CD33.
Peptide-mediated targeting.	Integrin $\alpha_v\beta_6$ (peptide H2009.1), integrin $\alpha_5\beta_1$ (peptide PR_b). Other targeting peptides: vasoactive intestinal peptide, EGFR1, CREKA, PH1, and LyP-1.
Aptamer-mediated targeting.	Extracellular domain of the prostate-specific membrane antigen (PSMA) (A10 2'-fluoropyrimidine RNA aptamer).
Folate-mediated targeting.	Folate-binding protein (FBP, a glycosylphosphatidylinositol surface folate cell receptor) (folate).
Transferrin-mediated targeting.	Transferrin receptor (transferrin, OX26 MAb, and TfRscFv MAb).

Ligand-receptor interactions typically lead to cell internalization (and cytosolic accumulation) of the nanomedicine by endocytosis.

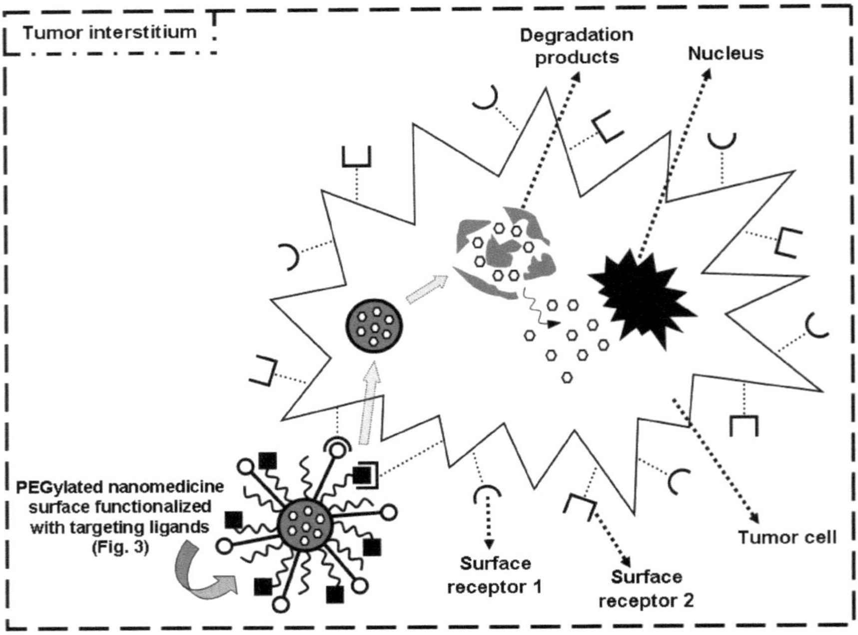

Fig. 4. Cell adhesion/internalization of a nanomedicine surface decorated with targeting moieties. As a consequence of the ligand-mediated cancer cell targeting, the drug will be specifically released into the tumor interstitium or into the cancer cell, upon nanoparticle biodegradation (disruption).

MAb-Mediated Targeting

MAbs can be directed against several surface antigens (or receptors) of tumor cells and vascular endothelium. These biomolecules can be either attached directly to the nanoparticle surface (e.g., by chemical conjugation to the phospholipid headgroup of liposomes), and to PEG endings previously incorporated onto the nanomedicine (Sapra and Allen 2003). Several investigations have reported an efficient delivery of the drug delivery nanoplatform to the tumor site when using MAbs (Arias 2011).

For instance, composite nanoparticles made of PLGA and montmorillonite were formulated for the delivery of the antitumor drug paclitaxel to malignant cells. Surface decoration of the nanomedicine with trastuzumab was investigated for enhanced cancer chemotherapy. Very interestingly, a greater cellular uptake by Caco-2 colon adenocarcinoma cells and SK-BR-3 breast cancer cells was described in comparison to the unfunctionalized nanomedicine. Consequently, *in vitro* cytotoxicity on SK-BR-3 cells was 12.7- and 13.1-fold higher than that of the bare nanoparticle and the free drug (Taxol®), respectively (Sun et al. 2008).

MAbs can be also used to formulate nanodrugs capable of overcoming MDR mechanisms. For example, a recent investigation has reported the formulation of liposomes conjugated to the MRK-16 MAb against P-glycoprotein (P-*gp*), and loaded with the chemotherapy molecule vincristine. It was described as an enhanced drug cytotoxicity against resistant human myelogenous leukemia cell lines, compared to non-functionalized vincristine-loaded liposomes. This enhanced anticancer effect was due to the inhibition of P-*gp* mediated vincristine efflux by MRK-16 MAb (Matsuo et al. 2001).

Aptamer-Mediated Targeting

Despite the fact that these biomolecules are directly used against cancer, aptamers can be also coupled to drug nanocarriers to maximize tumor cell targeting (Arias 2009). This strategy has been used to specifically deliver docetaxel to prostate cancer cells (Farokhzad et al. 2006). The antitumor drug was loaded to PLGA nanoparticles surface functionalized with PEG chains (passive targeting) and A10 2′-fluoropyrimidine RNA aptamers that recognize the extracellular domain of PSMA on prostate cancer cells (active targeting). It was reported a greater *in vitro* toxicity in LNCaP prostate epithelial cells, compared to the non-targeted drug nanoplatform ($p < 0.0004$). Interestingly, a single intratumoral injection of the nanomedicine induced total tumor reduction in ≈ 70% of LNCaP xenograft nude mice. This group exhibited a much higher survivability after 109 days compared to groups treated with non-functionalized nanomedicine, and free drug (100, 57, and 14%, respectively).

Peptide-Mediated Targeting

Integrins overexpressed on endothelial cells of tumor neovasculature can be targeted by peptides and peptidomimetics containing the RGD (arginine-glycine-aspartic acid) sequence. Thus, nanomedicine conjugation to these biomolecules has been described to improve their targetability to tumor vasculature, resulting in its destruction. In addition, nanomedicines surface functionalized with such peptides can be directed to malignant cells that overexpress the corresponding integrin. As a consequence, the antitumor effect of the chemotherapy agent will be significantly optimized (Arias 2009).

Gene therapy against cancer can also be enhanced by the use of gene nanocarriers surface conjugated to suitable peptides. For instance, the difficulties related to the systemic delivery of small interfering RNA (siRNA) to tumor tissues can be avoided (e.g., low penetration capability into the cell, and limited stability in blood). In a recent research report, siRNA-nanomedicines were prepared by using poly(propyleneimine) (PPI) dendrimers. The stabilization of the nanoparticles was possible thanks to the use of dithiol containing cross-linker molecules and a PEG coating. A synthetic analogue of the luteinizing hormone-releasing hormone (LHRH) peptide was incorporated to the distal end of PEG as the targeting ligand to cancer cells. *In vivo* results demonstrated the great stability in plasma and intracellular bioavailability of the siRNA-loaded nanoparticles. The strategy provides a specific tumor uptake, the concentration of siRNA into the cytoplasm, and an efficient gene silencing (Taratula et al. 2009).

Transferrin Receptor Targeting

Transferrin has been introduced in the formulation of advanced nanodrugs with greater cancer targeting efficiencies and prolonged contact with tumor cells. This is possible thanks to a selective accumulation of the drug nanocarrier into the tumor mass, even if the blood brain barrier (BBB) or MDR mechanisms must be overcame (Arias 2009, 2011). Moreover, transferrin can optimize cancer cell imaging and photothermal therapy (Li et al. 2009a). Unhappily, receptor saturation by endogenous plasma transferrin could happen. Intra-site administration of the nanomedicine, and/or the use of specific MAb to this receptor have been proposed for an efficient tumor targeting (Arias 2009).

For example, transferrin-conjugated PLGA nanoparticles loaded with paclitaxel can undergo three-fold higher *in vitro* uptake by human prostate cancer cells (PC3) than the unconjugated drug-loaded nanoplatform. *In vivo*, a single intratumoral injection of the functionalized nanomedicine (paclitaxel: 24 mg/Kg) in mice bearing PC3 subcutaneous xenografts resulted in complete cancer regression (Sahoo and Labhasetwar 2005). The use of liposomes surface functionalized with transferrin and loaded with both the anticancer drug doxorubicin and the P-*gp* inhibitor verapamil (to hinder drug efflux) has been suggested as a very insignificant approach to the problem of MDR in cancer cells (Wu et al. 2007).

Folate Receptor Targeting

Normally, tumor cells exhibit enhanced folate requirements for DNA synthesis. Hence, folate receptors are frequently overexpressed by them. Nanoplatforms surface decorated with folate moieties have been proposed for drug delivery to cancer (giving the chance to overcome MDR mechanisms), cancer phototherapy, and targeted imaging in cancer diagnosis (Arias 2009; Lai and Lee 2009; Low and Kularatne 2009).

Even anticancer prodrugs have been loaded to nanoplatforms surface functionalized with folate in order to obtain an increased cell uptake and cytotoxicity. SLNs surface decorated with folate molecules allowed the selective delivery of paclitaxel-2′-carbonylcholesterol to malignant cells. *In vivo*, this nanomedicine very efficiently increased the inhibition of tumor growth compared to non-targeted SLNs, and free paclitaxel (Stevens et al. 2004).

This active drug targeting strategy has been also applied to the problem of MDR tumor cells. For instance, doxorubicin-loaded pH-sensitive polymeric micelles has been prepared with two block copolymers poly(L-histidine)-β-PEG-folate and poly(D,L-lactide)-β-PEG-folate. The objective was to obtain folate-receptor targeting and dissolution at pH ≤ 6.8. It was concluded that the multifunctional nanoplatform exhibited higher cytotoxicity in MDR breast cancer cells at pH 6.8 (cell viability: 20%) compared to free drug (cell viability: 85%) (Lee et al. 2005).

Stimuli-Sensitive Nanocarriers

Stimuli-responsive, stimuli-sensitive, or environmental-sensitive nanocarriers are basically polymers or liposomes properly engineered to experience rapid changes in their structure and physical properties under exposure to small environmental modifications. As a consequence of the directed guidance, and disruption/aggregation, swelling/deswelling, etc. of the nanoplatform (which could be reversible), the drug will be accumulated and/or released exclusively into the targeted site (Fig. 5) (Brigger et al. 2002; Arias 2011). The stimuli may take place internally [e.g., pH changes in certain (non-healthy) tissues, or the presence of enzymatic systems], or could be externally produced (and controlled) with the help of different stimuli-generating devices (temperature increments, magnetic gradients, light, or ultrasounds) (Arias 2009). A proper activation of the stimulus could even induce a pulsatile drug release from the nanocarrier (Arias 2010). Commonly, passive targeting of the tumor takes place through the EPR effect, and the active drug targeting strategy assures cellular uptake of the nanomedicine and triggered drug release.

Enzyme-Triggered Drug Release

The strategy involves the formulation of nanodrugs susceptible to enzymes overexpressed by the tumor [i.e., secretory phospholipase A_2 (a lipid hydrolyzing enzyme significantly up-regulated in the extracellular tumor microenvironment), transglutaminase, alkaline phosphatase, metalloproteinase, and phosphatidylinositol-specific phospholipase C]. Under the influence of the enzymatic system, the nanocarrier is disrupted, leading to drug release (Andresen et al. 2005; Arias 2009). For example, the antitumor drug doxorubicin has been conjugated to albumin by

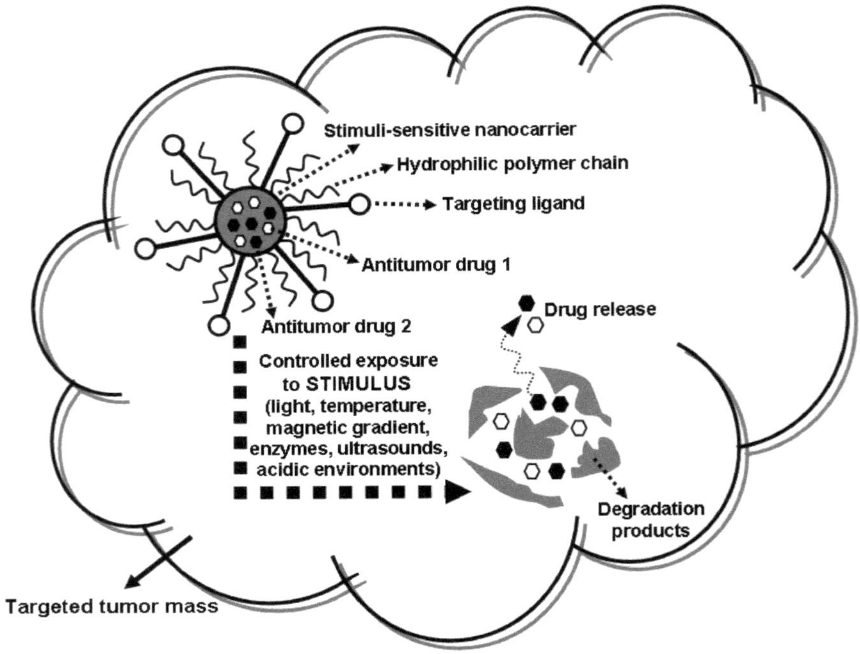

Fig. 5. Nanomedicine engineered to be sensitive to an external stimulus. Passive targeting (i.e., functionalization with PEG chains) and complementary active targeting (conjugation to ligands) strategies could be further introduced in the formulation of the nanomedicine. As a result, triggered drug release will happen exclusively within the tumor tissue when the nanoplatform is disrupted by the stimulus. The external stimulus could be also used to direct the nanodrug to the targeted site (e.g., magnetic gradient) after administration to the patient.

using a matrix metalloproteinase-2 specific octapeptide sequence. *In vitro*, it was reported a specific disruption of the conjugation by matrix metalloproteinase-2 (Cho et al. 2008).

Acid-Triggered Release

The slightly acid pH (≈ 6.6) that a nanomedicine faces inside the tumor mass differs from the physiological pH (≈ 7.4). This physiopathological condition can be used to trigger drug release (a consequence of nanoparticle degradation) exclusively into the tumor interstitium. Alternatively, this strategy involves lysosome targeting. After cellular uptake, the nanocarrier (principally, polymeric nanoparticles or fusogenic liposomes) will be disrupted into this compartment under a pH ≈ 4.5–5.0, and under contact with hydrolytic enzymes (i.e., cathepsin B) (Arias 2009). pH-sensitive polymeric nanoplatforms are habitually made of poly(acrylic acid), poly(methacrylic acid), poly(acrylamide), poly(dimethylaminoethyl methacrylate), poly(diethylaminoethyl methacrylate), and copolymers. These polymers contain basic (i.e., ammonium salts) or acidic (i.e., carboxylic, and sulfonic acids) groups that are capable of either releasing or accepting protons under certain pH values. This will provoke conformational modifications in the solubility or swelling behavior of the nanoparticle (Rapoport 2007; Arias 2011).

A recent research report have described the potential use of copolymers of poly(*N*-isopropylacrylamide) and chitosan in the formulation of pH-sensitive paclitaxel-based nanomedicines (Li et al. 2009b). Drug release from the copolymeric nanoparticle was significantly faster below pH 6.8, and decreased rapidly above pH 6.9. Fluorescence microscopy proved that paclitaxel release was promoted into tumor surroundings. S-180-bearing KM mice treated with this nanodrug exhibited almost null toxicity (negligible decrease in body weight), and significant anticancer activity (> 50% complete tumor regression). As a result, life span of treated mice was importantly extended.

Thermo-Sensitive Drug Delivery

Thermo-responsive nanoparticles are frequently made of liposomes, polymers, or conjugated (liposome-polymer, or inorganic-organic) composite materials with numerous hydrophobic groups in their chemical structure (e.g., methyl, ethyl, and propyl groups). These temperature-sensitive nanoplatforms are characterized by a critical solution temperature, at which their physical state is altered: the nanoparticle is solubilized and drug release takes place. Significant examples of these thermo-sensitive nanocarriers are poly(*N*-isopropylacrylamide), poly(*N*-(l)-1-hydroxymethyl-propylmethacrylamide), poly(2-carboxyisopropylacrylamide), poly(*N*-acryloyl-*N'*-alkylpiperazine), poly(*N*,*N'*-diethylacrylamide), and copolymers [e.g., poly(*N*-isopropylacrylamide-*co*-*N*,*N*-dimethylacrylamide)] (Rapoport 2007; Arias 2009).

This strategy also involves triggered drug release by hyperthermia which also induces selective tumor permeability. Nanomedicine uptake is the consequence of an increase in the microvascular pore size and in the tumor blood flow. Although hyperthermia can be cytotoxic to malignant cells (cancer treatment by hyperthermia), its main limitation comes from tumor localization and accessibility (Rapoport 2007; Arias 2011). Hyperthermia can also be externally controlled by using superparamagnetic iron oxides (Laurent et al. 2008; Arias 2010). When nanoplatforms based on these magnetic materials are subjected to an alternating magnetic gradient of high frequency (≈ 1 MHz), they produce heat due to magnetic hysteresis loss. This temperature increment will then facilitate drug desorption.

For instance, methotrexate-loaded thermo-sensitive magnetoliposomes have been synthesized by using cholesterol and 1,2-dipalmitoyl-*sn*-glycero-3-phosphocholine. Interestingly, more than 80% of the drug was released from the nanoparticles within 30 min when temperature was increased from 37 to 41°C. On the opposite, 60% of the drug remained inside the magnetic nanoplatform up to 24 hr at 37°C. Notably, the use of an external magnetic gradient and heating (up to 41°C) significantly enhanced methotrexate accumulation into the skeletal muscular tissue of mice (Zhu et al. 2009).

Light-Triggered Drug Release

A proper engineering of a liposomal or polymeric nanoplatform can make it sensitive to ultraviolet (UV) and/or visible light. The technique is highly advantageous since light stimulus can be delivered instantly to the targeted place in specific amounts with high accuracy (Bisby et al. 2000; Arias 2009). The strategy relies on the introduction into the nanoparticle structure of chemical groups that can be disrupted under exposure to UV light [i.e., the leuco derivative bis(4-dimethylamino)phenylmethyl leucocyanide], or visible light (e.g., light-sensitive chromophores: copper chlorophyllin trisodium salt). Visible light-responsive nanoplatforms are preferred in drug delivery because of their safety, low cost, and easy manipulation. Plasmalogen photooxidation is a very promising approach in the formulation of liposomal nanomedicines in which drug release results from enhanced membrane permeability upon plasmenylcholine photooxidative cleavage to single-chain surfactants (Shum et al. 2001). Despite the very interesting results previously obtained, this is maybe one of the areas of active drug targeting to cancer that needs more research to assure the viability of an *in vivo* targeted drug release into tumors.

Ultrasound-Mediated Drug Delivery

Ultrasounds can penetrate deep into the body, leading to localize and complete drug release from the nanocarrier. The mechanism of action relies on the induction of enhanced permeability of capillaries, the generation of thermal energy, and the alteration of cell membranes. The nanodrug is typically accumulated into the tumor tissue by passive targeting (EPR effect). Thereafter, the application of ultrasounds induces cellular uptake of the nanomedicine (thanks to the alteration of cell membrane permeability) and drug release (due to the nanocarrier disruption) (Rapoport et al. 2004; Arias 2009). It has been suggested that low frequency ultrasounds are best suited for superficial tumors (e.g., skin, gynaecological, and some head and neck cancers). On the other hand, high frequency ultrasounds are more suitable for deeper tumors (Schroeder et al. 2009).

As an example, low frequency ultrasounds (20 kHz) have been used to trigger cisplatin release from long-circulating liposomes. The drug-loaded nanoplatform (in combination with ultrasounds) demonstrated a very significant antitumor activity (tumor regression) upon administration to mice bearing J6456 murine lymphoma tumors in their peritoneal cavity, or C26 tumors in the footpad, compared to controls (untreated, ultrasounds alone, cisplatin solution with or without ultrasounds, and cisplatin-loaded nanoparticles without ultrasounds). Interestingly, ≈ 70% of cisplatin loaded into the nanocarrier was selectively released into tumors exposed to ultrasounds, compared to < 3% in non-exposed tumors (Schroeder et al. 2009).

Magnetic Drug Targeting

Magnetically responsive nanoparticles can specifically deliver anticancer drugs to targeted tumors under the guidance of magnetic gradients. Then, the magnetic force will focus the magnetic particle into the tumor interstitium (or tumor neovasculature) until the drug is completely released (Reddy et al. 2006; Arias 2010). Iron oxide nanoparticles (e.g., magnetite, maghemite) are principally used to formulate these nanoplatforms. However, due to the very poor drug loading capacity and uncontrollable drug release frequently exhibited by these magnetic nanoparticles, it is also needed to embed them into a biodegradable shell (mainly a polymer or a lipid vesicle). The shell [i.e., liposome, chitosan, PLGA, PCL, poly(alkylcyanoacrylate)] will play the role of improving the biocompatibility of the composite nanoparticle, and will transport the drug to the tumor tissue (controlling its release). At the same time, the iron oxide core will allow the magnetic guidance of the nanomedicine towards the site of action. In this way, the use of magnetic implants combined with externally applied magnetic gradients can optimize the delivery of magnetic nanomedicines to the tumor tissue (Arias 2010). The introduction of long-circulating capabilities along with complementary ligand-mediated targeting strategies in the formulation of magnetic nanoplatforms is currently revolutionizing cancer nano-chemotherapy (Lin et al. 2009).

Conclusions

The application of passive and active drug targeting strategies in the formulation of nanomedicines against cancer has contributed significantly to the enhancement of chemotherapy. An efficient nanocarrier must exhibit long-circulating properties to reach the tumor interstitium, surface functionalization with appropriate ligands to enter the cancer cell, and the capability of being disrupted exclusively inside the malignant cell (drug release in deep contact with the target). However, up to now very few formulations have been approved for clinical use. The future of nano-oncology still depends on improving the knowledge of tumor physiopathology, and on advances in nanomedicine engineering, especially regarding the relationship between nanoparticle structure and biological fate, nanotoxicity, and anticancer effect.

Summary Points

- Current chemotherapy is characterized by a significant lack of targetability for malignant cells which induces severe toxicity problems.
- Preclinical research reports have revealed the potential of nanocarriers in the improvement of drug accumulation into tumor interstitium.
- Polymeric and lipid-based nanoparticles are the most promising materials in the delivery of drugs to cancer. However, conventional nanocarriers can only be exploited against tumors located in the MPS.
- Nanomedicines against cancer must be formulated by using both passive and active drug targeting strategies.
- Drug delivery to brain cancers and MDR tumors is drastically optimized by multifunctional nanoplatforms.
- The future of nano-oncology relies on a complete clarification of tumor physiopathology to identify new targets for drug delivery, and on greater advances in the formulation, characterization, and nanoengineering of nanomedicines. Theragnostic nanoplatforms are expected to definitively defeat cancer.

Dictionary

Aptamer: DNA or RNA oligonucleotide capable of selectively bind to target antigens. These nucleic acid ligands are characterized by a small size and an easy chemical synthesis.

Immunonanoparticles: nanoparticles surface functionalized with MAbs for targeted drug delivery. For example, immunoliposomes.

Integrin: cell receptor that plays a key role in cell signaling (and determines cellular shape, mobility, and cell cycle).

Long-circulating nanoplatforms: nanoparticles surface functionalized with hydrophilic moieties (e.g., PEG chains).

Nanoparticle opsonization: a biological process that leads to rapid plasma clearance of the nanomedicine by the MPS. It is supported by high surface charges, hydrophobicity, and big dimensions (over 500 nm) of drug carriers. Even if the nanomedicine is small enough to escape from vascular filtration, it can be opsonised due to these properties. Opsonins will be adsorbed onto the nanoplatform surface, leading to rapid nanoparticle recognition and plasma clearance by Küpffer cells and other tissue macrophages.

Nanocapsule: vesicular nanosystems where the drug is incorporated into an oily or aqueous cavity, normally surrounded by a polymeric or lipidic shell.

Nanosphere: nanomatrix where the drug is distributed all along a continuous core.

Opsonin: plasma protein that encourages nanoparticle phagocytosis, e.g., fibronectin, immunoglobulin G, or complement system. They determine the biodistribution, biodegradation, and elimination processes of nanomedicines.

Transferrin: blood glycoprotein that controls the iron levels. When this biomolecule binds a receptor on cell surface, it is transported into the cell.

Key Facts

- Key facts of the biological fate of nanomedicines. Basic aspects in the engineering of nanomedicines are nanoparticle geometry, surface thermodynamics, and surface charge.
- Key facts of nanotoxicity in drug delivery to cancer. Nanoparticle biocompatibility is determined by both the toxicity of the nanocarrier, and the interaction of its biodegradation products with the immune system. The toxicity of drug delivery systems relies on the delivered dose (mass per cell or cm^3), dose of exposure (mass administered), cell dose (internalized mass), method of administration, biodegradability, pharmacokinetics, and biodistribution (Arias 2011). The degradation products must be cleared from the body in the shortest period of time. Further investigations are needed to establish predictive models for toxicity evaluations.
- Key facts of drug delivery to MDR cancer cells. Drug delivery by biodegradable nanoparticles must overcome both cellular-based, and non-cellular based drug resistance mechanisms. The former is associated to tumor physiology. Cell-based drug resistance

mechanisms consist of apoptosis blockage (i.e., decreased ceramide levels), enhanced drug efflux (e.g., up-regulated P-*gp*), minimized drug influx, and DNA repair activation and detoxification (thanks to specific enzymatic systems, e.g., topoisomerase) (Arias 2009).

- Key Facts of chemotherapy failure due to tumor physiology. Poor vascularization, heterogeneous blood supply, low microvascular pressure and high interstitial pressure, relatively long distances in tumor interstitium, and cellular heterogeneities contribute to an unsuccessful cancer cell targeting. For instance, uniform drug diffusion inside the tumor mass is unfeasible due to the very high hydrostatic pressure. The non-functional lymphatic system allows drug escaping from the tumor mass, and its dilution in the surroundings. Last but not least, the slightly acidic conditions of tumor mass (pH $\approx$ 6.6, due to a greater aerobic and anaerobic glycolisis, and spatial and temporal heterogeneities in blood flow) protect cancer cells from basic drugs which will be ionized (preventing their extravasation) (Arias 2009).

Applications to areas of health and disease

The use of nanoparticulate systems for the delivery of antitumor drugs has revolutionized current chemotherapy. Several preclinical investigations have demonstrated that this strategy can assure maximum drug concentration into the targeted place with almost null toxicity. Even more resistant malignant cells to anticancer agents (resulting in refractory tumors) can be treated by co-delivery of molecules that can regulate intracellular pH, resistance mechanisms (e.g., P-*gp* substrates), apoptotic threshold (e.g., ceramide), even combined with energy delivery (e.g., light or heat). Drug carriers can be also used to bypass BBB and reach brain cancers. The very promising preclinical results coming from this approach have introduced nanotechnology in the treatment of other diseases, e.g., inflammatory diseases (arthritis, etc.), and infectious diseases (caused by viruses, parasites, bacteria, or fungus).

The concept has been also used to increase the specificity and sensitivity of diagnostic tools. For example, luminophores and/or contrast agents for magnetic resonance imaging (e.g., gadolinium) loaded to biodegradable nanoparticles will give rise to improved signal detection. More interestingly, the concept of nanotheragnosis offers combined therapeutic and diagnostic activities. These multifunctional nanomedicines are intended to assure early disease detection, the recognition of disease biomarkers and signals for the choice of therapy, and an efficient (multi) drug delivery to the targeted site.

Abbreviations

AUC	:	area under the time-concentration curve
BBB	:	blood brain barrier
EPR effect	:	enhanced permeability and retention effect
FBP	:	folate-binding protein
human EGFR	:	human epidermal growth factor receptor
LHRH	:	luteinizing hormone-releasing hormone
MAb	:	monoclonal antibody
MDR	:	multi-drug resistance
MPS	:	mononuclear phagocyte system
PCL	:	poly(ε-caprolactone)
PEG	:	polyethylene glycol
PLGA	:	poly(D,L-lactide-*co*-glycolide)
P-*gp*	:	P-glycoprotein
PPI	:	poly(propyleneimine)
PSMA	:	prostate-specific membrane antigen
RGD sequence	:	arginine-glycine-aspartic acid sequence
siRNA	:	small interfering RNA
SLNs	:	solid lipid nanoparticles
$t_{1/2}$	:	plasma half life
UV	:	ultraviolet

References

Allen, T.M. 1994. Long-circulating (sterically stabilized) liposomes for targeted drug delivery. Trends Pharmacol. Sci. 15: 215–220.

Andresen, T.L., S.S. Jensen, T. Kaasgaard and K. Jørgensen. 2005. Triggered activation and release of liposomal prodrugs and drugs in cancer tissue by secretory phospholipase A2. Curr. Drug Deliv. 2: 353–362.

Arias, J.L. Micro- and nano-particulate drug delivery systems for cancer treatment. pp. 1–85. *In*: P. Spencer and W. Holt. [eds.] 2009. Anticancer Drugs: Design, Delivery and Pharmacology. Nova Science Publishers Inc., New York, USA.

Arias, J.L. 2010. Drug targeting by magnetically responsive colloids. Nova Science Publishers Inc., New York, USA.

Arias, J.L. 2011. Drug targeting strategies in cancer treatment: An overview. Mini-Rev. Med. Chem. 11: 1–17.

Bisby, R.H., C. Mead and C.G. Morgan. 2000. Active uptake of drugs into photosensitive liposomes and rapid release on UV photolysis. Photochem. Photobiol. 72: 57–61.

Brigger, I., C. Dubernet and P. Couvreur. 2002. Nanoparticles in cancer therapy and diagnosis. Adv. Drug Deliv. Rev. 54: 631–651.

Cho, K., X. Wang, S. Nie, Z. Chen and D.M. Shin. 2008. Therapeutic nanoparticles for drug delivery in cancer. Clin. Cancer Res. 14: 1310–1316.

Davis, M.E., Z. Chen and D.M. Shin. 2008. Nanoparticle therapeutics: An emerging treatment modality for cancer. Nat. Rev. Drug Dicov. 7: 771–782.

Decuzzi, P., R. Pasqualini, W. Arap and M. Ferrari. 2009. Intravascular delivery of particulate systems: Does geometry really matter? Pharm. Res. 26: 235–243.

Farokhzad, O.C., J. Cheng, B.A. Teply, I. Sherifi, S. Jon, P.W. Kantoff, J.P. Richie and R. Langer. 2006. Targeted nanoparticle-aptamer bioconjugates for cancer chemotherapy *in vivo*. Proc. Natl. Acad. Sci. USA 103: 6315–6320.

Lai, T.Y., and W.C. Lee. 2009. Killing of cancer cell line by photoexcitation of folic acid-modified titanium dioxide nanoparticles. J. Photochem. Photobiol. A: Chem. 204: 148–153.

Laurent, S., D. Forge, M. Port, A. Roch, C. Robic, L.V. Elst and R.N. Muller. 2008. Magnetic iron oxide nanoparticles: Synthesis, stabilization, vectorization, physicochemical characterizations, and biological applications. Chem. Rev. 108: 2064–2110.

Lee, E.S., K. Na and Y.H. Bae. 2005. Doxorubicin loaded pH-sensitive polymeric micelles for reversal of resistant MCF-7 tumor. J. Control Release 103: 405–418.

Li, J.L., L. Wang, X.Y. Liu, Z.P. Zhang, H.C. Guo, W.M. Liu and S.H. Tang. 2009a. *In vitro* cancer cell imaging and therapy using transferrin-conjugated gold nanoparticles. Cancer Lett. 274: 319–326.

Li, F., H. Wu, H. Zhang, F. Li, C.H. Gu and Q. Yang. 2009b. Antitumor drug Paclitaxel-loaded pH-sensitive nanoparticles targeting tumor extracellular pH. Carbohydr. Polym. 77: 773–778.

Lin, J.J., J.S. Chen, S.J. Huang, J.H. Ko, Y.M. Wang, T.L. Chen and L.F. Wang. 2009. Folic acid-Pluronic F127 magnetic nanoparticle clusters for combined targeting, diagnosis, and therapy applications. Biomaterials 30: 5114–5124.

Low, P.S., and S.A. Kularatne. 2009. Folate-targeted therapeutic and imaging agents for cancer. Cur. Opin. Chem. Biol. 13: 256–262.

Maeda, H., G.Y. Bharate and J. Daruwalla. 2009. Polymeric drugs for efficient tumor-targeted drug delivery based on EPR-effect. Eur. J. Pharm. Biopharm. 71: 409–419.

Matsuo, H., M. Wakasugi, H. Takanaga, H. Ohtani, M. Naito, T. Tsuruo and Y. Sawada. 2001. Possibility of the reversal of multidrug resistance and the avoidance of side effects by liposomes modified with MRK-16, a monoclonal antibody to P-glycoprotein. J. Control Release 77: 77–86.

Rapoport, N., D. Christensen, H.D. Fein, L. Barrows and Z. Gao. 2004. Ultrasound-triggered drug targeting to tumors *in vitro* and *in vivo*. Ultrasonics 42: 943–950.

Rapoport, N. 2007. Physical stimuli-responsive polymeric micelles for anti-cancer drug delivery. Prog. Polym. Sci. 32: 962–990.

Reddy, G.R., M.S. Bhojani, P. McConville, J. Moody, B.A. Moffat, D.E. Hall, G. Kim, Y.E.L. Koo, M.J. Woolliscroft, J.V. Sugai, T.D. Johnson, M.A. Philbert, R. Kopelman, A. Rehemtulla and B.D. Ross. 2006. Vascular targeted nanoparticles for imaging and treatment of brain tumors. Clin. Cancer Res. 12: 6677–6686.

Sahoo, S.K., and V. Labhasetwar. 2005. Enhanced antiproliferative activity of transferrin conjugated paclitaxel-loaded nanoparticles is mediated via sustained intracellular drug retention. Mol. Pharm. 2: 373–383.

Sapra, P., and T.M. Allen. 2003. Ligand-targeted liposomal anticancer drugs. Prog. Lipid Res. 42: 439–462.

Schroeder, A., R. Honen, K. Turjeman, A. Gabizon, J. Kost and Y. Barenholz. 2009. Ultrasound triggered release of cisplatin from liposomes in murine tumors. J. Control Release 137: 63–68.

Shenoy, D.B., and M.M. Amiji. 2005. Poly(ethylene oxide)-modified poly(epsilon-caprolactone) nanoparticles for targeted delivery of tamoxifen in breast cancer. Int. J. Pharm. 293: 261–270.

Shum, P., J.M. Kim and D.H. Thompson. 2001. Phototriggering of liposomal drug delivery systems. Adv. Drug Deliv. Rev. 53: 273–284.

Stevens, P.J., M. Sekido and R.J. Lee. 2004. A folate receptor-targeted lipid nanoparticle formulation for a lipophilic paclitaxel prodrug. Pharm. Res. 21: 2153–2157.

Sun, B., B. Ranganathan and S.S. Feng. 2008. Multifunctional poly(D,L-lactide-co-glycolide)/montmorillonite (PLGA/MMT) nanoparticles decorated by Trastuzumab for targeted chemotherapy of breast cancer. Biomaterials 29: 475–486.

Taratula, O., O.B. Garbuzenko, P. Kirkpatrick, I. Pandya, R. Savla, V.P. Pozharov, H. He and T. Minko. 2009. Surface-engineered targeted PPI dendrimer for efficient intracellular and intratumoral siRNA delivery. J. Control Release 140: 284–293.

van Verklen, L.E., Z. Duan, M.V. Seiden and M.M. Amiji. 2007. Modulation of intracellular ceramide using polymeric nanoparticles to overcome multidrug resistance in cancer. Cancer Res. 67: 4843–4850.

Wu, J., Y. Lu, A. Lee, X. Pan, X. Yang, X. Zhao and R.J. Lee. 2007. Reversal of multidrug resistance by transferrin-conjugated liposomes coencapsulating doxorubicin and verapamil. J. Pharm. Pharm. Sci. 10: 350–357.

Zhu, L., Z. Huo, L. Wang, X. Tong, Y. Xiao and K. Ni. 2009. Targeted delivery of methotrexate to skeletal muscular tissue by thermosensitive magnetoliposomes. Int. J. Pharm. 370: 136–143.

15

Gene Silencing with siRNA Encapsulated Nanoparticles to Overcome Tumor Multidrug Resistance

Mansoor M. Amiji,[1,*] *Francis Hornicek*[2,a] and *Zhenfeng Duan*[2,b]

ABSTRACT

Multidrug resistance (MDR) is a major obstacle in the successful treatment of human cancer. Experimental evidence from multiple laboratories implicates a wide range of mechanisms including over-expression of multidrug resistance 1 gene (*MDR-1*) and anti-apoptotic genes that contribute to the drug resistant phenotype. Reversing MDR has been a high priority goal of clinical and investigational oncology, but still remains elusive. RNA

[1]Department of Pharmaceutical Sciences, School of Pharmacy, Northeastern University, 360 Huntington Avenue, Boston, MA 02115; E-mail: m.amiji@neu.edu

[2]Department of Orthopedic Surgery, and Sarcoma Biology Laboratory, Massachusetts General Hospital, 55 Fruit Street, Boston, MA 02114.

[a]E-mail: fhornicek@partners.org

[b]E-mail: zduan@partners.org

*Corresponding author

List of abbreviations after the text.

interference (RNAi) has emerged as a powerful tool for sequence-specific gene silencing strategy. RNAi can be achieved by using small interfering RNA (siRNA) for target gene silencing in tumor cells. Synthetic, plasmid-based siRNA, and viral-based short hairpin RNA (shRNA) can significantly block *MDR-1* and anti-apoptotic gene expressions in drug resistant tumor cells. Although the RNAi technology is an excellent candidate to reverse MDR in cancer therapy, several serious challenges remain including unstable and rapid degradation by systemic and cellular nuclei and poor membrane permeability of siRNA for routine clinical application. Nanotechnology-based systems have been developed to deliver a variety of therapeutic agents, including nucleic acid constructs, into specific cells. Nanoparticulate formulations have recently emerged as alternatives to plasmid or viral vectors for the delivery of siRNA. This review summarizes current developments in drug resistant gene silencing with nanoparticle-encapsulated siRNA to overcome MDR in human cancer.

INTRODUCTION

The development of tumor multidrug resistance (MDR) is a significant clinical obstacle that often results in non-responsive, recurrent disease and eventual metastasis (Dong and Mumper 2010; Yusuf et al. 2003). MDR refers to a state of resilience against structurally and/or functionally unrelated drugs, and can be intrinsic or acquired due to initial exposure to chemotherapeutic agents. Intrinsic MDR may be genetically inherent or may develop as a response to selection pressures within the tumor microenvironment; a tumor cell confronted with survival challenges may undergo transformations resulting in MDR cells. Regardless of the mechanism of MDR, the clinical manifestation occurs due to inability of standard chemotherapeutic regimens to treat the disease effectively. Approaches that increase the dosage or combination of multiple agents in different therapeutic regimens generally do not lead to better clinical outcomes.

MDR is frequently due to expression of active transporters that pump a broad spectrum of chemically distinct, cytotoxic molecules out of tumor cells (Fig. 1). The classical form of MDR is mediated by P-glycoprotein (Pgp) that acts as a drug efflux pump. In humans the *MDR-1* (ABCB1) gene encodes Pgp, a 170 kDa membrane spanning protein with a highly conserved intracellular ATP-binding site. Pgp or *MDR-1* gene expression is frequently detected in both solid and hematologic cancers as well as in cancer stem cells and is a marker of both chemoresistance and decreased

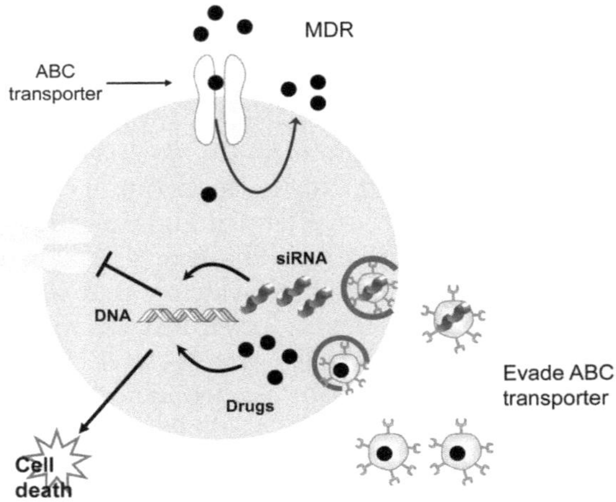

Fig. 1. Potential mechanism of overcoming multidrug resistance using nanoparticles. ABC transporters extrude chemotherapeutic drugs resulting in the survival of sarcoma cells. Conjugation of drugs and genes to nanoparticles results in increased accumulation of the drugs via non-specific endocytosis which could result in cell death. Drugs could be cytotoxic chemotherapeutic drugs, agents that could suppress the activity of ABC transporters or novel kinase inhibitors. Also, specific siRNA to the ABC transporters expressed in each sarcoma will result in the suppression of the ABC transporter.

Color image of this figure appears in the color plate section at the end of the book.

survival in lung cancer, ovarian cancer, leukemia, lymphoma, sarcoma and other malignancies (Yusuf et al. 2003). Pgp stands out among the ATP-binding cassette (ABC) family of transporters by conferring the strongest resistance to the widest variety of most chemotherapeutic agents. The multidrug resistance-associated protein 1 (MRP-1) or ABCC1 has also been identified. The MRP-1 gene encodes a 190 kDa N-glycosylated integral membrane phosphoprotein and is similar to Pgp in structure. MRP-1 confers resistance to several hydrophobic compounds that are also Pgp substrates. The discovery of MRP-1 stimulated a genomic search for homologues, leading to the discovery of eight additional members of the ABCC subfamily of transporters, of which six have been studied in some detail. Many MRP family members transport drugs in model systems and therefore have the potential to confer drug resistance (Teodori et al. 2006).

Altered regulation of apoptosis, or programmed cell death, has been demonstrated in a variety of human cancers (Table 1). The induction of apoptosis is believed to be the principal final step of chemotherapy associated cytotoxicity and failure to activate the intrinsic apoptotic program is a newly recognized mechanism of drug resistance. Several studies have demonstrated that the up-regulation of anti-apoptotic

Table 1. siRNA or shRNA silence genes to reverse MDR.

Target Gene	Type of RNAi	Major Function of the Gene
MDR-1 (ABCB1)	siRNA/shRNA	Efflux pump responsible for decreased drug accumulation in MDR cells
MRP-1 (ABCC1)	siRNA	Export of drugs from the cytoplasm and confers MDR
BCRP (ABCG2)	siRNA	Play a role in MDR for the drugs such as mitoxantrone,doxorubicin
Bcl-2	siRNA	Suppresses apoptosis
Bcl-XL	siRNA	Anti-apoptotic and inhibitor of cell death
Survivin	siRNA	Prevent apoptotic cell death
Stat3	siRNA/shRNA	Transcription factor and prevent apoptotic cell death
PLK1	siRNA	Regulators of cell cycle progression, mitosis, cytokinesis, and the DNA damage response
Cathepsin L	siRNA	Participate in intracellular degradation and turnover of proteins
MDM2	siRNA	Inhibits p53 mediated cell cycle arrest and apoptosis
GCS	shRNA	Involved in the production of glucosylceramide
RON	shRNA	Regulates expression of genes implicated in cancer-cell survival

proteins such as Bcl-2, Bcl-X_L, survivin, Stat3, IL-6, inhibitors of apoptosis (IAPs), PI3K, and AKT2 or down-regulation of pro-apoptotic proteins such as BAX and BAD decrease paclitaxel and cisplatin sensitivity *in vitro* (Duan et al. 2006; Yusuf et al. 2003). *In vivo* studies demonstrate that the expression of active PI3K renders ovarian cancer cells resistant to paclitaxel, a microtubule stabilizing agent, an effect that could be reversed by a PI3K inhibitor. The over-expression of epidermal growth factor receptor (EGFR) is also a hallmark of resistance and disease aggressiveness (Yusuf et al. 2003).

The development and discovery of agents that reverse MDR with high efficiency and low toxicity is an area of extensive research activity. Unfortunately, these compounds are often non-specific and have low efficiency and/or high toxicity; as such, phase 3 clinical trials of these agents are largely disappointing (Kaye 2008; Szakacs et al. 2006). Consequently, it is imperative to develop alternative, less toxic and more efficient strategies to overcome MDR.

One of the innovative approaches to addressing MDR is by inhibition of *MDR-1* mRNA expression by RNA interference (RNAi) (Table 2). RNAi is a technique that mimics and exploits endogenous silencing mechanisms resulting in post-transcriptional gene silencing of double-stranded RNA inside cells (Baker 2010; Lage 2006). The double stranded 21 to 23 nucleotide non-coding small interfering RNA (siRNA) can knock down expression of genes in a highly efficient and sequence specific manner. The efficiency of RNAi and its limited side effects have made this

Table 2. Summary of *MDR-1* RNAi Silencing to Reverse MDR.

Type of RNAi	Gene Targeted Site	Drugs Resistance Reversed
siRNA	508–528	Vinblastine, Doxorubicin, Paclitaxel and Hydroxyurea
siRNA	503–523	Daunorubicin
siRNA/shRNA	889-908	Paclitaxel
siRNA/shRNA	3495-3514	Doxorubicin, Paclitaxel and Colchicine
siRNA/shRNA	225-245	Doxorubicin
shRNA	88-110	Vincristine
siRNA/shRNA	499-517	Daunorubicin,Vincristine

technique an attractive alternative to the use of antisense oligonucleotides and ribozymes for therapies based on the inhibition of target genes. Both synthetic-based and plasmid-based siRNA can significantly block *MDR-1* expression in drug resistant tumor cells (Duan et al. 2004; Lage 2006; Nieth et al. 2003; Wu et al. 2003). Although the RNAi technology is an excellent candidate for cancer therapy, several challenges need to be addressed for clinical application. First, the poor membrane permeability of siRNA limits cellular uptake. Secondly, most of the reagents for delivery of siRNA that are currently available, such as Lipofectamine®, are cytotoxic and not amenable for systemic administration in patients. Thirdly, siRNA is unstable and rapidly degraded by nucleases (Baker, Baker 2010). Further use of siRNA as therapeutic agents will rely mostly on the development of more efficient delivery systems.

Nanotechnology offers solutions to overcome the adversity of other siRNA delivery systems. Several examples of nanoparticles can be used for encapsulation and delivery of siRNA. These include polymer- and lipid-based as well as self-assembling micelles and dendrimers (Fig. 2). The purpose of this review is to highlight the biological, therapeutic, and clinical role of nanoparticles encapsulated siRNA to overcome MDR in tumor cells.

GENE SILENCING STRATEGIES

RNA interference is an endogenous, well conserved mechanism that uses small non-coding RNAs to silence gene expression. When exogenous small double stranded siRNA is introduced into cells, it binds to the endogenous RNAi machinery to disrupt the expression of mRNAs containing sequences with high specificity. This pathway is initiated by the enzyme Dicer, which cleaves long double stranded RNA (dsRNA) into short fragments of ~20 nucleotides. One of the strands called the guide strand, is then incorporated into the RNA-induced silencing complex (RISC). The anti-sense RNA strand then guides RISC to homologous sequences on target mRNA and

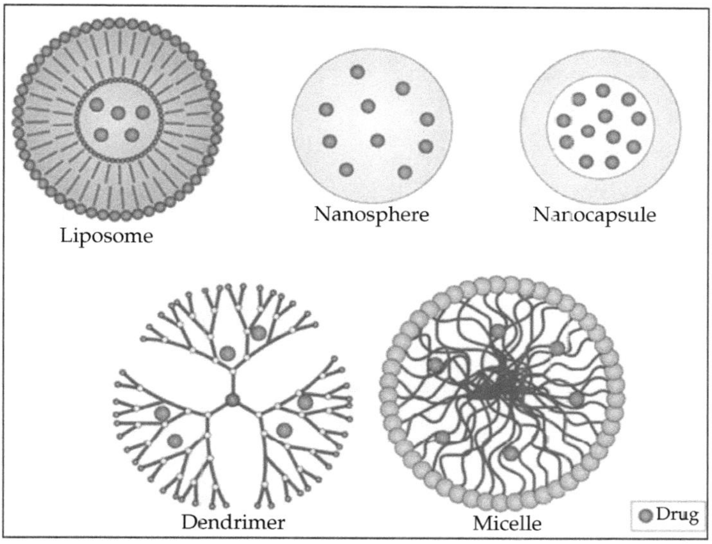

Fig. 2. Schematic illustration of various nanoparticle platforms. Nanoparticle systems are used for encapsulation and delivery of cancer therapeutics, including nucleic acid-based constructs. Examples include lipid vesicles or liposomes, polymeric nanospheres or nanocapsules, dendrimers, and self-assembling micelles. Each of these systems is engineered to protect the payload and deliver efficiently to disease tissues and cells.

Color image of this figure appears in the color plate section at the end of the book.

base-pairs with a complementary sequence of messenger RNA, inducing cleavage by Argonaute, the catalytic component of the RISC complex. The cleaved target mRNA is no longer capable of producing functional protein and therefore the mRNA is "silenced". This robust silencing effect of RNAi makes it a valuable research tool both in cell culture and in living organisms, although introducing siRNA into cells *in vivo* remains a significant obstacle (Baker 2010). Silencing of Pgp and down regulation of MDR or anti-apoptotic proteins could reverse the MDR phenotype, re-sensitizing MDR cells to cytotoxic agents (Duan et al. 2004; Duan et al. 2006; Wu et al. 2003) (Fig. 3).

MDR REVERSAL WITH SIRNA DELIVERY

Down-regulation of MDR transporters and anti-apoptotic genes by RNA interference (RNAi) has been suggested as a more specific alternative to overcome MDR than the use of conventional small molecule pharmacological inhibitors. Researchers believe that by using siRNA molecules, the mediators of RNAi, they can turn off the ability of cancer cells to produce

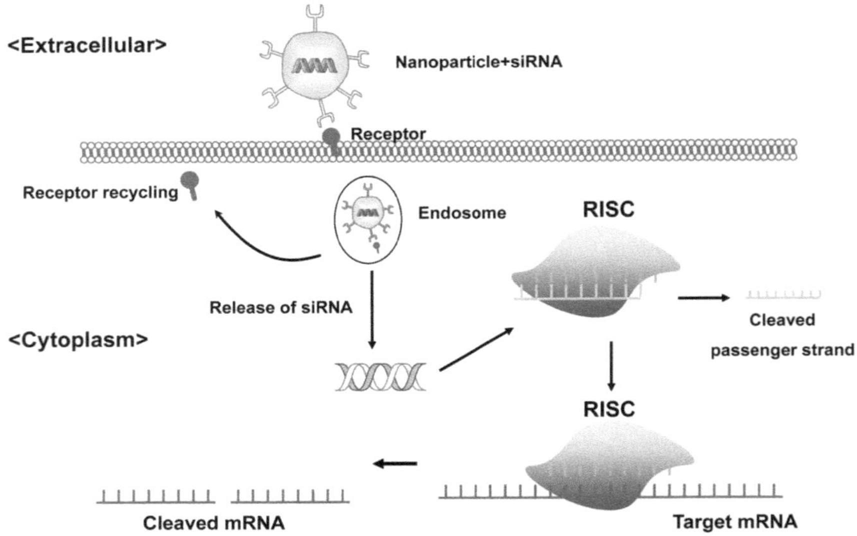

Fig. 3. RNA interference mechanisms utilizing siRNA-encapsulated nanoparticles. After internalization of nanoparticles via endocytosis, siRNA is released into the cytoplasm. siRNA is then incorporated into the RNA-induced silencing complex (RISC) which results in the cleavage of passenger strand of the siRNA by argonaute 2 (Ago2). Mature RISC complex binds to and degrades complementary mRNA resulting in the silencing of the target gene.

Color image of this figure appears in the color plate section at the end of the book.

the key proteins that induce malignancy. Transient RNAi can be attained by application of siRNA duplexes, whereas a stable RNAi-mediated gene-silencing can be achieved by transfection of mammalian cells with short hairpin RNA (shRNA) encoding expression cassettes localized on plasmid or viral vectors. Transient and stable RNAi strategies have been applied to overcome MDR-1/Pgp and MRP-1-mediated MDR in different *in vitro* models derived from various tumor types. Reversal of drug resistance by siRNA directed against the mRNA for MDR-1 has been reported for numerous cancer cells, including pancreatic, lung, ovarian, breast and leukemia (Duan et al. 2004; Nieth et al. 2003; Wu et al. 2003). For instance, RNAi against MDR-1 has been found to reverse etoposide, doxorubicin, and vincristine resistance in small cell lung cancer cells. RNAi may be a promising strategy to reverse MDR of human cancer cells. However, the human body is well-equipped to destroy double-stranded RNA circulating in the blood and prevent it from entering cells. The negative charge and hydrophilic macromolecular properties of siRNA further restrict its uptake into cells. Additionally, most of the reagents for delivery of siRNA such as Lipofectamine® are toxic. Furthermore, siRNA is unstable and rapidly degraded by nuclei. For *in vivo*, free siRNA are not ideal to produce efficient and predictable therapeutic effects. Within minutes after intravenous administration, the majority of the injected siRNA dose is removed from circulation by hepatic and renal clearance. Only a very small percentage of the administered siRNA molecules remain available for the target cells or tissues. Alternatively, nanotechnology-based siRNA delivery systems have resulted in gene silencing efficacy *in vitro* and *in vivo* in many different studies (Baker, Schroeder et al. 2010). Anti-MDR-1/Pgp strategies may show the highest clinical efficacy when administered in combination with conventional chemotherapeutic regimens.

NANOPARTICLE-ENCAPSULATED siRNA DELIVERY

Nanotechnology is an area of science that involves working with materials that are at the nanoscale level. A nanometre is billionth of a meter. That is about 1/80,000 of the diameter of a human hair. It manipulates the chemical and physical properties of a substance on molecular level (Jabr-Milane et al. 2008; Jain and Stylianopoulos 2010). Nano-biotechnology is the integration of biotechnology and nanotechnology. Nanoscale polymer particles can be designed to break down and release drugs at controlled rates, to allow differential release in certain environments, such as an acid medium and to promote uptake in tumors versus normal tissues. One of the key issues in bio-availability is cell transfection in DNA or siRNA gene therapy. Current methods have significant limitations, including the DNA or siRNA is unstable and rapidly degraded by cellular nucleases and risk of inadvertent transmission of disease by viral vectors. This has led researchers to explore polymer-DNA/siRNA complexes and liposome-DNA/siRNA complexes for gene delivery. It has also been shown that compacted DNA/siRNA in the form of nanoparticles can be used to transfect postmitotic cells (Liu et al. 2003).

Nanoparticles are useful for protecting labile therapeutics such as DNA and siRNA. These agents have poor stability in systemic circulation and are prone to extracellular and intracellular degradation; it is the protection and adequate delivery of gene therapy and RNAi that limits the clinical application of these agents. They can also be targeted to tumor mass either through passive or active targeting strategies (Fig. 4). Various nanocarrier platforms such as polymeric nanoparticles, cationic liposomes, and lipoplexes, have been shown to increase the stability of labile therapeutics. PEG- modified type B gelatin and thiolated gelatin nanoparticles have also been shown to increase the stability of therapeutic payloads (Kommareddy and Amiji 2007). Remarkably, the stability of plasmid DNA encapsulated in gelatin nanoparticles can be maintained

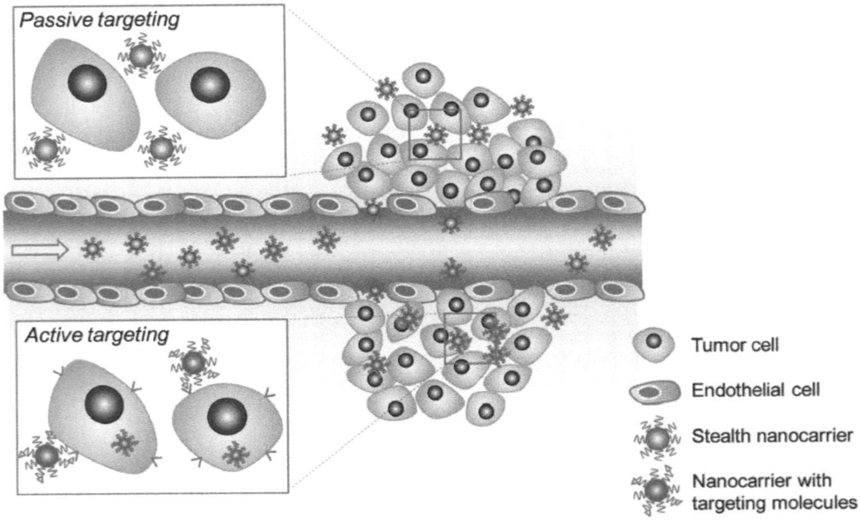

Fig. 4. Illustration of passive and active tumor targeted delivery with nanoparticles. Passive targeting relies on the abnormal tumor neovasculature that have greater fenestration allowing nanoparticles with up to 200 nm in diameter to extravasate and the properties of the delivery system. Active targeting is based on surface attachment of ligands that can specifically bind to over-expressed cellular targets.

Color image of this figure appears in the color plate section at the end of the book.

even in the presence of DNAs (Kommareddy and Amiji 2007). GFP plasmids are common reporter genes used to assess the transfection efficiency of different gene therapy strategies. Gelatin nanoparticles used to deliver a GFP plasmid have demonstrated transfection within 6 hrs of incubation, and the GFP plasmid expression was sustained until 96 hrs after transfection, indicating enhanced stability of the encapsulated plasmid compared to the naked plasmid (Kommareddy and Amiji 2007). PEO-PCL nanoparticles have also been used to simultaneously deliver paclitaxel and MDR-1 (Pgp) siRNA (Yadav et al. 2009). As Pgp (MDR-1) is responsible for the resistance phenotype of many MDR cells and is known to actively efflux paclitaxel, this strategy aimed to block paclitaxel efflux by silencing P-gp while increasing paclitaxel intracellular delivery using the nanocarrier system (Yadav et al. 2009). Nanoparticle delivery increased the intracellular concentration of paclitaxel relative to free drug administration (Devalapally et al. 2008; van Vlerken et al. 2007). When administered to $SKOV3_{TR}$ cells (established MDR ovarian cancer cells), the concentration of paclitaxel was 8.2 nM/mg of total cellular protein for cells treated with paclitaxel loaded PEO-PCL nanoparticles (no siRNA), this concentration increased to 11.5 nM/mg of total protein for cells treated with PEO-PCL nanoparticles loaded with both paclitaxel and Pgp siRNA (Yadav et al. 2009). This enhanced accumulation of paclitaxel due to P-gp down-regulation was confirmed with RT-PCR (Yadav et al. 2009). This strategy of combination drug delivery and Pgp silencing could be applied to other Pgp substrates to enhance efficacy.

The effectiveness of simultaneous and targeted delivery of paclitaxel, along with MDR siRNA has been investigated by using poly (D,L-lactide-co-glycolide) nanopartilces in MDR cells. Nanoparticles were surface functionalized with biotin for active tumor targeting. Dual agent nanoparticles encapsulating the combination of paclitaxel and MDR siRNA showed significantly higher cytotoxicity *in vitro* than nanoparticles loaded with paclitaxel alone. Increased therapeutic efficacy of dual agent nanoparticles could be correlated with effective silencing of the MDR-1 gene and with increased accumulation of paclitaxel in MDR cells. *In vivo* studies in a MDR mouse model demonstrated greater inhibition of tumor growth following treatment with biotin-functionalized nanoparticles encapsulating both paclitaxel and MDR siRNA at a paclitaxel dose that was ineffective in the absence of gene silencing (Patil et al. 2010).

In order to overcome both the dose-limiting side effects of conventional chemotherapeutic agents and the therapeutic failure resulting from MDR, A dextran-based nanoparticle was designed for *MDR-1* gene silencing siRNA delivery in doxorubicin resistant tumor cells. The efficacy of combination therapy with this system was evaluated. MDR cells were treated with the MDR-1 siRNA nanocarriers and Pgp expression, drug retention, and immunofluoresence were analyzed. Combination therapy of the *MDR-1* siRNA loaded nanocarriers with increasing concentrations of doxorubicin was also analyzed. The results showed that MDR-1 siRNA loaded dextran nanoparticles efficiently suppresses Pgp expression in the MDR cell lines (Susa et al. 2010) (Fig. 5). The results also demonstrated that this approach may be capable of reversing drug resistance by increasing the amount of drug accumulation in MDR cell lines. In addition, a liposome-based nanoparticle simultaneous co-delivery of doxorubicin as a cell-death inducer/anticancer agent with siRNA targeted to MRP-1 mRNA as a suppressor of drug resistance has also been reported in lung cancer drug resistant cells. The results showed nanoparticles provides an effective co-delivery of doxorubicin and siRNA as well as cell-death induction and suppression of cellular resistance in MDR lung cancer cells (Saad et al. 2008). A recent study shows mesoporous silica nanoparticles (MSNP) can be functionalized to effectively deliver doxorubicin as well as MDR-1 siRNA to MDR cell line to accomplish cell killing in an additive or synergistic fashion (Meng et al. 2010). The functionalization

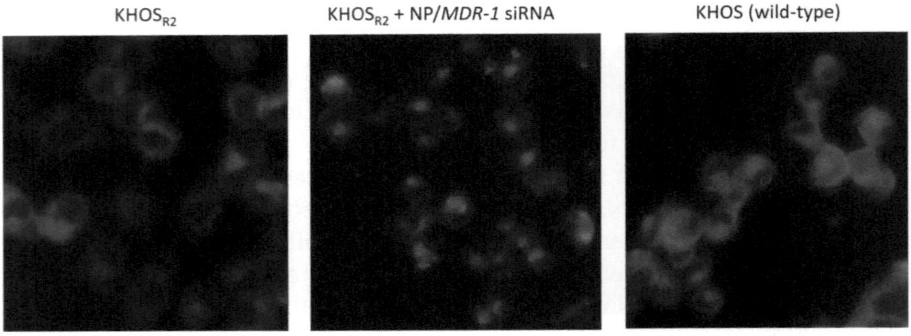

Fig. 5. Sub-cellular distribution of doxorubicin in drug sensitive (KHOS) and multidrug resistant (KHOS$_{R2}$) osteosarcoma cells. A prominent increase in fluorescence was observed in the nucleus when multidrug resistant cells KHOS$_{R2}$ were pre-treated with *MDR-1* gene silencing siRNA-loaded nanoparticles followed by administration with doxorubicin.

Color image of this figure appears in the color plate section at the end of the book.

of the particle surface with a phosphonate group allows electrostatic binding of doxorubicin to the porous interior, from where the drug could be released by acidification of the medium under abiotic and biotic conditions. In addition, phosphonate modification also allows exterior coating with the cationic polymer, polyethylenimine, which endows the MSNP to contemporaneously deliver MDR-1 siRNA. The dual delivery of doxorubicin and siRNA in MDR cells was capable of increasing the intracellular as well as intranuclear drug concentration to levels exceeding that of free doxorubicin or the drug being delivered by MSNP in the absence of siRNA codelivery (Meng et al.). Co-delivery of doxorubicin and anti-apoptotic gene Bcl-2 siRNA by nanoparticles has also been reported to overcome MDR in tumor cells (Chen et al. 2009).

Nanoparticles need to encapsulate siRNAs and protect them from clearance and degradation. Ideally they need to be functionalized with targeting ligands, which may potentially improve the delivery specificity and efficiency. Moreover, since siRNA molecules are hydrophilic and negatively charged, after the nanoparticles enter the cells through ligand-mediated endocytosis, the released siRNA cannot spontaneously diffuse across the plasma membrane. Instead, they may be trapped in endosomal/lysosome compartments and be subject to lysosomal degradation. Therefore, for efficient siRNA delivery, nanoparticles should incorporate certain mechanisms for siRNA to escape from the endosomal compartment and enter cytoplasm. Furthermore, once in the cytoplasm, the siRNA needs to be bioactive for efficient gene silencing. In order to meet these criteria for effective delivery of siRNA, combinatorial-designed multifunctional nanoparticles are needed.

MULTIFUNCTIONAL NANOPARTICLES FOR COMBINATION siRNA/DRUG DELIVERY

Nanoparticles hold tremendous potential for drug and siRNA delivery in tumor cells. Several varieties of nanoparticles are available including polymeric nanoparticles, dendrimers, inorganic/metal nanoparticles, quantum dots, liposomes, and micelles (Jabr-Milane et al. 2008; MacDiarmid et al. 2009; Susa et al. 2009; Susa et al.; van Vlerken et al. 2007). The advantages of nanoparticles including ease in surface modification, greater encapsulation efficiency of the payload, payload protection, large surface area-to-volume ratio and the ability to modify the rate of polymer erosion for temporal control over the release of nucleotides. These novel characteristics make nanoparticles an ideal system for the improved delivery of siRNAs. Nanoparticles need to encapsulate siRNAs and protect them from clearance and degradation. Ideally they need to be functionalized with targeting ligands, which may potentially improve the delivery specificity and efficiency. Furthermore, to achieve efficient drug and siRNA delivery, it is important to improve the interactions nanoparticles with the biological environment, targeting tumor cell-surface receptors (such as EGFR, IGF-1R), drug and siRNA release, stability of drug and siRNA. For efficient siRNA delivery, nanoparticles should incorporate certain mechanisms for siRNA to escape from the endosomal compartment and enter cytoplasm. Once in the cytoplasm, the siRNA needs to be bioactive for efficient gene silencing. The siRNA-containing nanoparticles (bacterially-derived minicells) targeted to the EGFR has been shown internalized by EGFR-mediated endosytosis and release siRNA into the cytoplasm (MacDiarmid et al. 2009). Similar results have been shown in plasmid-encoded shRNA directed against MDR-1. More recently, multifunctional liposome-polycation-DNA (LPD-II) nanoparticles which co-deliver siRNA and doxorubicin into the MDR tumor cells and trigger synergistic anti-cancer effect has been reported. siRNA and doxorubicin in the LPD-II nanoparticles via doxorubicin intercalation into the DNA in the nanoparticles. Both LPD and LPD-II nanoparticles were targeted specifically to the tumor cells by modification with anisamide (AA), a ligand of sigma receptor over-expressed in many human cancer cells. Two different siRNAs, VEGF and c-Myc siRNAs, are selected in this study to achieve the enhanced drug uptake and anti-cancer effect. siRNAs delivered to the MDR cells significantly down-regulate the target genes and sensitize the tumor cells to the co-delivered doxorubicin, resulting in an enhanced therapeutic activity. Similar results have been confirmed in a xenograft model of the doxorubicin resistant tumor (Chen et al. 2010).

Currently, application of RNAi in the clinical setting hinges on the low transfection efficiency, rapid degradation by serum nucleases, poor tissue penetration, and non-specific immune stimulation (Baker, Baker, Schroeder et al. 2010). Another problem when siRNA or shRNA delivery and tumor targeting fusion genes is that most fusion genes are transcription factors that are located in the nucleus, which makes it extremely difficult to introduce therapeutic tools to the destination. Nanoparticles have the potential to become the vehicle for stable, efficient, and tumor-specific delivery of RNAi. Currently, a phase I clinical trial is recruiting patients to test an intravenous nanoparticle-based siRNA treatment for solid tumors. The nanoparticles used are formulated using the three-part RONDEL technology, which combines a cyclodextrin polymer, an adamantine-modified stabilizer, and an adamantine-modified ligand targeted to the transferrin (an iron-binding protein) receptor on tumor cells. These nanoparticles were tested on both human beings and primates, and shown to be safe. Many other research projects are ongoing to improve the delivery of siRNA to the tumor site, but factors such as non-specific immune response by the host and off target effects of the siRNA also need to be addressed to improve the quality of siRNA. Novel chemical modification of the siRNA and smart multifunctional nanoparticulate complexes are needed to ensure an efficient and safe application of siRNA. Further work evaluating the efficacy of gene silencing with multifunctional nanoparticles encapsulated siRNA to overcome drug resistance in large animals is also needed.

Applications to Health and Diseases

The emergence of multidrug resistance (MDR) after chemotherapy is a major obstacle in cancer treatment. Both clinicians and basic oncology scientists have devoted considerable effort to developing therapies that can overcome MDR and deliver cytotoxic drugs.

In this review, we summarize the current developments in drug resistant gene silencing with nanoparticles encapsulated siRNA to overcome MDR in human cancer. Efficient siRNA encapsulation and subsequent release are fundamental requirements when using nanoparticles for siRNA delivery to overcome drug resistance. The nanoparticles need not only carry siRNA, but also protect siRNA from degradation. The nanoparticle-based combination strategies provide a promising example of engineering approaches that could potentially be used to overcome MDR in clinic.

Key Facts

- Multidrug resistance is a phenomenon after prolonged chemotherapy in cancer cells.
- Multidrug resistance is a major reason for the failure in the treatment of human cancer.
- Over-expression of specific genes such as *MDR-1* is the mechanism for cancer cell develop drug resistance.
- RNAi is a technique that can block the gene expression in cancer cells.
- Gene silencing with nanoparticles encapsulated siRNA could overcome multidrug resistance.
- Nanotechnology is an area of science that involves working with materials that are at the nanoscale level and named nanoparticles.
- Nanoparticles-based siRNA combinatorial approach hold promising potential to treat multidrug resistant tumors.

Key Terms

- **Multidrug resistance (MDR):** is a phenomenon in which tumor cells exposed to a single drug develop resistance to a wide variety of structurally and functionally unrelated drugs.
- **Nanoscience and nanotechnology:** is the study of the controlling matter on an atomic and molecular scale. Generally, nanotechnology deals with structures sized between 1 and 100 nanometer in at least one dimension and focuses on developing materials or devices within that size with unique engineered properties.
- **Nanoparticle:** a particle is defined as a small object that behaves as a whole unit in terms of its transport and properties. Nanoparticles are sized between 1 and 100 nanometers with unique engineered properties.
- **RNA interference (RNAi):** refers to the inhibition of gene expression by small double-stranded RNA molecules.
- **Small interfering RNA (siRNA):** is a class of double-stranded RNA molecules, 20–25 nucleotides in length that play a variety of roles in biology. Most notably, siRNA is involved in the RNA interference (RNAi) pathway, where it interferes with the expression of a specific gene.
- **Short hairpin RNA (shRNA):** is a sequence of RNA that makes a tight hairpin turn that can be used to silence gene expression via RNA interference.
- **Multidrug resistant gene 1 (MDR-1):** is the abbreviated name of a gene called Multi-Drug Resistance 1. MDR-1 is also called ABCB1, ATP-binding cassette sub-family B member 1.

- **P-glycoprotein (Pgp):** is a well-characterized ABC-transporter of the MDR/TAP subfamily protein.

Summary Points

- Multidrug resistance is a major obstacle for the successful treatment of human cancer.
- Over-expression of *MDR-1* gene and anti-apoptotic proteins are the major mechanisms for multidrug resistance in cancer cells.
- Gene silencing with nanoparticles encapsulated siRNA coould overcome multidrug resistance.
- The system includes siRNA targeted to *MDR-1*, MRP-1 and apoptosis mRNA.
- Next generation of nanoparticulate systems will be multifunctional nanoparticles capable of simultaneously achieving many goals.
- Multifunctional nanoparticles using siRNA, conventional chemotherapeutics, and active targeting hold great promise in the treatment of multidrug resistant tumors.

Abbreviations

RNAi	:	RNA interference
siRNA	:	Small interfering RNA
shRNA	:	Short hairpin RNA
MDR	:	Multidrug resistance
MDR-1	:	Multidrug resistant gene 1
Pgp	:	P-glycoprotein
MRP-1	:	Multidrug resistant associated protein 1

Acknowledgements

Our work in developing nanotechnology approaches to overcome tumor MDR is supported by the Alliance for Nanotechnology in Cancer at the National Cancer Institute, National Institutes of Health through grants R01-CA119617 and R01-CA119617S1 (ARRA Supplement).

References

Baker, M. 2010. RNA interference: From tools to therapies. Nature 464(7292): 1225.
Baker, M. 2010. RNA interference: Homing in on delivery. Nature 464(7292): 1225–8.
Chen, A.M., M. Zhang, D. Wei, D. Stueber, O. Taratula, T. Minko and H. He. 2009. Co-delivery of doxorubicin and Bcl-2 siRNA by mesoporous silica nanoparticles enhances the efficacy of chemotherapy in multidrug-resistant cancer cells. Small 5(23): 2673–7.

Chen, Y., S.R. Bathula, J. Li and L. Huang. 2010. Multifunctional nanoparticles delivering small interfering RNA and doxorubicin overcome drug resistance in cancer. J. Biol. Chem. 285(29): 22639–50.
Devalapally, H., Z. Duan, M.V. Seiden and M.M. Amiji. 2008. Modulation of drug resistance in ovarian adenocarcinoma by enhancing intracellular ceramide using tamoxifen-loaded biodegradable polymeric nanoparticles. Clin. Cancer Res. 14(10): 3193–203.
Dong, X., and R.J. Mumper. 2010. Nanomedicinal strategies to treat multidrug-resistant tumors: current progress. Nanomedicine (Lond) 5(4): 597–615.
Duan, Z., K.A. Brakora and M.V. Seiden. 2004. Inhibition of ABCB1 (MDR-1) and ABCB4 (MDR3) expression by small interfering RNA and reversal of paclitaxel resistance in human ovarian cancer cells. Mol. Cancer Ther. 3(7): 833–8.
Duan, Z., R. Foster, D.A. Bell, J. Mahoney, K. Wolak, A. Vaidya, C. Hampel, H. Lee and M.V. Seiden. 2006. Signal transducers and activators of transcription 3 pathway activation in drug-resistant ovarian cancer. Clin. Cancer Res. 12(17): 5055–63.
Jabr-Milane, L., L. van Vlerken, H. Devalapally, D. Shenoy, S. Komareddy, M. Bhavsar and M. Amiji. 2008. Multi-functional nanocarriers for targeted delivery of drugs and genes. J. Control Release 130(2): 121–8.
Jain, R.K. and T. Stylianopoulos. Delivering nanomedicine to solid tumors. Nat. Rev. Clin. Oncol. 2010 Nov; 7(11): 653–64.
Kaye, S.B. 2008. Reversal of drug resistance in ovarian cancer: where do we go from here? J. Clin. Oncol. 26(16): 2616–8.
Kommareddy, S., and M. Amiji. 2007. Poly(ethylene glycol)-modified thiolated gelatin nanoparticles for glutathione-responsive intracellular DNA delivery. Nanomedicine 3(1): 32–42.
Lage, H. 2006. MDR-1/P-glycoprotein (ABCB1) as target for RNA interference-mediated reversal of multidrug resistance. Curr. Drug Targets 7(7): 813–21.
Liu, G., D. Li, M.K. Pasumarthy, T.H. Kowalczyk, C.R. Gedeon, S.L. Hyatt, J.M. Payne, T.J. Miller, P. Brunovskis, T.L. Fink, O. Muhammad, R.C. Moen, R.W. Hanson and M.J. Cooper. 2003. Nanoparticles of compacted DNA transfect postmitotic cells. J. Biol. Chem. 278(35): 32578–86.
MacDiarmid, J.A., N.B. Amaro-Mugridge, J. Madrid-Weiss, I. S. Sedliarou, Wetzel, K. Kochar, V.N. Brahmbhatt, L. Phillips, S.T. Pattison, C. Petti, B. Stillman, R.M. Graham and H. Brahmbhatt. 2009. Sequential treatment of drug-resistant tumors with targeted minicells containing siRNA or a cytotoxic drug. Nat. Biotechnol. 27(7): 643–51.
Meng, H., M. Liong, T. Xia, Z. Li, Z. Ji, J.I. Zink and A.E. Nel. Engineered design of mesoporous silica nanoparticles to deliver doxorubicin and P-glycoprotein siRNA to overcome drug resistance in a cancer cell line. ACS Nano 4(8): 4539–50.
Nieth, C., A. Priebsch, A. Stege and H. Lage. 2003. Modulation of the classical multidrug resistance (MDR) phenotype by RNA interference (RNAi). FEBS Lett. 545(2-3): 144–50.
Patil, Y.B., S.K. Swaminathan, T. Sadhukha, L. Ma and J. Panyam. 2010. The use of nanoparticle-mediated targeted gene silencing and drug delivery to overcome tumor drug resistance. Biomaterials 31(2): 358–65.

Saad, M., O.B. Garbuzenko and T. Minko. 2008. Co-delivery of siRNA and an anticancer drug for treatment of multidrug-resistant cancer. Nanomedicine (Lond) 3(6): 761–76.

Schroeder, A., C.G. Levins, C. Cortez, R. Langer and D.G. Anderson. 2010. Lipid-based nanotherapeutics for siRNA delivery. J. Intern. Med. 267(1): 9–21.

Susa, M., A.K. Iyer, K. Ryu, F.J. Hornicek, H. Mankin, M.M. Amiji and Z. Duan. 2009. Doxorubicin loaded Polymeric Nanoparticulate Delivery System to overcome drug resistance in osteosarcoma. BMC Cancer 9: 399.

Susa, M., L. Milane, M.M. Amiji, F.J. Hornicek and Z. Duan. 2010. Nanoparticles: A Promising Modality in the Treatment of Sarcomas. Pharm. Res.

Susa, M., A.K. Iyer, K. Ryu, E. Choy, F.J. Hornicek, H. Mankin, L. Milane, M.M. Amiji and Z. Duan. 2010. Inhibition of ABCB1 (MDR-1) expression by an siRNA nanoparticulate delivery system to overcome drug resistance in osteosarcoma. PLoS One 5(5): e10764.

Szakacs, G., J.K. Paterson, J.A. Ludwig, C. Booth-Genthe and M.M. Gottesman. 2006. Targeting multidrug resistance in cancer. Nat. Rev. Drug Discov. 5(3): 219–34.

Teodori, E., S. Dei, C. Martelli, S. Scapecchi and F. Gualtieri 2006. The functions and structure of ABC transporters: implications for the design of new inhibitors of Pgp and MRP-1 to control multidrug resistance (MDR). Curr. Drug Targets 7(7): 893–909.

van, Vlerken L.E., Z. Duan, M.V. Seiden and M.M. Amiji. 2007. Modulation of intracellular ceramide using polymeric nanoparticles to overcome multidrug resistance in cancer. Cancer Res. 67(10): 4843–50.

Wu, H., W.N. Hait and J.M. Yang. 2003. Small interfering RNA-induced suppression of MDR-1 (P-glycoprotein) restores sensitivity to multidrug-resistant cancer cells. Cancer Res. 63(7): 1515–9.

Yadav, S., van L.E. Vlerken, S.R. Little and M.M. Amiji. 2009. Evaluations of combination MDR-1 gene silencing and paclitaxel administration in biodegradable polymeric nanoparticle formulations to overcome multidrug resistance in cancer cells. Cancer Chemother Pharmacol. 63(4): 711–22.

Yusuf, R.Z., Z. Duan, D.E. Lamendola, R.T. Penson and M.V. Seiden. 2003. Paclitaxel resistance: molecular mechanisms and pharmacologic manipulation. Curr. Cancer Drug Targets 3(1): 1–19.

16

Nano-flow Liquid Chromatography Mass Spectrometric Analysis of Glycans in Cancer

S. Hunter Walker[1,a] and *David C. Muddiman*[1,b,*]

ABSTRACT

Glycosylation is a post-translational modification frequently present in biological systems, regulating folding and function of proteins and cell-cell interactions, communications, and response. Because of the importance of glycosylation, aberrations in glycosylation patterns are potentially harmful, and it is these aberrations that are often studied in order to understand, diagnose, and treat diseases, including but not limited to cancer. However, there are no ubiquitous techniques or methodologies for the assay of glycans from tissues or biological fluids, and often, the sample preparation and analyses performed are unique to the individual lab, creating a barrier between researchers and limiting the reproducibility of the studies. Thus, the growth of

[1]W.M. Keck FT-ICR Mass Spectrometry Laboratory, Department of Chemistry, North Carolina State University, Raleigh, North Carolina 27695.

[a]E-mail: shwalke2@ncsu.edu

[b]E-mail: david_muddiman@ncsu.edu

*Corresponding author

List of abbreviations after the text.

glycomics as a whole is stunted because each researcher is often working on method development while profiling glycans and applying these methods to novel research. A well defined, high-throughput analysis for glycans is necessary for the progression of the field and to realize the potential for glycomics as a link to understanding the onset, progression, and treatment of cancer. Herein, the authors present various studies using nanoLC, nano-electrospray ionization, and mass spectrometric techniques for the analysis of glycans and application to studies in cancer related aberrant glycosylation. The techniques presented are often still being developed and have been included in order to provide a benchmark for the progression of the field to date and to show preliminary results using the assay of glycans as cancer biomarkers.

INTRODUCTION

This chapter discusses the relationship between the field of glycomics and cancer research and the benefits, progress, and obstacles in the analysis of glycans using nanoLC, nanospray MS toward the diagnosis, progression, and treatment of cancer. Numerous researchers have begun to make a comparative study of protein glycosylation patterns in healthy and cancerous patients, but the profiling of glycans in biological systems such as tissues or plasma has proven difficult. To date, there is no single methodology that has been developed and used as a paradigm for glycan analysis. This makes the field of glycomics extremely difficult when comparing data from one research lab to another, and without validation and reproduction of results by several laboratories, the forward progress of the field is significantly encumbered. Additionally, glycans are significantly more complex than nucleic acids and proteins, which now have well-known methodologies for the profiling of the genome and proteome, respectively. The complexity of glycans is due to different types of linkages (α and β) and the multiple branching possibilities at each monosaccharide unit, both of which are not trivial to discern and often involve laborious analyses for each individual glycan.

In order to fully profile the glycans in a given system, several properties must be determined including glycan composition (what monosaccharides and how many are present), the linkage of each monosaccharide (both glycosidic bond position and anomericity), the branching patterns (bi-, tri-antennary, bisecting, etc.), and the sequence of the monosaccharides. In order to accomplish this, often several different experiments are needed to determine the properties of even a single glycan including mass analysis, separation, endoglycosidase treatment, etc. Thus, in order to assay an entire glycome and possibly screen for changes in abundance of glycosylation in cancer studies, the development of a single high-throughput methodology is necessary in which specific glycans can be analyzed and quantified from a complex biological mixture. MS has become the instrument of choice in the analysis of glycans due to the fast analysis time, the ability for high resolution and mass accuracy, and the ability to acquire fragmentation spectra, which can often provide valuable information on sequence, linkage, and/or branching. Additionally, MS is compatible with online chromatography that can efficiently separate hundreds of compounds in complex mixtures and analytes can be analyzed in real-time as they are eluted from the column. While the MS methods are not perfect nor does MS currently provide all the information necessary for complete glycan analysis, certain MS/MS methods are able to elucidate some properties such as composition, linkage, branching, and/or sequence of the glycans, demonstrating the importance of MS and the need for continued MS research and development.

Herein, the relationship between protein glycosylation and cancer is discussed along with specific examples of aberrant glycosylation in cancer. Additionally, the movement to nano-flow LC and nanospray ionization and the advantages associated are presented. Several examples of the most recent glycan studies, both advances in glycan analysis and in cancer studies, employing these methods are described and the attributes of the techniques are highlighted. Though this is not a comprehensive overview of all the methods available, this chapter aims to sample the most recent and relevant techniques in glycan analysis and the profiling of aberrant protein glycosylation patterns of cancerous tissue samples or plasma.

ABERRANT GLYCOSYLATION AND CANCER

Glycosylation occurs throughout the secretory pathway in the endoplasmic reticulum and golgi apparatus and are grouped in two major classes, *N*- and *O*-linked. *N*-linked glycans are synthesized in a semi-template driven process where a core glycan is synthesized and transferred to the protein *via* a highly regulated process in the endoplasmic reticulum. However, once the precursor glycan is attached to the protein, trimming and addition of monosaccharide units occur in a non-regulated manner that is determined by substrate levels and the presence of specific glycosidases and glycotransferases (Taylor and Drickamer 2006). *O*-linked glycan synthesis differs from that of *N*-linked glycans in that there is no core glycan structure. Monosaccharides are added to proteins in the Golgi, much like the *N*-linked glycan elongation. The terminal elongation of

N-linked glycans and *O*-linked glycan synthesis both occur in the golgi apparatus and this is the origin of the diversity of both classes of glycans.

The glycan synthesis mechanism allows for a range of glycan structures, and at a single glycosylation site, numerous related glycan structures may be present. Thus, a copy of the same protein located in a different part of the cell or in a different environment may have significantly different glycosylation leading to different protein folding, function, and interaction. The elongation of glycans in the golgi apparatus provides the diversity of glycans and involves a specific glycosyltransferase enzyme for each linkage and monosaccharide unit. Thus, the final structure of the glycans attached to proteins is a function not only of the protein, other oligosaccharides, and substrate levels, but also the order and presence of the glycosyltransferase enzymes in the golgi apparatus of the specific cell. The presence or absence of glycosyltransferases is widely hypothesized to be a factor in the mechanism of aberrant glycosylation and cancer metastasis.

An example of aberrant glycosylation involves the terminal sugar residues on glycans which can serve as lectin binding sites. The glycans occupying these sites control cell-cell interactions including adhesion and communication, and the alteration of these residues has been implicated in the mechanism of cancer metastasis (Gorelik et al. 2001; Hakomori 1985). Sialyl Lewis X is a well known aberrant glycosylation pattern in which glycans in tumorous tissues have an over abundance of terminal sialic acid groups in comparison to healthy tissues (Mitsuoka et al. 1998). Another example of aberrant glycosylation in cancerous tissue is the increased abundance of β1-6 linkages in the profile of membrane *N*-linked glycans (Fernandes et al. 1991). This was later revised when it was shown that in mouse fibroblast cells, the β1-6 linkages are not up-regulated until they have become malignant, and it was further shown that the enhanced expression of β1-6 linkages correlates directly with an increase in malignancy, showing an overabundance of one of the six β-*N*-acetylglucosaminyltransferases (GnT-V) (Asada et al. 1997; Seelentag et al. 1998). These transferases have also been shown to increase the amount of branching in complex *N*-linked glycans in tumor cells (Kobata and Amano 2005). These examples have shown that glycosylation patterns in cancerous samples are often significantly different than those from healthy samples due to a number of possible reasons including the presence/absence of specific glycotransferase enzymes or the cell environment. Thus, changes in glycosylation patterns are important for the treatment and diagnosis of cancer but must be further studied in order to understand the mechanism of aberrant glycosylation and exploit glycans for the diagnosis or treatment of cancer.

MASS SPECTROMETRY AND NANOSPRAY IONIZATION

There is an abundance of different mass spectrometers made up of different ionization sources and mass analyzers that are capable of studying glycans. In biological studies where the analytes are typically large (> 1 kDa), the two ionization sources most often employed are ESI and MALDI. MALDI is an enticing choice for the ionization of glycans due to the short analysis times, the high throughput of samples, and the tolerance for contamination such as salts. However, ESI is often chosen in glycan analysis for 2 reasons: 1) MALDI imparts more internal energy into the molecules during ionization than ESI (known to cause in-source fragmentation of the glycosidic bonds primarily with sialic acid residues (Bereman et al. 2009a)), and 2) ESI can be directly coupled to liquid chromatography for the online fractionation of glycan samples just prior to injection into the MS. The primary reason that ESI is not ubiquitous for glycan analysis is the fact that ESI creates an inherent bias for the ionization of hydrophobic molecules (Fenn 1993). This is an extreme hurdle for glycan analysis due to the hydrophilic and polar nature of the sugar residues, but recent movements toward nano-flow ESI have significantly increased glycan detection and sensitivity.

Nanospray Fundamentals and Background. The invention of nano-flow electrospray ionization (Wilm and Mann 1994), has significantly enhanced the analysis of glycans in MS and allows significant increases in the ion abundances of glycans in MS. In nanospray, lower flow rates generate smaller droplets with a larger surface to volume ratio (Wilm et al. 1994), and this facilitates ionization of the more solvated, hydrophilic glycans, by increasing the surface activity of glycans. This increase in ion abundance can be rationalized by examining the mechanism of ESI. While still debated, generally it has been accepted that in ESI, the analytes are dissolved in small droplets that have a surplus of charge (H^+ in positive mode) on the surface due to the electrochemical reaction in the capillary floating at ~2 kV and are ejected from the electrospray emitter tip in the form of a "Taylor cone" (Taylor 1964). These droplets undergo a series of desolvation and fission processes and eventually produce individual gas-phase ions. The hydrophobic bias is introduced in the series of Coulombic fission events. As the droplets are being desolvated, they reach the Rayleigh limit, and when this occurs, several droplets (Kebarle and Tang 1993)) are ejected from the surface of the droplet (Duft et al. 2003). When these smaller progeny droplets are ejected, they are composed of solvent, analyte, and an abundance of charge from the surface of the parent droplet. Because the more hydrophobic molecules will be located closer to the surface of the parent droplet than the hydrophilic analytes, the hydrophobic molecules are significantly enriched in the progeny droplets.

Also, the hydrophilic molecules remain in the larger parent droplet that has a decreased charge to volume ratio, further decreasing the chance for the hydrophilic molecules to be ionized. Thus, when moving from micro-flow ESI to nano-flow ESI, the smaller droplet sizes allow the hydrophilic glycans to have a greater surface activity, allowing a higher probability for glycans to be ejected within the progeny droplets and subsequently ejected as gas-phase ions.

Nanospray and Glycan Analysis. Nanospray ionization was first developed and compared to ESI, both theoretically and experimentally, by Wilm and Mann (Wilm et al. 1994) and was shown to increase ionization and transmission efficiency into the MS. The authors showed that flow rate is directly proportional to the droplet size, and smaller droplet sizes allow for initial droplet size to analyte ratios an order of magnitude smaller than that of conventional ESI. These results have spurred many subsequent studies involving comparisons between the two and the increase in ion abundance, sensitivity, and decreased limits of detection in nanospray. Bahr et al. first demonstrated the advantages of using nanospray when analyzing oligosaccharides (Bahr et al. 1997). The authors show decreased detection limits and increased sensitivities when using ~30 nL/min flow rates compared to conventional 1 μL/min (Fig. 1). In the nanospray experiments, the intensities of the oligosaccharides were shown to be orders of magnitude larger than in the forced flow experiments and are able to be detected with a large signal-to-noise ratio at an order of magnitude lower concentration than the limit of detection of a 1 μL/min flow rate ESI. Additionally, proteins are more hydrophobic than glycans, and in mixtures of peptides and oligosaccharides, the peptides often suppress the signal of glycan. However, Bahr and coworkers show that in a mixture of peptides and oligosaccharides, nanospray reduces the suppression of oligosaccharides and achieves comparable sensitivities and intensities in an equimolar mixture. Another example by Karlsson and coworkers (Karlsson et al. 2004) describes the miniaturization of an entire nanoLC (reverse phase graphitized carbon capillary column) nanospray system and compares the results of negative ion glycan analysis when decreasing the flow rate an order of magnitude from 6 μL/min. to 600 nL/min. The authors report a 10-fold increase in sensitivity at the reduced flow rate and detection limits in the low femtomole range (Fig. 2). Additionally, it was shown that the absolute intensity of the nanospray ions in the MS was more than 100 times that of the higher flow rate ESI, allowing for high-quality fragmentation MS/MS spectra to be acquired.

Direct Infusion Nanospray Glycan Analysis. Several studies have shown the usefulness of direct infusion nanospray analysis of glycans without the necessity of online LC. Harvey and coworkers have recently demonstrated

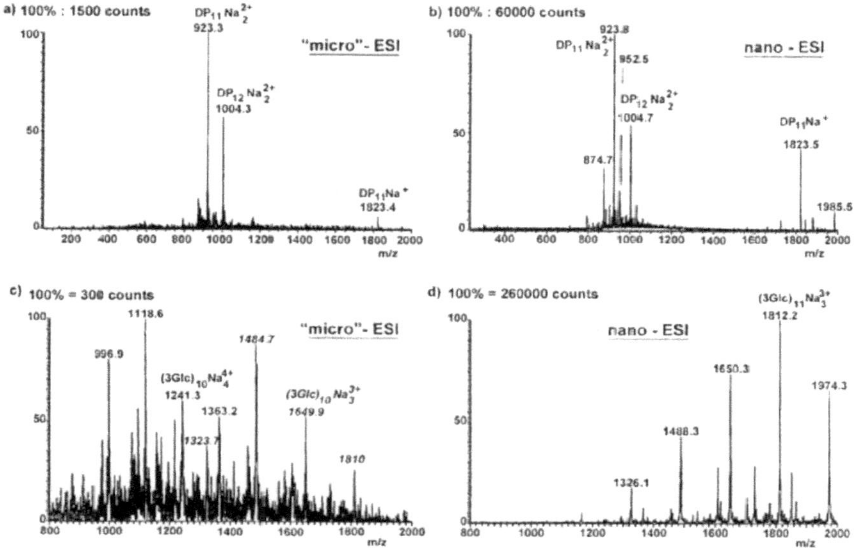

Fig. 1. Comparison between conventional ESI and nanospray. Spectra (a) and (c) are conventional ESI for dextran DP 11 and pullulan, respectfully, and spectra (b) and (d) are nanospray ionization for dextran DP 11 and pulluan, respectively. These spectra show the increased absolute abundance and signal to noise rations of oligosaccharides when moving toward nanospray ionization. Reprinted with permission from the publisher (Bahr et al. 1997).

the abundance of information that can be gained from performing direct infusion nanospray ionization *via* anion adduction and CID fragmentation of *N*-linked glycans in the negative ion mode (Harvey et al. 2008). The authors have shown that this method can elucidate glycan branching patterns, fucose location, and whether or not a bisecting GlcNAc is present, all of which are not possible by traditional positive ion mode analysis. Additionally, this study shows that different isomers from the same composition glycan can be determined using negative ion nanospray and fragmentation. Using this method produces different fragmentation masses due to the different linkages and positions of the monosaccharide units. An alternative study from Zaia and coworkers (Bowman and Zaia 2010) has also shown different fragmentation patterns from glycan isomers in negative ion nanospray. This additional information from fragmentation spectra is crucial for distinguishing the branching and sequence patterns of different isomers of glycans with the same chemical composition.

Recently, Reinhold and coworkers (Stumpo and Reinhold 2010) have analyzed the *N*-linked glycome of human plasma using only direct infusion nanospray ionization MS. The authors demonstrate that direct

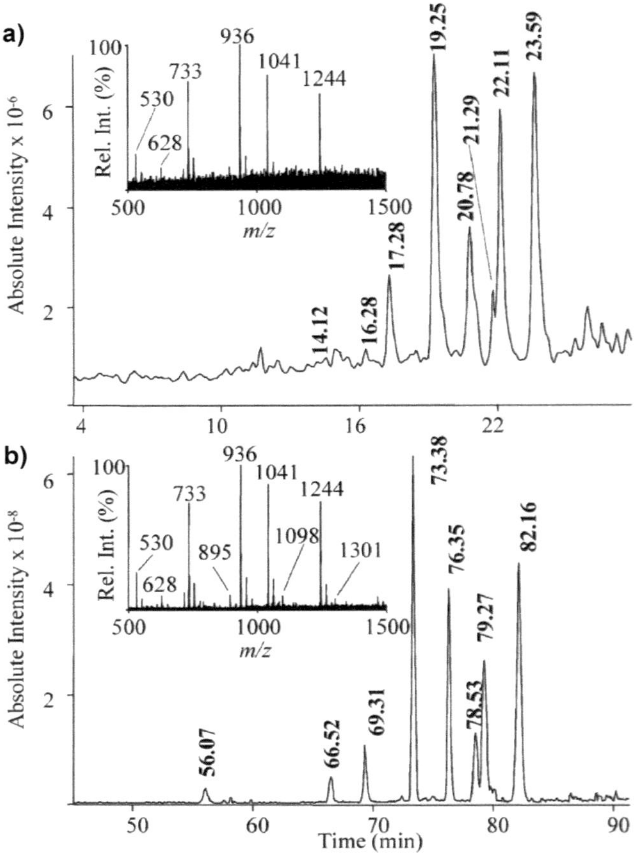

Fig. 2. Micro- vs. nano-flow LC ESI. The advantage of miniaturization from 6 μL/min (a) to 600 nL/min (b) flow rates in LC ESI IT MS. The signal to noise ratios are significantly increased and the absolute intensity of the chromatogram is increased by 100-fold in the lower flow LC ESI MS. Reprinted with permission from the publisher (Karlsson et al. 2004).

infusion nanospray of complex glycan mixtures can be achieved in a rapid analysis time. Additionally this work demonstrates the effectiveness of IgG protein depletion on the detection of lower abundance glycans and show a dramatic increase in the number of glycans detected. Using a combination of high- and low-resolution MS, the authors were able to observe 106 methylated *N*-linked glycans from human plasma after IgG depletion compared to only 53 before depletion (Fig. 3). The ability to detect >100 glycans without any prior fractionation is an impressive feat that has not been performed to date. Furthermore, the authors demonstrate that

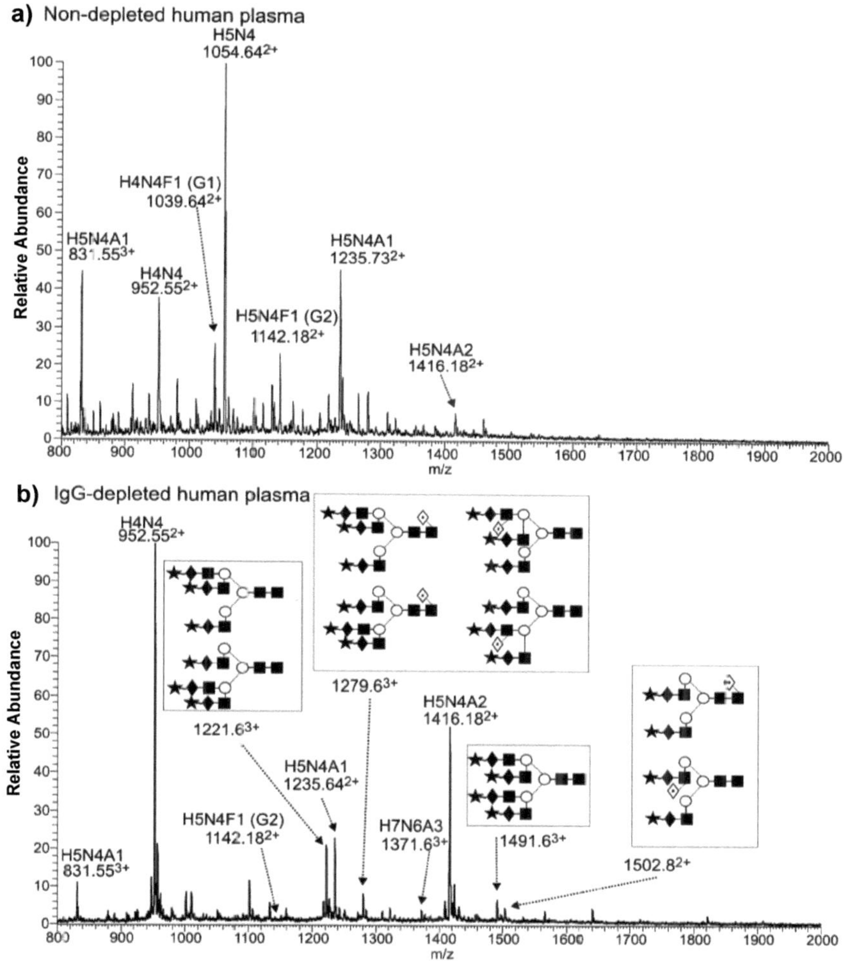

Fig. 3. Direct infusion nanospray entire *N*-glycome analysis. Nearly a 2-fold increase in the number of *N*-linked glycans was reported after the most abundant proteins were depleted. This method boasts rapid analysis times compared to methods with lengthy fractionation. Reprinted with permission from the publisher (Stumpo et al. 2010).

detailed characterization of the glycans can be achieved, including isomer and branching patterns from CID MS^3. These results were obtained on a healthy plasma sample and while this study does not have any implications toward cancer, these methods are capable of profiling glycans in a short amount of time, which is amenable to high-throughput glycan analysis necessary disease studies.

NANOLC MS ANALYSIS OF GLYCANS

An additional advantage for the analysis of glycans by nanospray MS is the capability of online liquid chromatography, and today, chromatography pumps are capable of pumping at or less than a few hundred nL/min reproducibly. This additional separation is often beneficial due to the multiple charging of glycans, which increases the clustering of glycans at low *m/z* and can cause overlapping MS distributions. Many different types of liquid chromatography have been used in the separation of glycans (Kirsch and Bindila 2009; Wuhrer et al. 2005) including both NP and RP, on- and off-line, and capillary and chip-based systems. RP chromatography is not practical for native glycan analysis, however, studies have shown the effectiveness in separating derivatized glycans using RP (Alley et al. 2010). More frequently used are HILIC and graphitized carbon separation techniques which are capable of efficiently separating native and derivatized oligosaccharides. A recent study compared these two nano-flow LC techniques using Amide-80 stationary phase (HILIC) and graphitized carbon stationary phase (Bereman et al. 2009a). The authors reported excellent performances that were comparable for both separation techniques. However, heavily sialylated glycans were permanently retained on the graphitized carbon stationary phase (a problem also reported in another study (Pabst and Altmann 2008)), and the life of the graphitized carbon column was much shorter than that of the HILIC column. Additionally, the development and miniaturization of HILIC stationary phases have exceeded that of graphitized carbon stationary phases (Ruhaak et al. 2009).

NanoLC equipment frequently uses capillaries for the trapping and analytical columns packed with the stationary phase of choice for a specific sample type. This setup is directly coupled to a nanospray source for the ionization and introduction into the MS. A more recent development in nanoLC is the use of microfluidic chip chromatography (Bindila and Peter-Katalinic 2009; Kirsch and Bindila 2009). Many researchers have begun to move toward chip-based systems because often the trap column, analytical column, and nanospray emitter can all be engineered onto one single chip, minimizing the number of connections and dead volume and increasing the reproducibility of the nanoLC-nanospray system from run to run and study to study (Kirsch and Bindila 2009). Though this technology is relatively new (< 5 years old), the chip system has been reported to be capable of separating most types of glycan species and is compatible with the three main stationary phases, RP, HILIC, and graphitized carbon.

Reverse Phase Chromatography. RP chromatography of glycans has been limited due to the polar nature of native glycans. However, increases in the hydrophobicity of glycans by derivatization such as reductive amination and permethylation have allowed for reverse phase separation of glycans. Novotny and coworkers (Alley et al. 2010) developed a strategy to profile the *N*-linked glycans from serum in order to profile the glycosylation in stage IV breast cancer proteins circulating in the blood. The authors cleave the glycans in solution using PNGase F, permethylate, and analyze the glycans by using a chip-based nano-reverse phase LC MS. Permethylation also increases the hydrophobicity of the glycans which increases the electrospray response. The separation of the *N*-linked glycans was performed on an Agilent ChipCube LC IT MS interface with a flow rate of 250 nL/min for separation and nanospray ionization.

The authors analyzed 15 healthy/control and 15 stage IV breast cancer samples of *N*-linked glycans cleaved from the glycoproteins circulating in human blood serum. The glycans were first grouped by glycan type (high mannose, total fucose, fucose only (no sialic acid), total sialic acid, and sialic acid only (no fucose)) in order to determine whether the class of the glycan is useful in predicting the health of the patient. Three of the different classes of glycans were shown to have statistical differences between late stage breast cancer and healthy glycan patterns. Total fucosylated glycans were significantly up-regulated in breast cancer sera, and glycans containing sialic acid residues were found to be statistically lower in abundance in late stage breast cancer. Also, each of the glycans detected was directly compared between healthy and diseased sera, and 5 glycans were determined to have statistically different ion abundances in the breast cancer samples than in the healthy control studies. Two glycans with the most dramatic changes in abundance between the cancerous and healthy samples were further studied, and box plots confirmed the statistical difference between the two states (Fig. 4). ROC curves were generated in order to determine the predictability of these glycans as disease markers. Large AUC values were found for the two glycans implying that these glycans are possible candidates for indicators of late stage breast cancer.

Graphitized Carbon Chromatography. Graphitized carbon has been a successful tool for the separation of native glycans. However, there has been little development in the stationary phase, such as bonded phases and the miniaturization of the stationary phases as in HILIC. Nonetheless, graphitized carbon separation of glycans is still a technique that is frequently used and can be an effective online fractionation method for LC. Barroso and coworkers developed a nanoLC Q-TOF MS analysis for profiling native oligosaccharides and reported low femtomole detection limits of separated *O*- and *N*-linked glycans (Barroso et al. 2002). This

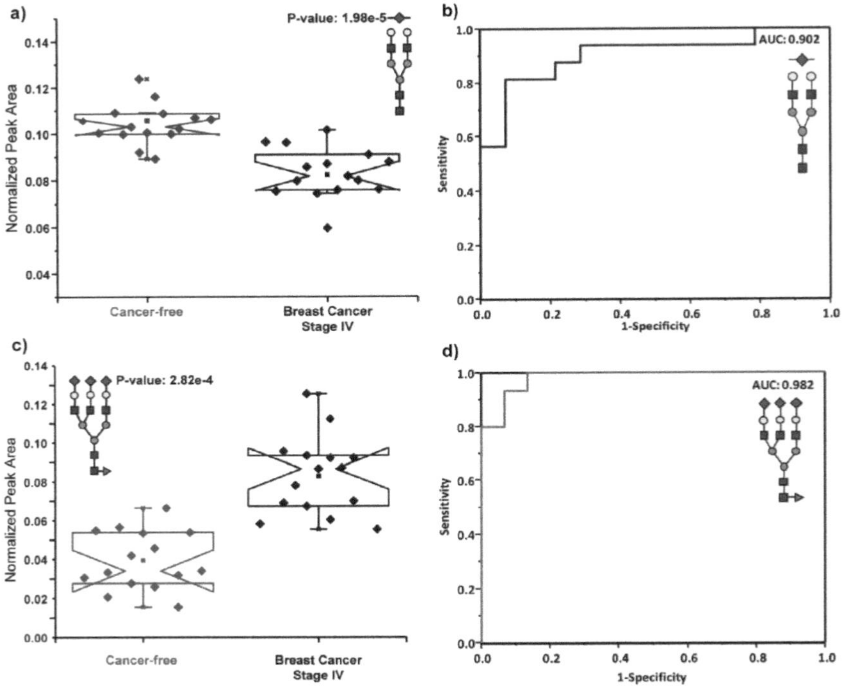

Fig. 4. Significant glycosylation changes in stage IV breast cancer. Two glycans found to have significantly different abundances in stage IV breast cancer vs. healthy samples. The box plots showing the statistical difference between the glycans in the two samples are displayed in (a) and (c), and the corresponding ROC plots are presented in (b) and (d). AUC values are >0.9 indicating that these glycans have a strong predictive value for stage IV breast cancer. Reprinted with permission from the publisher (Alley et al. 2010).

Color image of this figure appears in the color plate section at the end of the book.

study also shows the advantages of miniaturizing the LC and ESI source when performing analyses on glycans and demonstrates the analysis of glycans cleaved from a complex mixture of proteins.

Lebrilla and coworkers (Chu et al. 2009) have developed a chip-based graphitized carbon nanoLC TOF MS method that is capable of analyzing native, underivatized *N*-linked glycans. When profiling native glycans, the acidic glycans containing sialic acid residues are often suppressed, thus, the authors found LC to be necessary. This method has reduced the number of sample preparation steps by eliminating any chemical modification of the samples, saving time and reducing the variability of the sample. It was shown that the retention times are reproducible between samples and technical replicates and show a relative standard deviation of <1%.

Additionally, the authors reduced the glycans to their respective alditols in order to decrease the separation between α and β forms of the sugar, and it was found that the alditol samples have a better LC resolution. This study provides an alternative method to capillary nanoLC that is reproducible and commercially available. This method can be applied to the profiling of cancer samples and is a high throughput glycan analysis technique.

HILIC. Zaia and coworkers have developed a novel *N*-linked glycan derivatization method where stable isotopes are incorporated into the reductive amination reagents in order to perform relative quantification of glycans from different samples (Bowman et al. 2010). This method is the first to use tetraplex stable isotope reagents where four different glycan samples can each be tagged with the same reagent, but with different numbers of stable isotopes (0, 4, 8, 12) and quantified in the same mass spectrum (Fig. 5). Fractionation of the sample is crucial to this experiment so that the isotopic envelopes of different glycans do not overlap causing error in quantification. Thus, HILIC online separation was performed. This technique is extremely important to cancer research as the glycans can be quantified side-by-side for heathy and diseased samples. Because the samples are combined in an equimolar fashion into one sample after derivatization, the succeeding sample preparation steps are performed identically, reducing the amount of sample variability. Additionally, the novelty of incorporating tetraplex reagents will allow for time-course experiments where the glycan profile at 4 different time points can be compared in a single sample throughout the progression of cancer.

Muddiman and coworkers (Bereman et al. 2009a) demonstrate the use of capillary nanoLC (HILIC) MS in order to profile changes in glycosylation between sera from healthy, benign tumor, and epithelial ovarian cancer (OVC) patients. This preliminary study compares glycan spectra using MALDI and nanoLC/nanospray ionization and, as mentioned above, also compares glycan separation by HILIC (amide-80) and graphitized carbon. Due to in-source or post-source fragmentation of sialic acid containing glycans, ESI was chosen for glycan ionization and HILIC was found to be a superior separation mechanism (*vide infra*) (Bereman et al. 2009a). The authors use nanoLC and nanospray MS for glycan analysis and demonstrate the reproducibility in the LC retention times for 20 technical replicates over a 5 day experiment. This method was applied to the analysis of glycans in healthy, benign tumor, and OVC plasma samples. ROC curves were generated for two glycans (both biantennary and fucosylated) and it was found that these glycans were able to give good predictive value for OVC vs. healthy, but when comparing benign tumor sera to healthy or OVC, the glycans gave ambiguous predictions. This study was followed by the development of a high throughput procedure for the cleavage and nanoLC

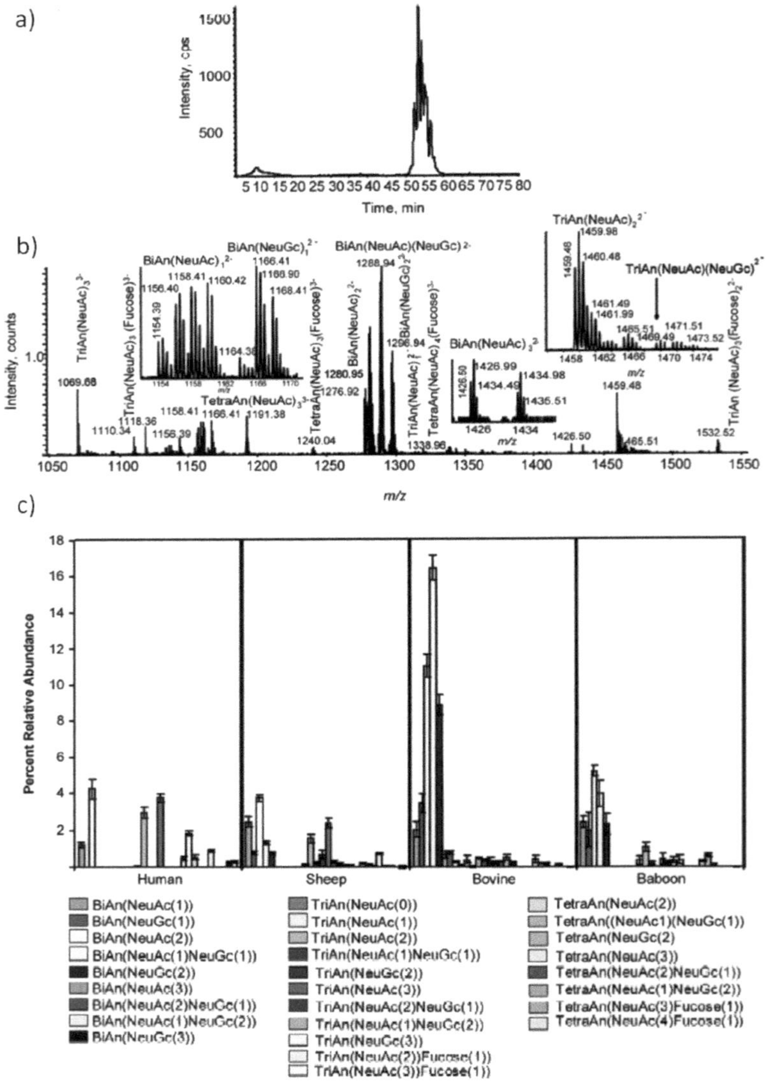

Fig. 5. Tetraplex stable isotope relative quantification strategy. (a) the base peak glycan chromatogram, (b) the MS showing the tetraplex isotopic envelopes for each glycan, and (c) the quantification of plasma *N*-linked glycans for 4 different species. This technique is relevant to cancer research as the relative quantification of glycans between several time-points in cancer can lead to the discovery of glycan biomarkers for the onset of cancer or possibly track the progression and metastasis of cancer. Reprinted with permission from the publisher (Bowman et al. 2010).

Color image of this figure appears in the color plate section at the end of the book.

MS analysis of *N*-linked glycans from plasma glycoproteins (Bereman et al. 2009b). Additionally, recent studies from this group have developed hydrazide tagging reagents in order to increase the hydrophobicity of the glycans (Bereman et al. 2010; Walker et al. 2010), which will increase the detection of glycans in nanospray MS, and in the future are capable of incorporating stable isotopes for relative quantification of glycans.

An additional aspect of the previous study is modeling the onset of OVC using a novel domestic hen model. Hens develop OVC spontaneously throughout years 2–4 of age and it has been shown that hen *N*-linked protein glycosylation in sera is nearly identical to that of human protein glycosylation (Dixon et al. 2010). Muddiman and coworkers have developed a repository of chicken OVC plasma and tissue at several different time points during the onset and development of OVC. The authors will be able to analyze the glycan patterns over longitudinal sampling and hypothesize that changes in glycosylation patterns will be indicative of the onset of OVC. A preliminary study has shown the importance of longitudinal sampling of glycans (Fig. 6) (Dixon et al. 2010). This study is an example of the combination of nanoLC MS technology with a novel animal model for the elucidation of aberrant glycosylation patterns with respect to OVC.

Key Facts of Glycosylation and Cancer

- Glycosylation is a post-translational modification of proteins and has been estimated to occur in more than 50% of all gene products (Apweiler et al. 1999). Additionally, the importance of glycosylation in biological processes is evident: of the entire translated genome, 0.5–1% of all gene products are involved in glycosylation (Begley 2009).
- Glycans take part in regulating numerous biological processes including cell-cell interactions, cellular recognition, adhesion, cell division, immune responses, and protein stability. The importance and ubiquity of glycosylation in biology also allows for frequent and potentially harmful aberrations.
- Aberrant glycosylation has been reported in a number of different diseases including many different types of cancer. It is hypothesized by many researchers that these changes in glycosylation can be quantified and used as biomarkers for the detection of the onset of cancer or metastasis.
- There is currently no ubiquitous method for the analysis of glycans, as each method has its disadvantages. However, mass spectrometry provides chemical information from which glycan composition and sequence information can often be deduced.

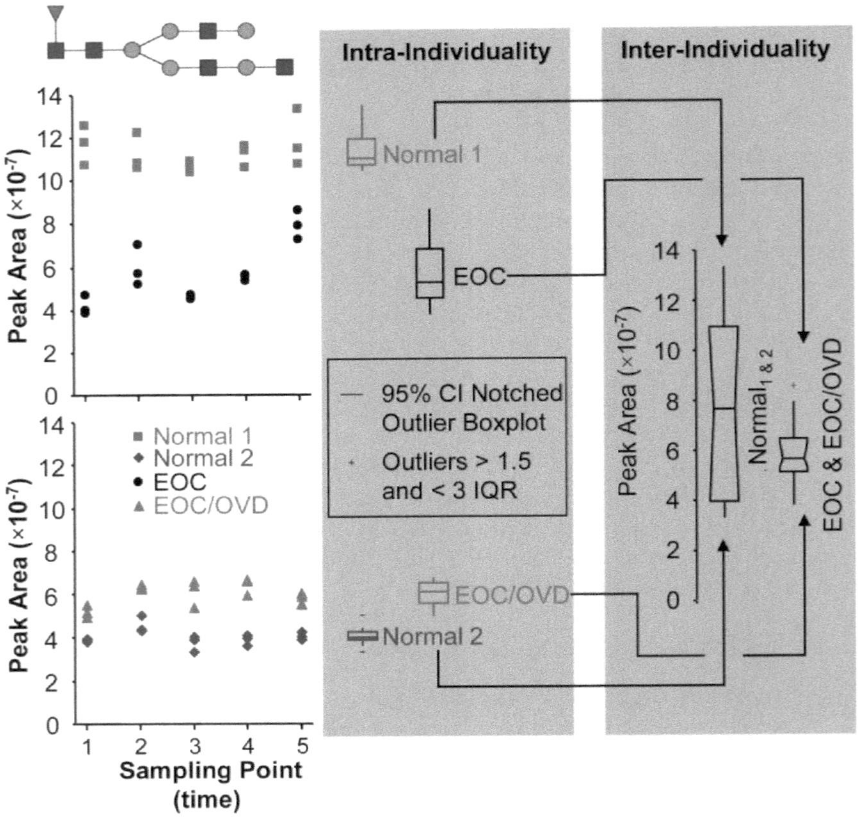

Fig. 6. Longitudinal glycosylation study of OVC in hens. This shows that biomarker discovery can be complicated as the levels of this glycan are shown to vary between the two normal chickens. Though there are distinct changes in the glycosylation between the OVC and healthy chickens, neither can be deemed statistically different due to the large disparity in the glycan abundance of the normal chickens. Thus, this study demonstrates the importance of longitudinal sampling and large sample sets of data in order to quantify the intra-individual and inter-individual variability of glycans in cancer. Reprinted with permission from the publisher (Dixon et al. 2010).

Color image of this figure appears in the color plate section at the end of the book.

- Nanospray ionization has significantly enhanced the abundances of glycans in MS. Reports show that glycans are ionized with an efficiency of >10-fold when compared to the efficiencies of traditional ESI.

Summary Points

- Many instances of aberrant glycosylation have been reported in cancer patients and often glycans change in abundance as the disease progresses, such as malignancy. It has been hypothesized that monitoring glycosylation of proteins in cancer tissue or blood can lead to biomarkers that can predict the occurrence of cancer. Additionally, it has been shown that glycans regulate many different cellular processes including communication, adhesion, recognition, and cell division. Thus, an aberration in glycosylation can affect any of these processes and be detrimental. Additionally, glycans are involved in the folding and structure of proteins. This can change the function of the proteins and affect the cell in a harmful manner.
- Glycans are often difficult to detect when using conventional ESI MS due to the hydrophobic bias of ESI and hydrophilicity of the glycans. However, miniaturization of the ESI source has significantly enhanced the detection of glycans. Often researchers use ESI because it is capable of online separation just prior to the MS. Thus, nanospray has allowed researchers to pursue this methodology. Today, nanoLC nanospray MS instruments are frequently used for glycan analyses.
- Nanospray sources were developed in order to facilitate ionization of analytes prior to introduction into the MS. Because nanospray creates smaller droplets from a smaller flow rate, the droplets are easier to evaporate. Also, hydrophilic analytes, such as glycans, are easier to ionize in nanospray. In conventional ESI, the glycans are deeply solvated in the center of the large droplets, whereas in nanospray, they are closer to the surface. The glycans are then more likely to become gas-phase ions and be detected.
- Nanospray is amenable to online LC separation just prior to ionization and MS analysis. This allows glycans extracted from a complex mixture such as blood and tissues to be analyzed in a simpler mixture. This makes the analysis much less complex. There are several different methods for separating glycans *via* LC including RP, graphitized carbon, and HILIC. Each of these can be used depending on what sample preparation has been performed.
- Much method development has been performed for the analysis of glycans by nanoLC MS, and recently, reports of application to cancer samples have been reported. Researchers are beginning to demonstrate glycan comparisons between healthy and cancerous samples and have found that certain glycans or classes of glycans are either up- or down-regulated. However, most of these studies are in the early stages and have shown only preliminary studies. The complexity of glycans has made the analysis of glycans more difficult than nucleic

acids and peptides, and the impact of the ubiquity of glycans in biology remains to be seen. Nevertheless, the assay of glycans for cancer research is gaining interest, and much effort will be needed to further develop analytical techniques in order to fully elucidate the correlation between aberrant glycosylation and cancer.

Applications to Areas of Health and Disease

The study of protein glycosylation has correlated aberrations in glycosylation patterns with numerous diseases, including cancer. Robbins and coworkers first described the disparity in the size of membrane glycoproteins between healthy and diseased fibroblasts (Wu et al. 1969), and aberrant glycosylation patterns were first linked to cancer by Penaloza and coworkers (Rostenberg et al. 1978), where it was shown that the glycosylation of α_1-antitrypsin is altered in lung, prostate, and gastrointestinal cancers. Recently, the number of studies providing evidence for the correlation of aberrant glycosylation and cancer has been increasing (Casey et al. 2003; Gorelik et al. 2001, Zhao et al. 2008), and it is hypothesized that glycans can fulfill the role of being sensitive and specific biomarkers for targeted screening (Packer et al. 2008). Additionally, recent studies have shown aberrant glycosylation to be present in numerous different types of cancers (Fuster and Esko 2005), including breast, prostate, liver, ovarian, pancreatic, etc. These changes in glycosylation patterns can be taken advantage of when probing for biological screening markers or possible treatment targets for various types of cancer.

Definitions

mass spectrometry: The separation and detection of different compounds, or analytes, based on the masses of each compound.

liquid chromatography: A method for separating compounds from each other in order to make a sample less complex. Different interactions between the analytes and the stationary phase and the mobile phase allow for compounds to be retained in the chromatography column for different periods of time, allowing for separation of the compounds.

electrospray ionization: Electrospray ionization is the process where analytes in liquid samples are desolvated and ionized in order to produce gas-phase ions. Analytes must be gas-phase ions in order to be introduced into and detected by MS.

permethylation: A glycan derivatization technique where all hydroxyl groups are converted to methyl esters. This technique increases the hydrophobicity and electrospray response of glycans, increasing the detection of glycans in ESI MS. Also, more informative fragmentation spectra can be acquired when permethylating glycans, in which branching patterns can be determined.

Peptide N-glycosidase F: An enzyme often used for the cleavage of N-linked glycans from proteins. This enzyme leaves an open reducing terminus which can be reduced to an alditol or derivatized using reductive amination or hydrazone formation.

HILIC: A form of normal phase LC which is able to retain polar analytes and the strong eluent is water. Several different types of stationary phases are used including bare silica and bonded stationary phases, such as amide-bonded.

Abbreviations

MS	:	mass spectrometer or mass spectrometry
LC	:	liquid chromatography
ESI	:	electrospray ionization
RP	:	reverse phase
NP	:	normal phase
HILIC	:	hydrophilic interaction liquid chromatography
MALDI	:	matrix assisted laser desorption/ionization
TOF	:	time of flight mass analyzer
Q-TOF	:	quadrupole time of flight mass analyzer
IT	:	ion trap mass analyzer
LTQ	:	linear ion trap mass analyzer
FTICR	:	Fourier transform ion cyclotron resonance mass analyzer
CID	:	Collision Induce Dissociation
MS/MS	:	tandem mass spectrometry
MS^n	:	mass spectrometry with 'n' successive mass detection events
OVC	:	ovarian cancer
IgG	:	Immunoglobulin G Protein
ROC	:	receiver operating characteristic
AUC	:	area under the curve
nanoLC	:	nL/min flowrates in LC

References

Alley, W.R., M. Madera, Y. Mechref and M.V. Novotny. 2010. Chip-based Reversed-phase Liquid Chromatography-Mass Spectrometry of Permethylated N-Linked Glycans: A Potential Methodology for Cancer-biomarker Discovery. Anal. Chem. 82: 5095–5106.

Apweiler, R., H. Hermjakob and N. Sharon. 1999. On the frequency of protein glycosylation, as deduced from analysis of the SWISS-PROT database. BBA-Gen. Subjects 1473: 4–8.

Asada, M., K. Furukawa, K. Segawa, T. Endo and A. Kobata. 1997. Increased expression of highly branched N-glycans at cell surface is correlated with the malignant phenotypes of mouse tumor cells. Cancer Res. 57: 1073–1080.

Bahr, U., A. Pfenninger, M. Karas and B. Stahl. 1997. High sensitivity analysis of neutral underivatized oligosaccharides by nanoelectrospray mass spectrometry. Anal. Chem. 69: 4530–4535.

Barroso, B., R. Dijkstra, M. Geerts, F. Lagerwerf, P. van Veelen and A. de Ru. 2002. On-line high-performance liquid chromatography/mass spectrometric characterization of native oligosaccharides from glycoproteins. Rapid Commun. Mass Spectrom. 16: 1320–1329.

Begley, T.P. 2009. Wiley Encyclopedia of Chemical Biology. 2: 785.

Bereman, M.S., D.L. Comins and D.C. Muddiman. 2010. Increasing the hydrophobicity and electrospray response of glycans through derivatization with novel cationic hydrazides. Chem. Comm. 46: 237–239.

Bereman, M.S., T.I. Williams and D.C. Muddiman. 2009a. Development of a nanoLC LTQ Orbitrap Mass Spectrometric Method for Profiling Glycans Derived from Plasma from Healthy, Benign Tumor Control, and Epithelial Ovarian Cancer Patients. Anal. Chem. 81: 1130–1136.

Bereman, M.S., D.D. Young, A. Deiters and D.C. Muddiman. 2009b. Development of a Robust and High Throughput Method for Profiling N-Linked Glycans Derived from Plasma Glycoproteins by NanoLC-FTICR Mass Spectrometry. J. Prot. Res. 8: 3764–3770.

Bindila, L., and J. Peter-Katalinic. 2009. Chip-Mass Spectrometry for Glycomic Studies. Mass Spectrom. Rev. 28: 223–253.

Bowman, M.J., and J. Zaia. 2010. Comparative Glycomics Using a Tetraplex Stable-Isotope Coded Tag. Anal. Chem. 82: 3023–3031.

Casey, R.C., T.R. Oegema, K.M. Skubitz, S.E. Pambuccian, S.M. Grindle and A.P.N. Skubitz. 2003. Cell membrane glycosylation mediates the adhesion, migration, and invasion of ovarian carcinoma cells. Clin. Exp. Metastasis 20: 143–152.

Chu, C.S., M.R. Ninonuevo, B.H. Clowers, P.D. Perkins, H.J. An, H.F. Yin, K. Killeen, S. Miyamoto, R. Grimm and C.B. Lebrilla. 2009. Profile of native N-linked glycan structures from human serum using high performance liquid chromatography on a microfluidic chip and time-of-flight mass spectrometry. Proteomics 9: 1939–1951.

Dixon, R.B., M.S. Bereman, J.N. Petitte, A.M. Hawkridge and D.C. Muddiman. 2010. One-year plasma *N*-linked glycome intra-individual and inter-individual variability in the chicken model of spontaneous ovarian adenocarcinoma. Int. J. Mass Spectrom. pub. on web: doi:10.1016/j.ijms.2010.05.023:

Duft, D., T. Achtzehn, R. Muller, B.A. Huber and T. Leisner. 2003. Coulomb fission —Rayleigh jets from levitated microdroplets. Nature 421: 128–128.

Fenn, J.B. 1993. Ion Formation from Charged Droplets-Roles of Geometry, Energy, and Time. J. Am. Soc. Mass Spectrom. 4: 524–535.

Fernandes, B., U. Sagman, M. Auger, M. Demetrio and J.W. Dennis. 1991. Beta-1–6 Branched Oligosaccharides as a Marker of Tumor Progression in Human Breast and Colon Neoplasia. Cancer Res. 51: 718–723.

Fuster, M.M., and J.D. Esko. 2005. The sweet and sour of cancer: Glycans as novel therapeutic targets. Nat. Rev. Cancer 5: 526–542.

Gorelik, E., U. Galili and A. Raz. 2001. On the role of cell surface carbohydrates and their binding proteins (lectins) in tumor metastasis. Cancer and Metastasis Reviews 20: 245–277.

Hakomori, S. 1985. Aberrant Glycosylation in Cancer Cell-Membranes as Focused on Glycolipids-Overview and Perspectives. Cancer Res. 45: 2405–2414.

Harvey, D.J., L. Royle, C.M. Radcliffe, P.M. Rudd and R.A. Dwek. 2008. Structural and quantitative analysis of N-linked glycans by matrix-assisted laser desorption ionization and negative ion nanospray mass spectrometry. Anal. Biochem. 376: 44–60.

Karlsson, N.G., N.L. Wilson, H.J. Wirth, P. Dawes, H. Joshi and N.H. Packer. 2004. Negative ion graphitised carbon nano-liquid chromatography/mass spectrometry increases sensitivity for glycoprotein oligosaccharide analysis. Rapid Commun. Mass Spectrom. 18: 2282–2292.

Kebarle, P., and L. Tang. 1993. From Ions in Solution to Ions in the Gas-Phase—the Mechanism of Electrospray Mass-Spectrometry. Anal. Chem. 65: A972–A986.

Kirsch, S., and L. Bindila. 2009. Nano-LC and HPLC-chip-ESI-MS: an emerging technique for glycobioanalysis. Bioanalysis 1: 1307–1327.

Kobata, A., and J. Amano. 2005. Altered glycosylation of proteins produced by malignant cells, and application for the diagnosis and immunotherapyof tumours. Immunol. Cell Biol. 83: 429–439.

Mitsuoka, C., M. Sawada-Kasugai, K. Ando-Furui, M. Izawa, H. Nakanishi, S. Nakamura, H. Ishida, M. Kiso and R. Kannagi. 1998. Identification of a major carbohydrate capping group of the L-selectin ligand on high endothelial venules in human lymph nodes as 6-sulfo sialyl Lewis X. J. Biol. Chem. 273: 11225–11233.

Pabst, M., and F. Altmann. 2008. Influence of electrosorption, solvent, temperature, and ion polarity on the performance of LC-ESI-MS using graphitic carbon for acidic oligosaccharides. Anal. Chem. 80: 7534–7542.

Packer, N.H., C.W. von der Lieth, K.F. Aoki-Kinoshita, C.B. Lebrilla, J.C. Paulson, R. Raman, P. Rudd, R. Sasisekharan, N. Taniguchi and W.S. York. 2008. Frontiers in glycomics: Bioinformatics and biomarkers in disease—An NIH White Paper prepared from discussions by the focus groups at a workshop on the NIH campus, Bethesda MD (September 11–13, 2006). Proteomics 8: 8–20.

Rostenberg, I., J. Guizarvazquez and R. Penaloza. 1978. Altered Carbohydrate Content of Alpha-1-Antitrypsin in Patients with Cancer. J. Natl. Cancer Inst. 61: 961–965.

Ruhaak, L.R., A.M. Deelder and M. Wuhrer. 2009. Oligosaccharide analysis by graphitized carbon liquid chromatography-mass spectrometry. Anal. Bioanal. Chem. 394: 163–174.

Seelentag, W.K.F., W.P. Li, S.F.H. Schmitz, U. Metzger, P. Aeberhard, P.U. Heitz and J. Roth. 1998. Prognostic value of beta 1,6-branched oligosaccharides in human colorectal carcinoma. Cancer Res. 58: 5559–5564.

Stumpo, K.A., and V.N. Reinhold. 2010. The N-Glycome of Human Plasma. J. Prot. Res. 9: 4823–4830.
Taylor, G. 1964. Proc. Royal Soc. London. Series a. Math and Phys. Sci. 280: 383–397.
Taylor, M.E. and K. Drickamer. 2006. Introduction to Glycobiology, 2nd ed., Oxford University Press Inc., New York.
Walker, S.H., B.N. Papas, D.L. Comins and D.C. Muddiman. 2010. Interplay of Permanent Charge and Hydrophobicity in the Electrospray Ionization of Glycans. Anal. Chem. 82: 6636–6642.
Wilm, M.S., and M. Mann. 1994. Electrospray and Taylor-Cone Theory, Doles Beam of Macromolecules at Last. Int. J. Mass Spectrom. Ion Processes 136: 167–180.
Wu, H.C., E. Meezan, P.H. Black and P.W. Robbins. 1969. Comparative Studies on Carbohydrate-Containing Membrane Components of Normal and Virus-Transformed Mouse Fibroblasts. I. Glucosamine-Labeling Patterns in 3t3 Spontaneously Transformed 3t3 and Sv-40-Transformed 3t3 Cells. Biochemistry 8: 2509–2517.
Wuhrer, M., A.M. Deelder and C.H. Hokke. 2005. Protein glycosylation analysis by liquid chromatography-mass spectrometry. J. Chromatogr. B: Anal. Technol. Biomed. Life Sci. 825: 124–133.
Zhao, Y.Y., M. Takahashi, J.G. Gu, E. Miyoshi, A. Matsumoto, S. Kitazume and N. Taniguchi. 2008. Functional roles of N-glycans in cell signaling and cell adhesion in cancer. Cancer Science 99: 1304–1310.

Section 2: Specific Cancers and Areas of Focus

17

Composite Nano-Sensor Arrays Technology in Lung Cancer

Silvano Dragonieri,[1,*] *Giorgio Pennazza*[2] and *Marco Santonico*[3]

ABSTRACT

Lung Cancer is the most common and most lethal cancer in many countries of the world. Diagnosing lung cancer is difficult, often requiring expensive and invasive methods. Exhaled breath contains thousands of volatile organic compounds (VOCs) that could be used as non-invasive markers of lung disease. According to several studies, sensors array technology seems to be able to discriminate exhaled breath of patients with lung cancer from that of individuals without it, discriminating different VOCs profiles. Nowadays, sensor arrays based on nanostructured materials with learning algorithms can be exploited to analyze VOCs mixtures by pattern recognition. Nanotechnology is used in commercial application, such as food industry and military field; medical applications and in particular breath analysis is the new scenario for this bottom-up strategy. Certain studies have shown that

[1]Department of Pulmonology, University of Bari, Bari, Italy; E-mail: sdragonieri@hotmail.com
[2]Faculty of Engineering, University Campus Bio-Medico di Roma, Italy;
E-mail: g.pennazza@unicampus.it
[3]Department of Electronic Engineering, University of Rome Tor Vergata, Rome, Italy;
E-mail: santonico@ing.uniroma2.it
*Corresponding author

List of abbreviations after the text.

gas sensor arrays based on nanotechnology can ameliorate the discrimination ability of VOCs pattern in the exhaled breath, increasing diagnostic performance also in a critical clinical case such as lung cancer. The aim of this chapter is to examine the principal literature on the different studies about non-invasive diagnosis of lung cancer by nanotechnology. Thus the use of these devices may potentially become an option in the diagnosis of lung cancer. However further studies with a larger number of subjects and with newly presented patients with various histology and severities of lung cancer are required.

INTRODUCTION

Lung cancer has an important clinical and socio-economical significance, as it is the most common and the most lethal cancer in many countries of the world.

The diagnosis of lung cancer is challenging. In the early stages lung cancer does not present evident symptoms, and when present those symptoms are not specific (Ferguson 1990). Therefore patients are often diagnosed at advanced stages, when curative treatment is no longer an option (Jemal et al. 2006). Diagnosis of lung cancer is generally made by expensive and often invasive tests which potentially lead to complications. For all these reasons, there is a need for new diagnostic tests which should be cheap, accurate and non-invasive.

Diseases diagnosis via 'breath smelling' is an old practice of ancient medicine. Physicians in the past knew that several diseases alter the odour of a patient's breath, for example diabetes, liver diseases and kidney diseases. This practice has been abandoned today because of the introduction of modern diagnostic technologies.

During the 1970s Pauling et al. detected more then 200 components in human breath by gas chromatography. (Pauling et al. 1971) Since then it is well-known nowadays that human exhaled breath contains thousands of volatile organic compounds (VOCs) in gas phase, which can be detected by gas chromatography and mass spectrometry (GC-MS) (Pauling et al. 1971; Moser et al. 2005). These include for example alkanes, aromatic compounds and benzene derivatives. It has been shown that VOCs analysis may potentially be used for a non-invasive marker of lung cancer (Gordon et al. 1985; Philips et al. 1999). Besides, GC-MS use in medical applications has been limited by a list of technical needs making this approach not optimal for volatiles profiling in breath oriented to pre-screening diagnostic activities.

A larger application of exhaled breath analysis for the diagnosis of lung cancer would be achieved by user-friendly, fast and less expensive technologies.

THE ROLE OF TECHNOLOGY IN BREATH ANALYSIS: PROBLEMS AND OPPORTUNITIES

A complex system or a simple device used as diagnostic instrument has, as first, the task of transforming measurable parameters related to the individual health-state into analytically useful signals able to support a diagnosis or suggest further investigations. These systems can be represented by a measure chain whose elements (the detector working principle, the chemical interface) give a peculiar contribution to the instrument performances: its sensitivity, selectivity, and limit of detection.

So far, the technological efforts contributing to the development of such diagnostic systems, allowing the transduction of the measuring input into the output signal, contemporary operates a filtration and sometimes a distortion of the signal itself.

Focusing on breath analysis, this transduction-filter-distortion action concerns with the interpretation of VOCs, based on two complementary aspects: a medical and an instrumental interpretation (Pennazza et al. 2010).

Considering exhaled breath as the biological input of a measure system its collection represents the first step, giving rise to a list of intermediate operations, which affect, amplify, distort and combine the information content of the exhaled breath (Pennazza et al. 2010).

Medical interpretation is beyond the scope of the present work. Instrumental interpretation, instead, concerns technological aspects: each different measurement technique with its proper sensing material and working principle operates an intrinsic selection among the VOCs composing the sample under analysis, so presenting a partial view of the VOCs totality (Pennazza et al. 2010).

Looking at the influence of the instrumental apparatus on the final result, the art of the matter concerns the sensors; sensing materials used for breath analysis mainly belong to two categories: organometallic compounds and polymers. The selection of such materials represents a constraint for the transduction mechanism, being organic compound scarcely conductive. Thus, transduction mechanism is mainly based on the variation of mass or on the modification of optical properties. An alternative is represented by a doping of the organic materials with conductive particles such as carbon black or gold nanoparticles (Peng et al. 2009).

General guidelines for sensing material selection in breath analysis ask for a bare sensitivity to O_2, N_2, CO_2 and water vapor; besides, peculiarity of different materials respect to key-VOCs families must be considered: metalloporphyrins (Di Natale et al. 2003; D'Amico et al. 2010) are more sensitive to alcohols, aromatic, and hydrazines while solvatochromic dyes (Mazzone et al. 2007) are not sensitive to alkanes and methylalkanes.

These differences in selectivity are based on different interaction characteristics, resulting in different performance for diagnosis of lung diseases. This fact suggests that, for lung cancer, a distinctive volatile signature is not based on a single compound, but it is revealed by the volatile compounds composition profile, a sort of the exhaled breath fingerprint.

THE ROLE OF NANOTECHNOLOGY IN BREATH ANALYSIS: BETTER POTENTIALITIES AND INCREASED DIFFICULTIES

Nanotechnology was firstly preconized by Richard Feynman in 1959, during a very fascinating lecture (Feynman 1960). He claimed that physics seems not to contradict the 'possibility of maneuvering things atom by atom': this is a fundamental starting point to discriminate nanotechnology with respect to merely using materials with nano-dimensional components without having a sort of control on them. On this basis we assume the following the definition given by N. Taniguchi in 1974: "the creation of useful materials, devices, and systems through the control matter at the nanometer length scale and the exploitation of novel properties and phenomena developed at that scale" (Taniguchi 1974).

What does it mean for sensor technology? And what for breath analysis using chemical sensors? It is useful to speak in term of better potentialities and increased difficulties. Actually, new properties of sensing materials used for chemical sensors could help to overcome some unresolved problems of breath analysis: humidity influence on sensors responses, cross-selectivity, detection of low-ppb concentration levels. Besides, the informative content of such measurements has to be reproducible, and this requirement pertains to the control of the new properties at the nano-scale level. The most challenging effort for this scope is a deep knowledge of the laws ruling these new phenomena.

In conclusion, can we claim that nowadays breath analysis is performed via nano-sensors? Looking at the literature many examples can be found of sensors based on nanostructured materials, but none performed with nanosensors. Really nanoscaled sensors, applied for breath analysis, could catch air samples directly inside the body, as a drug-nano-delivery task. This limitation acts as turning point between a current available list of opportunities offered by the use of nano-scaled sensing materials and the future perspective of catching air sample directly *in situ*, close to a cancerous lesion inside the lungs.

ELECTRONIC NOSE TECHNOLOGY

Progress in technology in the last few years has created chemical sensing and identification devices called "electronic noses".

Electronic nose technology aims to mimic mammalian olfactory smell strategy, using many receptors (sensors array) which are not specific for certain sensors. (Persaud and Dodd 1982).

The main part of the existing electronic noses is based on chemical gas sensors, but recently innovative working principles have been exploited to reproduce the functioning of the olfactive receptors. (Röck et al 2008).

In the last few years, electronic nose technology has been tested in every conceivable field dealing with odours and/or odourless volatiles and gases, in particular in food and beverage industry, environmental monitoring, military purposes and very recently for diagnosis of diseases. (Röck et al. 2008).

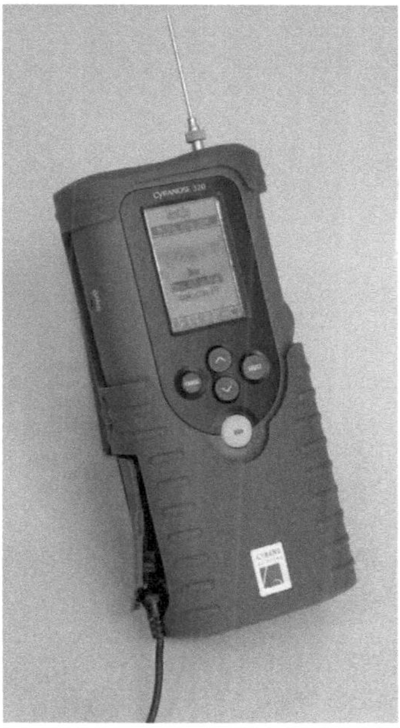

Fig. 1. One of the commercially available electronic noses (Dragonieri et al. 2007).

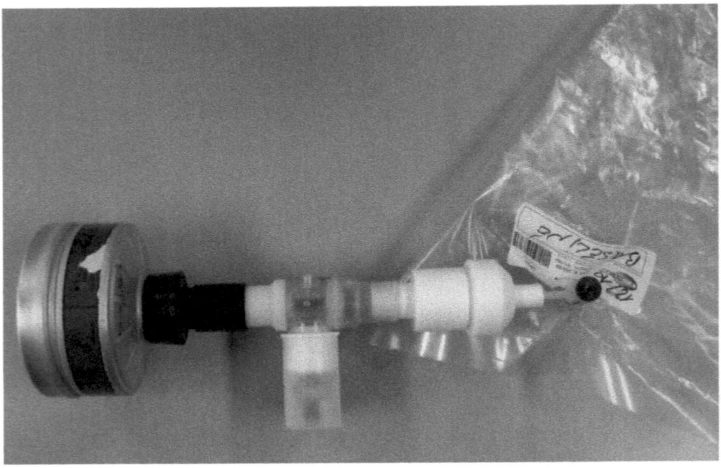

Fig. 2. An example of setup for exhaled breath collection (Dragonieri et al. 2007).

After the introduction of electronic noses, VOCs pattern of analysis of the exhaled breath has become easily available, due to their ability to perform on-board analysis and discrimination of "breathprints" by composite nano-sensors arrays (breatheomics). This is based on pattern recognition without providing information of the individual molecular components. (Briglin et al. 2002).

Fig. 3. Exhaled breath collection (Dragonieri et al. 2007).

Color image of this figure appears in the color plate section at the end of the book.

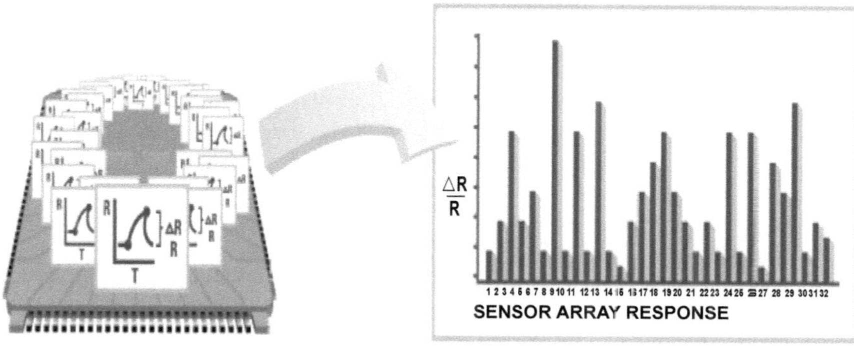

Fig. 4. The combination of the signal from all sensors generates the so-called "breathprint" (Dragonieri et al. 2007).

Color image of this figure appears in the color plate section at the end of the book.

To date, electronic nose technology has been applied in several medical fields, including the detection of sinusitis, cerebrospinal fluid leak, urinary tract infections, bacterial vaginosis and diabetes mellitus (Thaler and Hanson 2005). In respiratory medicine electronic noses have been used in the detection of asthma (Dragonieri et al. 2007; Fens et al. 2009; Montuschi et al. 2009), COPD (Fens et al. 2009) and in the detection of Mycobacterium Tuberculosis infections (Fend et al. 2006) and Ventilator-associated pneumonia (Hanson and Thaler 2005).

APPLICATIONS OF NANOTECHNOLOGY IN LUNG CANCER

In the last few years several groups have made efforts to detect lung cancer by nanotechnology with various sensor array systems.

Di Natale (Di Natale et al. 2003) showed 100% correct discrimination of patients with various forms of lung cancer, and 94% discrimination of controls by using quartz microbalance gas sensors.

Chen (Chen et al. 2007) used an electronic nose based on surface acoustic wave sensors and image recognition method showing promising results for the detection of lung cancer.

Machado (Machado et al. 2005) used a handheld portable chemical vapour analyzer, containing a nano-composite array with 32 polymer sensors and showed adequate discrimination between patients with lung cancer and those from other groups, including healthy controls and a mixed group of subjects with pulmonary diseases (α-1 AT deficiency, chronic beryllium disease and COPD). By building a validation model, these authors found 71.4% sensitivity and 91.9% specificity in the detection of bronchogenic carcinoma, with positive predictive values of 66.6% and 93.4%, respectively. Mazzone (Mazzone et al. 2007) by using a colorimetric sensor array showed the presence of non-small cell lung cancer (NSCLC) with a sensitivity of 73% and a specificity of 72,4%. Their control group was composed by patients with sarcoidosis, COPD, pulmonary arterial hypertension, idiopathic pulmonary fibrosis and healthy subjects. Dragonieri (Dragonieri et al. 2009) used the same electronic nose as Machado (Machado et al. 2005) showing that breathprints from NSCLC patients clustered distinctly from those of subjects with COPD (cross validation value: 85%; M-distance: 3.73). Moreover NSCLC could also be distinguished from healthy controls in duplicate measurements (cross validation value: 90% and 80%, respectively; M-distance: 2.96 and 2.26). These findings suggest that an electronic nose can distinguish between two separate smoking-related lung diseases, which may have pathophysiological and clinical implications. Peng (Peng et al. 2009) used an array of sensors based on gold nanoparticles showing that it can distinguish the breath of lung cancer patients from the breath of healthy individuals. Furthermore, these authors used GC-MS in combination with solid-phase microextraction to identify 42 volatile organic compounds that represent lung cancer biomarkers. Four of these were then used to train and optimize the sensors, achieving a good conformity between patient and simulated breath samples. Here it is worth remarking a brilliant case of nanotechnology used to develop an electronic nose for breath analysis which is showing great potential in lung cancer diagnosis; these nano-structured materials such as molecularly modified metal nanoparticles, metal oxide nanoparticles, carbon nanotubes, and semiconducting nanowires, ask for fabrication procedures relatively simple. Moreover they are so versatile that each sensing element can be tailored for specific adsorbing mechanisms. Eventually, they offer the opportunity to be easily integrated into established sensing platforms. The way they modify chemical sensors properties can be summarized by three simple features: the increased surface of the sensing element, the opportunity to functionalize adsorbing mechanism at molecular level, the ability to arrange many different sensing properties into an array of very small dimension (see Fig. 5). A last, but not least, characteristic of nanoscaled sensing material is the so called effect of quantum confinement (QC). QC effect consists in the disclosure of new properties of the used material emerged by its designing and utilization at dimension under the Bohr radius; considering that Bohr radius of Si is 4.9 nm, this effects is not so far to be exploited by the current technologies. QC effect can be also used to define different nano-structured materials: quantum-dots (materials whose nanoparticles three dimensions are under the Bohr radius); quantum-wires and quantum-belts (materials whose nano-particles length is under the Bohr radius); quantum-wells (materials whose nano-particles height is under the Bohr radius) (Di Francia et al. 2009).

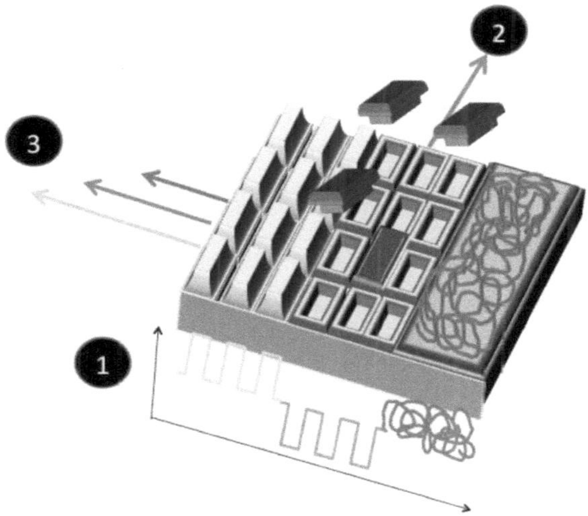

Fig. 5. Nanotechnology added value to chemical sensors: (1) the increased surface of the sensing element; (2) the opportunity to functionalize adsorbing mechanism at molecular level; (3) the ability to arrange many different sensing properties into an array of very small dimension.

Color image of this figure appears in the color plate section at the end of the book.

Finally, D'amico (D'Amico et al. 2010) used an array of metalloporphyrins coated QMB-sensors. By building three models these authors showed that their electronic nose was able to discriminate between 1) lung cancer and controls, 2) lung cancer and other lung diseases, 3) lung cancer from both lung diseases and control group. The complete classification obtained a global rate of success of 79.3% in the discrimination between the two different pathological conditions and controls and also showed a good performance in the discrimination between the two diseases (85.7%).

To conclude this brief survey it is useful to consider the pie-graphs reported in Fig. 6. Graph 6.a gives an overview of the study typologies, with a main part (75%) devoted to clinical studies on individuals; besides, researches on cell cultures and animals are growing in number, accounting for the needs to correlate the promising results obtained on breath analysis with findings on metabolism and its alteration. Graph 6.b shows that respiratory diseases (with or without lung cancer) cover 75% of the total number of studies. Graph 6.c shows that research groups developing also an electronic nose technology on their own conduct more than the half of the performed studies. Graph 6.d shows the incidence of nanotechnology in this field, which is now stated at the 30%, and is still growing.

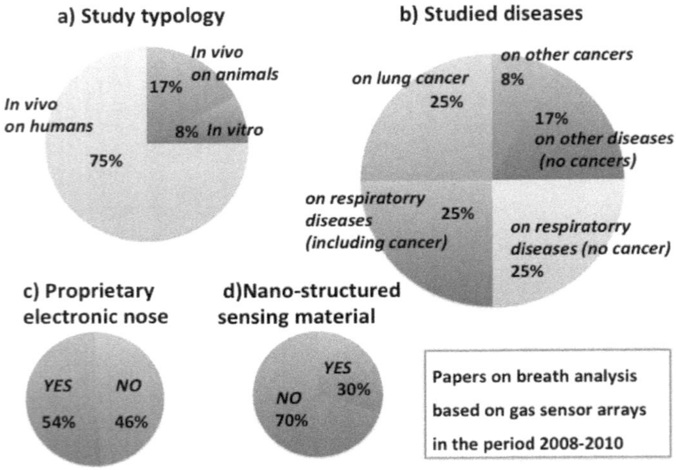

Fig. 6. Pie-graphs representing the different studies on breath analysis based on gas sensor arrays conducted in the period 2008–2010. (a) divided by study typology (*in vivo* on humans, *in vivo* on animals, in vitro on cell coltures) (b) divided by studied diseases (lung cancer, respiratory diseases including lung cancer, respiratory diseases not including lung cancer, other cancers, other diseases) (c) divided by research groups which are proprietary of the used electronic nose or not; d) divided by the study conducted with electronic noses based on nano-structured materials or not.

Color image of this figure appears in the color plate section at the end of the book.

VOCs IDENTIFICATION

Most of the devices based on nanotechnology measure VOCs profile in the exhaled breath by pattern recognition, which is sufficient for clinical assessment. However they do not provide information about the specific individual VOCs. Identification of critical VOCs by GC-MS is important for better understanding the pathophysiological pathways behind the disease and for helping to shape specific sensors for future clinical aims. Therefore at this stage GC-MS analysis of exhaled breath remains necessary and complementary to electronic nose investigations and further studies by GC-MS to identify the crucial VOCs are required.

APPLICATIONS TO AREAS OF HEALTH AND DISEASE

The first proof studies have shown promising results in the detection of lung cancer, indicating that if validated, nano-sensors array technology may have the potentials to become a convenient device for the physician for its cheapness, non-invasiveness, portability and ease to use. Devices based on nanotechnology may be used as diagnostic tool for selecting patients for additional diagnostic procedures or as a screening tool for excluding lung cancer in patients at risk.

Conclusion

According to several studies, devices based on nanotechnology seem to be able to discriminate exhaled breath of patients with lung cancer from control individuals, indicating that exhaled breath of patients with cancer has distinct characteristics in terms of VOCs profile.

Thus the use of nano-technology may potentially become an option in the diagnosis of lung cancer. However many steps still have to be taken before it can be possibly validated. It is therefore necessary to undertake further studies with a larger number of subjects and with newly presented patients with various histology and severities of lung cancer.

Key Terms

- Volatile Organic Compounds: any organic compound having an initial boiling point less than or equal to 250°C measured at a standard atmospheric pressure of 101.3 kPa.
- Nanotechnology: the engineering of functional systems at the molecular scale.
- Electronic nose: device developed in order to mimic human olfaction that functions as a non-separative mechanism: i.e., an odour/flavour is perceived as a global fingerprint.
- Pattern recognition analysis: assignment of some sort of output value to a given input value.

Summary Points

- Lung Cancer diagnosis is difficult, often requiring expensive and invasive methods.
- Exhaled breath contains thousands of volatile organic compounds (VOCs) that could be used as markers of lung disease.
- Devices based on nanotechnology analyze VOCs mixtures by pattern recognition.
- Several studies have shown that electronic noses discriminated the breath of patients with lung cancer from subjects without it.
- Nanotechnology may potentially become an option in the diagnosis of lung cancer.

Abbreviations

VOCs	:	volatile organic compounds
GC-MS	:	gas chromatography and mass spectrometry
NSCLC	:	non-small cell lung cancer

References

Briglin, S.M., M.S. Freund, P. Tokumaru, et al. 2002. Exploitation of spatio-temporal information and geometric optimization of signal/noise performance using arrays of carbon black-polymer composite vapor detectors. Sens. Actuators B. 82: 54–74.

Chen, X., M. Cao, Y. Li, et al. 2007. A study of an electronic nose for detection of lung cancer based on surface acoustic wave sensors and image recognition method. Meas Sci. Technol. 16: 1535–46.

D'Amico, A., G. Pennazza, M. Santonico, et al. 2010. An investigation on electronic nose diagnosis of lung cancer. Lung Cancer 2: 170–6.

Di Francia, G., B. Alfano and V. La Ferrara. 2009. Conductometric gas nanosensors. Journal of Sensors art ID 659275.

Di Natale, C., A. Macagnano, E. Martinelli, et al. 2003. Lung Cancer identification by the analysis of breath by means of an array of non-selective gas sensors. Biosens. Bioelectron. 18: 1209–1218.

Dragonieri, S., J.T. Annema, R. Schot, et al. 2009. An electronic nose in the discrimination of patients with non-small cell lung cancer and COPD. Lung Cancer 64: 166–70.

Dragonieri, S., R. Schot, J.A. Mertens, et al. 2007. An electronic nose in the discrimination of patients with asthma and controls. J. Allergy Clin. Immunol. 120: 856–62.

Fend, R., A.J. Kolk, C. Bessant, et al. 2006. Prospects for clinical application of electronic-nose technology to early detection of mycobacterium tuberculosis in culture and sputum. J. Clin. Microbiol. 44: 2039–45.

Fens, N., A.H. Zwinderman, M.P. van der Schee, et al. 2009. Exhaled Breath Profiling Enables Discrimination of Chronic Obstructive Pulmonary Disease and Asthma. Am. J. Respir Crit. Care Med. 180: 1076–82.

Ferguson, M.K. 1990. Diagnosing and staging of non-small cell lung cancer. Hematol Onco.l Clin. North Am. 4: 1053–68.

Gordon, S.M., J.P. Szidon and B.K. Krotoszynski. 1985. Volatile organic compounds in exhaled air from patients with lung cancer. Clin. Chem. 8: 1278–1282.

Hanson, C.W., and Thaler E.R. 2005. Electronic nose prediction of a clinical pneumonia score: biosensors and microbes. Anesthesiology 102: 63–8.

Jemal, A., R. Siegel, E. Ward, et al. Cancer statistics. 2006. CA Cancer J. Clin. 56: 106–30.

Machado, R.F., D. Laskowski, O. Deffenderfer, et al. 2005. Detection of lung cancer by sensor array analysis of exhaled breath. Am. J. Respir Crit. Care Med. 171: 1286–1291.

Mazzone, P.J., J. Hammel, R. Dweik, et al. 2007. Diagnosis of lung cancer by the analysis of exhaled breath with a colorimetric sensor array. Thorax 62: 565–568.

Montuschi, P., M. Santonico, G. Pennazza, C. Mondino, G. Mantini, E. Martinelli, R. Capuano, G. Ciabattoni, R. Paolesse, C. Di Natale, P. Barnes and A. D'Amico. 2009. Diagnostic performance of an exhaled nitric oxide and lung function testing in asthma. Chest 137(4): 190–796.

Moser, B., Bodrogi F., G. Eibl, et al. 2005. Mass spectrometric profile of exhaled breath: field study by PTR-MS. Respir. Physiol. Neurobiol. 145: 295–300.

Pauling, L., A.B. Robinson, R. Teranishi, et al. 1971. Quantitative analysis of urine vapor and breath by gas-liquid partition chromatography. Proc. Natl. Acad. Sci. USA 68: 2374–6.

Peng, G., U. Tisch, U. Adams, et al. 2009. Diagnosing lung cancer in exhaled breath using gold nanoparticles. Nature Nanotechnol. 4: 669–73.

Pennazza, G., E. Santonico, A. Martinelli, et al. September 2010. Interpretation of exhaled volatile organic compounds. An European Respiratory monograph: exhaled biomarkers. European Respiratory Society Ed. Number 24.

Persaud, K., and G. Dodd. 1982. Analysis of discrimination mechanisms in the mammalian olfactory system using a model nose. Nature 299, 352.

Phillips, M., K. Gleeson, J.M. Hughes, et al. 1999. Volatile organic compounds in breath as markers of lung cancer: a cross-sectional study. Lancet 353: 1930–1933.

Röck, F., N. Barsan and U. Weimar. 2008. Electronic Nose: Current Status and Future Trends. Chem. Rev. 108: 705–725.

Taniguchi, N., "On the Basic Concept of 'NanoTechnology'," Proc. Intl. Conf. Prod. Eng. Tokyo, Part II (Tokyo: Japan Society of Precision Engineering, 1974).

Thaler, E.R., and Hanson C.W. 2005. Medical applications of electronic nose technology. Exert. Rev. Med. Devices 2: 559–66.

18

Nano Arsenic Oxide (As_2O_3)/(Fe_2O_3) Iron Oxide Complexes for Liver Cancer

Ziyu Wang[1,a] and *Dongsheng Zhang*[1,b,*]

ABSTRACT

Clinically, thermotherapy, which includes heating certain organs or tissues to temperatures between 41°C and 46°C, also known as hyperthermia, has been long used in the treatment of tumors. With the development of nanotechnology, magnetic nanoparticles (MNPs), such as iron oxide nanoparticles, have been used for tumor hyperthermia, to absorb energy from an applied high-frequency alternating magnetic field (AMF) and transform it into heat at temperatures of 42–45°C, at which tumor cells are very sensitive. Arsenic trioxide (As_2O_3), a traditional Chinese medicine, is important for the research and treatment of human cancers. This chapter discusses the current status of the development of iron oxide MNPs and their application in the hyperthermic treatment of tumors, including their toxicity. Furthermore, we also review current studies using MNPs in combination with As_2O_3 for tumor therapy.

[1]School of Medicine, Southeast University, No.87 Ding Jia Qiao, Nanjing, China.
[a]E-mail: wangziyu717@163.com
[b]E-mail: b7712900@jlonline.com
*Corresponding author

List of abbreviations after the text.

INTRODUCTION

Over the last decade, nanotechnology has developed to such an extent that it has become possible to fabricate, characterize and specifically tailor the functional properties of nanoparticles for biomedical applications and diagnostics. Magnetic nanoparticles (MNPs) are a class of nanoparticles (i.e., engineered particulate materials of < 100 nm) that can be manipulated under the influence of an external magnetic field. MNPs are commonly composed of magnetic elements, such as iron, nickel, cobalt and their oxides. Based on their unique mesoscopic physical, chemical, thermal, and mechanical properties, superparamagnetic nanoparticles offer a high potential for several biomedical applications. These include: (a) cellular therapies, such as cell labeling and targeting, where they can also be used as a tool for cell biology research to separate and purify cell populations; (b) tissue repair; (c) drug delivery; (d) magnetic resonance imaging (MRI); (e) hyperthermic cancer therapy; and (f) magnetofection (Gupta and Gupta 2005). When further "functionalized" with drugs and bioactive agents, such as peptides and nucleic acids, MNPs form distinct particulate systems that penetrate cell and tissue barriers and offer organ-specific therapeutic and diagnostic modalities. The ability of MNPs to be functionalized and concurrently respond to a magnetic field has made them a useful tool for theragnostics (the fusion of therapeutic and diagnostic technologies to target and individualize medicine). However, variations in properties, such as MNP composition, shape, size, surface chemistry and dispersion, may influence their biodistribution and toxic potential.

The potential toxicity of nanosized materials relative to their bulk and molecular counterparts is owed largely to their enhanced reactive surface area, ability to cross cell and tissue barriers, and resistance to biodegradation. Activation of oxidative stress and inflammatory signaling pathways leads to increased apoptosis and genotoxicity, both of which are key parameters of nanotoxicity. In an *in vivo* setting, macrophages of the defense reticuloendothelial system (RES) quickly challenge and internalize MNPs, neutralizing their cytotoxic potential. Therefore, an integrative approach to improving MNP design and understanding their interface with specific organ systems, including their application and safety, are imperative to advancing nanomedicine.

Arsenic trioxide has been adopted from traditional Chinese medicine and used successfully to treat refractory or recurrent acute promyelocytic leukemia (APL). Arsenic acts on cells through a variety of mechanisms, influencing numerous signal transduction pathways and resulting in a vast range of cellular effects, including apoptotic induction, growth inhibition, promotion of differentiation, and inhibition of angiogenesis. However, its clinical application and use in the basic research of solid carcinomas is limited because of its poor dissolution and low bioavailability. When treated with a high dosage of As_2O_3, patients often suffer from acute and chronic side effects, such as gastrointestinal complications, some of which are often severe or fatal. Moreover, it is generally considered an extremely effective environmental co-carcinogen for some human malignancies, especially for cancers of the skin and lung. Investigations of epidemiology indicated that incidence of some tumors was associated with high levels of arsenic in drinking water. Therefore, it is extremely important to enhance the therapeutic effects and reduce the toxicity of As_2O_3, through chemical modifications. Developments in nanotechnology and increased applications in medicine have opened up a new realm for the modernization of traditional Chinese medicine. Studies on nanosized As_2O_3 were performed by several researchers to improve its bioavailability and therapeutic effect, while reducing the dosage and side effects.

IRON OXIDE MNPs FOR HYPERTHERMIA

Hyperthermia treatments for malignant tumors have been used for many years. At present, thermotherapy is commonly used clinically in applications such as those using radiofrequency, microwaves and lasers, all of which have many limitations for treatment of tumor hyperthermia. Microwave treatment of hyperthermia has a poor depth of penetration, making it unsuitable for the treatment of deep-seated tumors. Conversely, ultrasound has better depth of penetration and focusing abilities; however, it results in high energy absorption from bone and liquid-containing organs, with excessive reflection from air-filled cavities (Tasci et al. 2009). With the development of nanotechnology, MNPs have been used not only as drug carriers but also in tumor hyperthermia. In 1997, Hentschel discovered that a nanoscaled magnetic fluid could be absorbed with much higher power in an alternating magnetic field (AMF), and used to treat disease or tumors (Hentschel et al. 1997). This treatment was named 'magnetic fluid hyperthermia' (MFH). MNPs embedded around a tumor site, and placed within an AMF, will heat up to a temperature dependent on a number of properties, including the magnetic potential of the material, the strength of the magnetic field, the frequency of oscillation and the cooling capacity of the blood flow in the tumor site. Cancer cells are destroyed at temperatures above 43°C, whereas normal cells can survive under these conditions. MFH is more effective in the uniform heating of deeply situated tumors, with relatively good targeting. Generally, in this technique, magnetic fluids are dispersed into the target tissue, which are then heated through the application of an external AMF (Tasci et al. 2009). Magnetic nanoparticles may be effective thermoseeds for localized hyperthermia treatment of cancers (Zhao et al. 2006), thereby avoiding heat-related damage to normal tissue and thus overcoming the limitations of conventional heat treatments.

Iron oxide MNPs, such as magnetite (Fe_3O_4) or its oxidized and more stable form, maghemite (γ-Fe_2O_3), are superior to other metal oxide nanoparticles in terms of their biocompatibility and stability, making them the most commonly employed MNPs for biomedical applications. Typically, iron oxide MNPs are synthesized and dispersed into homogenous suspensions, called ferrofluids, composed of a large number of engineered composite nanoparticles. Each MNP consists of a magnetic core and a nonmagnetic coating, creating a different surface chemistry. Iron oxide MNPs used in hyperthermia were reported to have a significant therapeutic effect on xenograft models of liver cancer in nude mice (Yan et al. 2005). The Fe_2O_3 nanoparticles, prepared were spherical in shape, with diameters ranging from 13 to 20 nm, and could suspend stably in water, with good diffusibility (Fig. 1). The temperature of the magnetic fluid could rise from 37°C to 54°C, depending on the different concentrations, with a stable temperature being reached after 40 min of exposure to the magnetic field (Fig. 2). Results showed that Fe_2O_3 nanoparticles displayed good absorptive capabilities in the high-frequency alternating electromagnetic field, and strong magnetic responsiveness.

For MNPs to be useful for biomedical applications, they should first be stabilized against the absorption of plasma proteins and nonspecific uptake by components of the reticular-endothelial system (RES), such as macrophages (Xu and Sun 2009). The size, charge, surface chemistry and delivery route of MNPs, each influence their circulation time and biodistribution pattern in the body. Large (> 200 nm) particles are usually

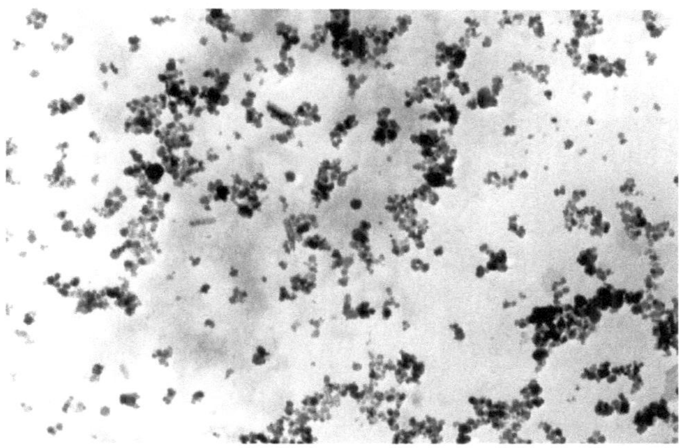

Fig. 1. TEM image of Fe_2O_3 nanoparticles (×50,000). The TEM image shows Fe_2O_3 nanoparticles are spherical in shape, with diameters about 20 nm (unpublished).

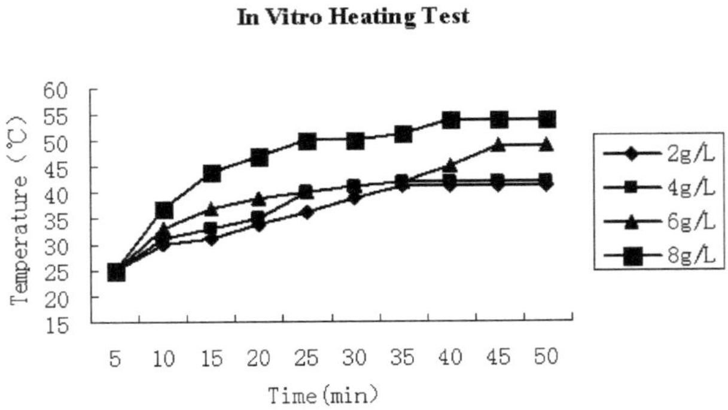

Fig. 2. Thermodynamic analysis of Various doses of Fe_2O_3. The temperature of Fe_2O_3 nanoparticles can rise from 37°C to 54°C, depending on the different Fe_2O_3 concentrations, with a stable temperature being reached after 40 min of exposure to the magnetic field (unpublished).

sequestered by the spleen via mechanical filtration, which results in their phagocytosis, whereas smaller (< 10 nm) particles are rapidly removed through extravasation and renal clearance. Therefore, particles in the range of 10–100 nm were believed to be optimal for *in vivo* delivery, as they escape both rapid renal clearance and sequestering by the RES of the spleen and liver.

However, the biomaterials would be in direct contact with tissues and cells when introduced into the body, so their biocompatibility needed to be evaluated before they could be applied in a clinical setting. Many studies have shown that some materials show signs of toxicity when their diameters are reduced to nanoscale (Lam et al. 2004). Therefore, the potential hazards and biosafety of ferromagnetic materials should be particularly noticeable when they are applied to tissues. Biomaterials must not only have long-term stability in biotic conditions, but also have no harmful effects on tissues, blood or the immune system. Fe_3O_4 MNPs were shown to exert no significant effects on the proliferation of HeLa cells and did not induce apoptosis. The activation of NF-kappa-B and the expressions of its typical target genes, IL-8, COX-2 and MCP-1, were not significantly changed under the various concentrations of Fe3O4 MNPs (Shen et al. 2009). To study the biocompatibility of Fe_2O_3 nanoparticles, Wang (Wang et al. 2009) evaluated the nanoparticles using *in vitro* cytotoxicity and hemolytic tests, a micronucleus experiment, the median lethal dose (LD50), and an *in vivo* study referring to ISO10993-1992 and other international standards. Experimental results indicated that Fe_2O_3 nanoparticles had no significant effects on cellular proliferation and did not cause any hemolytic reactions when in direct contact with blood, which was consistent with the requirement of hemolytic tests for biomaterials. Furthermore, they were confirmed to have no carcinogenic or mutagenic effects by the micronucleus assay. Fe_2O_3 also had a wide safety margin, with an LD50 of 5.75g/Kg in rats, rendering it in the 'nontoxic category'. No significant differences were found in serum AST, ALT, BUN and Cr levels, and no obvious pathological alterations were observed when MNPs were injected into the animals. Several reports found iron to accumulate in tissues, but with no remarkable histological changes in vital organs, the respective formulations were concluded as being safe to use. Although MNP deposits were detectable in the prostates of prostate cancer patients after a year of magnetic hyperthermia therapy, no signs of systemic toxicity were detectable (Johannsen et al. 2007). These results demonstrate that Fe_2O_3 nanoparticles are a highly biocompatible material and may be suitable for tumor hyperthermia.

To date, there has been much research into the use of MNPs for the hyperthermic treatment of several types of tumors by performing experiments in animals or cancerous cell lines. Heat-induced therapeutic gene expression is a highly desired gene therapy strategy, as it aims to minimize side effects. Furthermore, if the gene expression is triggered by heat stress, combined therapeutic effects of hyperthermia and gene therapy may be possible. Ito (Ito et al. 2001) combined *TNF-a* gene therapy driven by the stress-inducible promoter, gadd 153, with hyperthermia, using magnetite cationic liposomes (MCLs). In nude mice, MCLs induced cell death throughout much of the tumor area, when heated with an alternating magnetic field. This heat stress also resulted in a 3-fold increase in TNF-a gene expression driven, as compared with that of a non-heat-treated tumor. The combined treatment strongly arrested tumor growth in nude mice over a 30 day period, suggesting its potential for cancer treatment. In March 2003, the first clinical feasibility study on MNP hyperthermia treatment was performed with 14 patients suffering from glioblastoma multiforme (Maier-Hauff et al. 2007). All patients received a neuro-navigationally guided injection of the magnetic fluid into the tumor. The amount of fluid and the spatial distribution of the depots were calculated in advance by means of a specially designed treatment planning software, using MRI. The actual achieved magnetic fluid distribution was measured by computed tomography (CT).

ARSENIC FOR CANCER THERAPY

Arsenic is a well-documented carcinogen that also appears to be a valuable therapeutic tool in cancer treatment (Bode and Dong 2002). The first use of As_2O_3 in cancer therapy was in the treatment of APL (Soignet et al. 2001). As_2O_3 has now been approved by the Food and Drug Administration in the United States for the treatment of patients with APL that is refractory to Tretinoin. The results of *in vitro* research and clinical trials have shown that As_2O_3 is effective in inhibiting tumor growth, and in inducing the differentiation and apoptosis of APL cells. Owing to its significant anticancer effects, As_2O_3 has been tested in patients with various solid tumors, including gastric cancer (Xiao et al. 2007), neuroblastoma (Florea et al. 2007), esophageal carcinoma (Shen et al. 2000), and head and neck cancers (Seol et al. 2001). Moreover, As_2O_3 was shown to inhibit tumor metastasis by reducing the expression of metastasis-related genes (Lin et al. 2008). Overproduction of reactive oxygen species (ROS) was considered as an important mechanism of arsenic-dependent cell apoptosis (Kumagai and Sumi 2007). Accumulation of hydrogen peroxide (H_2O_2) leads to decreases in the mitochondrial membrane potential, resulting in cytochrome C release and activation of the caspase cascade. However, caspase-independent apoptosis pathways have also been reported to be activated by arsenic in myeloma cells and may mediate pro-apoptotic signals. Another recent study has implicated the JNK kinase as an essential component of As_2O_3-dependent apoptosis (Platanias 2009). Furthermore, As_2O_3 activates the pro-apoptotic Bcl-2 family member Bax and induces its translocation from the cytosol to the mitochondria. Bax engagement plays an important role in apoptosis, whereas its activation is suppressed by the anti-apoptotic protein Bcl-2 via inhibition of mitochondrial ROS generation (Zheng et al. 2005). As_2O_3 also up-regulates expression of other pro-apoptotic proteins and down-regulates Bcl-2 and a number of anti-apoptotic proteins.

However, there are many limitations to its use owing to its high toxicity and side effects. Patients treated with As_2O_3 suffer from acute and chronic side effects, such as gastrointestinal complications, which are often severe or fatal. It was reported that arsenic depressed the functions of the antioxidant defense system, leading to oxidative damage of cellular macromolecules. Moreover, it has been generally considered to be an extremely effective environmental co-carcinogen for certain human malignancies, especially renal, skin and lung cancers (Zhao et al. 2008; Lam et al. 2004; Platanias 2009; Li et al. 2009). Therefore, enhancing the selective curative effects and reducing the toxicity of As_2O_3, through modifications of its form or administration, is of great importance.

Deng first reported the size effect of the arsenic containing compound, realgar (As_2S_2), on the human umbilical vein endothelial cell line, ECV-304 (Deng et al. 2001). They found that As_2S_3 particles had barely any effect on cells; however, realgar nanoparticles of diameters between 100–150 nm could remarkably inhibit cell viability through the induction of apoptosis. Nanoparticles of another arsenic containing compound, orpiment (As_2S_3), were prepared successfully and had better effects on liver cancer cells than traditional As_2S_3, using the same concentration and incubation time reported by Lin (Lin et al. 2007). Other studies have shown that nanosized As_2O_3 (Fig. 3) could also produce increased cytotoxic effects on tumor cells than an As_2O_3 solution (Wang et al. 2005). Thus, there is an obvious size-dependent therapeutic effect of nanoparticles, with arsenic nanoparticles effectively inhibiting the growth of cancer cells by inducing apoptosis,

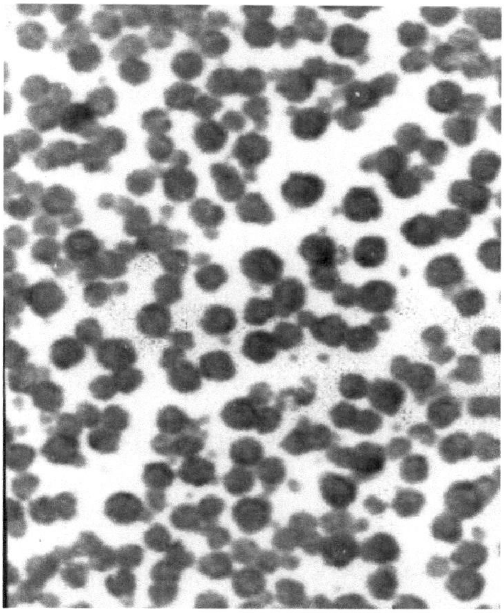

Fig. 3. TEM image of As_2O_3 nanoparticles (×50,000). The image shows As_2O_3 nanoparticles are spherical observed under TEM, with diameters about 80–100 nm (unpublished).

thereby being more therapeutically effective than traditional arsenic compounds. It is unknown why this occurs but may be a result of the physicochemical characteristics of nanomedicine. When normal particles are reduced to nanometer dimensions (1–100 nm), they exhibit novel physical, chemical and biological properties because of quantum, small size or surface effects, thereby creating new characteristics. These include the following: (1) Strong chemical activity whereby a particle of 10 nm has a 90 m^2/g surface area and a 5nm particle has a 180 m^2/g. The large surface area significantly increases the number of atoms on the exterior, thus increasing the chemical energy of the particle. (2) Increased bioadsorption. Drug incorporation is usually affected by its solubility. The increased surface area of nanoparticles increases their absorption. Furthermore, the effective contact area between the drug and particle is enlarged, thereby improving its adsorptive capacity. (3) Special pharmacodynamic effects. Numerous studies have also demonstrated the size of nanosphere particles is crucial for uptake and transport across the gastrointestinal mucosal barrier, thus modifying the expected pharmacodynamic activity. It is easier for nanoparticles than general particles to enter tumor cells via endocytosis, where they can contribute to the killing of tumor cells (Dobrovolskaia and McNeil 2007). Further research has documented that the uptake of cell microparticles significantly depends upon the microparticle diameter. In general, it can be said that the smaller the diameter, the greater the uptake of nanoparticles. Therefore, side effects and toxicity studies of arsenic could be diminished and the curative safety greatly improved. Nanotechnology may develop new applications of arsenic compound to treatment for solid tumors.

NANO-As_2O_3/Fe_2O_3 COMPLEXES FOR LIVER CANCER AND OTHER SOLID TUMORS

Primary hepatocellular carcinoma (HCC) is one of most common malignant tumors in China, and the incidence of HCC reported has apparently increased in recent years. Despite a variety of therapeutic strategies, HCC remains a significant cause of cancer morbidity (Zhang and Xia 2008). Current therapeutic options remain unsatisfactory for most patients. In the past years, surgical resection has been recognized as the most effective method for the treatment of HCC; however, it is only indicated for a small number of HCC patients. Therefore, it is crucial to identify a new method to treat hepatocarcinoma. For patients where resection is not possible, regional interventional therapies have led to major breakthroughs. These include transarterial chemoembolization, percutaneous ethanol injection, radiofrequency ablation, microwave coagulation therapy, laser-induced thermotherapy, and MFH. In comparison to traditional hyperthermia, MFH has many advantages, including strong magnetic responsiveness, automatic

Table 1. Key Features of Liver Cancer (Yip et al. 1999; Sun and Tang 2003).

1.	Liver cancer is a relatively rare form of cancer but has a high mortality rate.
2.	Liver cancers can be classified into two types. They are either primary, when the cancer starts in the liver itself, or metastatic, when the cancer has spread to the liver from some other part of the body.
3.	Primary hepatocellular carcinoma (HCC) is common in Africa and Southeast Asia.
4.	HCC is associated with cirrhosis of the liver in 70% of the cases. Other risk factors include hemochromatosis, hepatitis, schistosomiasis, exposure to vinyl chloride, and possibly nutritional deficiencies.
5.	Surgical resection has been recognized as the most effective method for the treatment of HCC. Other therapeutic approaches include: chemotherapy, radiation therapy, hepatic artery embolization with chemotherapy, liver transplantation etc.
6.	Current therapeutic options remain unsatisfactory for most patients.

This table lists the key facts of liver cancer.

temperature control in a high-frequency alternating electromagnetic field, and specific targeting of the therapy. Iron oxide MNPs, used in MFH, of xenograft liver cancer in nude mice were reported to have a significant therapeutic effect (Yan et al. 2005). They found that Fe_2O_3 nanoparticles combined with MFH could significantly inhibit the proliferation of the SMMC-7721 human liver cancer cells in a dose-dependent manner. *In vivo* experiments showed that Fe_2O_3 nanoparticles combined with MFH have a significant inhibitory effect in weight and volume of the xenograft liver cancer.

A number of researchers attempted to prepare a new kind of magnetic material containing As_2O_3 which had double functions: the chemotherapeutic function of As_2O_3 and thermotherapeutic function of iron MNPs. Nanosized As_2O_3/Fe_2O_3 complexes were prepared and its therapeutic effect on liver cancer were observed both *in vitro* and *in vivo* (Wang et al. 2009). The As_2O_3/Fe_2O_3 complexes were also observed using TEM, where they were shown to be round or elliptical in shape, well-dispersed and of approximately 100 nm in diameter, which is suitable for *in vivo* delivery (Fig. 4). Upon the application of an AMF, the temperature of As_2O_3/ Fe_2O_3 could rapidly reach 46°C within 5 min, which could kill tumor cells, while having little effect on normal cells. Fe_2O_3 and As_2O_3/ Fe_2O_3 nanoparticles were injected directly into the tumor tissue, instead of the boundary between normal tissue and tumor. This method allows thermogenic action to be administered locally, even in tumors located deep inside bodies, while minimizing heating of the normal tissue surrounding the tumor. Compared with As_2O_3/ Fe_2O_3 groups, As_2O_3/ Fe_2O_3 combined with MFH had an increased inhibitory effect on xenograft liver tumors,

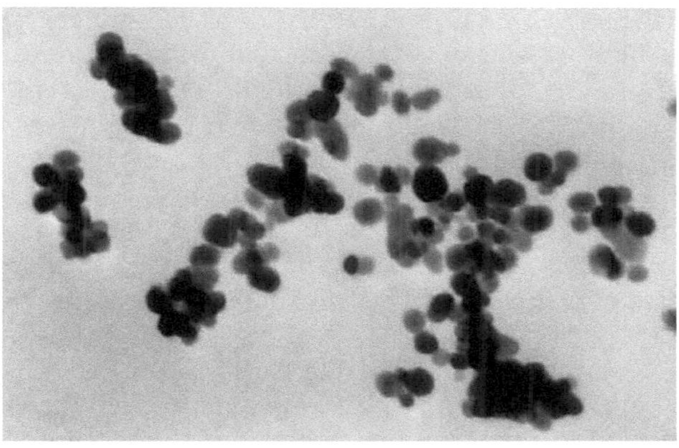

Fig. 4. TEM image of the As_2O_3/Fe_2O_3 complexes. (×50,000). The image shows As_2O_3/Fe_2O_3 complexes are round or elliptical in shape observed under TEM, with diameters about 100 nm (unpublished).

indicating that the MFH had a significant therapeutic effect. Moreover, the group of As_2O_3/ Fe_2O_3 combined with MFH group was the most effective therapeutic agent among all the groups tested. These findings suggest that the As_2O_3/Fe_2O_3 composite nanoparticles may be useful as thermoseeds for localized hyperthermia treatment of cancers, without damaging normal tissue, thus overcoming the limitations of conventional heat treatment. Moreover, by using these complexes, As_2O_3 can be localized to tumor tissue, thus reducing the As_2O_3 toxicity to normal tissue. These particles are suitable for treating superficial tumors by direct injection and for deep tumors by using different injection techniques (CT, ultrasound- or X-fluoroscopy-guided, intraoperatively under visual control), as they absorb energy from high-frequency AMF and also have chemotherapeutic effects.

Other studies have shown that As_2O_3 combined with MFH had good therapeutic effects on other solid tumors. For example, As_2O_3/Fe_3O_4 nanoparticles were investigated on xenograft HeLa cells in nude mice by Du (Du et al. 2009). By adjusting the concentration of Fe_3O_4, a suitable temperature (42°C–47°C) for tumor hyperthermia could be selected. At a concentration of 1 mg/mL Fe_3O_4 in the As_2O_3/Fe_3O_4 complex, the temperature of the magnetic fluid rose to 47°C after exposure to high-frequency AMF for 30 min. Following intratumor injection of simple Fe_3O_4 magnetic nanoparticles or nanosized As_2O_3/ Fe_3O_4 complexes, and exposure to high-frequency AMF (f = 230 kHz, I = 20 A) for 30 min, the entire tumor was almost heated by the nanoparticles and the temperature of tumors rose to 44°C–45°C. The mass and volume inhibition rates in the As_2O_3/ Fe_3O_4 group were significantly higher than what was observed in the control, As_2O_3, and Fe_3O_4 groups. Histological examination revealed that, in both the As_2O_3/ Fe_3O_4 and Fe_3O_4 groups, many black nanoparticles accumulated in the stroma of the tumors, with widespread tumor necrosis surrounding the nanoparticles. Although the mechanism by which nanoparticles enter tumor cells is not yet known, pinocytosis or penetration may occur, owing to the small diameter of the nanoparticles. Moreover, hyperthermia may increase the permeability of biological membranes. They also measured the effect of As_2O_3/Fe_3O_4 nanoparticles on the expression of cluster of differentiation-44 (CD44)-v6, vascular endothelial growth factor (VEGF)-C, and matrix metalloproteinases (MMP)-9, all of which were related to cancer and/or metastasis. The results showed As_2O_3 dose-dependently inhibited the expression of CD44v6, VEGF-C, and MMP-9 mRNA and thermotherapy alone inhibited the expression of VEGF-C. In addition, thermochemotherapy with the nanosized As_2O_3/ Fe_3O_4 complex may inhibit tumor metastasis to a certain extent. Other MNPs containing As_2O_3 forms have also been prepared, including As_2O_3/

$Mn_{0.5}Zn_{0.5}Fe_2O_4$ microspheres, nanomagnetoliposomes containing As_2O_3, which were used to treat other solid tumors, such as human breast and ovarian cancers.

CONCLUSION

Iron oxide MNPs are a suitable biomaterial for hyperthermia treatment with low toxicity. As_2O_3 combined with MFH is a promising technique for the minimally invasive elimination of solid tumors. However, to date, the majority of research in this field has only been evaluated in preclinical studies. Despite a small number of phase I clinical trials, it remains too early to claim therapeutic advantages for the applied method as survival rates and regressive time have not yet been defined, due to feasibility studies not yet having reached completion. Although there is still a long way to go before this technology can be applied in a clinical setting, it has the potential to develop into a new approach for the treatment of liver cancer and other solid tumors.

Applications to Areas of Health and Disease

Magnetic nanoparticles have been attractive for medical diagnostics and therapeutics. These NPs can be used for drug delivery, magnetic separation, magnetic resonance imaging, magnetofection and magnetic fluid hyperthermia. Hyperthermia is a very promising approach in the treatment of cancer. Compared with available hyperthermia modalities, magnetic fluid hyperthermia yields better results in uniform heating of the deeply situated tumors.

Key Terms

- **Median lethal dosage (LD50):** the dose that will kill 50% of the tested group.
- **Arsenic:** a shiny gray element with atomic symbol As, atomic number 33, and atomic weight 75. It occurs throughout the universe, mostly in the form of metallic arsenides. Most forms are toxic.
- **Acute promyelocytic leukemia (APL):** one type of acute myelogenous leukemia in which more than half the cells are malignant promyelocytes.
- **ROS:** molecules or ions formed by the incomplete one-electron reduction of oxygen. These reactive oxygen intermediates include singlet oxygen, superoxides, peroxides, hydroxyl radical and hypochlorous acid.

- **Apoptosis:** a natural process of self-destruction in certain cells that is determined by the genes and can be initiated by a stimulus or by removal of a repressor agent. Also called programmed cell death.

Summary Points

- Magnetic fluid hyperthermia (MFH) has become a potential method to treat tumors without injuring the normal tissue and cells.
- Iron oxide nanopartitcles (MNPs) are a suitable biomaterial for hyperthermia treatment with low toxicity.
- Arsenic is a traditional Chinese medicine that appears to be a valuable therapeutic tool in leukemia.
- With the development of nanotechnology, arsenic compounds have been prepared to nanoparticles with high therapeutic effect and low toxicity.
- As_2O_3 combined with MFH is a promising technique for the therapy of solid tumors because of its two functions: chemotherapy of As_2O_3 and thermotherapy of MNPs.

Abbreviations

AMF	:	alternating magnetic field
APL	:	acute promyelocytic leukemia
As_2O_3	:	arsenic trioxide
CD44	:	cluster of differentiation44
CT	:	computed tomography
HCC	:	primary hepatocellular carcinoma
LD50	:	median lethal dose
LITT	:	laser-induced thermotherapy
MCLs	:	magnetite cationic liposomes
MCT	:	microwave coagulation therapy
MFH	:	magnetic fluid hyperthermia
MMPs	:	matrix metalloproteinases
MNPs	:	magnetic nanoparticles
MRI	:	magnetic resonance imaging
PEI	:	percutaneous ethanol injection
RFA	:	radiofrequency ablation
RES	:	reticular-endothelial system
ROS	:	reactive oxygen species
TACE	:	transarterial chemoembolization
TEM	:	transmission electron microscope
VEGF	:	vascular endothelial growth factor

References

Bode, A.M., and Z. Dong. 2002. The paradox of arsenic: molecular mechanisms of cell transformation and chemotherapeutic effects. Crit. Rev. Oncol. Hematol. 42: p. 5–24.

Deng, Y., H. Xu, K. Huang, X. Yang, C. Xie and J. Wu. 2001. Size effects of realgar particles on apoptosis in a human umbilical vein endothelial cell line: ECV-304. Pharmacol. Res. 44: p. 513–8.

Dobrovolskaia, M.A., and S.E. McNeil. 2007. Immunological properties of engineered nanomaterials. Nat. Nanotechnol. 2: 469–478.

Du, Y., D. Zhang, H. Liu and R. Lai. 2009. Thermochemotherapy effect of nanosized As2O3/Fe3O4 complex on experimental mouse tumors and its influence on the expression of CD44v6, VEGF-C and MMP-9. BMC Biotechnol. 9: 84

Florea, A.M., F. Splettstoesser and D. Busselberg. 2007. Arsenic trioxide (As2O3) induced calcium signals and cytotoxicity in two human cell lines: SY-5Y neuroblastoma and 293 embryonic kidney (HEK). Toxicol. Appl. Pharmacol. 220: 292–301.

Gupta, A.K., and M. Gupta. 2005. Synthesis and surface engineering of iron oxide nanoparticles for biomedical applications. Biomaterials 26: 3995–4021

Hentschel, M., S. Mirtsch, A. Jordan, P. Wust, T. Vogl, W. Semmler, K.J. Wolf and R. Felix. 1997. Heat response of HT29 cells depends strongly on perfusion—a 31P NMR spectroscopy, HPLC and cell survival analysis. Int. J. Hyperthermia 13: 69–82.

Ito, A., M. Shinkai, H. Honda and T. Kobayashi. 2001. Heat-inducible TNF-alpha gene therapy combined with hyperthermia using magnetic nanoparticles as a novel tumor-targeted therapy. Cancer Gene Ther. 8: 649–654.

Johannsen, M., U. Gneveckow, K. Taymoorian, B. Thiesen, N. Waldofner, R. Scholz, K. Jung, A. Jordan, P. Wust and S.A. Loening. 2007. Morbidity and quality of life during thermotherapy using magnetic nanoparticles in locally recurrent prostate cancer: results of a prospective phase I trial. Int. J. Hyperthermia 23: 315–323.

Kumagai, Y., and D. Sumi. 2007. Arsenic: signal transduction, transcription factor, and biotransformation involved in cellular response and toxicity. Annu. Rev. Pharmacol. Toxicol. 47: 243–262.

Lam, C.W., J.T. James, R. McCluskey and R.L. Hunter. 2004. Pulmonary toxicity of single-wall carbon nanotubes in mice 7 and 90 days after intratracheal instillation. Toxicol. Sci. 77: 126–134.

Li, Z., F. Piao, S. Liu, L. Shen, N. Sun, B. Li and S. Qu. 2009. Preventive effects of taurine and vitamin C on renal DNA damage of mice exposed to arsenic. J. Occup. Health 51:169–172.

Lin, M., Z. Wang and D. Zhang. 2007. Preparation of orpiment nanoparticles and their cytotoxic effect on cultured leukemia K562 cells. J. Nanosci. Nanotechnol. 7: 1–7.

Lin, T.H., H.C. Kuo, F.P. Chou and F.J. Lu. 2008. Berberine enhances inhibition of glioma tumor cell migration and invasiveness mediated by arsenic trioxide. BMC Cancer 8: 58.

Maier-Hauff, K., R. Rothe, R. Scholz, U. Gneveckow, P. Wust, B. Thiesen, A. Feussner, A. von Deimling, N. Waldoefner, R. Felix and A. Jordan. 2007. Intracranial thermotherapy using magnetic nanoparticles combined with external beam radiotherapy: results of a feasibility study on patients with glioblastoma multiforme. J. Neurooncol. 81: 53–60.

Platanias, L.C. 2009. Biological responses to arsenic compounds. J. Biol. Chem. 284: 18583–7.

Seol, J.G., W.H. Park, E.S. Kim, C.W. Jung, J.M. Hyun, Y.Y. Lee and B.K. Kim. 2001. Potential role of caspase-3 and -9 in arsenic trioxide-mediated apoptosis in PCI-1 head and neck cancer cells. Int. J. Oncol. 18: 249–255.

Shen, S., Y. Liu, P. Huang and J. Wang. 2009. *In vitro* cellular uptake and effects of Fe3O4 magnetic nanoparticles on HeLa cells. J. Nanosci. Nanotechnol. 9: 2866–2871.

Shen, Z.Y., J. Shen, W.J. Cai, C. Hong, M.H. Zheng. 2000. The alteration of mitochondria is an early event of arsenic trioxide induced apoptosis in esophageal carcinoma cells. Int. J. Mol. Med. 5:155–158.

Soignet, S.L., S.R. Frankel, D. Douer, M.S. Tallman, H. Kantarjian, E. Calleja, R.M. Stone, M. Kalaycio, D.A. Scheinberg, P. Steinherz, E.L. Sievers, S. Coutre, S. Dahlberg, R. Ellison, R.P. Warrell, Jr. 2001. United States multicenter study of arsenic trioxide in relapsed acute promyelocytic leukemia. J. Clin. Oncol. 19: 3852–3860.

Sun, H.C. and Z.Y. Tang. 2003. Preventive treatments for recurrence after curative resection of hepatocellular carcinoma—a literature review of randomized control trials. World J. Gastroenterol. 9: 635–40.

Tasci, T.O., I. Vargel, A. Arat, E. Guzel, P. Korkusuz and E. Atalar. 2009. Focused RF hyperthermia using magnetic fluids. Med. Phys. 36: 1906–1912.

Wang, Z.Y., Dongsheng Zhang, Ning Gu, Xinli Lu and Shiyan Yan. 2005. Preparation and evalua tion of As2O3 nanoparticles for treatment of human liver cancer cells *in vitro*. Journal of Southeast University (English Edition) 21: 58–62.

Wang, Z.Y., J. Song and D.S. Zhang. 2009. Nanosized As2O3/Fe2O3 complexes combined with magnetic fluid hyperthermia selectively target liver cancer cells. World J. Gastroenterol. 15: 2995–3002.

Xiao, Y.F., D.D. Wu, S.X. Liu, X. Chen and L.F. Ren. 2007. Effect of arsenic trioxide on vascular endothelial cell proliferation and expression of vascular endothelial growth factor receptors Flt-1 and KDR in gastric cancer in nude mice. World J. Gastroenterol. 13: 6498–6505.

Xu, C., and S. Sun. 2009. Superparamagnetic nanoparticles as targeted probes for diagnostic and therapeutic applications. Dalton. Trans. 7: 5583–5591.

Yan, S., D. Zhang, N. Gu, J. Zheng, A. Ding, Z. Wang, B. Xing, M. Ma, Y. Zhang. 2005. Therapeutic effect of Fe2O3 nanoparticles combined with magnetic fluid hyperthermia on cultured liver cancer cells and xenograft liver cancers. J. Nanosci. Nanotechnol. 5: 1185–1192.

Yip, D., M. Findlay, M. Boyer and M.H. Tattersall. 1999. Hepatocellular carcinoma in central Sydney: a 10-year review of patients seen in a medical oncology department. World J. Gastroenterol. 5: 483–487.

Zhang, Y.Y., and H.H. Xia. 2008. Novel therapeutic approaches for hepatocellular carcinoma: fact and fiction. World J. Gastroenterol. 14: 1641–1642.
Zhao, D.L., H.L. Zhang, X.W. Zeng, Q.S. Xia, J.T. Tang. 2006. Inductive heat property of Fe3O4/polymer composite nanoparticles in an ac magnetic field for localized hyperthermia. Biomed. Mater. 1: 198–201.
Zhao, X., T. Feng, H. Chen, H. Shan, Y. Zhang, Y. Lu and B. Yang. 2008. Arsenic trioxide-induced apoptosis in H9c2 cardiomyocytes: implications in cardiotoxicity. Basic Clin. Pharmacol. Toxicol. 102: 419–425.
Zheng, Y., H. Yamaguchi, C. Tian, M.W. Lee, H. Tang, H.G. Wang, Q. Chen. 2005. Arsenic trioxide (As(2)O(3)) induces apoptosis through activation of Bax in hematopoietic cells. Oncogene 24: 3339–3347.

19

Hepatocellular Carcinoma and Midkine-antisense Oligonucleotide-loaded Nanoparticles

Licheng Dai,[1,a,*] *Xing Yao*[1,b] and *Jing Zhong*[1,c]

ABSTRACT

It has been demonstrated that ASODNs (antisense oligonucleotides) to selectively target genes had enormous potential for treatment of several human diseases including cancer. We recently found that MK-ASODNs (antisense oligonucleotides targeting midkine) could inhibit growth of HCC (hepatocellular carcinoma) cells by especially down-regulating expression of MK. However, the instability in blood and poor penetration into cells of MK-ASODNs limit their widespread using in the clinic. In order to overcome these impediments, we employ the nanoparticles as delivery tools by encapsulating MK-ASODNs within nanoparticles. It is observed that the NANO-ASODNs (MK-ASODNs packing with nanoparticles) were stable in the blood and absorbable into

[1]Huzhou Key Laboratory of Molecular Medicine, Huzhou Central Hospital, Huzhou, Zhejiang Province, China.
[a]E-mail: dlc@hzhospital.com
[b]E-mail: yaoy@mail.huptt.zj.cn
[c]E-mail: zhongjing103@yahoo.com.cn
*Corresponding author

List of abbreviations after the text.

HCC cells. Furthermore, results showed the growth of HCC cells was significantly inhibited by the NANO-ASODNs both *in vitro* and *in vivo* with fewer side effects. In summary, NANO-ASODNs might be used as potential drugs for HCC therapy in the near future by reducing the side effects and improving the therapeutic efficacy.

INTRODUCTION

HCC (Hepatocellular carcinoma) is one of the most malignant diseases in the world with a high incidence and mortality. Surgical resection treatment offers the best prognosis for long-term survival, but unfortunately only 10–15% of patients are suitable for surgical resection (El-Serag et al. 2008). Recently, we reported that antisense oligonucleotides targeting MK (Midkine) suppressed the growth of HCC tumors in nude mice significantly. MK, a heparin-binding growth factor or cytokine, has been reported to be generally overexpressed in malignant tumors and exhibits several cancer-related activities, including fibrinolytic, anti-apoptotic, mitogenic, transforming, angiogenic and chemotactic functions (Calinescu et al. 2009; Ezquerra et al. 2008; Garver et al. 1994; Kato et al. 2000; Maeda et al. 2007; Murasugi and Tohma-Aiba 2001; Rha et al. 1997).

As we known, the ASODNs (antisense oligonucleotides) are instable in the physical fluids and difficult to penetrate into cells. At present, the delivery tools of ASODNs are ineffective and toxic. Although lentiviral technology is a proven tool in the laboratory setting, it has several adverse effects, such as toxic immune responses and genetic alterations. So, the effect delivery system for ASODNs is urgently needed. In our recent studies, we used the systemic delivery of NANO-ASODNs (nanoparticles incorporated with MK-ASODNs). The results showed that the nanoparticles delivery approach is much safer and more effective for the treatment of HCC. NANO-ASODNs have been demonstrated to play an important role in the suppression of HCC growth significantly *in vitro* and *in vivo* and provide insight into their future clinical application to HCC therapy.

ANTISENSE OLIGONUCLEOTIDES ARE APPLIED TO AREAS OF HEALTH AND DISEASE

Antisense therapy has been investigated extensively over the decades because its ease of design and low cost. In general, antisense oligonucleotides are single-stranded DNA molecules with 13–25 nucleotides. Antisense oligonucleotides could inhibit target gene expression by binding the starting location of mRNA translation inside cells to block translation of target mRNA into protein (Crooke 2004). Antisense drugs are being applied to treat several kinds of human disease including cancers, diabetes, asthma and arthritis. One antisense drug Vitravene has been approved by the US Food and Drug Administration (FDA) as a treatment for cytomegalovirus retinitis.

In cancers, a number of cellular genes (proto-oncogenes) are altered. Selective inhibition by antisense oligonucleotides directed against several oncogenes has been investigated. In chronic myeloid leukemia, antisense oligonucleotides is targeted to the mRNAs of the bcr-abl fusion gene or c-myb suppressed leukemic cell proliferation (Ratajczak et al. 1992). In several tumors, the Ha-ras gene is activated by a single nucleotide mutation. Antisense oligomers can selectively inactivate the mutated but not the normal Ha-ras gene, showing the highly selective action of antisense oligonucleotides. Further, growth of these tumor cells is suppressed *in vitro* and *in vivo*. We recently found antisense oligonucleotides targeting MK, an oncogene overexpressed in various tumors, suppressed the growth of HCC tumors (L.C. Dai et al. 2006; Dai et al. 2009). However, to date, rapid degradation and poor diffusion across the cell membrane remains challenging. To overcome these bottlenecks, a variety of vectors such as lipids, polymers, peptides and nanoparticles have been explored.

KEY FACTS OF HEPATOCELLULAR CARCINOMA

Hepatocellular carcinoma (HCC, also called malignant hepatoma) is a primary malignancy of the liver and is usually caused by hepatitis B or C infection. HCC is one of the most common tumors worldwide. The epidemiology of HCC exhibits two main patterns, one in North America and Western Europe and another in non-Western countries. Usually males are affected more than females and it is most common between the age of 30 to 50. (Parkin et al. 2005). The mortality rate from HCC is the third highest worldwide for any cancer-related diseases, and since the 1990s, HCC has been the cause of the second highest mortality rate due to cancer in China. Resection, transplantation or percutaneous and transarterial interventions are common therapies for HCC, which are usually diagnosed at advanced stage (El-Serag et al. 2008). However, these therapeutic results are not satisfied. It is accepted that there is no satisfactory treatment available for patients with HCC and chemotherapy has been extremely disappointing (El-Serag et al. 2008). The poor prognosis of patients with HCC and the lack of satisfactory therapy for advanced cases indicate a need for more effective therapeutic options.

THE ONCOGENE OF MIDKINE

Midkine (MK), is a 13-kDa heparin-binding growth factor with a high affinity for heparin, which shares a 50% homology in amino acid sequence with pleiotrophin (PTN), another member of MK family. MK is a heparin-binding growth factor with its gene first identified in embryonal carcinoma cells at early stages of retinoic acid-induced differentiation. MK is frequently and highly expressed in a variety of human carcinomas, such as breast, prostate, grastric, colon, hepatocellular and urinary bladder carcinomas, neuroblastoma, glioma and Wilms' tumor. Furthermore, the blood MK level is frequently elevated with advance of human carcinomas, decreased after surgical removal of the tumors (Chung et al. 2002; Ezquerra et al. 2008; Mitsiadis et al. 1995; Nakamura et al. 1998; Obama et al. 1995; Tao et al. 2007). Thus, it is expected to become a promising marker for evaluating the progress of carcinomas. There is mounting evidence that MK plays a significant role in carcinogenesis-related activities, such as proliferation, migration, anti-apoptosis, mitogenesis, transforming, and angiogenesis. (Fujita et al. 2008; Nakamoto et al. 1992; Owada et al. 1999; Qi et al. 2000) In addition, siRNA and antisense oligonucleotides for MK have yielded great effects in anti-tumor activities. (L.C. Dai et al. 2006; Dai Wang et al. 2007a). Therefore, MK appears to be a potential candidate molecular target of therapy for human carcinomas.

ANTISENSE OLIGONUCLEOTIDES TARGETING MIDKINE ARE POTENTIAL THERAPEUTIC AGENTS FOR CURING HEPATOCELLULAR CARCINOMA CANCER

HCC (Hepatocellular carcinoma) is one of the most common malignancies with high mortality. There are several mechanisms of carcinogenesis which have been proposed for HCC. Down-regulation of apoptosis and accelerated proliferative activity in hepatic cells had been regarded as malignant factors leading to HCC. Recently, HCC tumor cells were found to overexpress MK (Kato et al. 2000), whereas it is low or undetectable in normal liver tissue. It is suggested that MK could be a potential target for hepatocellular carcinoma therapy.

In our previous studies, we developed potential ASODNs to repress MK expression in hepatocellular carcinoma cells and examine their antiproliferation activity. In our study, we designed and synthesized ten ASODNs targeting MK according to a computational neural network model (Dai et al. 2009) (Table1). It is found that ten ASODNs showed various inhibitory activities on the 3 lines of liver cancer cells, BEL7402, HepG2, and SMMC-7721. The inhibition rate ranged from 20% to 85%, positively related to the ASODNs concentrations. As measured by RT-PCR (Real time-PCR) and Western blotting, the antisense compounds efficiently downregulated the MK expression level in HepG2 cells in a dose-dependent manner. However, the transfection of mismatched and sense compounds showed little effect on MK expression in HepG2 cells. Our data suggested that the proliferation inhibition by ASODNs was due to the downregulation of MK expression.

Drug tolerance is often developed during combination chemotherapy of HCC. Recently, we analyzed the combination of MK-ASODNs with chemotherapeutic drugs. The results showed that MK-ASODNs could increase the anti-cancer effect of chemotherapeutic drugs (cisplatin (DDP), 5-fluorouracil (5-FU) and adriamycin (ADM)) (L.C. Dai et al. 2006; Dai Wang et al. 2007b; Dai et al. 2007). Furthermore, we also observed the synergism at a lower concentration of all these chemotherapeutic drugs. This result implied that combination of MK-ASODNs with chemotherapeutic drugs will possibly provide a more marked therapeutic effect.

In brief, our studies suggest that MK-ASODNs can inhibit the proliferation of HCC cells significantly. Moreover, MK-ASODNs could increase the therapeutic effect of chemotherapeutic drugs. However, rapid degradation and poor penetration of cell membrane remains challenging.

MIDKINE-ANTISENSE OLIGONUCLEOTIDE-LOADED NANOPARTICLES INHIBIT GROWTH OF HEPATOCELLULAR CARCINOMA TUMOR BOTH *IN VITRO* AND *VIVO*

It is demonstrated that naked oligonucleotides are internalized poorly by cells whether or not they are negatively charged (Crooke 2004). More specifically, naked oligonucleotides tend to localize in endosomes/ lysosomes, where they are unavailable for antisense purposes. To improve cellular uptake and oligonucleotide spatial and temporal activity, a range of techniques and transporters have been developed. In our recent studies, we packaged the MK-ASODNs with nanoparticles to enhance the MK-ASODNs penetrating into cells and stabilizing them.(H. Dai et al. 2006). Evidence from our experiments showed that nanoliposomes packaged with MK-ASODNs could suppress HCC growth both *in vitro* and *in vivo*. In addition, the data also indicated that nanoliposomes effectively delivered MK-ASODNs and showed less systematic toxicity. Consequently, nanoliposomes incorporated with MK-ASODNs should represent an effective and less toxic approach for treatment of HCC, and potentially, other tumor types.

Nanoparticles Liposome Packageding and Features

The produced nanometer liposomes were used to package the MK-ASODNs using a ratio of 1.8 μl of the nanometer liposomes to 1 μg MK-ASODNs (Dai et al. 2009) (Fig. 1). The nanoparticles packaged with MK-ASODNs were stained with 1% uranyl acetate and examined with TEM (transmission electron microscope). The size of the micelles was determined with a Zetasizer 5000 (Malvern Instruments, Malvern, and Worcestershire, UK). The morphology of the nanoparticles packaged with MK-ASODNs was examined using dynamic light scattering and TEM (Fig. 1). In order to determine the transduction of NANO-ASODNs, the ASODNs were conjugated with carboxyfluorescein (FAM) and the rate of transduction under the confocal microscope at different time points was observed. It is indicated that the NANO-ASODNs were effectively transduced into the HepG2 cells at the indicated times (6, 12 and 18 h) (Fig. 2A). *In vivo*, Fig. 3 showed that NANO-ASODNs mainly targeted the liver after injection through the tail vein. We also found that the concentration of the NANO-ASODNs reached a peak 90 min after the injection, and then slowly decreased. (Dai et al. 2009).

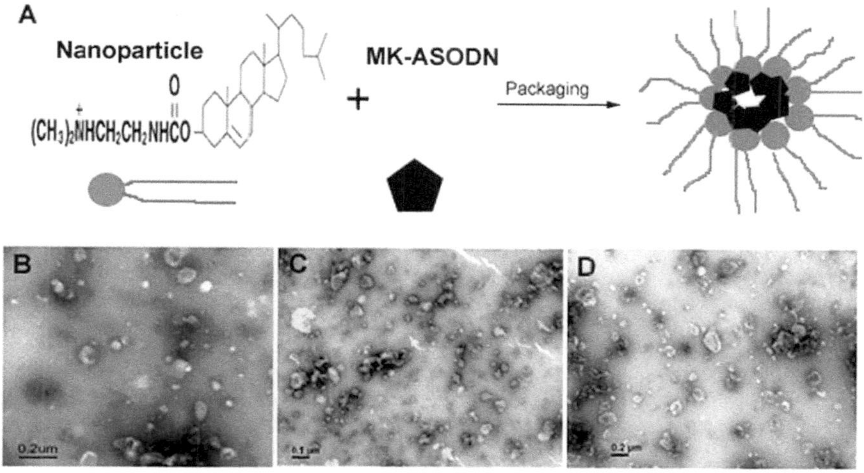

Fig. 1. Nano-assembly of MK-ASODNs and nanoliposomes and characterization of NANO-ASODNs. Characteristics of NANO-ASODNs. A: Schematic illustration of the self-assembly of MKASODN and nanoliposomes; B: TEM image of the empty nanoliposomes stained with 1% uranyl acetate; C, D: TEM image of the NANO-ASODNs. Data are from Dai et al. (2009), with permission from the Publishers.

MK, midkine; ASODNs, antisense oligonucleotides; MK-ASODNs, antisense oligonucleotides targeting midkine; NANO-ASODNs, MK-ASODNs packaged with nanoparticles; TEM, Transmission electron microscopy

Inhibition of Growth of HCC *in vitro* and *in vivo*

In vitro, we demonstrated that the NANO-ASODNs (0.1, 0.2, 0.4 and 0.8 μg/ml) could significantly down-regulate the MK mRNA levels (Fig. 2B). Moreover, the MTS (3-(4,5-dimethylthiazol-2-yl)-5-(3-carboxymethoxyphenyl)-2-(4-sulfophenyl)-2H-tetrazolium), assay showed that NANO-ASODNs could inhibit HCC cell proliferation significantly. Figure. 2C showed that the inhibition rates ranged from 20% to 80%, which correlated with the NANO-ASODNs concentrations.

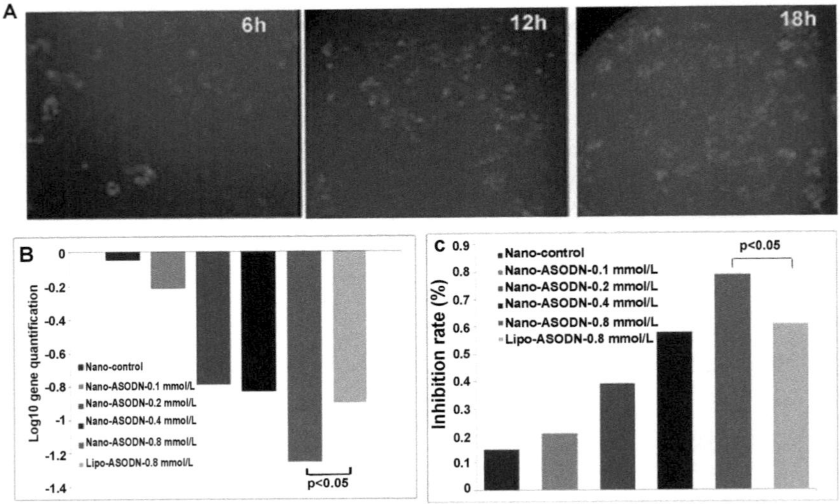

Fig. 2. Transduction and function of NANO-ASODNs *in vitro*. Analysis of NANO-ASODNs function. A: 0.2 mmol/L FAM-conjugated NANO-ASODNs transduced into HepG2 cells. The results were observed under a confocal microscope at indicted times of 6, 12 and 18 h; B: NANO-ASODNs down-regulated expression of MK mRNA; C: The proliferation of HepG2 cells was significantly inhibited by NANO-ASODNs (Data are from Dai et al. 2009), with permission from the Publishers.

MK, midkine; ASODNs, antisense oligonucleotides; MK-ASODNs, antisense oligonucleotides targeting midkine; NANO-ASODNs, MK-ASODNs packaged with nanoparticles; TEM, Transmission electron microscopy

Color image of this figure appears in the color plate section at the end of the book.

Later, we used an *in situ* mouse HCC model to evaluate the antitumor activity of NANO-ASODNs *in vivo*. After establishing the mouse HCC model for 2 d, PBS, free nanoparticles, 10 mg/kg per day fluorouracil (5-FU), various doses of NANO-ASODNs or ASODNs (25, 50 and 100 mg/kg per day) were administered through the tail vein for 20 d. The tumors were removed after sacrificing the mice. The tumors were measured and weighed. Table 2 and Fig. 4, 5 show the final tumor volumes and

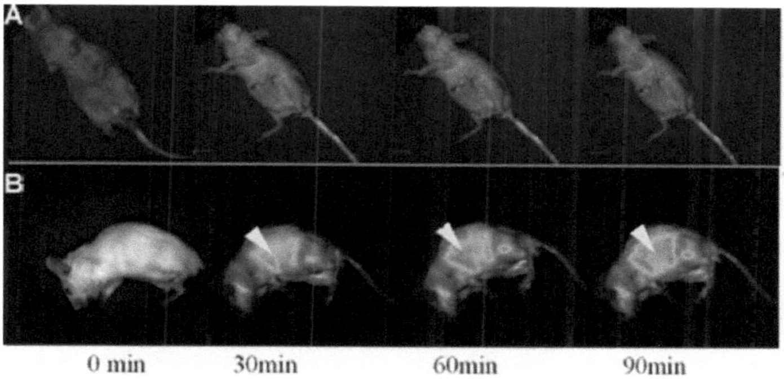

Fig. 3. NANO-ASODNs targeted to liver. The kinetic results of the NANO-ASODNs were observed through in vivo imaging systems at indicated times after NANO-ASODNs were injected through the tail vein. A: Free ASODNs did not concentrate within the liver and these ASODNs disappeared quickly; B: NANO-ASODNs were found to mainly target the liver (the arrow represents the NANO-ASODNs) (Data are from Dai et al. 2009), with permission from the Publishers.
MK, midkine; ASODNs, antisense oligonucleotides; MK-ASODNs, antisense oligonucleotides targeting midkine; NANO-ASODNs, MK-ASODNs packaged with nanoparticles; TEM, Transmission electron microscopy

Color image of this figure appears in the color plate section at the end of the book.

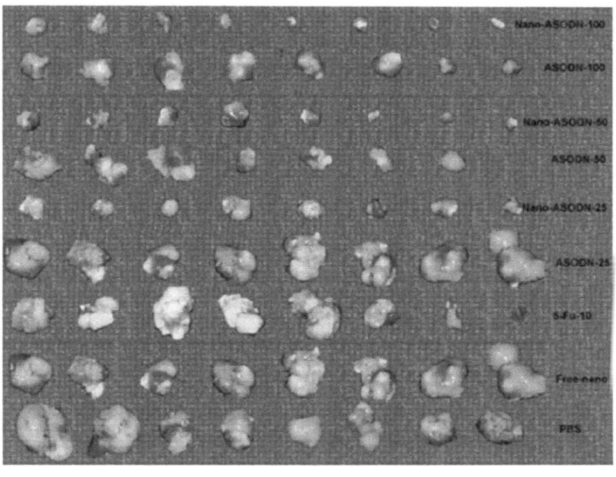

Fig. 4. Morphological changes of HCC following treatment with NANO-ASODNs. The volume of HCC decreased significantly following treatment with 100, 50 and 25 mg/kg per day of NANO-ASODNs for 20 d. MK-ASODNs were the positive control. The PBS or free nanoparticles represent the negative controls. (Data are from Dai et al. 2009), with permission from the Publishers.
MK, midkine; ASODNs, antisense oligonucleotides; MK-ASODNs, antisense oligonucleotides targeting midkine; NANO-ASODNs, MK-ASODNs packaged with nanoparticles; TEM, Transmission electron microscopy

Color image of this figure appears in the color plate section at the end of the book.

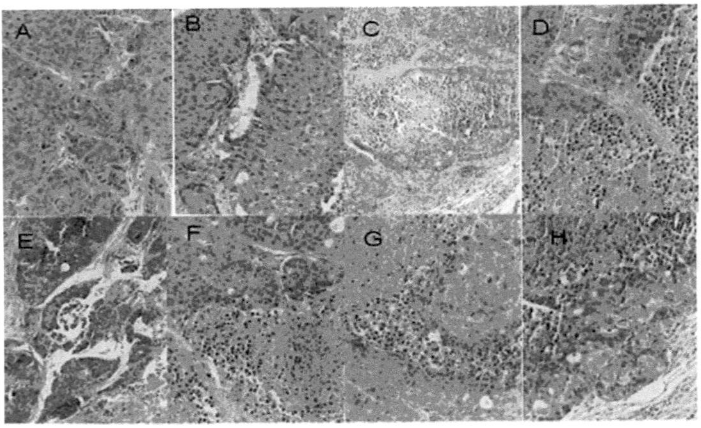

Fig. 5. Histopathological analysis. Analysis changes of histopathology. A: Tissue sections of the tumors from in situ xenograft HCC; B: Tissue sections of the tumors from nanoparticles; C: Tissue sections of the tumors from 5-FU (10 mg/kg per day); D: Tissue sections of the tumors treated with ASODNs 50 mg/kg per day; E-G: Tissue sections of tumors treated with NANO-ASODNs 100, 50 and 25 mg/kg per day of NANO-ASODN treated tumors, respectively; H: Tissue sections of the tumors treated with 5-FU (10 mg/kg per day) and 50 mg/kg per day of NANO-ASODNs. Regions showing an increase of necrosis and fibrosis were observed in the 5-FU or ASODN treatment groups (C-H, × 200) compared with the free nanoparticles and untreated groups (A and B, × 200) (Data are from Dai et al. 2009), with permission from the Publishers.
MK, midkine; ASODNs, antisense oligonucleotides; MK-ASODNs, antisense oligonucleotides targeting midkine; NANO-ASODNs, MK-ASODNs packaged with nanoparticles; TEM, Transmission electron microscopy

Color image of this figure appears in the color plate section at the end of the book.

weights after 20 d of treatment. The results showed that the tumor volumes decreased in both free ASODNs and NANO-ASODNs treated groups compared with the PBS control group ($p < 0.01$). Additionally, the effect of NANO-ASODNs on tumor growth inhibition was superior to the free ASODNs ($p < 0.05$). Moreover, the effect of NANO-ASODNs on inhibiting tumor proliferation was dose-dependent (Table 2). In addition, the NANO-ASODNs treatment also resulted in a significant inhibition of tumor weight compared with the PBS and free-nanoparticle-treated mice. In contrast to the PBS group, it had the highest inhibitory efficacy for the tumor weight which was 81.77% and 10 mg/kg 5-FU had an inhibitory efficacy of 48.96%; however, for the free nanoparticles treatment, the inhibitory rate was only 17.18% ($p < 0.01$) (Table 1) (Dai et al. 2009).

Table 1. The sequence of ASODNs.

NO	target position	size	sequence
ASODN 1	629–648	20	ATCCCTTGTCCCTCCCCAC
ASODN 2	178–197	20	CGCGGA AACCCACGCCGCAA
ASODN 3	1–20	20	TGCCCGCGCTGCTTGGCCCG
ASODN 4	428–447	20	CCCTTCCCTTTCTTGGCTTT
ASODN 5	108–127	20	CCCCGGGCCGCCCTTCTTCA
ASODN 6	296–315	20	CCCCATCACACGCACCCCAG
ASODN 7	435–454	20	GTCCTTTCCCTTCCCTTTCT
ASODN 8	636–655	20	TCCCAGAATCCCTTGTCCCT
ASODN 9	61–80	20	AGGAGGGTGAGGAGGAGGAA
ASODN 10	511–530	20	CGGGCCTGGGAGAGGGAGGG
Sense	–	20	TGA AGA AGGGCGGCCCGGGG
Mismatch	–	20	CCGGCCGCGCGGCAAGATCA

In this table, it showed several sequences of ASODNs. (Data is from Dai et al. 2006), with permission from the Publishers.
MK, midkine; ASODNs, antisense oligonucleotides; MK-ASODNs, antisense oligonucleotides targeting midkine; NANO-ASODNs, MK-ASODNs packaged with nanoparticles; TEM, Transmission electron microscopy

Table 2. NANO-ASODNs inhibit growth of HCC *in vivo*.

Group (mg/kg)	Tumor volume (mm3)	Tumor weight (g)	Tumor inhibition (%)
Control (PBS)	1633.38 ± 525.93	1.92 ± 0.45	
ASODN-100	321.56 ± 85.55b	0.57 ± 0.16b	70.31
ASODN-50	509.29 ± 300.85b	0.73 ± 0.22b	61.98
ASODN-25	835.25 ± 263.33a	1.14 ± 0.12b	40.63
5-FU10	717.19 ± 281.25b	0.98 ± 0.16b	48.96
Nano-ASODN-100	225.81 ± 128.75b	0.35 ± 0.17b	81.77
Nano-ASODN-50	457.88 ± 249.29b	0.52 ± 0.21b	72.92
Nano-ASODN-25	584.00 ± 261.92b	0.83 ± 0.20b	56.77
Nano control	1319.25 ± 340.70	1.59 ± 0.18	17.18

aP < 0.05, bP < 0.01 vs control (PBS).
The tumor growth was inhibited significantly by NANO-ASODNs. a: p < 0.05, b : p < 0.01 vs control (PBS). (Data are from Dai et al. 2009), with permission from the Publishers.
MK, midkine; ASODNs, antisense oligonucleotides; MK-ASODNs, antisense oligonucleotides targeting midkine; NANO-ASODNs, MK-ASODNs packaged with nanoparticles; TEM, Transmission electron microscopy

SUMMARY AND PROSPECT

In our reports, we used more effective and less toxic nanoparticle liposomes that have previously been used effectively to deliver siRNA for the treatment of lymphoma and ovarian cancer, as well as colorectal carcinoma. (Takei et al. 2006). Liposomes, the first nanotechnology to benefit cancer patients, are continuing to evolve as tools for delivering potentially useful therapies for the treatment against tumors. MK-ASODNs incorporated into nanoliposomes have been used effectively *in vivo*. Evidence from these experiments showed that nanoliposomes packaged with MK-ASODNs could suppress HCC growth both *in vitro* and *in vivo*. In addition, the data also indicated that nanoliposomes could effectively deliver MK-ASODNs and showed less systematic toxicity. Consequently, nanoliposomes incorporated with MK-ASODNs should represent an effective and less toxic approach for treatment of HCC, and potentially, other tumor types.

In summary, our reports suggested that nanoliposomes packaged with MK-ASODNs can increase the therapeutic effect of MK-ASODNs, both *in vivo* and *in vitro*, for the treatment of HCC. The combination of nanoliposomes and MK-ASODNs showed a more effective and less toxic tool for therapy against HCC and should provide a novel strategy for cancer treatment.

Summary Points

- Hepatocellular carcinoma is one of the most common tumors worldwide, caused thousands of deaths worldwide per year, about half of them in China.
- Midkine was involved in progression of hepatocellular carcinoma, which has been demonstrated to be a promising therapeutic target for hepatocellular carcinoma.
- Oligonucleotides are unmodified or chemically modified single-stranded DNA molecules, which hybridize to a unique sequence in the total pool of targets present in cells.
- Liposome-nanoparticle hybrid constructs present great opportunities in terms of nanoscale delivery system engineering for combinatory therapeutic-imaging modalities.
- Antisense oligonucleotides of midkine delivering by nanoparticles inhibit growth of hepatocellular carcinoma significantly.

Abbreviations

HCC	:	hepatocellular carcinoma
ASODNs	:	antisense oligonucleotides
MK	:	midkine
MK-ASODNs	:	antisense oligonucleotides against to midkine
NANO	:	nanoparticles

References

Calinescu, A.A., T.S. Vihtelic, D.R. Hyde and P.F. Hitchcock. 2009. Cellular expression of midkine-a and midkine-b during retinal development and photoreceptor regeneration in zebrafish. J. Comp. Neurol. 1: 1–10.

Chung, H.W., Y. Wen, E.A. Choi, L. Hao, H.S. Moon, H.K. Yu, et al. 2002. Pleiotrophin (PTN) and midkine (MK) mRNA expression in eutopic and ectopic endometrium in advanced stage endometriosis. Mol. Hum. Reprod 4: 350–355.

Crooke, S.T. 2004. Progress in antisense technology. Annu. Rev. Med. 55: 61–95.

Dai, H., X. Jiang, G.C. Tan, Y. Chen, M. Torbenson, K.W. Leong, et al. 2006 Chitosan-DNA nanoparticles delivered by intrabiliary infusion enhance liver-targeted gene delivery. Int. J. Nanomedicine 4: 507–522.

Dai, L.C., X. Wang, X. Yao, Y.L. Lu, J.L. Ping and J.F. He. 2006. Antisense oligonucleotides targeting midkine induced apoptosis and increased chemosensitivity in hepatocellular carcinoma cells. Acta. Pharmacol. Sin. 12: 1630–1636.

Dai, L.C., X. Wang, X. Yao, Y.L. Lu, J.L. Ping and J.F. He. 2007a. Antisense oligonucleotide targeting midkine suppresses *in vivo* angiogenesis. World J. Gastroenterol. 8: 1208–1213.

Dai, L.C., X. Wang, X. Yao, Y.L. Lu, J.L. Ping and J.F. He. 2007b. Enhanced therapeutic effects of combined chemotherapeutic drugs and midkine antisense oligonucleotides for hepatocellular carcinoma. World J. Gastroenterol. 13: 1989–1994.

Dai, L.C., X. Wang, X. Yao, L.S. Min, J.L. Ping and J.F. He. 2007. Antisense oligonucleotides targeting midkine inhibit tumor growth in an in situ human hepatocellular carcinoma model. Acta. Pharmacol. Sin. 3: 453–458.

Dai, L.C., X. Yao, X. Wang, S.Q. Niu, L.F. Zhou, F.F. Fu, et al. 2009. *In vitro* and *in vivo* suppression of hepatocellular carcinoma growth by midkine-antisense oligonucleotide-loaded nanoparticles. World J. Gastroenterol. 16: 1966–1972.

El-Serag, H.B., J.A. Marrero, L. Rudolph and K.R. Reddy. 2008. Diagnosis and treatment of hepatocellular carcinoma. Gastroenterology 6: 1752–1763.

Ezquerra, L., L.F. Alguacil, T. Nguyen, T.F. Deuel, I. Silos-Santiago and G. Herradon. 2008. Different pattern of pleiotrophin and midkine expression in neuropathic pain: correlation between changes in pleiotrophin gene expression and rat strain differences in neuropathic pain. Growth Factors 1: 44–48.

Fujita, S., S. Seki, M. Fujiwara and T. Ikeda. 2008. Midkine expression correlating with growth activity and tooth morphogenesis in odontogenic tumors. Hum. Pathol. 5: 694–700.

Garver, R.I., Jr., D.M. Radford, H. Donis-Keller, M.R. Wick and P.G. Milner. 1994. Midkine and pleiotrophin expression in normal and malignant breast tissue. Cancer 5: 1584–1590.

Kato, M., T. Shinozawa, S. Kato, A. Awaya and T. Terada. 2000. Increased midkine expression in hepatocellular carcinoma. Arch. Pathol. Lab Med. 6: 848–852.

Maeda, S., H. Shinchi, H. Kurahara, Y. Mataki, H. Noma, K. Maemura, et al. 2007. Clinical significance of midkine expression in pancreatic head carcinoma. Br. J. Cancer 3: 405–411.

Mitsiadis, T.A., M. Salmivirta, T. Muramatsu, H. Muramatsu, H. Rauvala, E. Lehtonen, et al. 1995. Expression of the heparin-binding cytokines, midkine (MK) and HB-GAM (pleiotrophin) is associated with epithelial-mesenchymal interactions during fetal development and organogenesis. Development, 1: 37–51.

Murasugi, A., and Y. Tohma-Aiba. 2001. Comparison of three signals for secretory expression of recombinant human midkine in Pichia pastoris. Biosci. Biotechnol. Biochem. 10: 2291–2293.

Nakamoto, M., S. Matsubara, T. Miyauchi, H. Obama, M. Ozawa and T. Muramatsu. 1992. A new family of heparin binding growth/differentiation factors: differential expression of the midkine (MK) and HB-GAM genes during mouse development. J. Biochem. 3: 346–349.

Nakamura, E., K. Kadomatsu, S. Yuasa, H. Muramatsu, T. Mamiya, T. Nabeshima, et al. 1998. Disruption of the midkine gene (Mdk) resulted in altered expression of a calcium binding protein in the hippocampus of infant mice and their abnormal behaviour. Genes Cells 12: 811–822.

Obama, H., J. Tsutsui, M. Ozawa, H. Yoshida, Y. Yoshida, M. Osame, et al. 1995. Midkine (MK) expression in extraembryonic tissues, amniotic fluid, and cerebrospinal fluid during mouse embryogenesis. J. Biochem. 1: 88–93.

Owada, K., N. Sanjo, T. Kobayashi, H. Mizusawa, H. Muramatsu, T. Muramatsu, et al. 1999. Midkine inhibits caspase-dependent apoptosis via the activation of mitogen-activated protein kinase and phosphatidylinositol 3-kinase in cultured neurons. J. Neurochem. 5: 2084–2092.

Parkin, D.M., F. Bray, J. Ferlay and P. Pisani. 2005. Global cancer statistics, 2002. CA Cancer J. Clin. 2: 74–108.

Qi, M., S. Ikematsu, K. Ichihara-Tanaka, S. Sakuma, T. Muramatsu and K. Kadomatsu. 2000. Midkine rescues Wilms' tumor cells from cisplatin-induced apoptosis: regulation of Bcl-2 expression by Midkine. J. Biochem. 2: 269–277.

Ratajczak, M.Z., J.A. Kant, S.M. Luger, N. Hijiya, J. Zhang, G. Zon, et al. 1992. *In vivo* treatment of human leukemia in a scid mouse model with c-myb antisense oligodeoxynucleotides. Proc. Natl. Acad. Sci. USA 24: 11823–11827.

Rha, S.Y., S.H. Noh, H.J. Kwak, A. Wellstein, J.H. Kim, J.K. Roh, et al. 1997. Comparison of biological phenotypes according to midkine expression in gastric cancer cells and their autocrine activities could be modulated by pentosan polysulfate. Cancer Lett. 1: 37–46.

Takei, Y., K. Kadomatsu, T. Goto and T. Muramatsu. 2006. Combinational antitumor effect of siRNA against midkine and paclitaxel on growth of human prostate cancer xenografts. Cancer 4: 864–873.

Tao, P., D. Xu, S. Lin, G.L. Ouyang, Y. Chang, Q. Chen, et al. 2007. Abnormal expression, highly efficient detection and novel truncations of midkine in human tumors, cancers and cell lines. Cancer Lett. 1: 60–67.

20

Nanotechnology and Nanostructures Applied to Head and Neck Cancer

Erik C. Dreaden,[1,a] *Mostafa A. El-Sayed*[1,b] and *Ivan H. El-Sayed*[2,*]

ABSTRACT

Global incidence and mortality from cancer is projected to more than double between 2008 and 2030, resulting in an additional 8.7 million new cases and 5.6 million deaths annually (United Nations, 2009). Head and neck cancers (HNCs) account for approximately 12 percent of current incidence and mortality worldwide. HNC is more prevalent in less developed regions than in economically developed countries (8 versus 15 percent) and most arise from the squamous cell lining of the upper aerodigestive tract. Despite advances in diagnostic imaging and treatment, there has been

[1]Laser Dynamics Laboratory, Department of Chemistry and Biochemistry, Georgia Institute of Technology, 901 Atlantic Drive NW, Atlanta, GA 30332-0400, USA.
[a]E-mail: edreaden@gatech.edu
[b]E-mail: melsayed@gatech.edu
[2]Otolaryngology Minimally Invasive Skull Base Program, Department of Otolaryngology-Head and Neck Surgery, University of California, San Francisco, 2380 Sutter Street, San Francisco, CA 94115, USA; E-mail: ielsayed@ohns.ucsf.edu
*Corresponding author

List of abbreviations after the text.

little change in survival rates over the past 50 years. There is a dire need to improve early detection of HNC and therapeutic outcomes. Nanoscience refers to the study and application of structures ranging from 1 to 100 nm in size. Materials of these dimensions differ dramatically from their atomic, molecular, or bulk counterparts. At the nanoscale, new physical properties emerge as interfacial properties dominate and electrons become restricted. The equally matched size between nanoparticles and cellular organelles and biomolecules allows potential interactions of these materials within cells. This chapter discusses several promising diagnostic/therapeutic nanotechnologies with special emphasis on rational design, photonic properties/applications, and relevance to head and neck cancer.

INTRODUCTION

Cancer is a leading cause of death in the United States for those under 85, with two to six percent resulting from HNC (El-Sayed 2010; Jemal et al. 2010). The most common HNC tumor type is squamous cell carcinoma of the upper aerodigestive tract. Roughly 60 percent of those newly diagnosed and 70 percent of all killed by HNC globally are male, with half of all deaths occurring in patients over 65. Diagnosis and treatment of these tumors is subsite specific and can be hampered by late presentation of tumors due to their ambiguous symptoms at early stages and difficulty in identifying deep-seated tumors within body cavities such as the paranasal sinuses or throat. Mortality rates for HNC have changed little in the past 50 years and are still dismal for certain types. For instance, the five year relative survival rate for Stage 1–3 cancer of the pharynx is reported 30 to 40 percent (Edge et al. 2010). Current treatment methods for HNC are still limited, consisting of surgical excision, radiation therapy, chemotherapy, and antibody-blocking therapy. Tumors however, may be unresectable or can develop resistance to noninvasive treatments.

Improved technologies are necessary for prevention/early detection, less invasive therapies, circumvention of treatment resistance, and monitoring for recurrence. Current imaging modalities offer a lower limit detection of nearly 4 mm, but are expensive and may lead to false positive results. *Ex vivo* assessment of tissues and conventional immunosorbent/blotting can be difficult to interpret, at times based on subjective visual assessment. Moreover, due to the proximity of HNC to vital functional organs (tongue, vocal cords, eyes, etc.) and nerves, as well as cosmetic concerns, less invasive treatment methods are also needed.

Several potential molecular-level diagnostic/therapeutic targets for HNC exist based on the selective overexpression of markers related to malignant progression such as, cell-surface growth signaling receptors (EGFR/ErbB-1/HER1;CD44), regulators of apoptosis, metabolic regulators (folate receptor, FR), proteolytic enzymes (matrix metalloproteinases, MMPs; urokinase plasminogen receptor, uPAR), tumor necrosis factor receptors (TNFRs), and proangiogenesic receptors (vascular endothelial growth factor receptor, VEGFR). Functionalized biomedical nanoparticles can allow tailored targeting of relevant malignant processes, personalized treatments, and sensitive *ex vivo* detection and *in vivo* imaging techniques. Methods involving systemic and local administration of nanoparticle conjugates are currently in clinical trials with research exploring multivalent, cancer-selective targeting schemes, as well as both intrinsic and extrinsic activation/release.

Nanobiotechnology platforms are attractive due to their inherent multifunctionality. These particles exhibit novel electronic/optical properties, synthetic tunability over a wide range, and facile conjugation to an array of biologically-active molecules for tailored diagnostic/therapeutic applications. For instance, a human immunoglobulin G (IgG) antibody contains two antigen binding sites and roughly 20 primary amine sites available for conjugation to a drug, fluorophore, etc. In contrast, a gold nanoparticle roughly half the size of an IgG contains 2,300 available surface sites for binding any combination of targeting ligand, drug molecule, or imaging contrast agent. While IgG antibodies exhibit increased antigen binding affinity due to the proximity of two binding sites, nanoparticles can present thousands of binding sites over the same volume. Poorly soluble molecules or those which normally enter the cell through passive diffusion can be rapidly delivered across the cell membrane by receptor-mediated nanoparticle endocytosis, delivering thousands of additional molecules into the targeted cell (Dreaden et al. 2009). The pharmacokinetics and pharmacodynamics of nanoparticles can be tailored by the size, shape and composition of the nanoparticle and may serve useful for drug delivery of chemotherapeutic agents. This chapter broadly discusses some of the properties and design considerations for HNC-relevant medicinal nanoparticles of varying composition and highlights some especially promising applications.

NANOPARTICLE TUMOR-TARGETING/CLEARANCE, STABILITY, AND BIOCOMPATIBILITY

One of the most promising applications of anti-cancer nanotechnologies involves their size-selective accumulation in tumors due to the so-called enhanced permeability and retention (EPR) effect (Maeda 2010). Two factors allow accumulation within tumors: increased penetration and decreased clearance of nanoparticles. Malignant cells typically utilize significant amounts of nutrients due to their fast growth/division and rapidly initiate angiogenesis when they reach 2–3 mm in size. The endothelium of these vessels is highly disordered compared to normal tissues, containing abnormally enlarged fenestrations into the extravasculature compartment. As a result, these vessels allow for the extravasation of high molecular weight compounds such as plasma proteins, micelles, liposomes, IgG, DNA complexes, and bacteria. Diminished lymphatic drainage from the tumor interstitium further increases the retention of these compounds at the tumor site. Obviously, the optimal hydrodynamic diameter for EPR of a nanoconjugate depends on the tumor type, location, and stage. It is generally considered that molecules larger than the renal clearance threshold (ca. 6 nm dia, 50 kDa) and up to 2 μm, exhibit preferential tumor accumulation, however optimal tumor targeting parameters are still being defined. In cases where nuclear penetration is desired, particles should be smaller than the diameter of the nuclear pore complex (ca. < 40–50 nm).

Table 1. Key Features of Tumor-Targeted Nanoparticle Delivery, Clearance, Stability, and Biocompatibility.

1. Systemically administered nanoparticles ≤6–8 nm in hydrodynamic diameter are efficiently cleared in the urine.
2. The EPR effect allows particle accumulation inside the interstitium of tumors larger than 2–3 mm in size. Increased circulation time of nanoparticles in the blood increases EPR tumor accumulation.
3. Stable biomedical nanoparticles must account for the physiologic acidity, ionic concentration, and adsorption by proteins and macromolecules.
4. PEG is a common biocompatible polymer used to coat nanoparticles due to its exceptional affinity for aqueous environments and resistance to protein adhesion.
5. The RES is the portion of the immune system responsible for scavenging and discarding foreign/superfluous objects from the body.

This table lists key concepts pertaining to anti-cancer nanoparticle pharmacokinetics and design. EPR: enhanced permeability and retention; PEG: poly(ethylene glycol); RES: reticuloendothelial system (Dreaden et al. 2011b).

The EPR effect may be exploited to deliver diagnostic or therapeutic nanoprobes to diseased tissues while also decreasing their dose to healthy tissues. In the head and neck, many tumors are also accessible for direct inspection, topical application, or direct injection. Nanoparticles can be loaded with a range of agents such as chemotherapuetics, targeting molecules, or gene therapy agents for possible multimodal therapies. Gold nanoparticles have already completed Phase I human trials using gold nanospheres conjugated with tumor necrosis factor (TNF) and appear well tolerated with no serious adverse side effects and are capable of systemically delivering TNF to tumors in humans at concentrations far higher than those attained in previous investigations (Libutti et al. 2009).

Pharmacokinetic clearance of nanoparticle conjugates is size dependent (Longmire et al. 2008). Kidney filtration from the bloodstream (i.e., renal clearance) is based largely on size and charge. Only 1% of serum albumin, for example, (69 kDa, 7.10 nm eff dia) is filtered into the urine, whereas 75% of myoglobin (16.9 kDa, 3.76 nm eff dia) is renally excreted. Renal excretion of nanoparticles is also largely determined by hydrodynamic diameter (HD). Because nanoparticles can exhibit slightly wider ranges in net charge versus endogenous proteins, cationic particles with HDs as large as 8 nm can still be excreted in the urine due to the net negative charge of the capillaries which separate circulating blood from urinary filtrates, however these particles typically exhibit low circulatory half lives due to adsorption to the negatively charged laminar surface of blood vessels. Polyanions, conversely, exhibit decreased cell penetration due to electrostatic repulsion with cell-surface proteoglycans. Nanoparticles that avoid the kidneys may be transported to the liver where they can be processed and excreted into the bile by hepatocytes (and removed) or intracellularly degraded by Kupffer cells. Since many inorganic or metallic nanoparticles cannot be degraded by these processes and can remain in the cells, inert reactivity and potential toxicity are important considerations.

Designing nanoparticles with *in vivo/in vitro* stability is critically important since physiologic conditions can significantly affect the activity of nanoparticle conjugates. Irreversible aggregation of nanoparticles can occur due to attractive forces, such as London Van der Waals forces, hydrophobic-hydrophobic interactions, and complementary charge pairing (electrostatic binding). Because physiological environments are aqueous and charged, increasing particle stability generally relies on enhanced hydrophilicity and/or electrostatic repulsion between the particles. Serum proteins *in vitro* and *in vivo* can adsorb the nanoparticles and block targeting groups, increase hydrodynamic diameter and uptake by the reticuloendothelial system (RES). High ionic concentrations *in vivo* can also screen the apparent charge at the surface of the particles (i.e., Debye screening), reducing their repulsion and inducing aggregation. Coating nanoparticles with immunopassivating agents such as poly(ethylene glycol) (PEG) increases the *in vivo/in vitro* stability of nanoparticles (Harris and Chess 2003) by preventing opsonin protein adsorption and subsequent phagogcytic uptake of nanoparticle by the RES. PEG consists of poly(ethylene oxide) repeat units, which can be linear, or preferably branched. Other stabilizer ligands include N-oleoyl sarcosine, alginate, chitosan, dextran, poly(lysine), poly(cyclodextrin), poly(lactic acid), poly(glycolic acid), poly(vinyl alcohol) (PVA), starches, poly(methyl methacrylate) (PMMA), and oleic acid (Xu et al. 2004;

Xu et al. 2005). Immunopassifying molecules also serve to increase hydrophilicity/circulatory half life, thereby maximizing tumor-selective EPR accumulation.

FUNDAMENTAL PROPERTIES OF NANOMATERIALS

At the nanoscale, particles possess properties unique from their individual atoms/molecules and bulk assemblies. In the case of free electron metals, the collective motions of electrons at the nanoparticle surface dominate those at its interior. Photons in the ultraviolet (UV), optical, near-infrared (NIR), and infrared (IR) wavelength regions can couple with these surface electron oscillations, becoming absorbed, depending on the metal's composition, size, shape, and dielectric environment. Confinement and absorption of these photons results in highly intense local electric fields on the particle's surface, a process known as surface plasmon resonance (SPR). The SPR field can enhance photon interactions in its vicinity over several orders of magnitude and create optical effects with potential biomedical applications such as SPR absorption/scattering, fluorescence enhancement, and the so-called electromagnetic surface-enhanced Raman scattering (SERS) effect (Dreaden et al. 2011a; Huang et al. 2007a; Qian et al. 2008).

Table 2. Fundamental Properties of Anti-cancer Nanoparticle Platforms*.

1. Nanoparticles may allow for targeted drug deliver vectors, contrast probes based on scattering/absorption/photoacoustic properties, or selective tumor ablation using their photothermal properties.
2. Semiconductor nanoparticles exhibit absorption/emission properties superior to molecular fluorophores, allowing for stable, high-intensity detection in multiple labeling schemes.
3. Polymeric nanoparticles allow for the controlled and variable delivery of a variety of drugs with high *in vivo* stability, biocompatibility, and biodegradability.
4. Carbon nanoparticles exhibit photoacoustic, Raman, photothermal, and photoluminescent properties amenable to a variety of diagnostic, delivery, and treatment strategies.
5. Liposomes can be used to deliver hydrophilic drugs in high quantity to tumor sites while micelles can be used in cases of poorly soluble drugs.

This table describes some of the most common platforms applied in anti-cancer nanotechnologies and their advantages/multifunctionality (Dreaden et al. 2011b).

In the case of spherical particles, light incident on a noble metal nanoparticle can be absorbed or scattered in a manner first described by Mie (Mie 1908). An interesting aspect of the SPR effect is the intensity of the absorbed and scattered light that is wavelength-dependent. SPR absorption and scattering can be used for diagnostic imaging of nanoparticles bound to cells (El-Sayed et al. 2005b) over the visible and near-infrared spectrum (Huang et al. 2006). Sensitive molecular and spatial information may be discerned from changes in the peak absorption/scattering wavelengths of nanoparticles as their surface is modified by molecular binding interactions or brought close to other nanoparticles. Further, nanoparticles can be directly quantified by their high optical extinction cross section.

SPR can also result in the transfer of energy to molecules adsorbed to the surface of the nanoparticle and enhance the Raman spectra of molecules on the nanoparticles by orders of magnitude. Because SPR fields enhance both the incident and Raman scattered photons, molecules in close proximity produce strong SERS signals characterized by high signal-to-noise ratios, allowing for the sensitive, non-invasive detection of Raman reporters in tumors. Intense signals from gold nanorod SERS may identify a molecular signature of EGFR overexpression in oral cancer cells not witnessed in benign epithelial controls (Huang et al. 2007a). Further, strong gold nanoparticle SERS signals have been detected noninvasivley in malignant tumor models *in vivo* after intravascular injection (Qian et al. 2008).

In addition to imaging, plasmonics offers the possibility of photothermal therapeutic applications. When light is absorbed, its energy dissipates through vibrations in the particle's atomic lattice (i.e., phonons) and can generate heat sufficient to perforate cell membranes, facilitate the melting of duplex DNA, and elicit hyperthermic cellular responses (Bert et al. 2002; Huang et al. 2007b). Wavelength-dependent optical properties of plasmonic nanoparticles, particularly those made of gold due to its inert reactivity, can also be used in a variety of diagnostic and sensing applications based on scattering (El-Sayed et al. 2005a) and wavelength shift (Huang et al. 2007b). In a similar manner, gold nanoparticles can also enhance the local effect of ionizing radiation nearly 200 percent for potential therapeutic effect.

Where the energetic separation between ground and excited electronic states in an insulator are far, and those in metals overlap, semiconductors exhibit energy gaps in the optical and NIR region. The spacing of this energy gap can be increased due to quantum confinement when the particle size is decreased below the semiconductor's characteristic Bohr radius. Following excitation, a semiconductor dissipates energy in the form of longer wavelength photons equivalent in energy to the minimum gap between the ground and excited state. As a result, semiconductor nanoparticles and quantum dots exhibit intense, spectrally broad absorption and narrow emission compared to photo-oxidation prone molecular fluorophores. Because of their structural tenability and the narrowness, intensity and temporal stability of their emission, multiple quantum dots can identified (multiplexed) from a single *in vivo* image or *in vitro* tissue sample. By targeting quantum dots to cell surface growth receptors, angiogenic proteins, or other cancer-specific molecules, high sensitivity imaging and high throughput diagnostics can be performed. At the moment, *in vivo* human applications of semiconductor nanoparticles and quantum dots have been hampered due to potential systemic toxicity, however (Geys et al. 2008).

Polymeric nanoparticles represent one of the most versatile and biocompatible platforms for the targeted delivery of therapeutic drugs, nucleic acids and/or imaging contrast agents. Nanoparticle polymers generally encapsulate or intercalate their active compounds and rely on hydrophilic functionalization for stability/RES avoidance, surface ligands for targeting, and size-dependent EPR at tumor sites. Such systems are often comprised of extrusions, condensates, or emulsions of biocompatible polymers previously discussed in the stabilization of solid nanoparticles and include liposomes, hydrogels (nanogels), micelles, and dendrimers among others. Using varying side chains, polymeric nanoparticle conjugates can be designed to release compounds in response to internal stimuli, such as changes in temperature, pH, or enzymatic degradation, as well as external stimuli such as the application of electromagnetic radiation. Hydrophyllic polymeric nanoparticles can also facilitate the intravenous delivery of therapeutic hydrophobic agents and the intracellular transport of nucleic acids for gene therapy applications. The greatest advantage for the *in vivo* use of polymeric nanoparticles is likely their hepatobiliary degradability.

Carbon nanostructures such as carbon nanotubes (CNTs) and fullerenes (C_{60}) represent another interesting nanoparticle platform for therapy and imaging with the advantage of low cost, scalable production, and potential for biodegradation. Like most hydrophobic particles, carbon nanostructures can be surface-functionalized with a variety of amphiphilic molecules, often a phospholipid derivative or biocompatible polymer such as chitosan. CNTs exhibit strong, highly Stokes-shifted NIR photoluminescence, multiple intrinsic Raman-active scattering vibrational modes, and remarkably high NIR extinction cross section. Potential uses include actively- and passively- targeted delivery of CNTs to delineate tumor margins or metastatic spread by photoluminescence and/or Raman imaging. Because of their high NIR absorption cross section, pulsed photoexcitation results in high acoustic signal generation which can be discriminated from tissues and biological fluids using an ultrasonic transducer, yielding three dimensional images with high lateral resolution. Fullerenes can be similarly used in photoacoustic imaging applications, as well as photothermal ablation techniques, hydrophobic anti-cancer drug delivery, and gene therapy applications. Because of their small size (< 2 nm), CNT and C_{60} renal clearance can also be achieved in some cases.

PHOTONIC IMAGING/DIAGNOSTICS/THERAPEUTICS AND CONTROLLED RELEASE

Due to their inherent multifunctionality, potential nanoparticle platforms often consist of an imaging contrast agent or are co-conjugated with some such compound. Examples include fludeoxyglucose- (FDG) labeling for three dimensional positron emission tomograpic (PET) imaging, as well as iron oxide- or gadolinium-labeling for magnetic resonance imaging (MRI). In contrast, molecular fluorophores, gold nanorods/shells, quantum dots, and CNTs most often make use of NIR laser excitation/ emission due to the low attenuation from biological fluids and high tissue penetration depths observed in the 650–900 nm window region (Fig. 1). Microwatt NIR lasers (FDA class 1), for example, have exhibited 10 cm penetration depths through breast tissue and 4 cm through skull/brain and deep muscle tissue. Higher power lasers (FDA class 3) have exhibited up to 7 cm penetration through muscle and neonatal skull/brain tissue (Weissleder 2001). In combination with fiber optic guidance, imaging

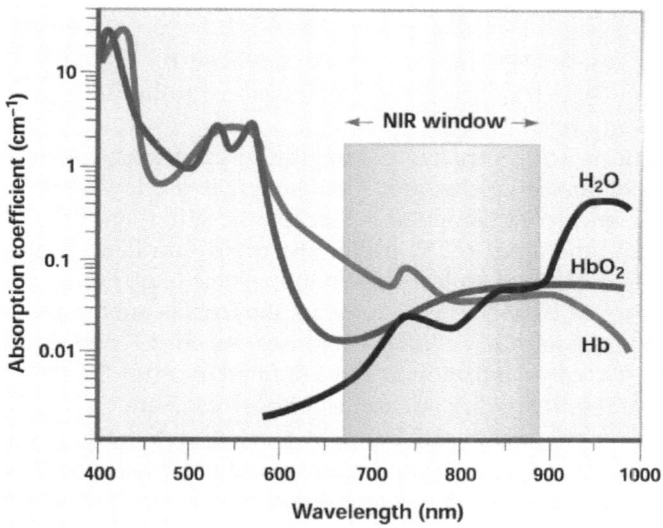

Fig. 1. Wavelength range of the near-infrared (NIR) tissue transmission window. In the 650–900 nm range, absorption by oxygenated/de-oxygenated hemoglobin and water is minimal, allowing up to 10 cm penetration through breast tissue using FDA class 1 NIR lasers and 7 cm penetration through muscle and neonatal skull/brain tissue using FDA class 3 NIR lasers. Photoexcitation of fluorescent compounds, plasmonic particles, and Raman-labels is most efficient in this region. Reprinted with permission from R. Weissleder. Copyright 2001 Nature Publishing Group/Macmillan Publishers Ltd.

Color image of this figure appears in the color plate section at the end of the book.

Table 3. Head and Neck Cancer Relevant Photonic Imaging/Diagnostics/Therapeutics and Controlled Release Using Nanotechnologies.

1.	Photonic-based imaging can allow treatment guidance, quantitative protein diagnostics, ablative therapeutics, and post-treatment monitoring.
2.	Biological tissues exhibit minimal attenuation in the 650–900 nm near-infrared (NIR) window.
3.	PTT tumor injury and ablation can complement, substitute for, or guide surgical tumor excision.
4.	The photonic scattering and emission properties of metal and semiconductor nanoparticles offer high-throughput, low cost, quantitative, multiplexed detection of cancer-associated proteins with sensitivity/reproducibility superior to conventional methods.
5.	Successful gene therapies require efficient shielding of nucleic acids, cell penetration, and endosomal escape which can be provided by nanotechnologies.
6.	Interference in the transcription of genes necessary for the production of growth-promoting proteins can decrease malignant progression and increase sensitivity to chemotherapy.
7.	Chemical reactions specific to extra- and intra-cellular tumor environments can be used for the selective delivery and activation of therapeutic/diagnostic nanoparticles.

This table discusses the rational design/application of anti-cancer nanotechnologies relevant to head and neck cancer with emphasis on photonic properties and controlled release/delivery. PTT: photothermal therapy.

of contrast-labeled HN tumors can be achieved by excitation with NIR radiation and spectrally resolved from tissue autofluorescence with high lateral resolution.

In addition to *in vivo* HNC imaging applications, newly emerging properties of nanoparticles and their conjugates allow for a variety of alternative and complementary therapeutic interventions. Plasmonic photothermal therapies (PTT) utilize the rapid, localized heat generated following excitation of a plasmonic nanoparticle (e.g., gold, silver etc.) to ablate the tumor tissues to which the nanoparticles are targeted by either illiciting hyrperthermic cellular responses, or direct perforation of cell membranes. In cases of tumors with ill-defined margins, PTT might ensure complete removal of malignant tissues, either by systemic administration of targeted plasmonic nanoparticles, or by local administration and washing treatments to dislodge non-specifically bound nanoparticles. PTT ablation of surgically-excised tumor margins can allow for increasing preservation of non-malignant HN tissues which can be of significant functional and/or cosmetic importance. In instances of poorly accessible HNCs, PTT can be used in place of surgical excision. While researchers such as Halas and West evaluated dielectric-metal core-shell nanoparticle geometries to facilitate cancer-targeted PTT, rod-shaped gold nanoparticles appear to be increasingly efficient contrast agents for PTT. El-Sayed and coworkers used both locally and systemically administered PEGylated gold nanorods to target and treat oral squamous cell carcinoma tumor models in mice using a small, portable, inexpensive NIR diode laser (Fig. 2). They observed inhibition of average tumor growth for both delivery methods over a 13 day period following a single treatment, with resorption of >57 percent of the directly-injected tumors and 25 percent of the intravenously-treated tumors (Dickerson et al. 2008).

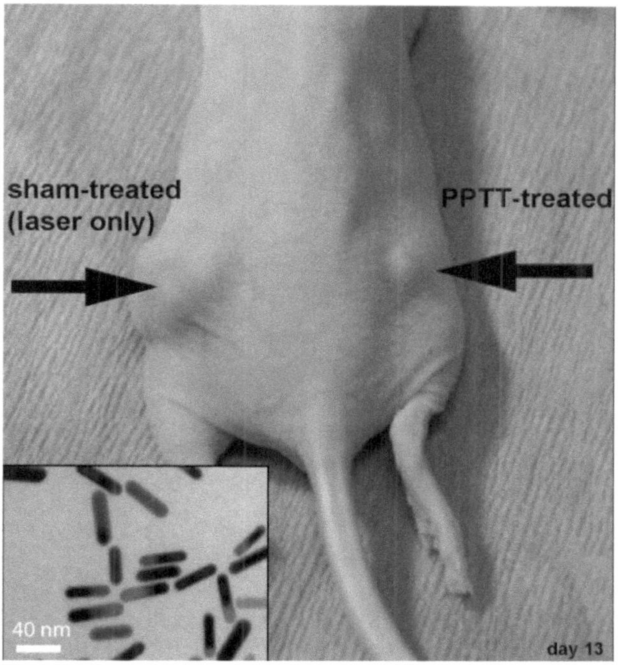

Fig. 2. *In vivo* near-infrared laser photothermal cancer therapy of oral squamous cell carcinoma tumor models using gold nanorods in mice. Representative HSC-3 tumor resorption and growth inhibition in female nu/nu mice at day 13 following direct injection of PEGylated gold nanorods (15 µL, $OD_{\lambda=800}=40$) and a single near-infrared PPTT treatment (10 min, 0.9–1.1 W/cm^2, 6 mm dia) versus sham/NIR treatment (15 µL, 10 mM PBS) at the same NIR laser exposure power density and duration. Inset shows transmission electron micrographs of the nanorods (unpublished data). Reprinted with permission from Dickerson et al. Copyright 2008 Elsevier.

Bhatia and coworkers improved these results by exploiting the high x-ray absorption cross section of gold nanorods (which is two times greater than clinical iodine standards) using x-ray computed tomographic (CT) imaging-guidance. By three-dimensionally reconstructing tumor margins from EPR-accumulated gold nanorods, they were able to computationally guide laser PTT treatments, achieving apparent complete tumor resorption over a >50 day period (von Maltzahn et al. 2009).

Improved *ex vivo* analysis beyond current bioanalytical techniques can be achieved with increasing sensitivity and selectivity using nanotechnologic platforms. Diagnostic enzyme-linked immunosorbent assays (ELISAs) and immunoblotting techniques for example, can be performed with increasing sensitivity/selectivity, lower cost, time and skill required for processing/interpretation. Mirkin and coworkers demonstrated the use of antibody-gold nanoparticle conjugates in the multiplexed detection of cancer protein markers from both buffer and serum solutions using a microarray-based sandwich assay (Fig. 3). In this scheme, antibody-labeled glass slides were incubated with a complex antigen solutions followed by gold nanoparticle-labeled IgG. Subsequent electroless deposition of gold allowed for simultaneous light scattering detection of prostate specific antigen (PSA), a prostate cancer marker, human chorionic gonadotropin (hCG), a testicular cancer marker, and R-fetoprotein (AFP), a hepatic cancer marker from 1.4 pM buffered solutions. Non-multiplexed detection of PSA was achieved from buffered solutions at 300 aM (ca. 9000 copies)

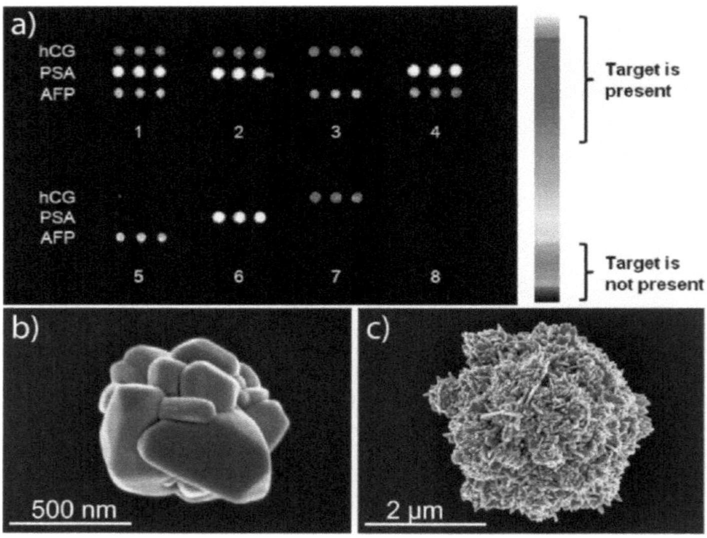

Fig. 3. Multiplexed detection of protein cancer markers based on light scattering using an antibody-gold nanoparticle sandwich assay. Prostate specific antigen (PSA), a prostate cancer marker, human chorionic gonadotropin (hCG), a testicular cancer marker, and R-fetoprotein (AFP), a hepatic cancer marker were correctly identified at 1.4 pM concentrations in buffer (a) following scattering amplification by silver (b) and gold (c) deposition. (1) All targets present; (2) hCG and PSA; (3) hCG and AFP; (4) PSA and AFP; (5) AFP; (6) PSA; (7) hCG; (8) no targets present. Reprinted with permission from Kim et al. Copyright 2009 American Chemical Society.

Color image of this figure appears in the color plate section at the end of the book.

and from 10 percent serum solutions at a concentration (3 fM) three orders of magnitude lower than that typically capable by ELISA (Kim et al. 2009). HNC-specific antigens do not exist, but HNC can be identified by a variety of molecules that tend to be overexpressed on the cell-surface or in the cell. Multiplexed detection may provide a means to delineate HNC-specific antigen expression profiles with high throughput and low cost.

Semiconductor quantum dots also provide a means by which highly sensitive, highly throughput, multiplexed cancer protein detection can be achieved. Due to their intense, spectrally broad absorption, narrow emission, and low background signal, quantum dots allow for simultaneous detection of multiple cancer-associated proteins using existing fluorescence-based imaging/spectroscopic techniques. Chan and coworkers developed a multiplexed tissue microarray screening/staging assessment method based on IgG-quantum dot conjugates which can quantify cancer-associated protein levels with less subjectivity than current immunohistochemical scoring methods (Fig. 4). Tissue microarrays were assembled from needle biopsies of tumor xenografts using established techniques. Fluorescence intensity from spectrally-resolved quantum dot-labeled antibodies of epidermal growth factor receptor (EGFR), pan cytokeratin, and E-cadherin was quantified for each sample and autofluorescence from its correspondingly unlabeled microarray sample subtracted. Moreover, detected protein levels agreed highly with RNA expression levels determined by more cumbersome and time-consuming quantitative real time polymerase chain reaction (Q-RT-PCR) methods (Ghazani et al. 2006).

Improved early detection, identification of small tumors/lymphatic metastases, monitoring for recurrence, and detection/removal of circulating tumor cells (CTCs) are significant goals in cancer prevention/treatment strategies. To this end, Zharov and coworkers devised a non-invasive, *in vivo* cytometric method to detect and sequester CTCs by dual-labeling them with photoacoustically-active plasmonic and magnetic nanoparticles (Fig. 5). Iron oxide nanoparticles labeled with a fragment of urokinase plasminogen activator and gold-coated CNTs labeled with folic acid actively target cell-surface receptors commonly upregulated in malignant cells, but are absent in red blood cells. CTCs were detected *in vitro* at a sensitivity limit of 35 gold- and 720 magnetic-nanoparticles per cell with high targeting efficiency (96 ± 2.1 percent) over red blood cells. *In vivo*, the ratio of CTCs detected in the vicinity of tumor xenograft models to those in distant vessels was found to correlate highly with the stage of the primary tumor progression. Further, a small-rare earth magnet could be used to sequester the circulating particle-labeled CTCs (Galanzha et al. 2009).

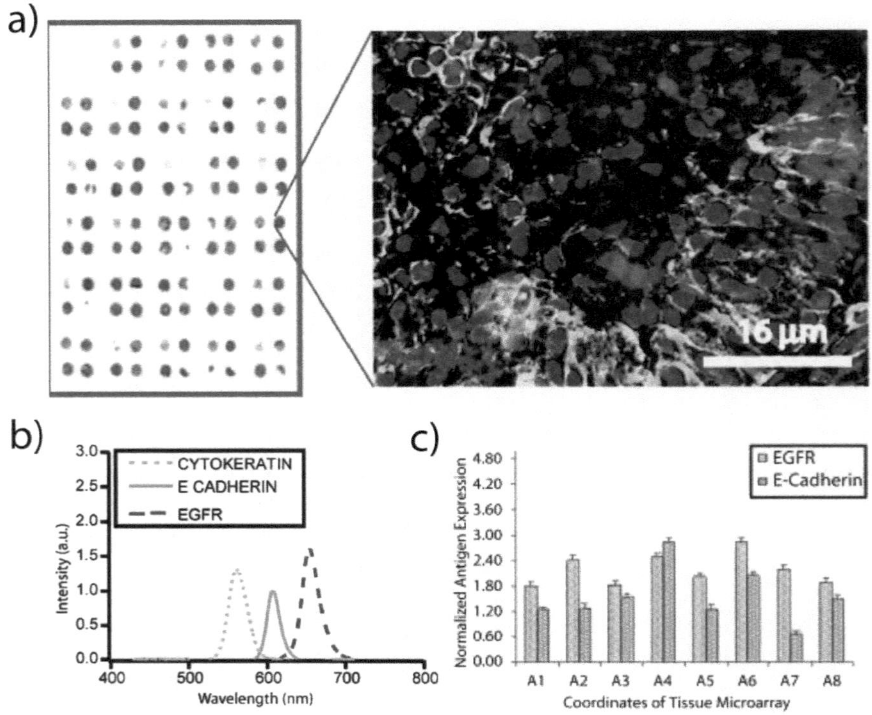

Fig. 4. High throughput, multiplexed quantification of HNC cancer protein markers based on quantum dot-antibody labeling of tissue microarrays. Fluorescence intensity from spectrally-resolved quantum dot-labeled antibodies of epidermal growth factor receptor (EGFR), pan cytokeratin, and E-cadherin was quantified for each tissue microarray sample. (a) Using cytokeratin as an internal standard (b) cancer protein expression levels (c) could be compared over multiple tumor phenotypes. Adapted with permission from Ghazani et al. Copyright 2006 American Chemical Society.

Color image of this figure appears in the color plate section at the end of the book.

Efficient penetration of the cell membrane is one of the greatest impediments to current applications of gene therapy. In HNC, the ability to decrease the production of growth promoting proteins by silencing gene transcription would be of tremendous value by altering the progression of their growth, and their sensitivity to chemo- and radio-therapy. For example, the overexpression of EGFR, commonly observed in HNC, is believed to promote resistance to taxane-based chemotherapeutics. Lyon and McDonald explored the effects of EGFR RNA interference (RNAi) on the sensitivity of cancer cells to docetaxel treatment (Fig. 6). Cancer cells treated with small interfering RNA- (siRNA) loaded core-shell hydrogel nanoparticles targeted with a YSA-peptide ligand

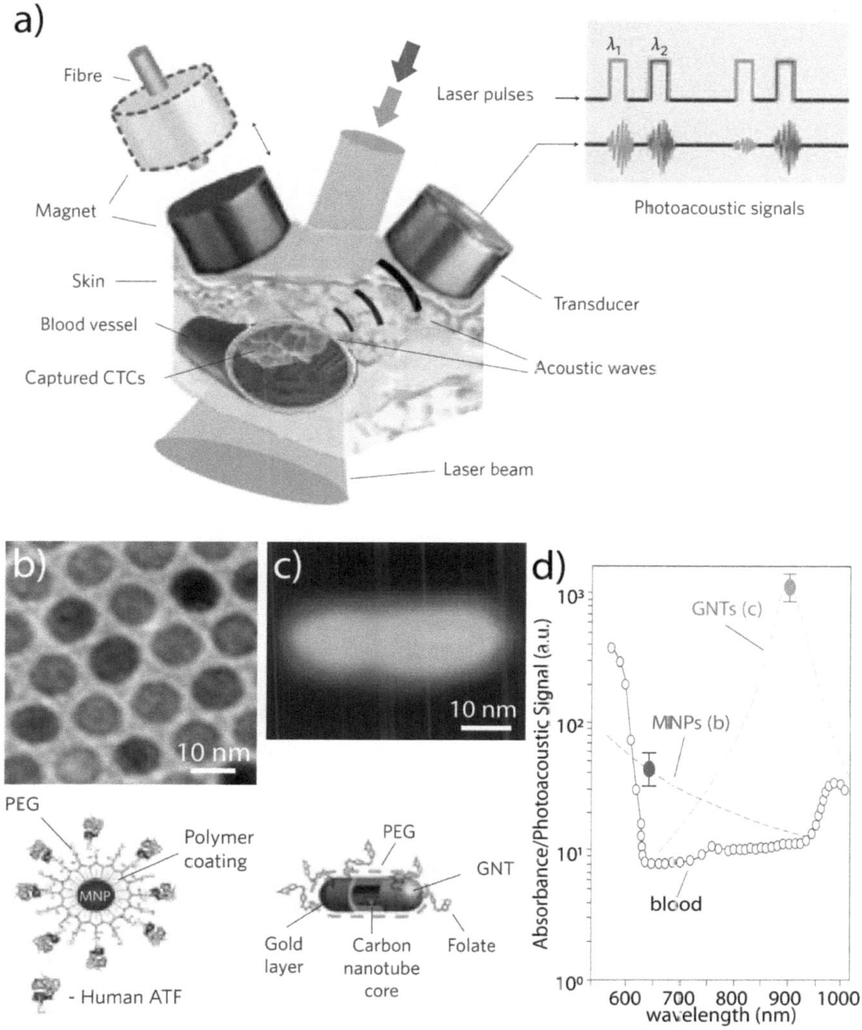

Fig. 5. Schematic of *in vivo* detection and magnetic sequestration of circulating tumor cells (CTCs) by dual nanoparticle labeling and photoacoustic cytometry. Dual-labeled CTCs were non-invasively detected (a) by their enhanced photoacoustic response and sequestered with a small rare-earth magnet. Photoacoustic response from magnetic iron oxide (b) and gold-coated carbon nanotube particles (c) labeled with ligands of cell-surface receptors upregulated in malignant cells, but absent in red blood cells, was significantly higher than that from blood and red blood cells (d) Error bars represent standard deviation. Adapted with permission from Galanzha et al. Copyright 2009 Nature Publishing Group/Macmillan Publishers Ltd.

Color image of this figure appears in the color plate section at the end of the book.

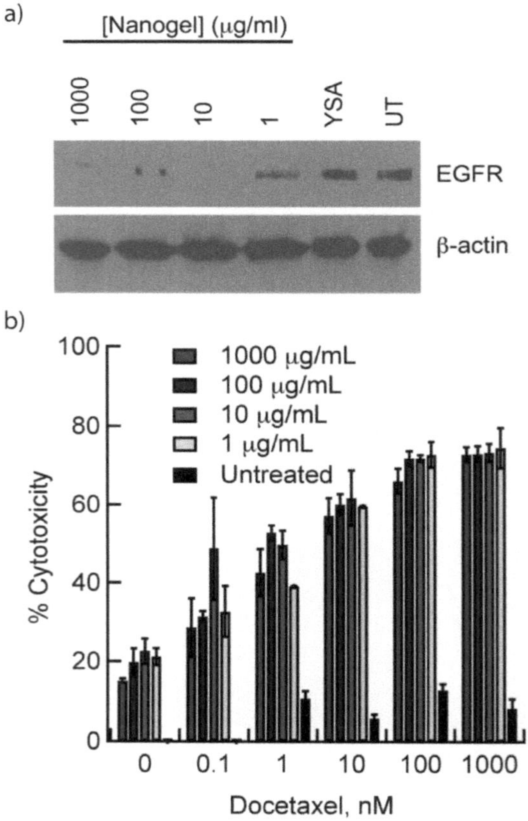

Fig. 6. Targeted RNA interference of epidermal growth factor receptor expression and chemosensitization to docetaxel by treatment with YSA-targeted small interfering RNA-nanogel particles. Selective reduction of EGFR expression (a) in EphA2 receptor (+) cancer cells was achieved by YSA peptide conjugation to siRNA-loaded core-shell hydrogel nanoparticles and incubation for 48 hr, followed by 96 hr incubation with docetaxel (b) Error bars represent standard deviation. Reprinted with permission from Dickerson et al. Copyright 2010 retained by the authors.

Color image of this figure appears in the color plate section at the end of the book.

were able to selectively and efficiently reduce EGFR expression *in vitro*, dramatically increasing sensitivity to docetaxel treatment. They used siRNA-loaded ca. 100 nm diameter nanogel particles (composed of poly(N-isopropylmethacrylamide) (pNIPMAm) crosslinked with N,N'-methylenebisacrylamide conjugated to a YSA peptide ligand of EphA2, a cell-surface receptor commonly upregulated in malignant cells (Dickerson et al. 2010).

Nanoparticle constructs may take advantage of tumor microenvironments by pH-responsive functionality. Because the extracellular environment of tumors is more acidic (pH 7.0–7.2) than normal tissue, imaging contrast agents which activate upon a drop in pH can be advantageous. A cancer-targeted nanoparticle-fluorophore conjugate, for example, which completely dissociates from a quencher group at pH 7.2, but not pH 7.4, would function well in diagnostic imaging methods. Polymeric nanoparticles which swell and release encapsulated drug molecules only at tumor pH would also function effectively in tumor-selective drug delivery applications.

Changes in intracellular pH are also important in the release of nanoparticle conjugates from intracellular organelles. Although endocytotic uptake of nanoparticle conjugates is often desired, their sequestration in acidic endosomal/lysosomal compartments can be challenging to overcome. Cationic polymer nanoparticles often used to transport (anionic) nucleic acids into cells can become sequestered in endosomes and degraded in the lysosomes. Because the early/late endosomes and the lysosomes typically maintain pH in the 5.9–6.0, 5.0–6.0, and 5.5–6.0 range, respectively, co-functionalization of nanoparticle vectors with groups which facilitate endosomal escape, or endosomolysis, in response to a drop in pH would greatly improve the efficacy of particle delivery. Groups which become highly protonated in the pH environment of the endosome/lysosome, such as the nitrogens in polyethyleneimine (PEI), have been used to escape sequestration by increasing Cl^- influx in response to protonation at acidic pH, thus increasing osmotic pressure and swelling which leads to organellar burst—the so-called "proton sponge effect"—and successful nuclear delivery of its conjugate.

Mesoporous silica nanoparticles have been designed which are capped by pH-responsive molecular valves capable of releasing anti-cancer drugs at endosomal pH (Fig. 7). The particles were synthesized by Stöber hydrolysis, yielding ca. 100 nm diameter silica particles comprised of 2 nm wide hexagonally-arranged cylindrical pores. An aromatic amine, N-methylbenzimidazole, with a pKa (5.67) comparable to the endosomal pH range was silanized and surface functionalized onto the silica nanoparticles for use as the stalk of the molecular valve. A solution of doxorubicin, an anti-cancer drug and DNA transcription/replication inhibitor, was subsequently loaded into the porous nanoparticles for delivery. A cyclic oligosaccharide, β-cyclodextrin, was pyrene-appended for use as the valve cap, as it is both biocompatible and large enough to block doxorubicin release from the 2 nm silica pores. At physiological pH (7.4), the pyrene-modified β-cyclodextrin cap remains non-covalently associated to the unprotonated aryl amine stalk. At pH < 6 however, the aryl amine becomes protonated and the cap dissociates from the nanoparticle, releasing its drug payload. The authors demonstrated *in vitro* cellular uptake and endosomal colocalization of the silica nanoparticle conjugates,

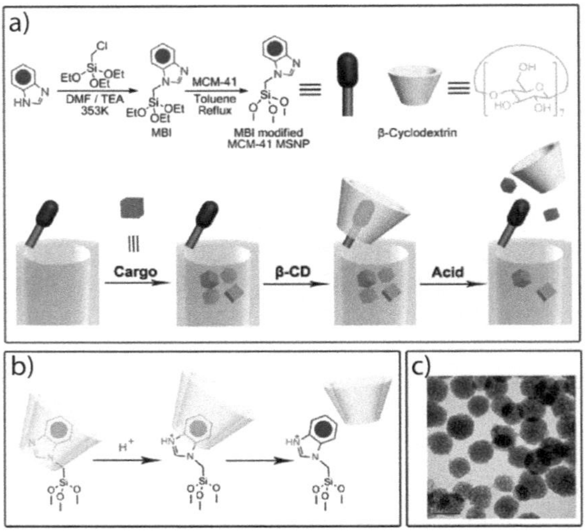

Fig. 7. Schematic design of a pH-responsive molecular valve for the intracellularly-selective delivery of anti-cancer drugs loaded into porous nanoparticles. The molecular valve (a) consists of an aromatic amine stalk which non-covalently binds to a bulky, cyclic oligosaccharide cap at pH > 6, preventing the release of drugs from the nanoparticle pores to which the stalks are attached. Following endosomal uptake (b) the stalk becomes protonated, the cap dissociates, and drugs loaded into the nanoparticle pores are released. Here, doxorubicin was loaded into mesoporous silica nanoparticles (c) for endosomal pH-dependent cytotoxic delivery. Reproduced with permission from Meng et al. Copyright 2010, American Chemical Society.

Color image of this figure appears in the color plate section at the end of the book.

as well as pH-dependent release and drug cytotoxicity. Interestingly, the efficiency of the (cationic) doxorubicin release was optimized by (cationic) ammonium-functionalization of the particle's (anionic) interior surface, further demonstrating the importance of pH in the rational design of biomedical nanoparticles (Meng et al. 2010).

OUTLOOK

Clinical applications of therapeutic nanoparticles are extremely promising to improve detection and decrease treatment morbidity in HNC. *Ex vivo* diagnostic applications will no doubt become commonplace. *In vivo* methods offer potential multimodal platforms for combined diagnostic/ therapeutic strategies, however methods of delivery, targeting, and intracellular processing require optimization. Toxicity to short and long term exposure will warrant further investigation. Emphasis should be directed toward systematic investigations involving details of absorption, distribution, metabolism, and excretion based on well-characterized nanoparticle size, shape, composition, and surface ligands.

Summary Points

- Nanoscale materials exhibit properties wholly different from their individual atomic, molecular, and bulk forms which can be exploited in both conventional and unconventional diagnostic/treatment strategies.
- Multifunctional biomedical nanotechnologies allow for combined therapeutics/diagnostics.
- Nanoparticle conjugates are relatively easily prepared and may allow patient-specific tailoring.
- The increased size of nanoparticles versus small-molecules greatly improves tumor accumulation but also greatly affects the way they are trafficked throughout the body.
- Photonic-based nanotechnologies represent a new modality of cancer detection, assessment, and treatment to complement current technologies.
- Nanotechnological platforms can facilitate externally- and internally-controlled release of cancer-targeted drug conjugates and gene therapies.

Definitions

Ablation: physical destruction of cellular components by heat or locally produced pressure.

Angiogenesis: the growth of new blood vessels.

Carbon nanotubes and fullerenes: CNTs and C_{60}s are atomic sheets of bonded carbon atoms which are rolled into tubular and spherical structures, respectively.

Cation/anion: an ion with a positive/negative charge, respectively.

Circulatory half life: the time in which it takes for one half of an intravenously administered substance to be removed from circulation.

Cross section: a measure of how much an incident electromagnetic field is perturbed by a fixed quantity of an atom, molecule, or material.

Cytokeratin: now termed keratin, is the fibrous, structural protein containing "intermediate" filaments and found on the intracellular portion of the cytoskeleton in epithelial tissue; keratin expression levels are generally independent to malignancy.

E-cadherin: a calcium-dependent cell-cell adhesion glycoprotein found in epithelial cells which is commonly downregulated in malignant cells.

enzyme-linked immunosorbent assay: a bioanalytical method used to quantify the concentration of a molecule based on its affinity for an antibody and the subsequent localized change in color intensity as enzymes attached to the antibody react.

gene therapy: the use and delivery of nucleic acids to increase the production of desired genes/proteins or to disrupt the production undesirable genes/proteins.

hydrodynamic diameter: the diameter of a particle which includes contribution from its surface ligands and associated water molecules.

hyperthermia: the cellular response to gentle heating 5–10 °C above normal temperature, to which malignant cells are increasingly sensitive.

immunoblot: a bioanalytical method used to quantify proteins electrophoretically separated and transferred (blotted) onto another surface, based on the intensity of their labeling by fluorescent or colored antibodies of the protein.

ligand: in the case of a nanoparticle, any molecule which is chemically or physically bonded to it.

liposome/micelle: here, a liposome is a spherical bilayer of amphiphilic molecules which present both interior and exterior hydrophilic groups, while a micelle is a similarly composed monolayer with a hydrophilic exterior and lipophilic interior.

opsonins: any molecule which binds a substance in order to facilitate its removal from the body by the RES.

pH: the molar concentration of protons, as represented in log base ten.

phagocytosis: the process by which cells of the immune system engulf foreign bodies, cellular debris, and/or pathogens for transport or degradation.

pKa: the pH at which a proton can be added/removed by an increase/decrease in proton concentration.

quantitative real time polymerase chain reaction: a bioanalytical method used to amplify and quantify nucleic acid sequences.

quantum dot: a semiconductor whose dimensions are comparable to or below the material's Bohr radius, resulting in an increase in absorption/emission energy gap with decreasing size.

renal clearance: the process by which small molecules and ion are filtered from the blood and excreted in the urine.

silica: the most abundant material in the earth's crust, composed on one silicon atom for every two oxygen atoms; in biomedical applications it is often amorphous and exhibits low toxicity.

taxane: a polyaromatic hydrocarbon molecule naturally produced by coniferous trees/shrubs of the yew family which acts as an inhibitor of cell division by disruption of microtubule formation necessary for mitosis.

YSA peptide: the 12 amino acid peptide abbreviated YSAYPDSVPMMS which mimics the ephrin-A1 ligand of EphA2 receptor, a receptor which regulates cell migration and vascular assembly and is upregulated in many malignancies.

Abbreviations

CNT	:	carbon nanotubes
CTC	:	circulating tumor cell
CT	:	computed tomography
DLS	:	dynamic light scattering
EPR	:	enhanced permeability and retention
EGFR ErbB-1, HER1	:	epidermal growth factor receptor
FDG	:	fludeoxyglucose
FR	:	folate receptor
C_{60}	:	fullerene
HNC	:	head and neck cancer
hCG	:	human chorionic gonadotropin
HD	:	hydrodynamic diameter
IgG	:	immunoglobulin G
IR	:	infrared
MRI	:	magnetic resonance imaging
MMP	:	matrix metalloproteinase
NIR	:	near-infrared
PTT	:	photothermal therapy
PEG	:	poly(ethylene glycol)
PEI	:	poly(ethyleneimine)
PMMA	:	poly(methyl methacrylate)
pNIPMAm	:	poly(N-isopropylmethacrylamide)
PVA	:	poly(vinyl alcohol)
PET	:	positron emission tomography
PSA	:	prostate specific antigen
Q-RT-PCR	:	quantitative real time polymerase chain reaction
RES	:	reticuloendothelial system
AFP	:	R-fetoprotein
RNAi	:	RNA interference
SERS	:	surface-enhanced Raman scattering
TNFR	:	tumor necrosis factor receptor
UV	:	ultraviolet
uPAR	:	urokinase plasminogen activator receptor
VEGFR	:	vascular endothelial growth factor receptor

References

Bert, H., W. Peter, A. Olaf, D. Annette, S. Geetha, K. Thoralf, F. Roland and R. Hanno. 2002. The cellular and molecular basis of hyperthermia. Clin. Rev. Oncol./Hematol. 43: 33–56.

Dickerson, E., W. Blackburn, M. Smith, L. Kapa, L.A. Lyon and J. McDonald. 2010. Chemosensitization of cancer cells by siRNA using targeted nanogel delivery. BMC Cancer 10: 10.

Dickerson, E.B., E.C. Dreaden, X. Huang, I.H. El-Sayed, H. Chu, S. Pushpanketh, J.F. McDonald and M.A. El-Sayed. 2008. Gold nanorod assisted near-infrared plasmonic photothermal therapy (PPTT) of squamous cell carcinoma in mice. Cancer Lett. 269: 57–66.

Dreaden, E.C., S.C. Mwakwari, Q.H. Sodji, A.K. Oyelere and M.A. El-Sayed. 2009. Tamoxifen-Poly(ethylene glycol)-Thiol Gold Nanoparticle Conjugates: Enhanced Potency and Selective Delivery for Breast Cancer Treatment. Bioconjugate Chem. 20: 2247–2253.

Dreaden, E.C., R.D. Near, T. Abdallah, M.H. Talaat and M.A. El-Sayed. 2011a. Multimodal plasmon coupling in low symmetry gold nanoparticle pairs detected in surface-enhanced Raman scattering. Appl. Phys. Lett. 98: 183115–183113.

Dreaden, E.C., M.A. Mackey, X. Huang, B. Kang and M.A. El-Sayed. 2011b. Beating cancer in multiple ways using nanogold. Chem. Soc. Rev. 40: 3391–3404.

Edge, S.B., D.R. Byrd, C.C. Compton, A.G. Fritz, F.L. Greene and A. Trotti. 2010. AJCC Cancer Staging Handbook: From the AJCC Cancer Staging Manual. *In*. Springer, New York.

El-Sayed, I. 2010. Nanotechnology in Head and Neck Cancer: The Race Is On. Current Oncol. Rep. 12: 121–128.

El-Sayed, I.H., X. Huang and M.A. El-Sayed. 2005a. Surface Plasmon Resonance Scattering and Absorption of anti-EGFR Antibody Conjugated Gold Nanoparticles in Cancer Diagnostics: Applications in Oral Cancer. Nano Lett. 5: 829–834.

El-Sayed, I.H., X. Huang and M.A. El-Sayed. 2005b. Surface Plasmon Resonance Scattering and Absorption of anti-EGFR Antibody Conjugated Gold Nanoparticles in Cancer Diagnostics: Applications in Oral Cancer. Nano Lett. 5: 829–834.

Galanzha, E.I., E.V. Shashkov, T. Kelly, J.-W. Kim, L. Yang and V.P. Zharov. 2009. *In vivo* magnetic enrichment and multiplex photoacoustic detection of circulating tumour cells. Nat. Nano 4: 855–860.

Geys, J., A. Nemmar, E. Verbeken, E. Smolders, M. Ratoi, M.F. Hoylaerts, B. Nemery and P.H.M. Hoet. 2008. Acute toxicity and prothrombotic effects of quantum dots: impact of surface charge. Environ. Health Perspect. 116: 1607.

Ghazani, A.A., J.A. Lee, J. Klostranec, Q. Xiang, R.S. Dacosta, B.C. Wilson, M.S. Tsao and W.C.W. Chan. 2006. High Throughput Quantification of Protein Expression of Cancer Antigens in Tissue Microarray Using Quantum Dot Nanocrystals. Nano Lett. 6: 2881–2886.

Harris, J.M., and R.B. Chess. 2003. Effect of pegylation on pharmaceuticals. Nat. Rev. Drug Discov. 2: 214–221.

Huang, X., I.H. El-Sayed, W. Qian and M.A. El-Sayed. 2006. Cancer Cell Imaging and Photothermal Therapy in the Near-Infrared Region by Using Gold Nanorods. J. Am. Chem. Soc. 128: 2115–2120.

Huang, X., I.H. El-Sayed, W. Qian and M.A. El-Sayed. 2007a. Cancer Cells Assemble and Align Gold Nanorods Conjugated to Antibodies to Produce Highly Enhanced, Sharp, and Polarized Surface Raman Spectra: A Potential Cancer Diagnostic Marker. Nano Lett. 7: 1591–1597.

Huang, X., P.K. Jain, I.H. El-Sayed and M.A. El-Sayed. 2007b. Gold nanoparticles: interesting optical properties and recent applications in cancer diagnostics and therapy. Future Med. Nanomed. 2: 681–693.

Jemal, A., R. Siegel, J. Xu and E. Ward. 2010. Cancer Statistics, 2010. CA Cancer J. Clin. 60: 277–300.

Kim, D., W.L. Daniel and C.A. Mirkin. 2009. Microarray-Based Multiplexed Scanometric Immunoassay for Protein Cancer Markers Using Gold Nanoparticle Probes. Anal. Chem. 81: 9183–9187.

Libutti, S.K., G.F. Paciotti, L. Myer, R. Haynes, W. Gannon, M. Walker, G. Seidel, A. Byrnes, N. Yuldasheva and L. Tamarkin. 2009. Results of a completed phase I clinical trial of CYT-6091: A pegylated colloidal gold-TNF nanomedicine. J. Clin. Oncol. 27: 3586.

Longmire, M., P.L. Choyke and H. Kobayashi. 2008. Clearance properties of nano-sized particles and molecules as imaging agents: considerations and caveats. Future Med. - Nanomed. 3: 703–717.

Maeda, H. 2010. Tumor-Selective Delivery of Macromolecular Drugs via the EPR Effect: Background and Future Prospects. Bioconjugate Chem. 21: 797–802.

Meng, H., M. Xue, T. Xia, Y.-L. Zhao, F. Tamanoi, J.F. Stoddart, J.I. Zink and A.E. Nel. 2010. Autonomous *in vitro* Anticancer Drug Release from Mesoporous Silica Nanoparticles by pH-Sensitive Nanovalves. J. Am. Chem. Soc. 132: 12690–12697.

Mie, G. 1908. Contributions to the Optics of Turbid Media, Especially Colloidal Metal Solutions. Ann. Phys. 25: 377–445.

United Nations, Department of Economic and Social Affairs, Population Division. 2009. World Population Prospects: The 2008 Revision: Volume I: Comprehensive Tables.(U.N. publication)

Qian, X., X.-H. Peng, D.O. Ansari, Q. Yin-Goen, G.Z. Chen, D.M. Shin, L. Yang, A.N. Young, M.D. Wang and S. Nie. 2008. *In vivo* tumor targeting and spectroscopic detection with surface-enhanced Raman nanoparticle tags. Nat. Biotech. 26: 83–90.

von Maltzahn, G., J.-H. Park, A. Agrawal, N.K. Bandaru, S.K. Das, M.J. Sailor and S.N. Bhatia. 2009. Computationally Guided Photothermal Tumor Therapy Using Long-Circulating Gold Nanorod Antennas. Cancer Res. 69: 3892–3900.

Weissleder, R. 2001. A clearer vision for *in vivo* imaging. Nat. Biotechnol. 19: 316–317.

Xu, X.Q., H. Shen, J.R. Xu and X.J. Li. 2004. Aqueous-based magnetite magnetic fluids stabilized by surface small micelles of oleolysarcosine. App. Surf. Sci. 221: 430–436.

Xu, X.Q., H. Shen, J.R. Xu, J. Xu, X.J. Li and X.M. Xiong. 2005. Core-shell structure and magnetic properties of magnetite magnetic fluids stabilized with dextran. App. Surf. Sci. 252: 494–500.

Index

About the Editors

Rajaventhan Srirajaskanthan is a consultant physician at University Hospital Lewisham and Kings College Hospital, London. He graduated in 2001 from Guys and St. Thomas' medical school and his postgraduate training was predominantly in prestigious teaching hospitals in London. In 2010 he was awarded a MD for his thesis on biomarkers and receptor expression in Neuroendocrine tumours. He has published over 30 peer review papers and authored 2 books and numerous book chapters. His research interests include biomarkers in cancer, nutrition and liver disease.

Victor R. Preedy BSc, PhD, DSc, FSB, FIBiol, FRCPath, FRSPH is Professor of Nutritional Biochemistry, King's College London, Professor of Clinical Biochemistry, Kings College Hospital (Hon) and Director of the Genomics Centre, King's College London. Presently he is a member of the King's College London School of Medicine. Professor Preedy graduated in 1974 with an Honours Degree in Biology and Physiology with Pharmacology. He gained his University of London PhD in 1981 when he was based at the Hospital for Tropical Disease and The London School of Hygiene and Tropical Medicine. In 1992, he received his Membership of the Royal College of Pathologists and in 1993 he gained his second doctoral degree, i.e. DSc, for his outstanding contribution to protein metabolism in health and disease. Professor Preedy was elected as a Fellow to the Institute of Biology in 1995 and to the Royal College of Pathologists in 2000. Since then he has been elected as a Fellow to the Royal Society for the Promotion of Health (2004) and The Royal Institute of Public Health (2004). In 2009, Professor Preedy became a Fellow of the Royal Society for Public Health. In his career Professor Preedy has carried out research at the National Heart Hospital (part of Imperial College London) and the MRC Centre at Northwick Park Hospital. He is visiting lecture at University College London (UCL) and has collaborated with research groups in Finland, Japan, Australia, USA and Germany. He is a leading expert on the pathology of disease and has lectured nationally and internationally. He has published over 570 articles, which includes over 165 peer-reviewed manuscripts based on original research, 90 reviews and numerous books.

Color Plate Section

Chapter 1

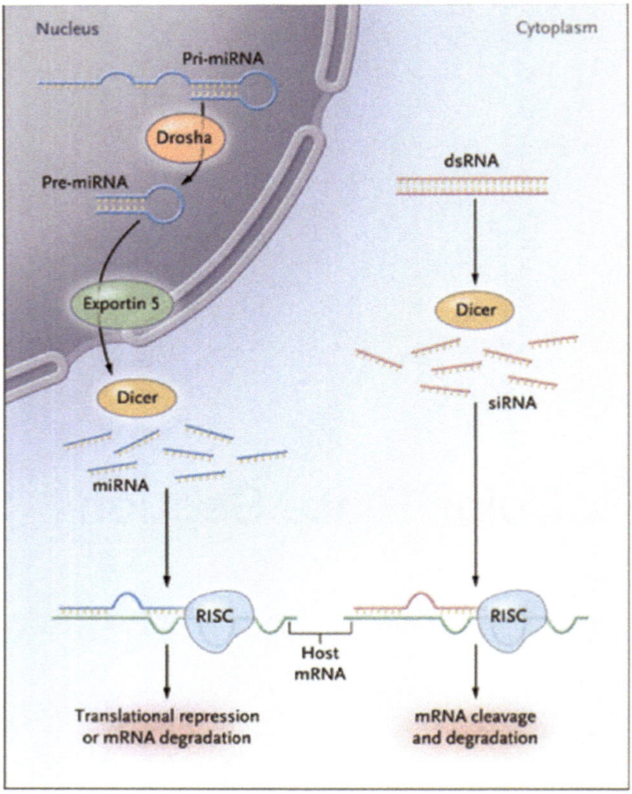

Fig. 1. Mechanism of RNA interference. Long precursor microRNA (miRNA) segments, called pri-miRNA, are first cleaved in the nucleus by Drosha, an RNase III endonuclease, into segments of approximately 70 nucleotides each (called pre-miRNA). Transportation into the cytoplasm by means of exportin 5 leads to cleavage by Dicer, another RNase III endonuclease, which produces mature miRNA segments. Host degradation of messenger RNA (mRNA) and translational repression occurs after miRNA binds to the RNA-induced silencing complex (RISC). Cytoplasmic long double-stranded RNA (dsRNA) is cleaved by Dicer into small interfering RNA (siRNA), which is incorporated into RISC, resulting in the cleavage and degradation of specific target mRNA. (Reprinted with permission from Merritt et al. Copyright 2008a Massachusetts Medical Society. All rights reserved).

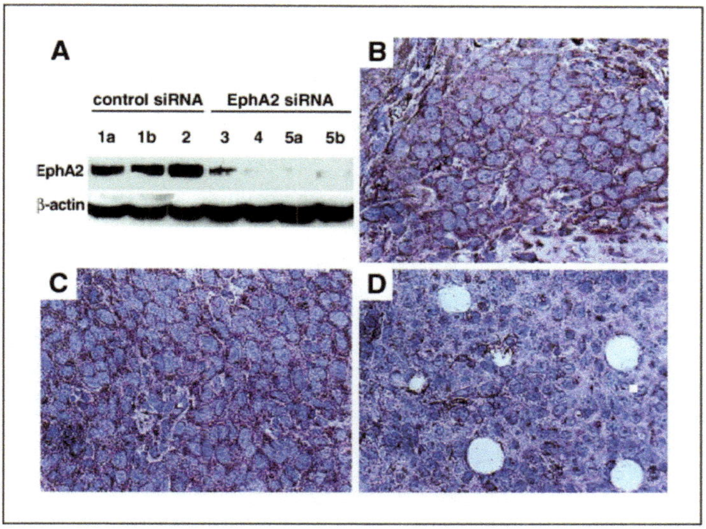

Fig. 2. ***In vivo*** **down-regulation of EphA2 by siRNA/DOPC after systemic administration.** (a) Western blot of lysate from orthotopic tumors collected 48 hours after a single administration of control siRNA (*lanes 1* and *2*) or EphA2-targeting (*lanes 3–5*) siRNA, each complexed within DOPC. To control for sampling error, *lanes 1a* and *1b* are separate preparations from the same tumor treated with control siRNA. Similarly, *lanes 5a* and *5b* are separate preparations from the same tumor treated with EphA2-targeting siRNA. *Lanes 2* to *4*, additional tumor-bearing mice treated with control or EphA2-targeting siRNA-DOPC. Adjacent sections were stained by H&E to confirm the presence of tumor. (b) Immunohistochemical staining for EphA2 of tissue treated with control siRNA/DOPC. The typical cobblestone appearance of this overexpressed membrane-bound protein is noted. (c) Immunohistochemistry 48 hours after a single treatment of EphA2-targeting siRNA without a transfection agent ("naked") is shown and had no detectable affect on EphA2 expression. (d) Treatment of EphA2-targeting siRNA encapsulated within DOPC effectively down-regulated EphA2 expression 48 hours after a single dose. EphA2 expression is restored 1 week after a single treatment (not pictured). (b-d) Original magnification, ×400. (EphA2: Brown). (Reprinted with permission from the American Association for Cancer Research, Inc., from Landen et al. 2005 Fig. 4).

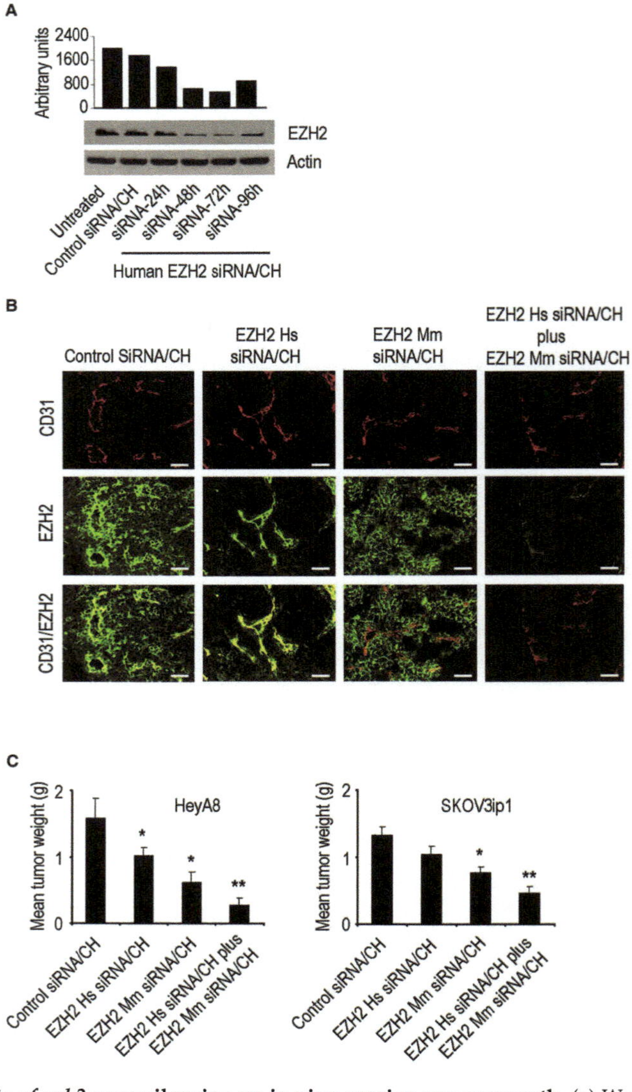

Fig. 3. Effects of *ezh2* gene silencing on in vivo ovarian cancer growth. (a) Western blot of lysates from orthotopic tumors collected 24, 48, 72, and 96 hr after a single injection of control siRNA/CH or human (*ezh2* Hs siRNA/CH). (b) *Ezh2* gene silencing in HeyA8 tumor as well as tumor endothelial cells. Tumors collected after 48 hr of single injection of control siRNA/CH, *ezh2* Hs siRNA/CH, or *ezh2* Mm siRNA/CH and stained for EZH2 (green) and CD31 (red). The scale bar represents 50 µm. (c) Effects of *ezh2* Hs siRNA/CH or *ezh2* Mm siRNA/CH on tumor weight in orthotopic mouse models of ovarian cancer. Error bars indicate SEM. $^{*}p < 0.05$; $^{**}p < 0.001$. (Reprinted with permission from Elsevier, Inc., from Lu et al. 2010. Fig. 5).

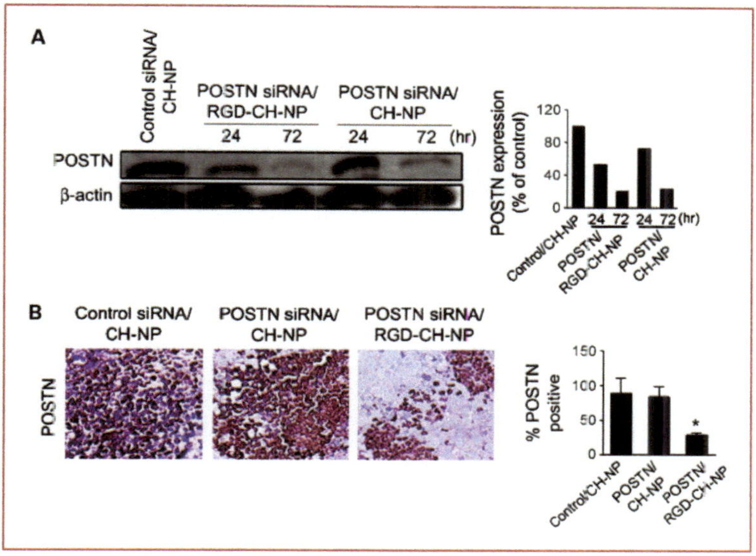

Fig. 4. Effect of *POSTN* downregulation following i.v. injection of POSTN siRNA/RGD-CH-NP into SKOV3ip1-bearing mice. (a) Western blot analysis was done for *POSTN* expression in tumor tissue (20 µg of protein used). Quantitative differences were determined by densitometry analysis. (b) *POSTN* expression in tumor tissues was assessed by immunohistochemistry at 24 hours. All of these analyses were recorded in five random fields for each slide and quantitative difference was determined by positive/negative expression of cells for staining (magnification, ×100). Error bars, SE; *, $P < 0.05$. (Reprinted with permission from the American Association for Cancer Research, Inc., from Han et al. 2010. Fig. 4).

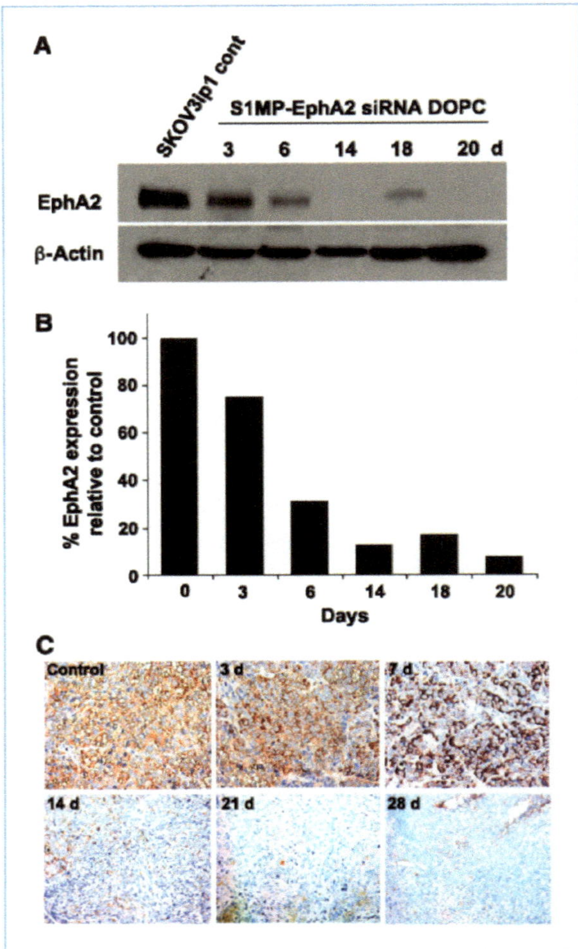

Fig. 5. Systemic delivery of siRNA-DOPC using S1MP results in long-lasting *in vivo* gene silencing. The mice (three mice per time point) bearing SKOV3ip1 orthotopic ovarian tumors were injected with S1MP-EphA2-siRNA-DOPC or left untreated. (a) The tumors were harvested at the indicated time points for Western blot to measure EphA2 expression levels. Thirty micrograms of tumor lysate were separated on a 10% SDS-PAGE and transferred on to a polyvinylidene difluoride membrane. The membrane was incubated with anti-EphA2 antibody overnight at 4°C. The membrane was tested for β-actin to confirm equal loading. (b) Densitometric analysis was performed to normalize EphA2 expression by β-actin. Data were expressed as % of normalized value to the untreated group. (c) Immunohistochemical analysis of EphA2 expression in the SKOV1ip3 tumor. Images were taken at original magnification of ×400. (Reprinted with permission from the American Association for Cancer Research, Inc., from Tanaka et al. 2010. Fig. 2).

Chapter 5

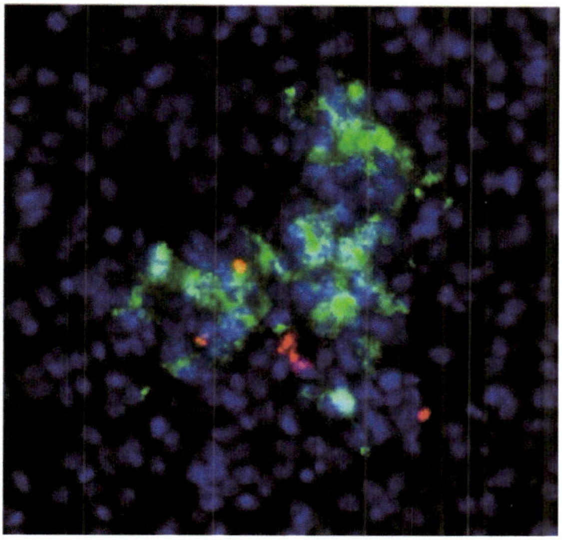

Fig. 4. Umbilical cord matrix stem cells (UCMSC) home to melanoma and deliver therapeutic proteins. Dye loaded human UCMSC (red or orange), engineered to express β-interferon, home to metastatic human breast cancer tumors in SCID mouse lungs (green) and attenuate the tumors (unpublished).

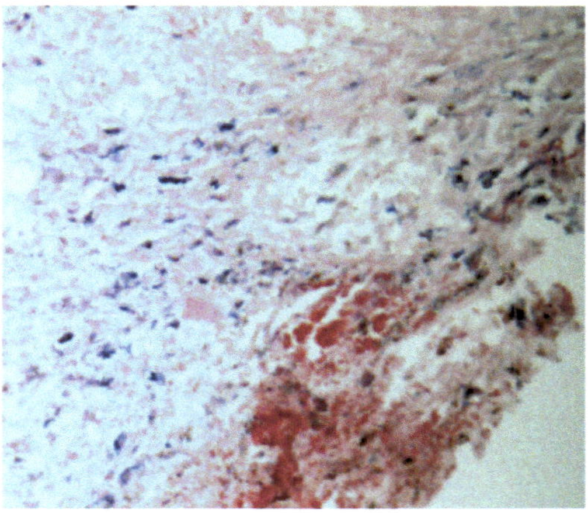

Fig. 5. Murine Neural Progenitor Cells (NPCs) (blue) home to melanomas. Image was captured four days after intravenous administration of stem cells into mice bearing subcutaneous B16F10 melanoma. (Reproduced with permission from (Rachakatla et al. 2010) Copyright: American Chemical Society 2010).

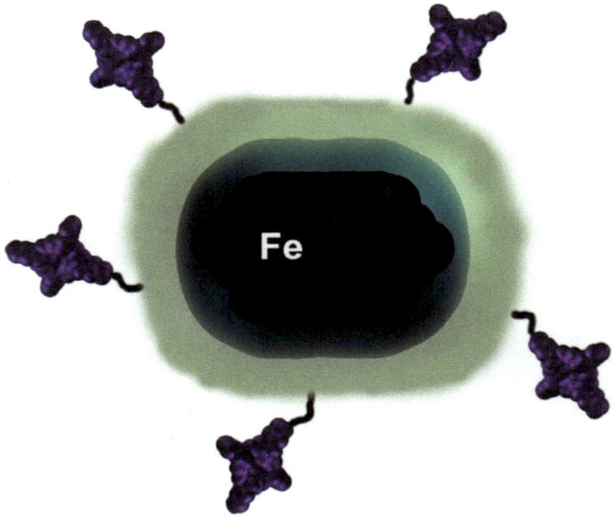

Fig. 7. Aminosiloxane (blue layer)-stabilized Fe/Fe_3O_4 composite core/shell nanoparticles featuring a dopamine-anchored organic stealth-layer (green layer) and tetrakis-carboxyphenyl-porphyrin (TCPP) labels for the enhanced uptake of the nanoparticle by the LDL (low density lipid) receptor, which is overexpressed in numerous cancer cell lines (Reproduced with permission from (Rachakatla et al. 2010) Copyright: American Chemical Society 2010).

Chapter 8

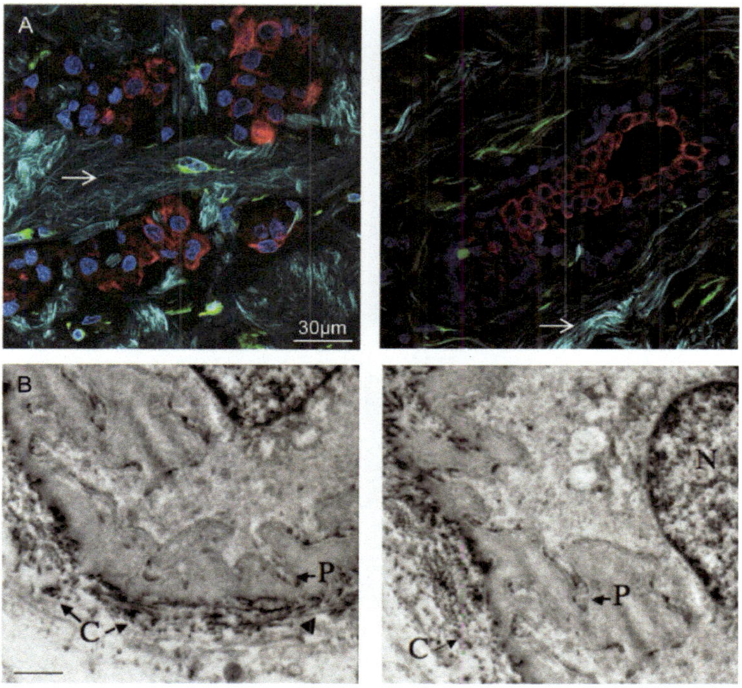

Fig. 1. Multiphoton, second harmonic and TEM imaging of tumour cells and collagen. A. Cohorts of tumour cells from the acini of the breast are stained by cytokeratin-19 antibodies (red). The tumour mass is surrounded by dense collagen areas imaged by second harmonic (cyan) (arrows). **B.** The protrusions (P) of tumour cells extend into the extracellular matrix and appear to be surrounded by collagen fibrils (C) that are mostly imaged in cross-section.

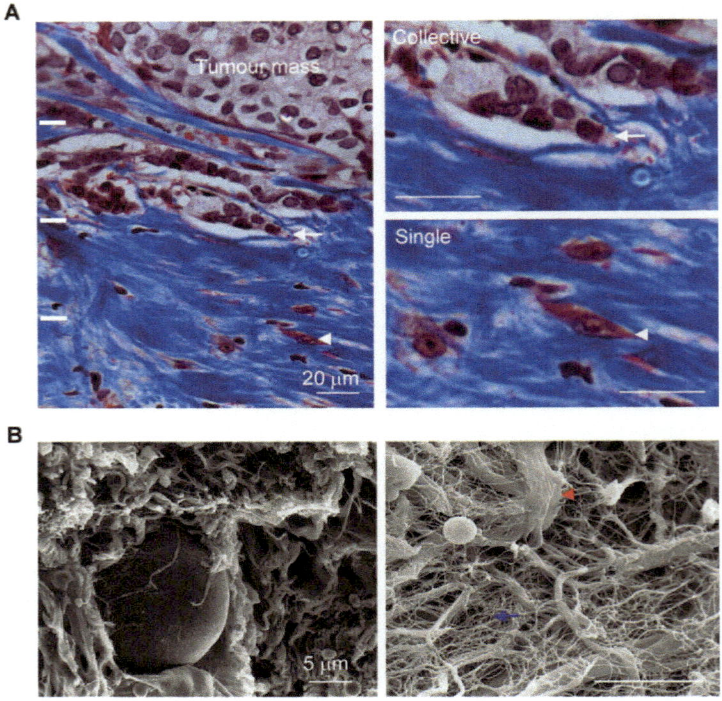

Fig. 2. Collagen fibrils surrounding invasive tumours in HMD matrices. A. Masson's Trichrome staining revealed infiltration of DCIS cells in HMD matrices occurring in regions abundant with fibrillar collagen (blue). Collective migratory cells occur close to the primary tumour mass (arrow). Distance between white markers on left = 50 µm. B. SEM demonstrating close apposition of fibrillar collagen and an epithelial cell. Mesh-like organisation of fibrillar collagen (blue arrow) and bundles of larger collagen fibrils are present throughout the HMD matrix.

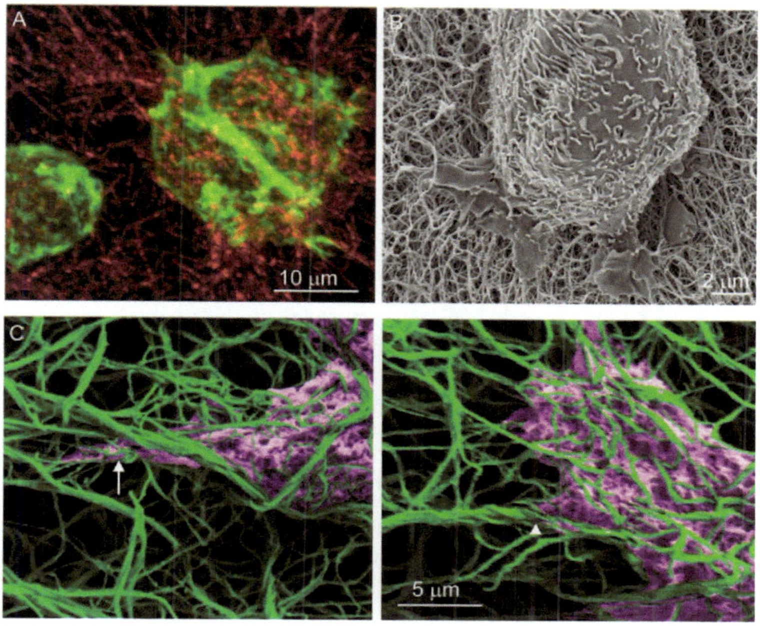

Fig. 5. Confocal and reflection microscopies, and FESEM, illustrating cell-matrix interactions. MTLn3 breast carcinoma cells are seeded onto dense collagen matrices for 48 hours prior to fixation and treatment. **A.** Tumour cells staining for phalloidin actin-Alexa 488 (green) have migrated into the matrices (red). Tumour cell fluorescence was captured by confocal microscopy and the matrix fibrils were imaged using reflection microscopy. **B.** A low resolution FESEM image illustrating a tumour cell extending protrusions into collagen matrix. **C.** Higher resolution FESEM images demonstrate the interactions between the protrusions with matrix fibrils. Arrow shows how the protrusion is enmeshed within tight fibrillar networks. Images have been pseudocoloured using Photoshop according to differences in the intensity of objects.

Chapter 9

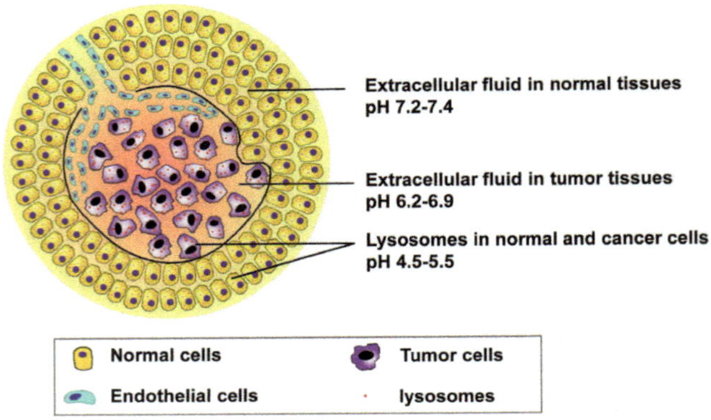

Fig. 1. A schematic of the acidic compartment distribution in normal and tumor tissues. The acidic compartments include the extracellular fluid in the tumor (pH 6.2–6.9), and the lysosomal luments (pH 4.5–5.5) in the normal as well as cancer cells. The pH of the blood, extracellular fluid of the normal tissues and cytosol in both normal and cancer cells stay in the neutral range (7.2–7.4).

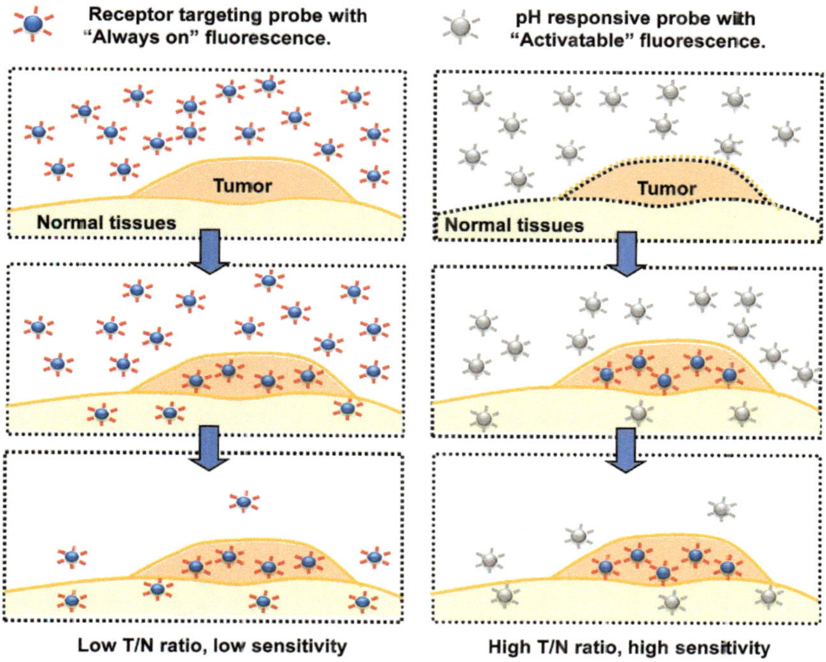

Fig. 2. pH activatable probe significantly increasing the sensitivity of tumor imaging. (A) Schematics of conventional receptor-associated tumor imaging strategy. The probe signal remains at the "On" level during the whole imaging process. (B) Schematics of pH activatable tumor imaging strategy. In the normal tissues, the signal of the probe stays at the "Off" level, but it increases significantly in the tumor acidic microenvironment.

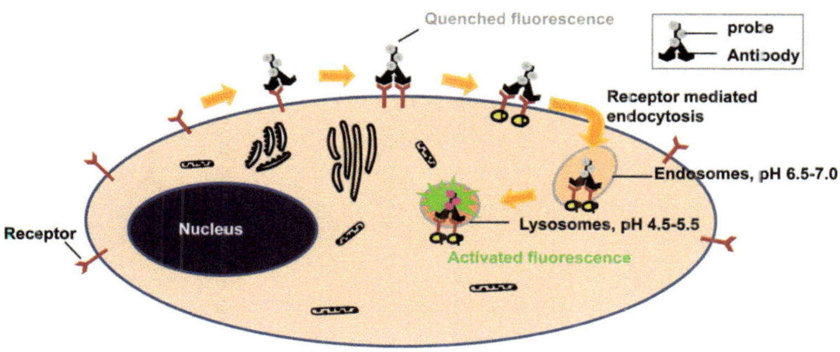

Fig. 3. Probes imaging tumor by sensing the acidic endosome/lysosome lument. The probes labeled in the antibody are quenched in neutral pH. After internalization *via* the receptor-mediated endocytosis, the probes are delivered into acidic endosomes followed by lysosomes, where the probes are activated concomitantly with significant signal enhancement. The picture is reproduced with permission from Ref. 18 © the Nature Publishing Group.

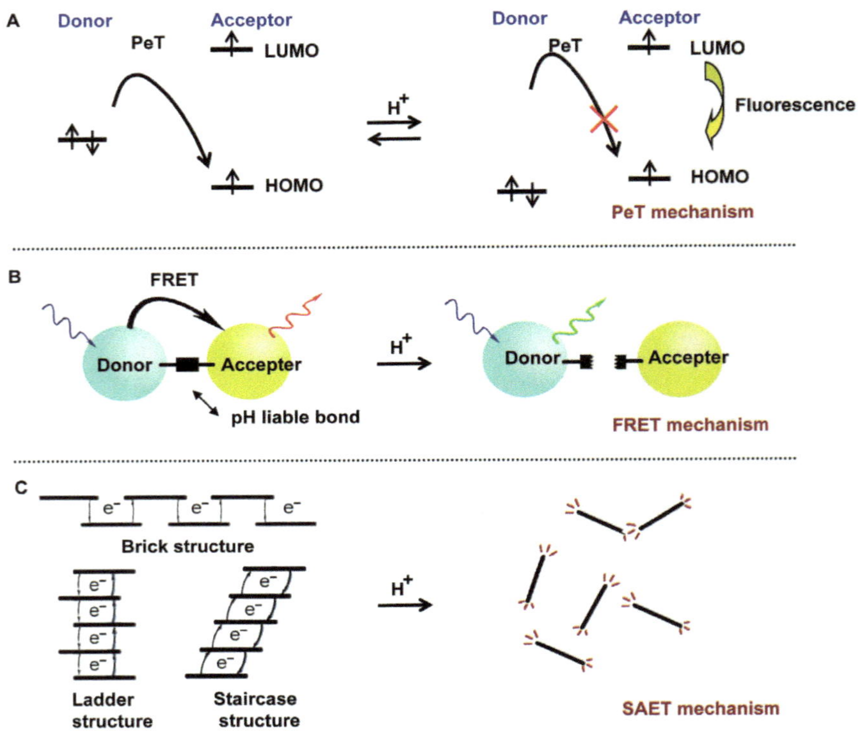

Fig. 4. Schematics of the mechanisms inducing pH responsive fluorescence. (A) The pH activated photon-induced electron transfer (PeT) effect. (B) The pH activated fluorescence resonance energy transfer (FERT) effect. (C) The pH activated self-aggregation associated energy transfer (SAET) effect.

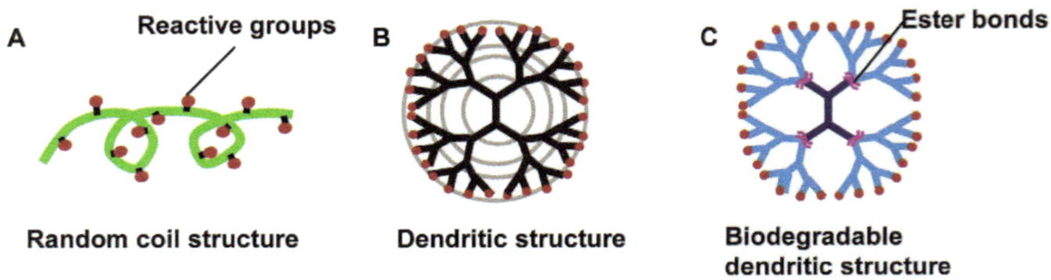

Fig. 5. Structures of the polymers used for the nanoprobe platform. (A) The linear or branched polymers with a random coil-like structure. (B) The dendrimer-like polymer with a unique molecular weight, spherical shape and well-defined topology of the peripheral reactive groups on the particle surface. (C) The biodegradable dendrimer-like polymer possesses the advantages of a dendrimer but with low systemic toxicity.

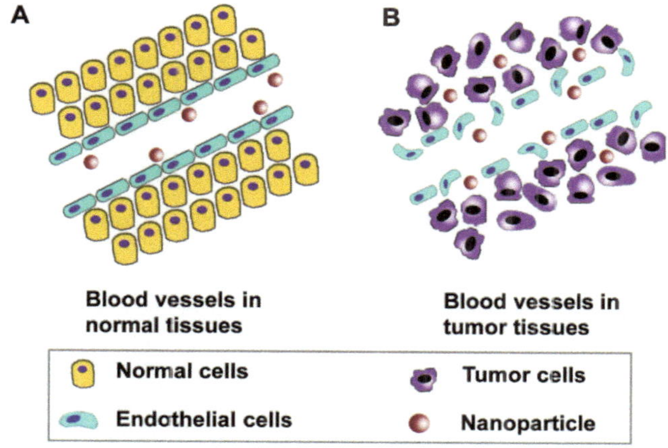

Fig. 6. A schematic of the enhanced permeable and retention (EPR) effect. (A) The nanoparticles are restricted in the normal vasculatures. (B) Due to the high permeability of tumor vasculatures, the nanoparticles penetrate the vessel walls and enter the tumor interstitium.

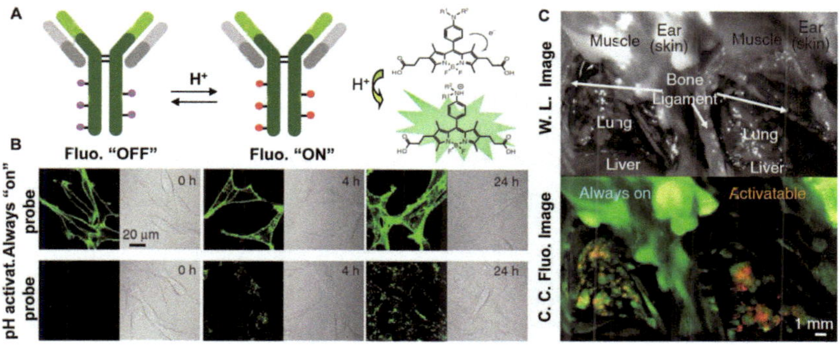

Fig. 8. Nanoprobe imaging tumors *via* the pH induced PeT effect. (A) The schematic of the antibody-based nanoprobe, in which the pH activatable fluorophores were labeled to the HER2 antibody. (B) Confocal fluorescence microscope images of NIH3T3 HER2$^+$ cells at a selected time after the treatment of "always-on" or the pH activatable nanoprobe. (C) White light (top) and color coded fluorescence image (bottom) of the lung metastases after injection of "always-on" (left) or pH-activatable (right) nanoprobes. The activatable probe (bottom right) readily distinguishes between HER2$^+$ (green) and RFP+HER2$^-$ (red) tumors. In contrast, the "always on" nanoprobe cannot distinguish HER2$^+$ or HER2$^-$ tumors. Pictures in panel A–C are reproduced with permission from Ref. 18 © the Nature Publishing Group.

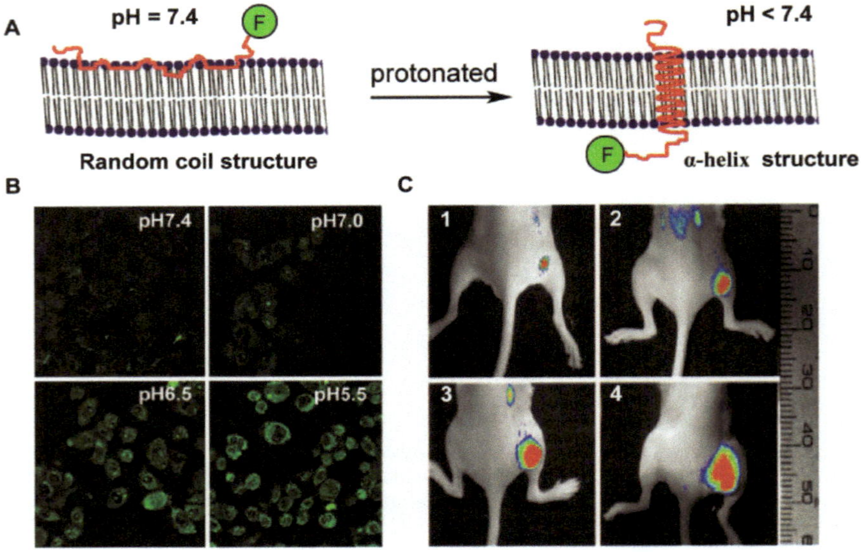

Fig. 9. Nanoprobes imaging tumor by changing the topological structure in the acidic environment. (A) Schematic of the up-regulated nanoprobe uptake induced by its pH responsive topological structure variation. (B) Microscopic fluorescence images of HeLa cells incubated with the pH responsive nanoprobe at pH 5.5, 6.5, 7.0, and 7.4 for 15 min. (C) *In vivo* overlapped white light and NIR fluorescent images of mice bearing tumors in right flanks. Panel 1–4 show a tumor undetected by the naked eye, 3–4 mm, 5–6 mm, and 8–9 mm. Pictures in panel A–C are reproduced with permission from Ref. 21 © the National Academy of Sciences, USA.

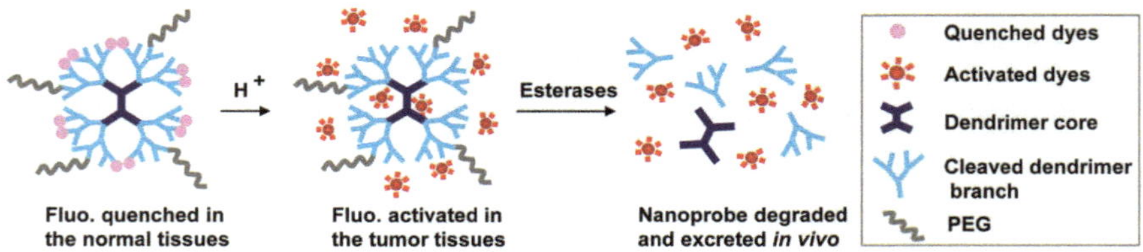

Fig. 10. Schematic of the pH responsive biodegradable nanoprobe. In the physiological pH, the NIR fluorophores labeled on the nanoprobe quench with each other *via* the SAET effect. In the tumor acidic environment, the cleavage of the pH liable bonds results in the pH activated fluorescence. The nanoprobe can be further degraded by endogenous esterases following the excretion *in vivo*.

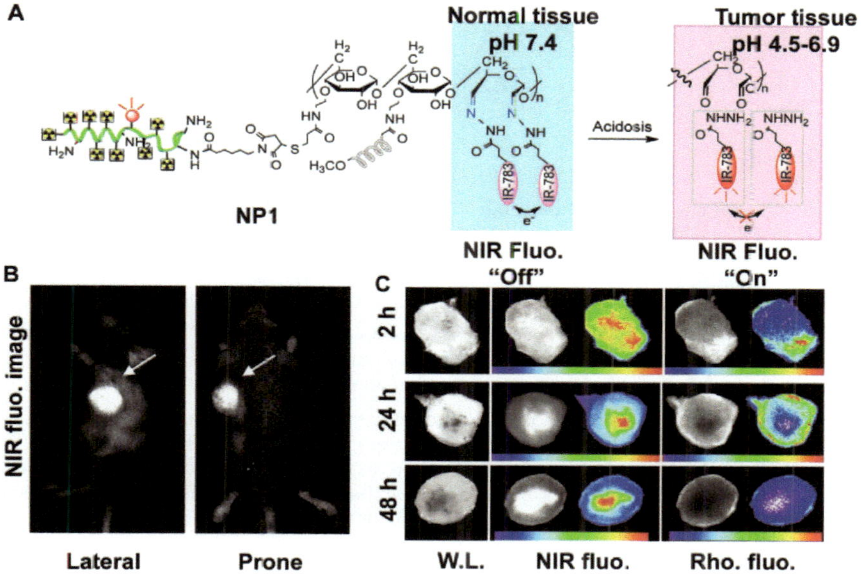

Fig. 11. Nanoprobe imaging tumor *via* the pH induced SAET effect *in vivo*. (A) The schematic of the dextran based nanoprobe, in which the NIR fluorophores are labeled into dextran *via* the pH liable hydrazone bonds. (B) Representative *in vivo* NIR fluorescence images of a mouse bearing a U87MG tumor xenograft at 48 h post-injection of **NP1** *via i.v.* The arrow points to the position of the tumor. (C) Representative white light, NIR and rhodamine fluorescence images of *ex vivo* U87MG tumor sections at 2, 24 and 48 h post-injection of **NP1**. WL: white light; CC: color coded.

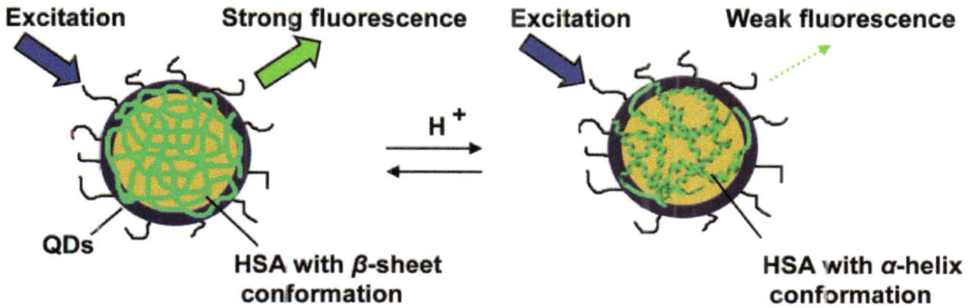

Fig. 12. Quantum dot based pH responsive fluorescence nanoprobe. In the acidic pH, the denatured HSA protein coated on the surface of QDs forms an α-helix conformation, but it turns into the β-sheet conformation in the basic environment. The pH dependent conformation of the protein coating leads to a fluorescence intensity change of the QD.

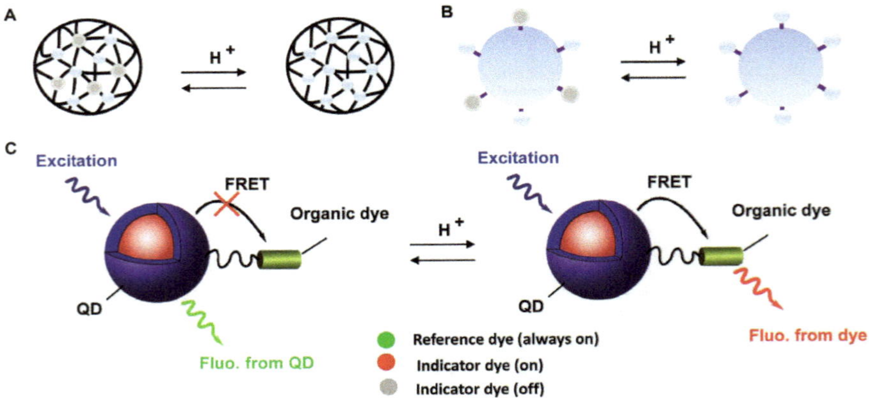

Fig. 13. Schematics of pH responsive ratiometric fluorescence nanoprobes. (A) The nanoprobe in which both reference and indicator dyes are embedded inside the polymer. (B) The nanoprobe in which reference and indicator dyes are labeled covalently on the surface. (C) The inorganic nanoprobe measures the pH *via* the intramolecular FRET effect between the quantum dot and the organic dyes modified on its surface.

Chapter 10

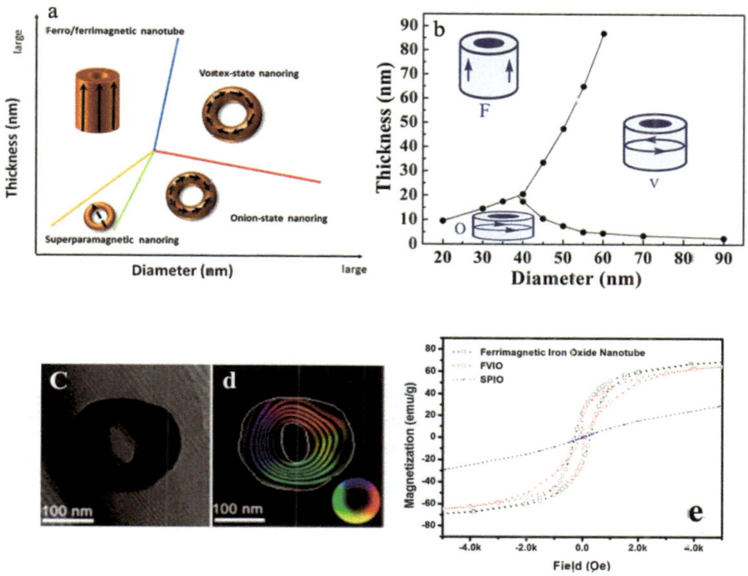

Fig. 1. Unique properties of magnetic nanorings and nanotubes. (a) Schematically demonstrates the shape-dependent magnetic properties of nanorings and nanotubes. (b) The shape-dependent magnetic configurations of single crystalline Fe_3O_4 nanorings and nanotubes obtained from LLG Micromagnetics simulation. (Reprinted with permission from Yang et al. unpublished data, provided by Dr. Jun Ding in Materials Science and Engineering, National University of Singapore) (c) Off-axis electron hologram of a single magnetite ring with an average thickness of 50 nm. In the form of bending interference fringes, the phase changes can be seen. (d) Direction of the magnetic induction under field-free conditions following magnetisation, indicated by colour as shown in the colour wheel in the inset (red = right, yellow= down, green=left, blue = up). (e) The hysteresis loops of a ferrimagnetic magnetite nanotube (370 nm in length and 80 nm in outer diameter), FVIO (50 nm in thickness and 70nm in outer diameter), and SPIO (10 nm) at 300K. (Reprinted with permission from Jia et al. 2008 and Fan et al. 2009, ©2008 and 2009 ACS Publishing Group.)

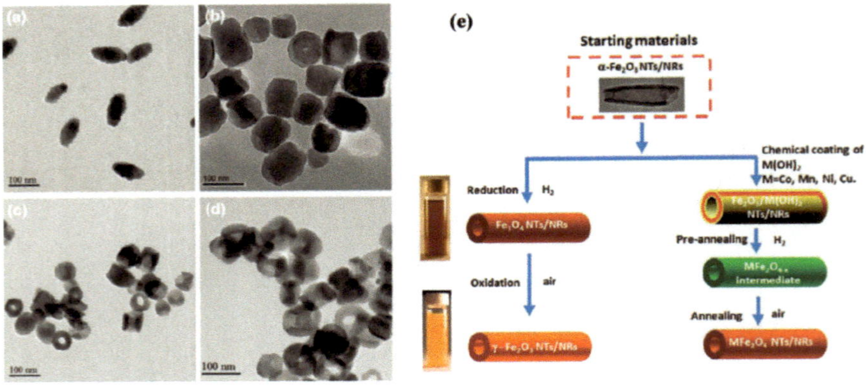

Fig. 3. Fabrications of iron oxide nanorings and nanotubes through coordination-assisted dissolution process. (a)-(d) Morphological evolutions of the α-Fe_2O_3 nanorings with different reaction times of (a) 3 h, (b) 10 h, (c) 24 h, and (d) 48 h, as revealed by TEM images. (e) The overall scheme for the synthesis of spinel MFe_2O_4 (M=Co, Mn, Ni, Cu) nanotubes/nanorings. The insets are the optical images of Fe_3O_4 and γ-Fe_2O_3 nanotubes water dispersion after surface modification by citric acid. (Reprinted with permission from Fan et al. 2008 and Fan et al. 2009, ©2009 American Chemical Society.)

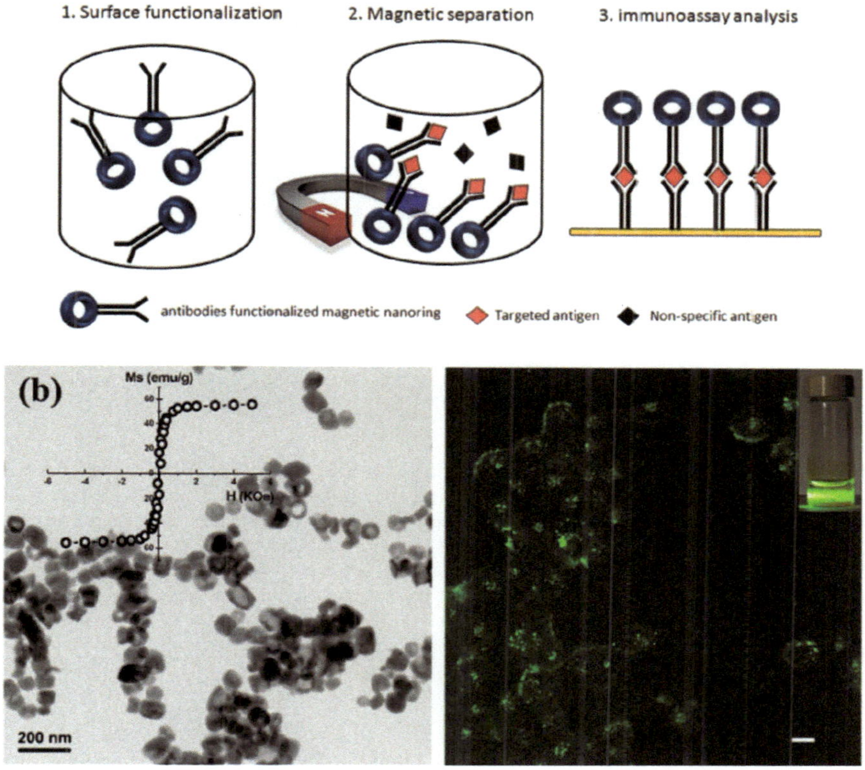

Fig. 4. Magnetic nanoring for cell separation and immunoassay. (a) Schematic illustration of a magnetic cell separation and capture-detection immunoassay. (b) TEM images of luminescent/magnetic nanorings. The insert is the hysteresis loop (luminescent/magnetic) at 300K. (c) Fluorescence images of lung cancer cells labelled with the luminescent/magnetic nanorings. The insert is the luminescent image of luminescent/magnetic nanoring dispersion in water. (Reprinted with permission from Fan et al. 2009, ©2009 American Chemical Society.)

Chapter 11

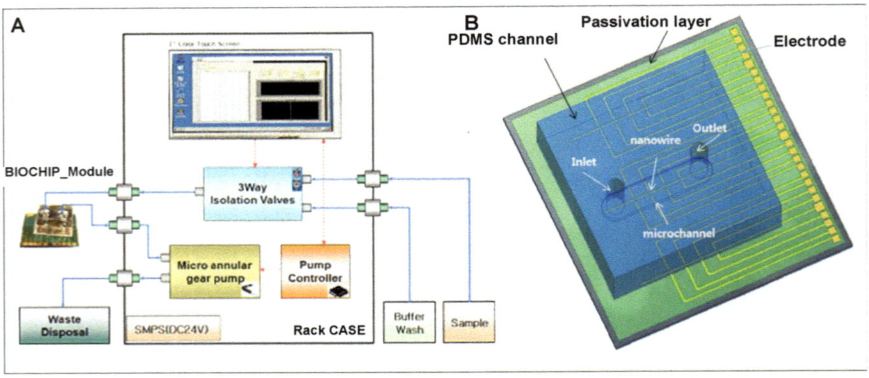

Fig. 3. (A) Schematic of automated fluidic control system. Three-way isolation valves were used—one for buffer, one for target samples, and one for delivery to the biochip module. The waste solution was directly pushed into waste disposal by a microannular gear pump. All pumps and times were carefully controlled by a custom-made lab view–based program. (B) PDMS microfluidic channel through which sample solution flows, and where SiNW FETs sensing took place. (Lee et al 2010) (Reprinted with permission from Elseveir).

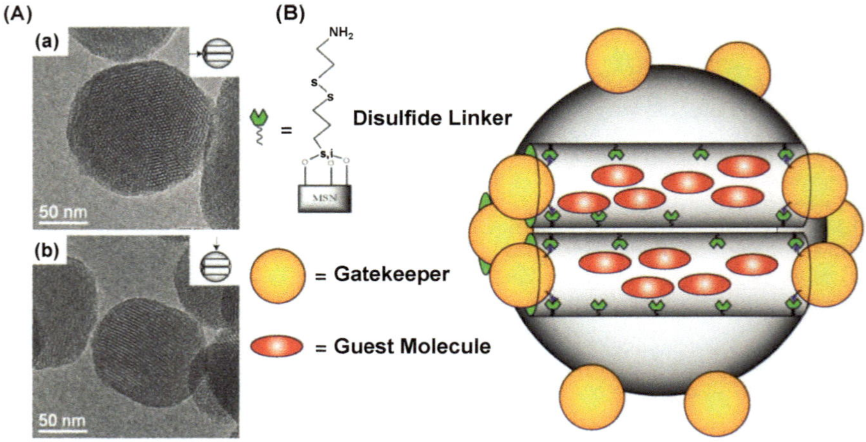

Fig. 5. (A) TEM images of MSNs materials recording from the direction (a) parallel or (b) perpendicular to the long axis of the mesochannels. ((B) Representation of an MSN loaded with guest moleculesand end-capped with a general gatekeeper. (Slowing et al. 2008) (Reprinted with permission from Elsevier).

Chapter 12

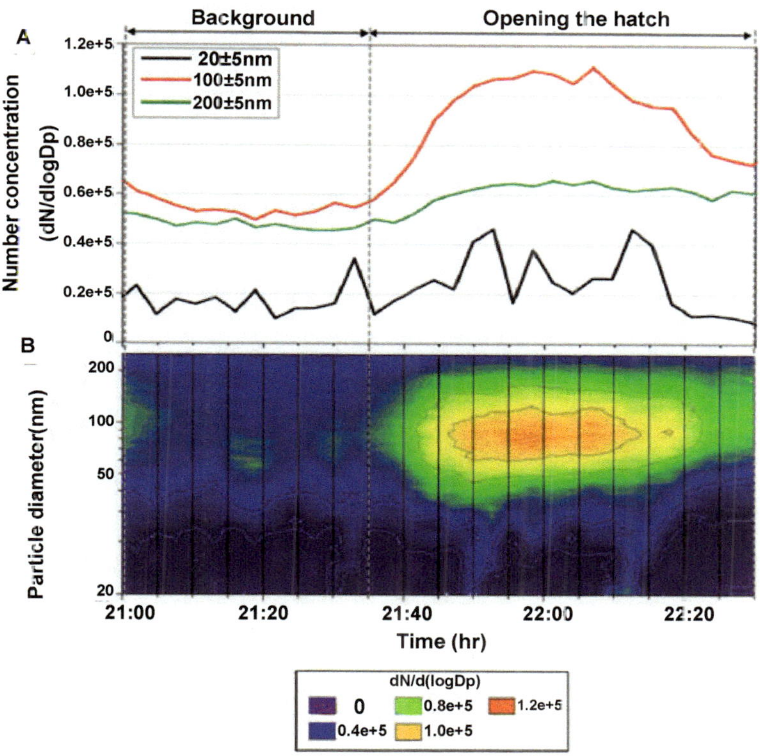

Fig. 1. Time series for the total silver nanoparticle number concentration measurement at a manufacturing facility. The particle concentration increased over time, with 100 nm silver nanoparticles predominating when the reactor hatch was opened for 1.5 hr. (A) Change in the number concentrations of 20 ± 5, 100 ± 5, and 200 ± 5 nm diametre particles and (B) change in particle number concentrations between 20 and 250 nm with time is shown as a filled contour plot (Adapted from Park et al 2009 with permission).

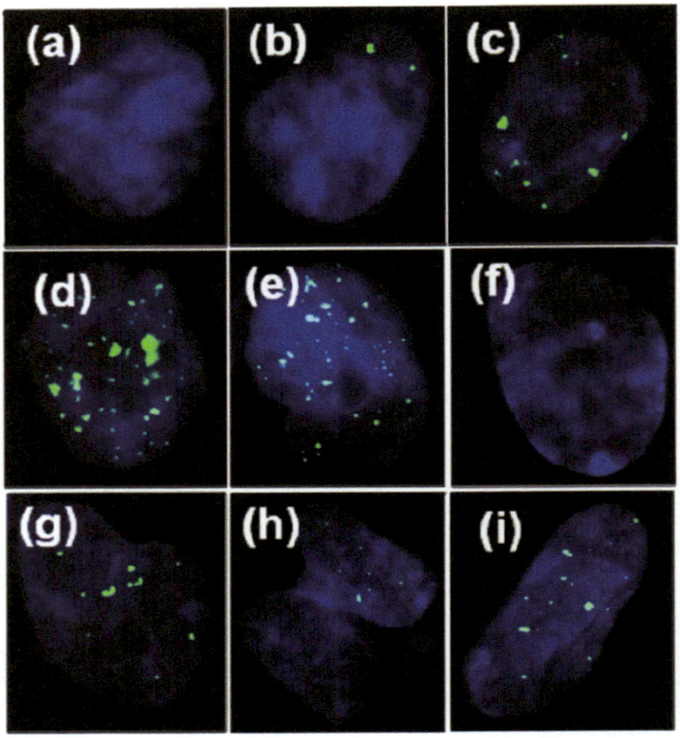

Fig. 3. Gamma H2AX staining; Silver nanoparticles induce DNA damage mainly through oxidative stress. γH2AX staining of untreated human cancer (U251) cells (a) showing minimal foci, (b) cells exposed to 25 μg/mL of Ag-np-3 (c) 50 μg/mL (d) 100 μg/mL and (e) 10 μM H_2O_2, showing multiple foci. (f) Untreated IMR-90 cells shows no foci whereas, treatment with (g) 25 μg/mL of Ag-np-3 (h) 50 μg/mL (i) 100 μg/mL and (K) 10 μM H_2O_2, shows multiple foci.

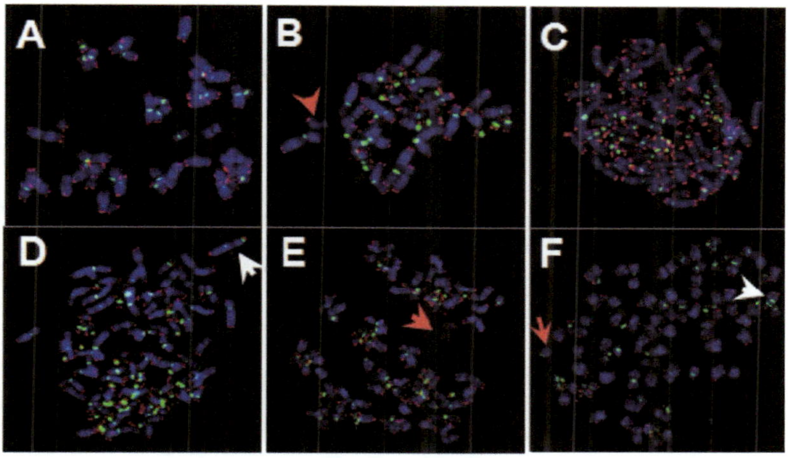

Fig. 4. Chromosomal aberrations in silver nanoparticle treated human cells. Metaphase spreads from the untreated cells show no apparent damage in the chromosomes (A). Silver nanoparticle treated human normal fibroblast cells (IMR-90) show acentric and centric fragments (B). Arrow indicates acentric fragments. (C) Untreated cancer cells with no aberrations, metaphases show dicentric chromosomes in untreated cells (D) and treated cells. White arrow points to a dicentric chromosome. Cancer cells treated with silver nanoparticles also show acentric fragments (E) and centric fragments (F). Red arrow points to a chromosome fragment. (G) Summary of the frequency of aberrations observed in silver nanoparticle treated cells. A minimum of 50 metaphases per sample was scored for chromosome analysis (Reproduced from Asharani et al. 2009a).

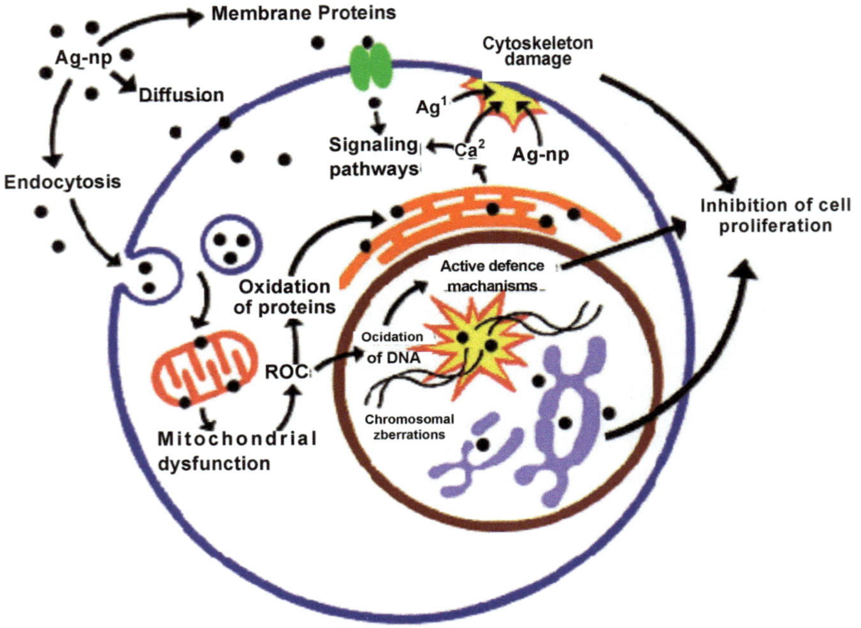

Fig. 5. The proposed mechanism of Ag-np toxicity based on the experimental data obtained from our studies (Reproduced from Asharani et al. 2009a).

Chapter 13

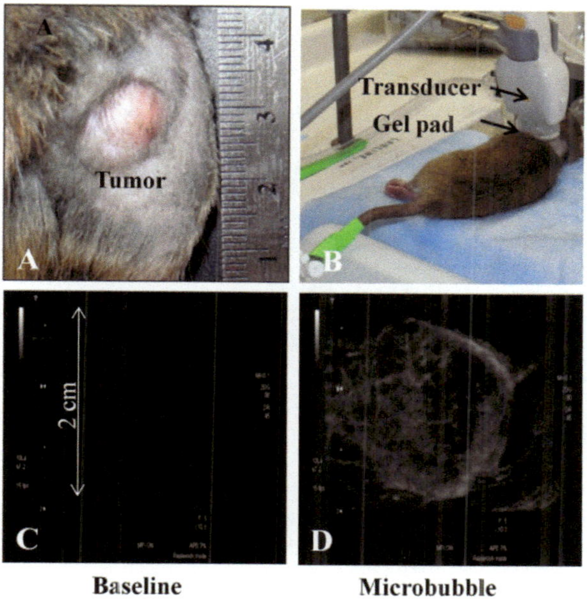

Fig. 1. Contrast enhanced ultrasound imaging of blood flow with microbubbles. A: representative subcutaneous rat tumor; B: tumor imaging setup; C: baseline image of rat tumor without contrast; D: microbubble enhanced ultrasound of tumor vasculature *(unpublished original illustration).*

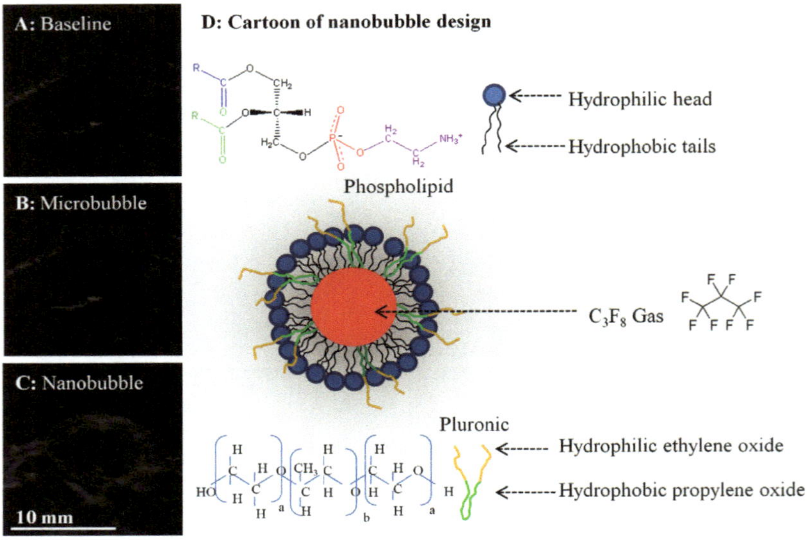

Fig. 4. Ultrasound imaging enhanced with Pluronic nanobubbles. A: Baseline image of subcutaneous rat tumor; B: microbubble enhanced image; C: nanobubble enhanced image; D: Pluronic nanobubbles (C_3F_8 gas stabilized by phospholipid shell and Pluronic, a triblock amphiphilic polymer) (*unpublished data*).

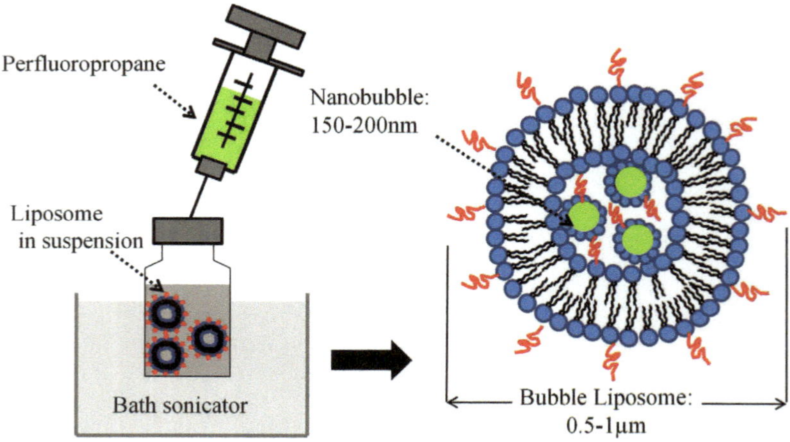

Fig. 5. Schematic of bubble liposome formulation. Sonication of C_3F_8 and suspension of liposomes in a sealed vial lead to the production of nanobubbles, with diameters of 150–200 nm *(unpublished original illustration).*

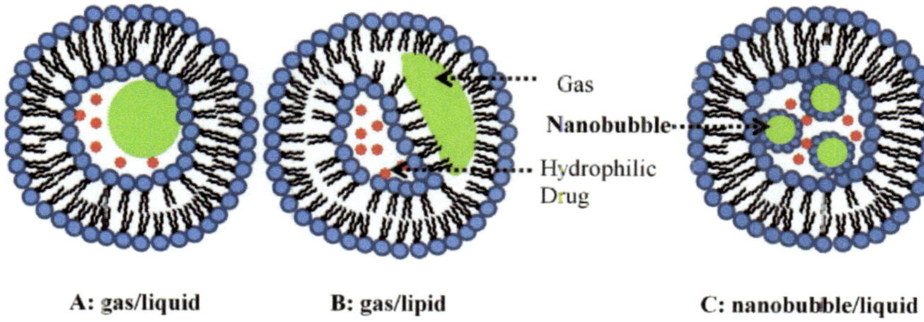

Fig. 6. Potential configurations of acoustically active liposomes. A: gas and hydrophilic drug floating in the aqueous core of the liposome without stabilizing shell; B: gas is encapsulated between the lipid bilayer of the liposome shell; C: bubbles stabilized by a single lipid shell are suspended in the aqueous core of the liposomes *(unpublished original illustration).*

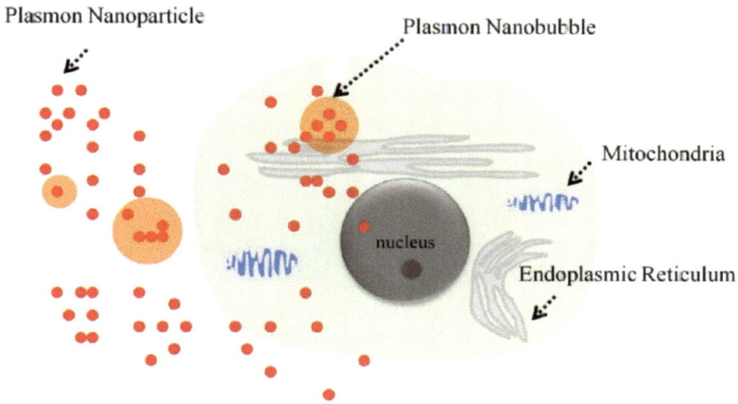

Fig. 7. Plasmon nanobubble formation. Laser pulse at appropriate fluence causes heating of plasmon nanoparticles. Heat accumulation causes evaporation of surrounding medium either around a single particle or clusters of nanoparticles. Evaporated medium forms nanobubbles either in the cell or around the cell and subsequent nanobubbles cavitation leads to cell damage *(unpublished original illustration).*

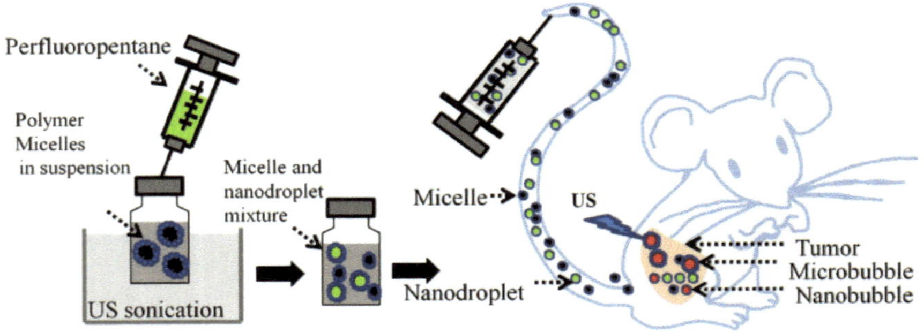

Fig. 8. Schematic of formulation, gene/drug delivery and imaging of PEG-PLLA (PCL) nanobubble system. A mixture of micelle carrying therapeutics and perfluoropentane (PFP) nanodroplets is administered systemically. While PFP is susceptible for vaporization at body temperature, Laplace pressure maintains them in the droplet state. At the tumor site, upon insonation, acoustic droplet vaporization takes place triggering drug release, sonoporation and nanodroplet to bubble phase transition. In the tumor, nanobubbles coalesce forming large microbubbles producing signals that are more suitable for clinically relevant ultrasound imaging *(unpublished original illustration).*

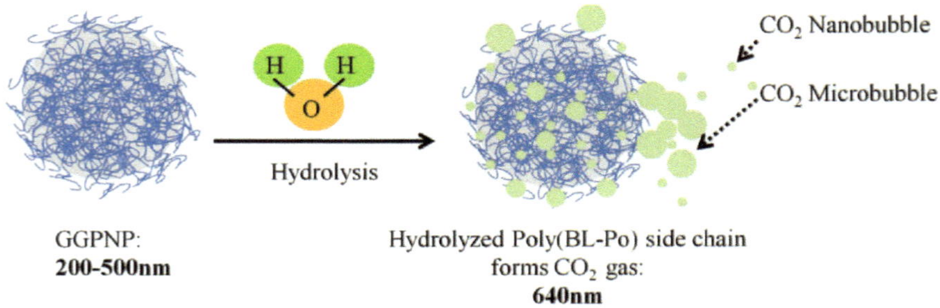

Fig. 9. Schematic of CO_2 nanobubble drug delivery system. Inward diffusion of H_2O molecules into the Poly (BL-PO) nanoparticle cleaves the carbonate side chains via hydrolysis. One product of this chemical process is CO_2. CO_2 accumulation lead to nanobubble formation on the surface of the Poly (BL-PO) nanoparticle and the coalescence of these nanobubbles forms CO_2 microbubbles *(unpublished original illustration).*

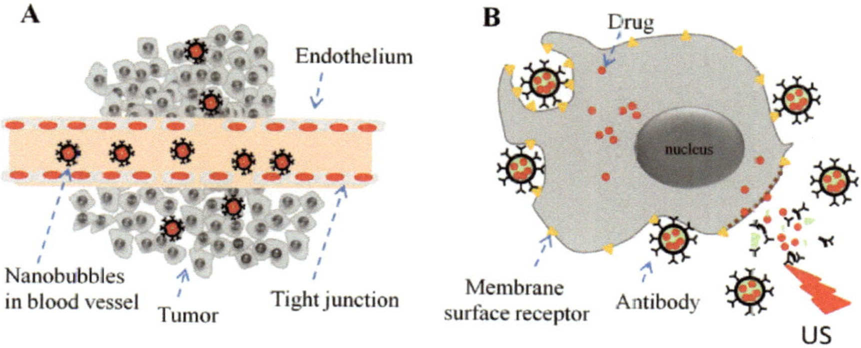

Fig. 10. Nanobubble tumor targeting. A: Passive tumor targeting: nanoparticles circulating in the blood stream leaks out vasculature through defective fenestra of the vessel wall and remains in tumor tissue due to defective lymphatic drainage; B: Active and mechanical targeting: Nanoparticles are conjugated with antibodies that recognize cancer specific cell surface receptors following extravasation. Ultrasound disrupts nanobubble, increases cell membrane permeability creating transient pores, enhancing drug diffusion into tumor cells *(unpublished original illustration).*

Chapter 15

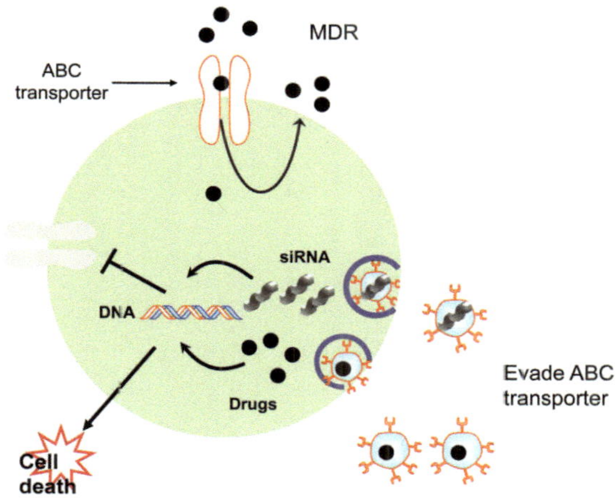

Fig. 1. Potential mechanism of overcoming multidrug resistance using nanoparticles. ABC transporters extrude chemotherapeutic drugs resulting in the survival of sarcoma cells. Conjugation of drugs and genes to nanoparticles results in increased accumulation of the drugs via non-specific endocytosis which could result in cell death. Drugs could be cytotoxic chemotherapeutic drugs, agents that could suppress the activity of ABC transporters or novel kinase inhibitors. Also, specific siRNA to the ABC transporters expressed in each sarcoma will result in the suppression of the ABC transporter.

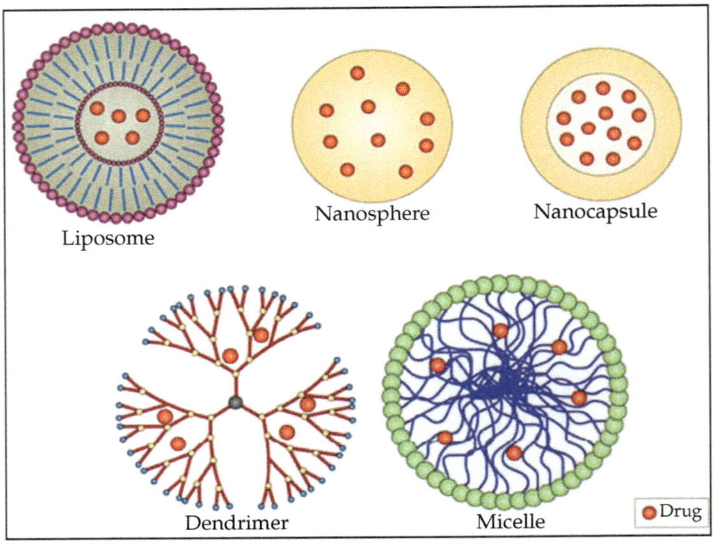

Fig. 2. Schematic illustration of various nanoparticle platforms. Nanoparticle systems are used for encapsulation and delivery of cancer therapeutics, including nucleic acid-based constructs. Examples include lipid vesicles or liposomes, polymeric nanospheres or nanocapsules, dendrimers, and self-assembling micelles. Each of these systems is engineered to protect the payload and deliver efficiently to disease tissues and cells.

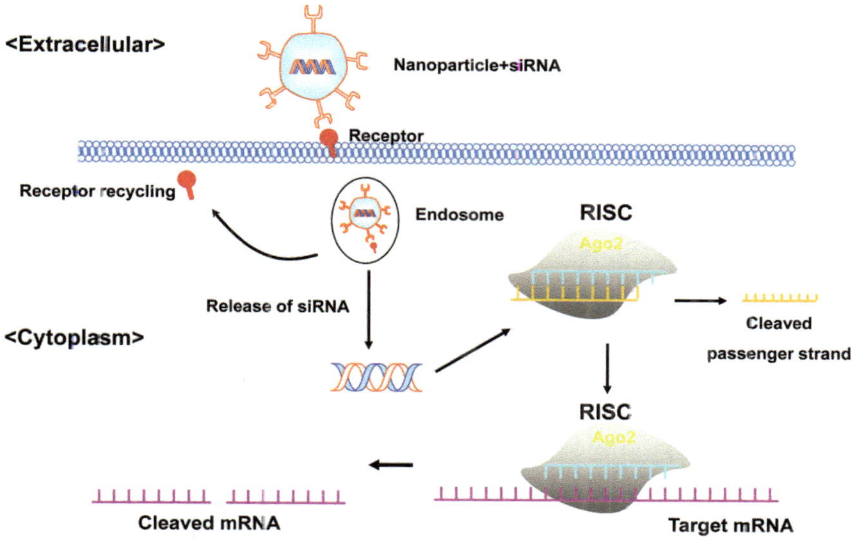

Fig. 3. RNA interference mechanisms utilizing siRNA-encapsulated nanoparticles. After internalization of nanoparticles via endocytosis, siRNA is released into the cytoplasm. siRNA is then incorporated into the RNA-induced silencing complex (RISC) which results in the cleavage of passenger strand of the siRNA by argonaute 2 (Ago2). Mature RISC complex binds to and degrades complementary mRNA resulting in the silencing of the target gene.

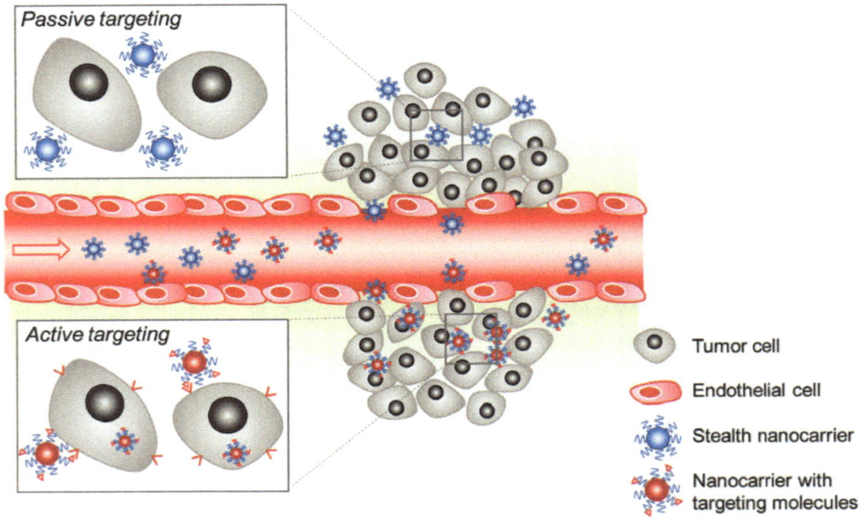

Fig. 4. Illustration of passive and active tumor targeted delivery with nanoparticles. Passive targeting relies on the abnormal tumor neovasculature that have greater fenestration allowing nanoparticles with up to 200 nm in diameter to extravasate and the properties of the delivery system. Active targeting is based on surface attachment of ligands that can specifically bind to over-expressed cellular targets.

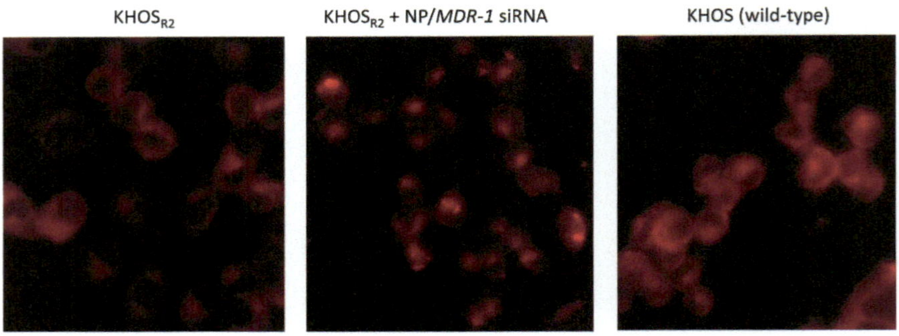

Fig. 5. Sub-cellular distribution of doxorubicin in drug sensitive (KHOS) and multidrug resistant ($KHOS_{R2}$) osteosarcoma cells. A prominent increase in fluorescence was observed in the nucleus when multidrug resistant cells $KHOS_{R2}$ were pre-treated with *MDR-1* gene silencing siRNA-loaded nanoparticles followed by administration with doxorubicin.

Chapter 16

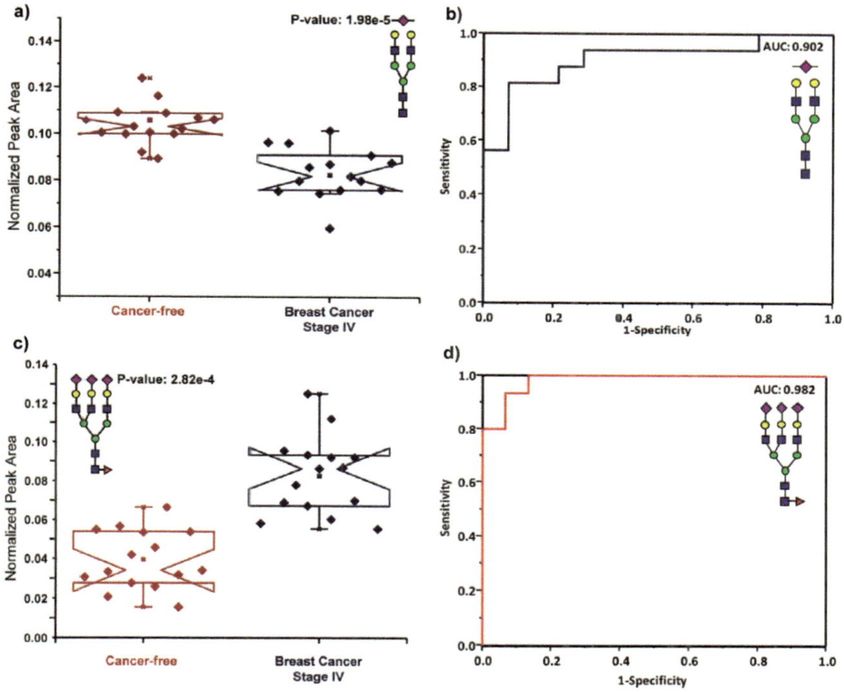

Fig. 4. Significant glycosylation changes in stage IV breast cancer. Two glycans found to have significantly different abundances in stage IV breast cancer vs. healthy samples. The box plots showing the statistical difference between the glycans in the two samples are displayed in (a) and (c), and the corresponding ROC plots are presented in (b) and (d). AUC values are >0.9 indicating that these glycans have a strong predictive value for stage IV breast cancer. Reprinted with permission from the publisher (Alley et al. 2010).

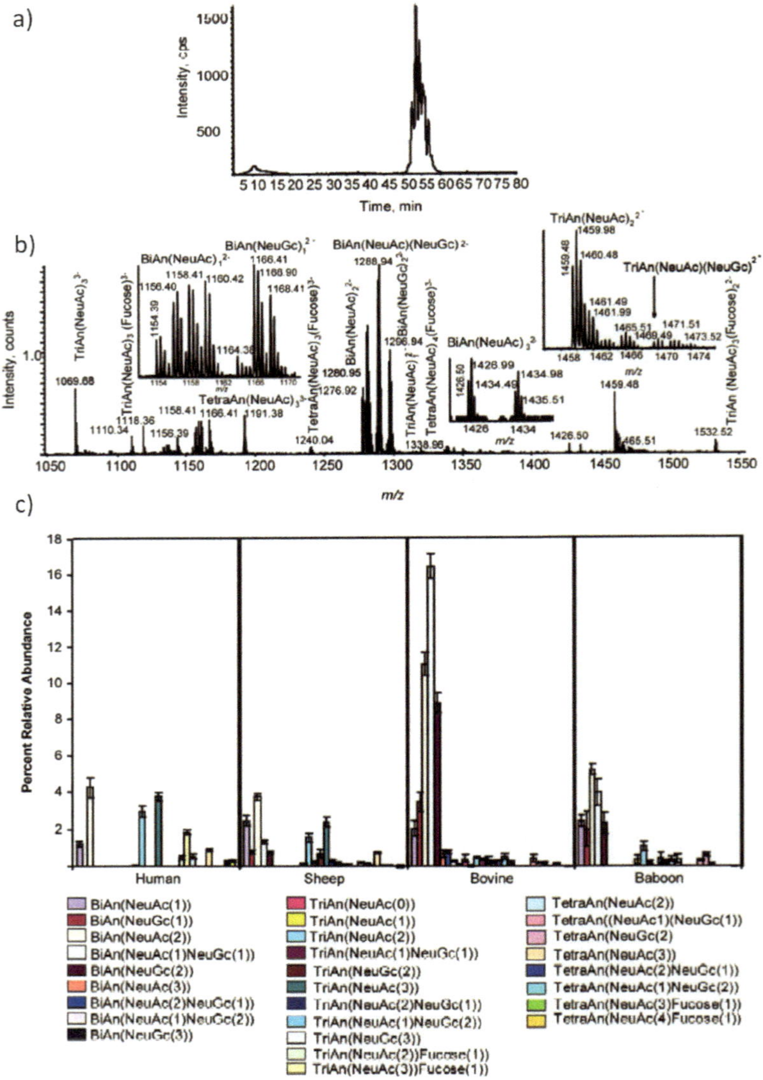

Fig. 5. Tetraplex stable isotope relative quantification strategy. (a) the base peak glycan chromatogram, (b) the MS showing the tetraplex isotopic envelopes for each glycan, and (c) the quantification of plasma *N*-linked glycans for 4 different species. This technique is relevant to cancer research as the relative quantification of glycans between several time-points in cancer can lead to the discovery of glycan biomarkers for the onset of cancer or possibly track the progression and metastasis of cancer. Reprinted with permission from the publisher (Bowman et al. 2010).

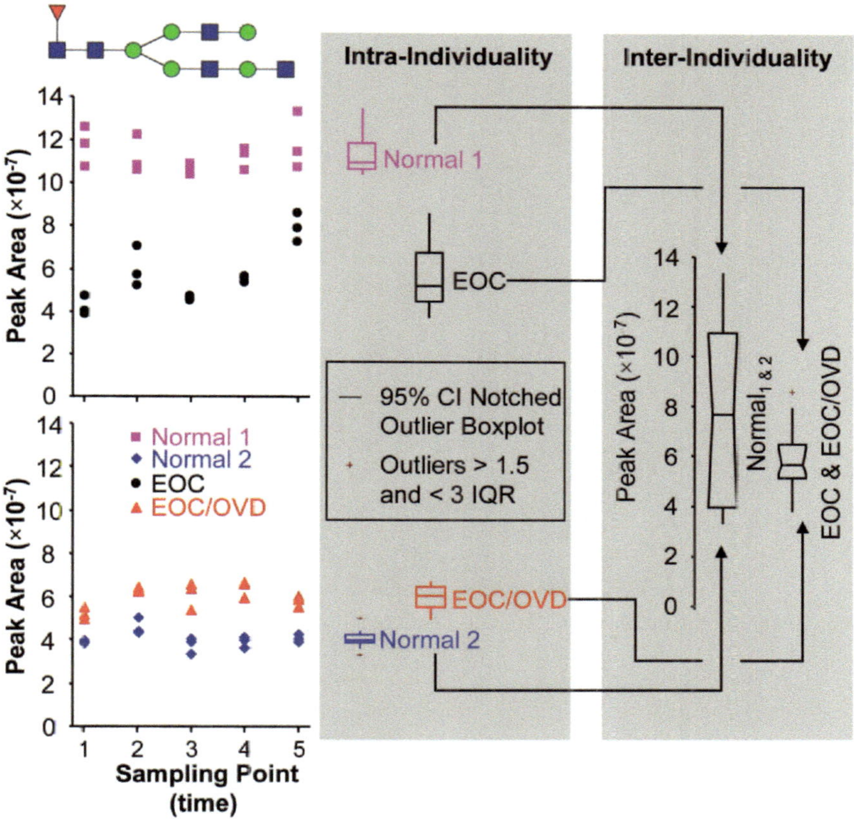

Fig. 6. Longitudinal glycosylation study of OVC in hens. This shows that biomarker discovery can be complicated as the levels of this glycan are shown to vary between the two normal chickens. Though there are distinct changes in the glycosylation between the OVC and healthy chickens, neither can be deemed statistically different due to the large disparity in the glycan abundance of the normal chickens. Thus, this study demonstrates the importance of longitudinal sampling and large sample sets of data in order to quantify the intra-individual and inter-individual variability of glycans in cancer. Reprinted with permission from the publisher (Dixon et al. 2010).

Fig. 3. Exhaled breath collection Dragonieri et al. 2007.

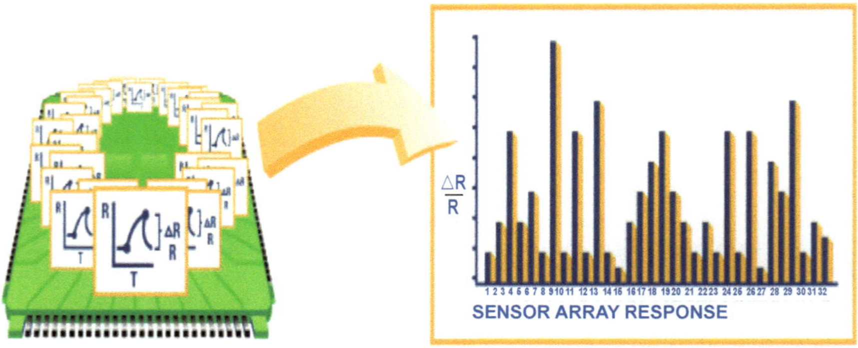

Fig. 4. The combination of the signal from all sensors generates the so-called "breathprint" (Dragonieri et al. 2007).

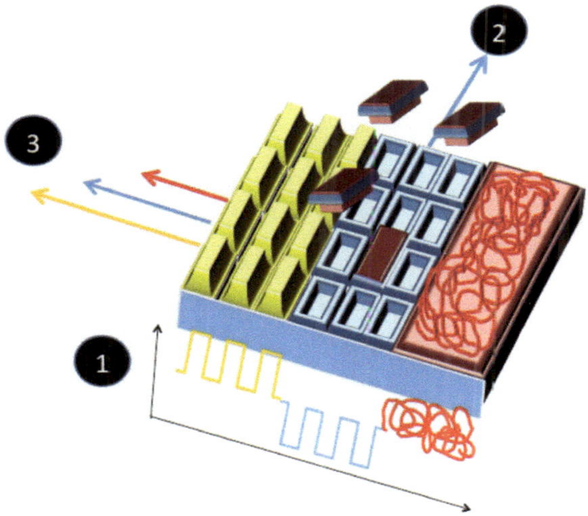

Fig. 5. Nanotechnology added value to chemical sensors: (1) the increased surface of the sensing element; (2) the opportunity to functionalize adsorbing mechanism at molecular level; (3) the ability to arrange many different sensing properties into an array of very small dimension.

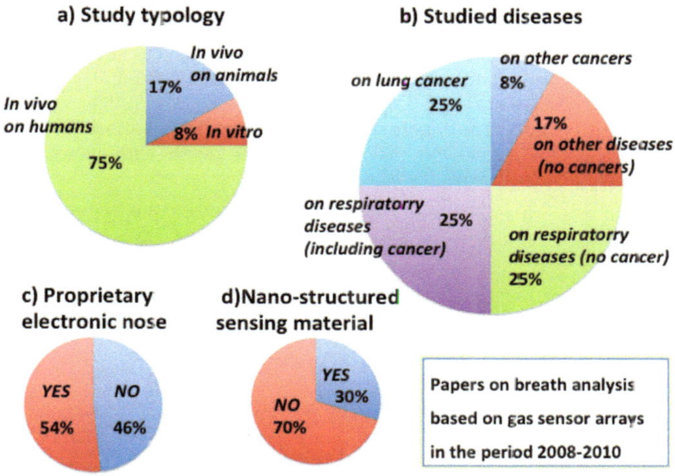

Fig. 6. Pie-graphs representing the different studies on breath analysis based on gas sensor arrays conducted in the period 2008–2010. a) divided by study typology (in vivo on humans, in vivo on animals, in vitro on cell coltures); b) divided by studied diseases (lung cancer, respiratory diseases including lung cancer, respiratory diseases not including lung cancer, other cancers, other diseases); c) divided by research groups which are proprietary of the used electronic nose or not; d) divided by the study conducted with electronic noses based on nano-structured materials or not.

Chapter 19

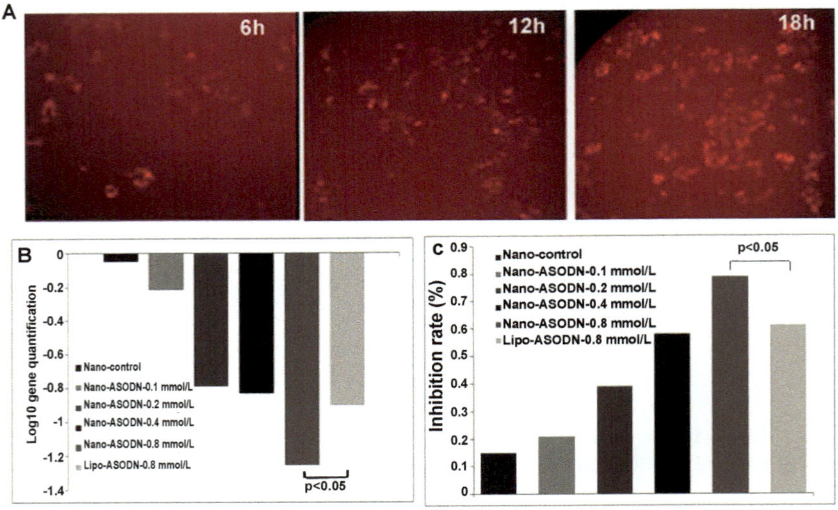

Fig. 2. Transduction and function of NANO-ASODNs in vitro. Analysis of NANO-ASODNs function. A: 0.2 mmol/L FAM-conjugated NANO-ASODNs transduced into HepG2 cells. The results were observed under a confocal microscope at indicted times of 6, 12 and 18 h; B: NANO-ASODNs down-regulated expression of MK mRNA; C: The proliferation of HepG2 cells was significantly inhibited by NANO-ASODNs (Data are from Dai et al. 2009), with permission from the Publishers.

MK, midkine; ASODNs, antisense oligonucleotides; MK-ASODNs, antisense oligonucleotides targeting midkine; NANO-ASODNs, MK-ASODNs packaged with nanoparticles; TEM, Transmission electron microscopy

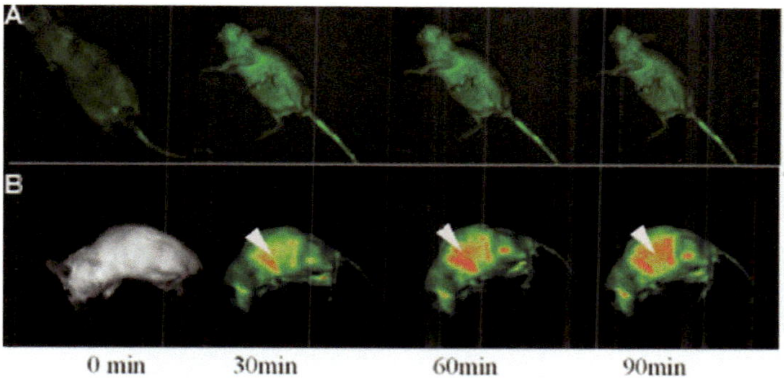

Fig. 3. NANO-ASODNs targeted to liver. The kinetic results of the NANO-ASODNs were observed through in vivo imaging systems at indicated times after NANO-ASODNs were injected through the tail vein. A: Free ASODNs did not concentrate within the liver and these ASODNs disappeared quickly; B: NANO-ASODNs were found to mainly target the liver (the arrow represents the NANO-ASODNs) (Data are from Dai et al. 2009), with permission from the Publishers.
MK, midkine; ASODNs, antisense oligonucleotides; MK-ASODNs, antisense oligonucleotides targeting midkine; NANO-ASODNs, MK-ASODNs packaged with nanoparticles; TEM, Transmission electron microscopy

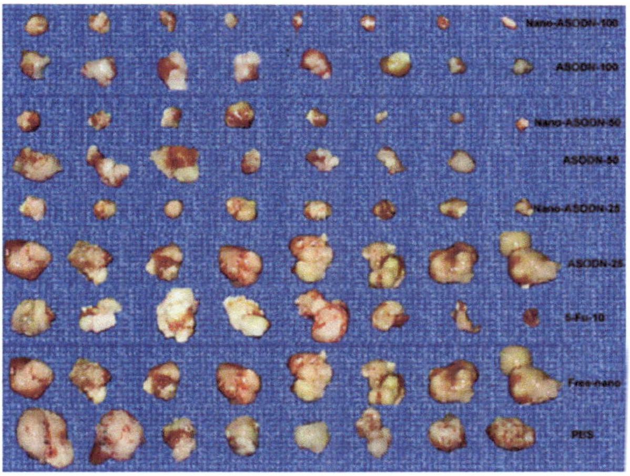

Fig. 4. Morphological changes of HCC following treatment with NANO-ASODNs. The volume of HCC decreased significantly following treatment with 100, 50 and 25 mg/kg per day of NANO-ASODNs for 20 d. MK-ASODNs were the positive control. The PBS or free nanoparticles represent the negative controls. (Data are from Dai et al. 2009), with permission from the Publishers.
MK, midkine; ASODNs, antisense oligonucleotides; MK-ASODNs, antisense oligonucleotides targeting midkine; NANO-ASODNs, MK-ASODNs packaged with nanoparticles; TEM, Transmission electron microscopy

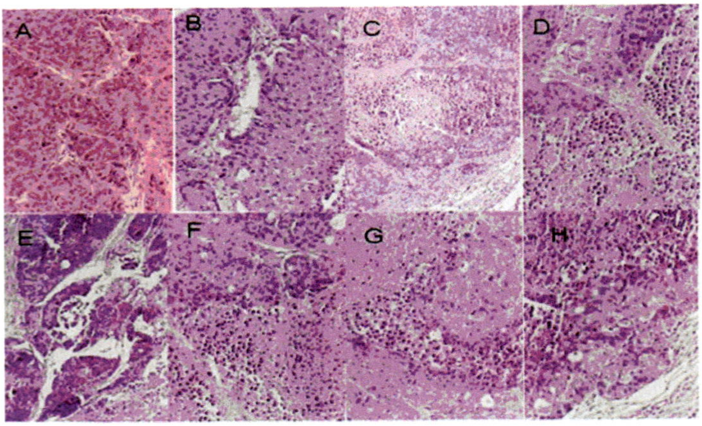

Fig. 5. Histopathological analysis. Analysis changes of histopathology. A: Tissue sections of the tumors from in situ xenograft HCC; B: Tissue sections of the tumors from nanoparticles; C: Tissue sections of the tumors from 5-FU (10 mg/kg per day); D: Tissue sections of the tumors treated with ASODNs 50 mg/kg per day; E-G: Tissue sections of tumors treated with NANO-ASODNs 100, 50 and 25 mg/kg per day of NANO-ASODN treated tumors, respectively; H: Tissue sections of the tumors treated with 5-FU (10 mg/kg per day) and 50 mg/kg per day of NANO-ASODNs. Regions showing an increase of necrosis and fibrosis were observed in the 5-FU or ASODN treatment groups (C-H, × 200) compared with the free nanoparticles and untreated groups (A and B, × 200) (Data are from Dai et al. 2009), with permission from the Publishers.

MK, midkine; ASODNs, antisense oligonucleotides; MK-ASODNs, antisense oligonucleotides targeting midkine; NANO-ASODNs, MK-ASODNs packaged with nanoparticles; TEM, Transmission electron microscopy

Chapter 20

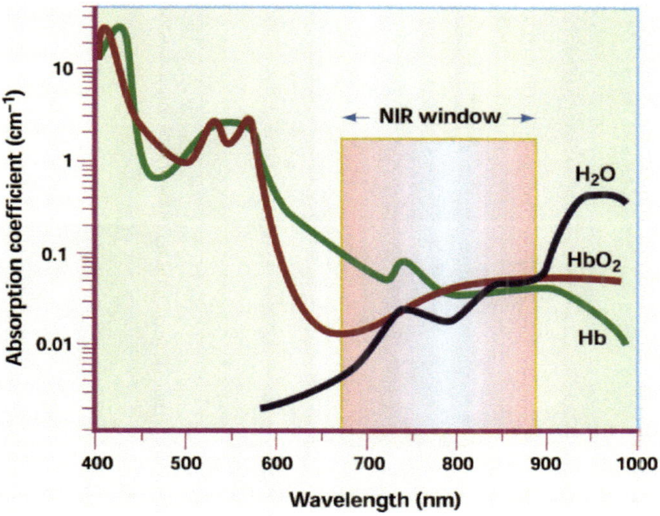

Fig. 1. Wavelength range of the near-infrared (NIR) tissue transmission window. In the 650–900 nm range, absorption by oxygenated/de-oxygenated hemoglobin and water is minimal, allowing up to 10 cm penetration through breast tissue using FDA class 1 NIR lasers and 7 cm penetration through muscle and neonatal skull/brain tissue using FDA class 3 NIR lasers. Photoexcitation of fluorescent compounds, plasmonic particles, and Raman-labels is most efficient in this region. Reprinted with permission from R. Weissleder. Copyright 2001 Nature Publishing Group/Macmillan Publishers Ltd.

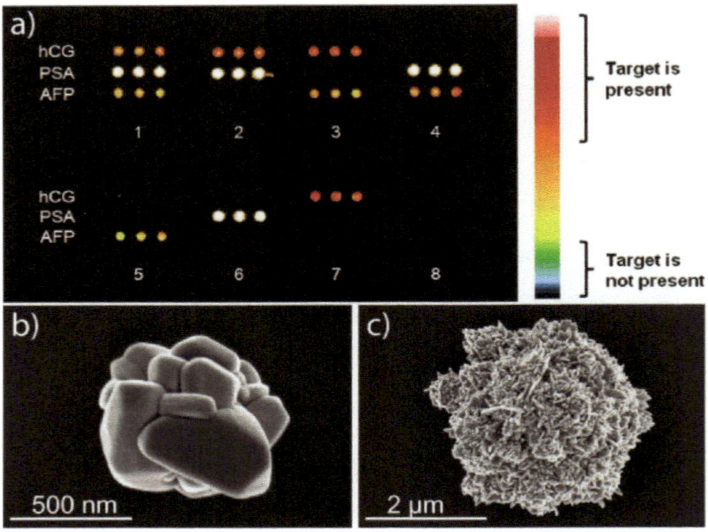

Fig. 3. Multiplexed detection of protein cancer markers based on light scattering using an antibody-gold nanoparticle sandwich assay. Prostate specific antigen (PSA), a prostate cancer marker, human chorionic gonadotropin (hCG), a testicular cancer marker, and R-fetoprotein (AFP), a hepatic cancer marker were correctly identified at 1.4 pM concentrations in buffer (a) following scattering amplification by silver (b) and gold (c) deposition. (1) All targets present; (2) hCG and PSA; (3) hCG and AFP; (4) PSA and AFP; (5) AFP; (6) PSA; (7) hCG; (8) no targets present. Reprinted with permission from Kim et al. Copyright 2009 American Chemical Society.

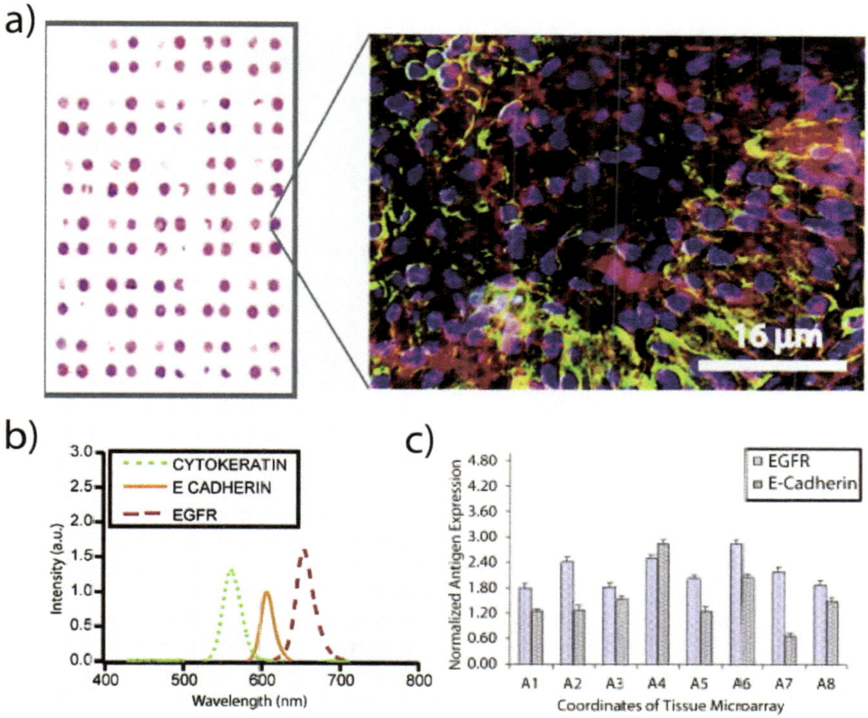

Fig. 4. High throughput, multiplexed quantification of HNC cancer protein markers based on quantum dot-antibody labeling of tissue microarrays. Fluorescence intensity from spectrally-resolved quantum dot-labeled antibodies of epidermal growth factor receptor (EGFR), pan cytokeratin, and E-cadherin was quantified for each tissue microarray sample (a). Using cytokeratin as an internal standard (b), cancer protein expression levels (c) could be compared over multiple tumor phenotypes. Adapted with permission from Ghazani et al. Copyright 2006 American Chemical Society.

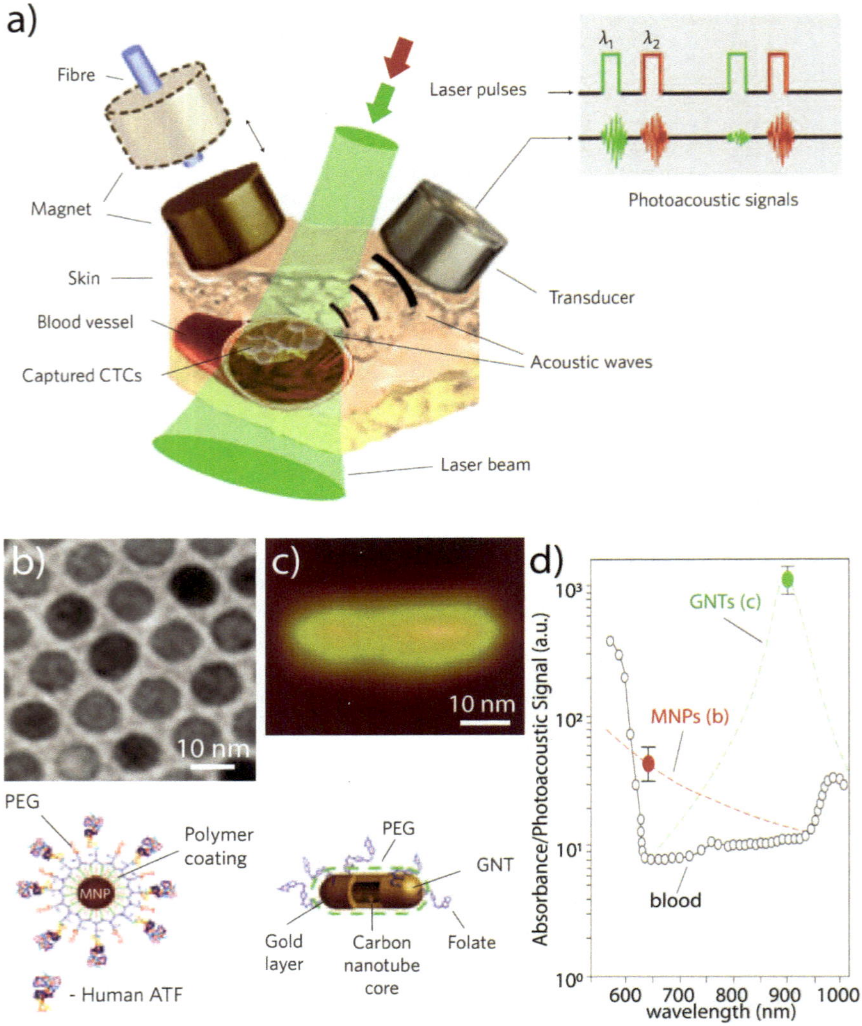

Fig. 5. Schematic of *in vivo* detection and magnetic sequestration of circulating tumor cells (CTCs) by dual nanoparticle labeling and photoacoustic cytometry. Dual-labeled CTCs were non-invasively detected (a) by their enhanced photoacoustic response and sequestered with a small rare-earth magnet. Photoacoustic response from magnetic iron oxide (b) and gold-coated carbon nanotube particles (c) labeled with ligands of cell-surface receptors upregulated in malignant cells, but absent in red blood cells, was significantly higher than that from blood and red blood cells (d). Error bars represent standard deviation. Adapted with permission from Galanzha et al. Copyright 2009 Nature Publishing Group/Macmillan Publishers Ltd.

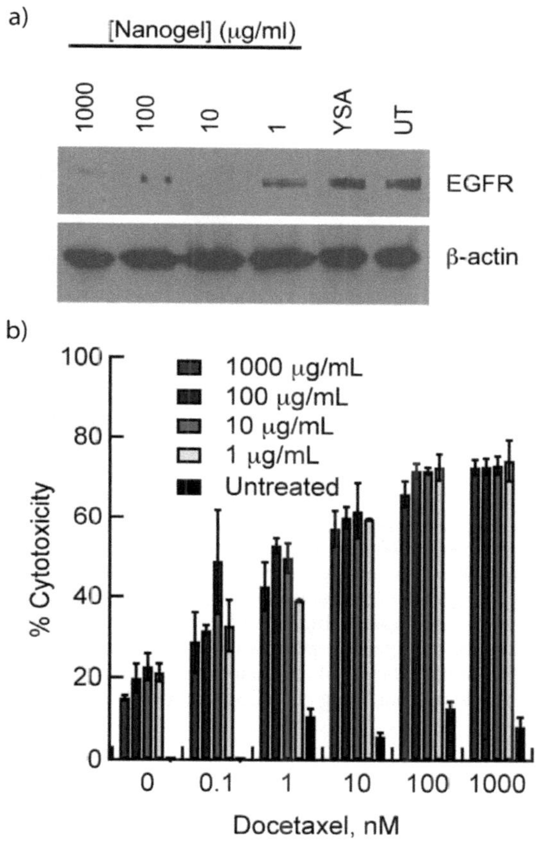

Fig. 6. Targeted RNA interference of epidermal growth factor receptor expression and chemosensitization to docetaxel by treatment with YSA-targeted small interfering RNA-nanogel particles. Selective reduction of EGFR expression (a) in EphA2 receptor (+) cancer cells was achieved by YSA peptide conjugation to siRNA-loaded core-shell hydrogel nanoparticles and incubation for 48 hr, followed by 96 hr incubation with docetaxel (b). Error bars represent standard deviation. Reprinted with permission from Dickerson et al. Copyright 2010 retained by the authors.